$F=ma$
$E=mc^2$

BARRON'S
E–Z
PHYSICS

Robert L. Lehrman
Former Science Department Supervisor
Roslyn High School
Roslyn, New York

BARRON'S

Better Grades or Your Money Back!

As a leader in educational publishing, Barron's has helped millions of students reach their academic goals. Our E-Z series of books is designed to help students master a variety of subjects. We are so confident that completing all the review material and exercises in this book will help you, that if your grades don't improve within 30 days, we will give you a full refund.

To qualify for a refund, simply return the book within 90 days of purchase and include your store receipt. Refunds will not include sales tax or postage. Offer available only to U.S. residents. Void where prohibited. Send books to **Barron's Educational Series, Inc., Attn: Customer Service** at the address on this page.

Credit: Figure 8.1 from FOUNDATIONS OF PHYSICS by Robert L. Lehrman, copyright © 1965, renewed 1993 by Holt, Rinehart & Winston, reproduced by permission of the publisher.

All inquiries should be addressed to:
Barron's Educational Series, Inc.
250 Wireless Boulevard
Hauppauge, New York 11788
www.barronseduc.com

ISBN-13: 978-0-7641-4126-3
ISBN-10: 0-7641-4126-0

Library of Congress Control Number: 2008053112

Library of Congress Cataloging-in-Publication Data

Lehrman, Robert L.
    E-Z physics / Robert L. Lehrman.—4th ed.
        p. cm.
    Rev. ed. of: Physics the easy way. 3rd ed. 1998.
    Includes index.
    ISBN-13: 978-0-7641-4126-3
    ISBN-10: 0-7641-4126-0
  1. Physics.   I. Lehrman, Robert L. Physics the easy way.   II. Title.

    QC23.2.L44 2009
    530—dc22

                                                                          2008053112

Printed in the United States of America
9 8 7 6 5 4 3 2

CONTENTS

iv   Contents

Preface

So you have decided to study physics. Anyone who would like to understand the mechanism of the universe and the ways in which we investigate it must come to this decision sooner or later. Physics is the most fundamental of the sciences, the place where the most ubiquitous question—WHY?—comes to a stop and peters out into the murky waters of speculative philosophy.

Physics represents the human urge to make sense of the universe, to write the rules of its workings in exact and comprehensible terms. Professional physicists find that they must resort to higher mathematics in order to formulate their theories and explain their findings. For this reason, most of the new developments in physics, the advancing front of scientific knowledge, are well beyond the comprehension of the beginning student. But the basics on which the great superstructure of physical theory is built are accessible to anyone with a reasonable grasp of simple algebra. These fundamental laws have, for the most part, been established for many years, and most of the basic theory in this book could have been written a century ago.

This is by no means the case with the practical applications of these theories. Technology is advancing so rapidly that some of the statements made here may soon be out of date. This is no cause for alarm; properly armed with the fundamentals, you should be able to keep up fairly well with many of the technological advances based on a particular body of theory.

Although this book limits itself to simple mathematics, a knowledge of algebra and geometry is indispensable. To understand any part of physics, you must solve problems. You will have to be able to manipulate and solve simple linear equations, to set up and solve proportionalities, and to take square roots. You will have to know how to deal with the sines, cosines, and tangents of angles. Scientific notation will be useful throughout and is absolutely essential in Chapters 16 and 17. In this day, it would be folly to do all this arithmetic without a calculator; a simple, four-function type is adequate, and a scientific calculator, with trigonometric functions and scientific notation, will be even more useful.

In your pursuit of this work, I can wish you nothing better than the uplift in spirit that flows from perceiving the beauty inherent in the patterns according to which the universe behaves.

Robert L. Lehrman

ATTENTION!

The science of physics encompasses a wide range of natural phenomena. The selected topics covered in introductory high school and college courses vary greatly. Selection of material in this book was made in consultation with high school and college teachers. No modest-sized book can incorporate everybody's favorite topic; an effort has been made to deal with those topics generally found in curricula in use.

The tedium of problem solving can be greatly reduced by using a scientific calculator with the capacity to handle trigonometric functions and scientific notation.

Questions labeled "Try this" are scattered throughout the text. The answers will be found in Appendix 5.

Each chapter ends with a series of problems. The answers will be found in Appendix 6.

Many important equations are introduced throughout the book and will be required for problem solving. Appendix 1 is a list of all such equations used in this book.

Problem solving will often call for use of various physical constants, such as the speed of light and the mass of a proton. A table of these important constants can be found in Appendix 2.

The Art of Measurement

WHAT YOU WILL LEARN

All of science is a method of finding out how things work by observing and measuring what we see. Measurement is always limited in accuracy, and this chapter deals with techniques we use to assure that our conclusions are consistent with the accuracy of our measurements.

SECTIONS IN THIS CHAPTER

- Why Measure?
- How Certain Is a Measurement?
- Significant Digits
- Big Numbers, Little Numbers
- Arithmetic in Scientific Notation
- Labels
- Dimensionality
- New Dimensionalities
- Le Système Internationale d'Unités

Why Measure?

Fifty thousand years ago, people lived in much of Europe. They left us a legacy of their art, mostly paintings of animals, on the walls of the inner recesses of their caves. These are probably religious paintings, appeals to their gods or spirits to favor them in the hunt.

As long as people have been on earth, they have strived to understand and control the world around them. This need spawned an endless variety of myths, superstitions, and religions. In our era, sophisticated religion and philosophy still pursue this same end. Science, dedicated to this goal, has created a worldwide professional coterie of men and women who pursue knowledge according to a set of rules that make them strictly accountable to each other.

Among all branches of knowledge, science is unique in one respect. Its conclusions are testable. Philosophers and theologians may theorize endlessly about good and evil; they can never reach universal agreement because no decisive test will distinguish truth from fallacy. Scientists have such a test. A scientific theory is a proposed description of things that happen in the real, observable world. The theory will stand if it describes and predicts events that can be observed and measured. If it fails this test, the theory will be rejected. Erroneous theories are an important part of the process of making science; they are tested against reality and eventually discarded, clearing the way for ideas that work. The unique feature of the scientific enterprise is that it is self-correcting.

Measurement is at the heart of reality testing. It is the bridge between the real world and the theory that describes it. Measurement is a crucial skill for any experimental scientist. It is also a scientific enterprise in its own right, with specialists who set up standards and techniques that allow scientists to communicate with each other all over the world.

CORE CONCEPT
Standard and careful methods of measurement are needed in all scientific work.

TRY THIS
— 1 —

Why was it necessary to establish an international treaty to define the meter?

How Certain Is a Measurement?

Consider the following two statements. First, your friend tells you, "I'll pick you up at your place at 7:15." Second, the daily newspaper tells you, "The moon will rise tonight at 7:15."

While these two time values look the same, they are really very different. You would not be surprised if your friend showed up at 7:10 or 7:20. On the other hand, if the moon peeked over the horizon at 7:16, an astronomer might soon be looking for a job. The *magnitudes* of these two measurement statements are the same, but their uncertainties are very different.

No measurement is perfect. Consider the measurement of the height of a cylinder, shown in Figure 1.1. The centimeter scale, even if you paid a lot for it, is not perfect. Whoever uses it does not have infinitely perfect eyesight. It might not be perfectly aligned with the cylinder, and the cylinder might not be perfectly uniform all the way around. These pitfalls in the process of measurement can be reduced, but they can never be completely eliminated. Furthermore, measurements are necessarily inexact because of the nature of the thing being measured. What is the diameter of a tennis ball? It is so fuzzy that no sharp boundary exists between the ball and the space it is in. Ultimately, on the atomic scale, all surfaces are fuzzy.

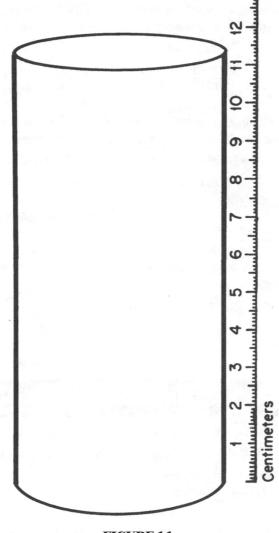

FIGURE 1.1

In scientific work, every measured value must be accompanied by a statement of its uncertainty. The height of the cylinder in Figure 1.1 would be given as 11.4 centimeters ±0.1 centimeter; the ±0.1 centimeter is the *limits of uncertainty* in the measurement. This means that the actual height of the cylinder is *probably* between 11.3 centimeters and 11.5 centimeters. These limits are not absolute but only probabilistic; even the uncertainty has its uncertainty.

If the limits of uncertainty are ±0.1 centimeter, is the measurement highly accurate? It depends. Finding the height of the cylinder within 0.1 centimeters is not difficult.

However, in measuring the distance to the moon, an uncertainty of 0.1 centimeter would be an extraordinary level af accuracy. It would require a lot of complex and expensive equipment. The accuracy of a measurement is its limits of uncertainty *compared with the measurement itself.*

Accuracy is the ratio of the limits of uncertainty to the measurement, usually expressed as a percent. For the measurement of the cylinder, the accuracy (percent uncertainty) is

$$\frac{\pm 0.1 \text{ cm}}{11.4 \text{ cm}} \times 100\% = \pm 0.9\%$$

If the percent uncertainty is known, it is simple to find the limits of uncertainty. Just multiply the magnitude of the measurement by its percent uncertainty. For example, what are the limits of uncertainty of a mass measurement given as 232 grams ± 2%?

$$(232 \text{ g}) \times (\pm 0.02) = \pm 5 \text{ g}$$

CORE CONCEPT

All measurements are limited in accuracy because of errors introduced by the nature of the measuring instrument and the object being measured.

TRY THIS
—2—

What is the percent uncertainty in a measurement given as 55 ± 2 kilograms?

Significant Digits

The equations of physics tell us what kind of arithmetic to use in processing measurement data. The accuracy of the answer depends on the uncertainties in the measurement and on the kind of arithmetic. A rough-and-ready method of evaluating the accuracy of an answer is to keep track of the number of *significant digits*.

Significant digits are those digits of a measurement that represent meaningful measurement data. The more accuracy there is in the measurement, the more significant digits there will be in its record. If you guess that a room is 24 feet long, you are giving the measurement accurate to 2 significant digits. It is understood that the final digit (the 4) could be wrong; the last significant digit in any measurement is always the *estimated digit*. If you measure the room with care, accurate to the nearest millimeter, you might find that its length is 6.097 meters. The 7 (underlined) is the estimated digit, and there are four significant digits in all.

Zeroes are a problem because they serve two functions. They may be significant, but they are also used merely to indicate the location of the decimal point. You could measure the thickness of a pencil and find it to be 8.2 millimeters—a two-digit measurement. Expressed in meters, this would be 0.0082 meters. There is no change in accuracy, and this is still a two-digit number. The rule is this: *Initial zeroes are never significant.*

Final zeroes may or may not be significant. If you drive a golf ball, and say, "It went 250 yards," you could be off by 10 or 20 yards. The estimated digit is the 5. The final zero is not significant but is needed because of the way our system of numbers is designed. If, on the other hand, a surveyor laying out the course measures a distance and finds it to be 250 yards, it is probably not 251 yards. In the surveyor's measurement, the final zero is the estimated digit, and there are three significant digits in the number.

If there is a decimal point, all zeroes are significant. A surveyor's 250.0 yards is a measurement made accurate to the nearest 0.1 yard, and there are four significant digits. Without a decimal point, there is often no way to tell whether a final zero is significant unless you know how the measurement was made.

In adding or subtracting measurement data, there can be no significant digit in any place (except the first) unless all of the numbers have significant digits in that same place. For example:

$$
\begin{array}{r}
227.31 \\
14.516 \\
+3\ 155 \\
\hline
3\ 396.826
\end{array}
$$

However, you have no idea what digit belongs in the tenths place because there is no tenths digit in 3 155. Therefore, the answer must be rounded off to the units place and given as 3 397.

The rule for propagation of error in multiplication and division is different. Count the number of significant digits in the numbers; the number of significant digits in the answer must be no more than the smallest number of significant digits in the problem. For example (using a calculator):

$$7.0660 \div 0.022 = 321.18182$$

This is a five-digit number divided by a two-digit number, so the answer must have only two digits; it is 320 (the final zero is not significant).

There are two exceptions to these rules. First, it does not make sense to treat 101 as a three-digit number and 99 as a two-digit number, since both imply accuracy to about 1 part in 100. Treat 101 as a two-digit number. A handy rule is this: Do not count initial 1's as significant unless the second digit is more than 5.

A second useful technique, in doing a series of calculations, is to keep one extra, nonsignificant digit until the very end, and to get rid of it when you round off the answer.

**EXAMPLE
1.1**

Subtract 6.9 from 226.151, and multiply the result by 35.

SOLUTION

226.151
– 6.9
219.251   Four significant digits; round off to 219.25.

219.25
× 35
7 673.75   Since 35 has only two digits, round off the answer to 7 700.

CORE CONCEPT

A rough statement of accuracy in measurement data and in values derived from them is given by careful use of the correct number of significant digits.

**TRY THIS
—3—**

Add 87.0 to 9.136, and divide the result by 22.

Big Numbers, Little Numbers

If you want to express all distances in the SI (Système Internationale; see Page 13) unit, the meter, you will run into trouble when you write the distance to the nearest star or the diameter of an atomic nucleus. In either case, you will need 15 zeroes—at the end of the number for the star and at the beginning of the number for the nucleus.

The only function of all these zeroes is to indicate where the decimal point belongs. But there is a better way to do this. It is called *scientific notation*. We write every number with the decimal point just after the first nonzero digit and then multiply by some power of 10 to show where it really belongs. For example, instead of writing a number like 15 200, we write

$$15\ 200 = 1.52 \times 10\ 000 = 1.52 \times 10^4$$

This may seem like a lot of trouble, but it is extremely convenient once you get used to it, and it is really the only way to deal with very large and very small numbers.

To put a number in scientific notation, just count the number of places to the left you have to move the decimal point to place it after the first digit. That number becomes the exponent of 10. In the example given above, the decimal point was moved four places to the left.

If the decimal point is somewhere else, and powers of 10 are already given, just move it to the left and add the number of places to the exponent of 10. For an example, see Example 1.2.

EXAMPLE 1.2 Put the number 351.7×10^4 into standard form.

SOLUTION

The decimal point must be moved two places to the left, which amounts to dividing by 100. Therefore, you must also multiply by 100 by increasing the exponent of 10 by 2; the answer is 3.517×10^6.

For small numbers—less than 1—you will have to move the decimal point to the right to get it where you want it. Then you *subtract* the number of places moved from the exponent of 10. For example:

$$0.0000162 = 1.62 \times 10^{-5}$$

If the number already has a power-of-ten multiplier, follow the same rule:

$$0.135 \times 10^4 = 1.35 \times 10^3$$

One big advantage of scientific notation is that when it is used, all zeroes are significant. Thus, 4 700 is written 4.700×10^3 if all zeroes are significant, or 4.70×10^3 if only the first zero is significant, or 4.7×10^3 if neither zero is significant.

CORE CONCEPT

To put a number into scientific notation, write it with the decimal point after the first nonzero digit and multiply by the appropriate power of 10.

TRY THESE
—4—

Put the following numbers into the standard form of scientific notation:
1. 14 million
2. 234×10^{-3}
3. 4 190
4. 0.000000398

Arithmetic in Scientific Notation

One benefit of scientific notation is that it is easier to add exponents than to count zeroes. If you have to multiply 23 400 by 17 000 000 you can greatly simplify the process by first putting the numbers into scientific notation. Then the problem looks like this:

$$(2.34 \times 10^4)(1.7 \times 10^7)$$

Then all you have to do is multiply the numbers in the usual way and add the exponents of 10. The answer is therefore 3.978×10^{11}. Round off to 4.0.

EXAMPLE 1.3   How much is $(5.6 \times 10^4)(7.9 \times 10^2)$?

SOLUTION

Multiply the numbers and add the exponents to get 44.2×10^6; shift the decimal point, and the answer is 4.4×10^7.

The same rule holds if the exponents are negative:

$$(2.91 \times 10^{-3})(5.0 \times 10^{-6}) = 14.55 \times 10^{-9}$$

Now shift the decimal point to its proper position, remembering that a move to the left requires you to increase the exponent of 10. Was your answer 1.46×10^{-8}?

To divide, just divide the numbers in the usual way and *subtract* the exponent of 10 in the denominator from that in the numerator:

$$\frac{4.5 \times 10^8}{1.7 \times 10^3} = 2.6 \times 10^5$$

When the exponents are negative, watch out! Remember that subtracting a negative number is like adding!

$$\frac{2.8 \times 10^{-2}}{1.1 \times 10^{-6}} = 2.5 \times 10^4$$

If that surprises you, perhaps you have forgotten that $-2 - (-6) = +4$.

You can add and subtract numbers in scientific notation only if they have the same exponent; the exponent of the sum or difference is then the same as that of the numbers you are combining. For example, you know that 4.5 million plus 2.2 million is 6.7 million. In scientific notation, it looks like this:

$$(4.5 \times 10^6) + (2.2 \times 10^6) = 6.7 \times 10^6$$

If you have to add numbers that do not have the same exponent of 10, you first have to shift the decimal point.

EXAMPLE 1.4   Subtract 4.17×10^{-8} from 2.25×10^{-7}.

SOLUTION

Shift the decimal point of one of the numbers until both have the same exponent of 10; then subtract:

$$(2.25 \times 10^{-7}) - (0.417 \times 10^{-7}) = 1.83 \times 10^{-7}$$

CORE CONCEPT

When multiplying numbers in scientific notation, add the exponents of 10; when dividing, subtract the exponents of 10.

TRY THESE
—5—

1. Multiply 3.5×10^4 by 4.6×10^2.
2. Multiply 5.7×10^{-5} by 6.0×10^8.
3. Divide 2.6×10^8 by 1.3×10^{-12}.
4. Add 3.75×10^4 to 1.91×10^3.
5. Subtract 5.17×10^5 from 3.95×10^4.

Labels

Every measurement has three parts—magnitude, uncertainty, and a label. If you tell someone you ran a distance of 50, you are conveying no information unless you say whether that measurement was in meters, yards, kilometers, or miles. The label says that your measurement is a ratio, based on some standard. Fifty meters is fifty times as long as the standard meter.

If your height is 160 centimeters, are you taller or shorter than a friend who is 64 inches tall? To find out, you have to convert either your measurement to inches or your friend's to centimeters. You can go either way if you know the proper conversion factor. This is it:

$$2.54 \text{ centimeters} = 1 \text{ inch}$$

Now, let's see, do you multiply the inches or divide them to change them to centimeters? Making the wrong choice in this situation is a very common error, which can be avoided absolutely by following a simple rule: *Treat the labels exactly like algebraic quantities.*

Here's how it works: Since 2.54 centimeters = 1 inch, it follows that

$$\frac{2.54 \text{ cm}}{1 \text{ inch}} = 1 = \frac{1 \text{ inch}}{2.54 \text{ cm}}$$

Now everyone knows that you can multiply anything by 1 without changing it. So you can multiply your friend's height in inches by either of the fractions above without breaking any rules. Choose the one that eliminates the inches and leaves your answer in centimeters:

$$64 \text{ inches} \times \frac{2.54 \text{ cm}}{1 \text{ inch}} = 163 \text{ cm}$$

Don't antagonize the guy—he's bigger than you are. And it does not help to convert your height to inches instead:

$$160 \ \cancel{cm} \times \frac{1 \ \text{inch}}{2.54 \ \cancel{cm}} = 63 \ \text{inches}$$

EXAMPLE 1.5   How many pounds is 45 ounces? There are 16 ounces to a pound.

SOLUTION

$$(45 \ \cancel{oz})(\frac{1 \ \text{lb}}{16 \ \cancel{oz}}) = 2.8 \ \text{lb}$$

Note that conversion of units does not affect accuracy. Accuracy is a property of measurement data. Twelve inches to the foot is not a measurement but a definition, and it is thus exact. When you multiply 22.17 feet by 12 inches/foot, the answer is 266.04. The least accurate measurement is 22.17 feet, which has four significant digits. The answer must also have four significant digits; it is 266.0 inches.

CORE CONCEPT
Every measurement has a label, comparing it to some standard.

TRY THESE
—6—

1. There are 16 ounces to a pound. What is the sum of 3 pounds and 27 ounces, in ounces?
2. A board is 57 inches long. If you cut off 65 centimeters, how many inches are left?

Dimensionality

How much is 6 pounds plus 3 feet? This is a meaningless question; pounds is a measure of weight, and feet is a measure of length. You can convert feet to centimeters and add any two lengths together. If your friend stands on your head, the combined height is 160 centimeters + 163 centimeters = 323 centimeters, or 63 inches + 64 inches = 127 inches. However, you cannot convert pounds to feet, and you cannot add them together.

The property of a measurement that expresses the convertability of its units is called the *dimensionality* of the measurement. Feet, inches, miles, centimeters, kilometers, fathoms, rods, and microns all have the dimensionality of length, and any of these units can be converted into any other. Units with the dimensionality of weight include pounds, ounces, tons, newtons, slugs, and (for our British readers) stone. (Note that grams and kilograms are not on this list for reasons you will learn in Chapter 3.)

We may measure area in square inches, not inches. Area has a different dimensionality than length, and you cannot add an area to a length.

For example, suppose you have a rectangular table with a top measuring 30 inches long and 18 inches wide. What is its surface area, A? The area of a rectangle is the length times the width. Treating the labels like algebraic quantities, you find that the solution looks like this:

$$A = (30 \text{ in})(18 \text{ in}) = 540 \text{ in}^2$$

The answer is read "five hundred forty square inches." The square inch (in^2) is a new label that cannot be compared with, added to, or subtracted from inches. It has the dimensionality of area and can be added to square centimeters, acres, or any other unit of area.

Similarly, the cubic inch (in^3) is a measure of volume and can be converted to cubic centimeters, quarts, and so on.

EXAMPLE 1.6 How many gallons of water are there in a rectangular aquarium 20 inches long and 12 inches wide if the water is 9.5 inches deep? There are 231 cubic inches to a gallon.

SOLUTION

Volume is length × width × height:

$$(20 \text{ in})(12 \text{ in})(9.5 \text{ in}) = 2\,280 \text{ in}^3$$

Now convert the units:

$$(2\,280 \text{ in}^3)\left(\frac{1 \text{ gal}}{231 \text{ in}^3}\right) = 9.9 \text{ gal}$$

CORE CONCEPT

Any labeled quantities can be multiplied or divided, and the process creates new labels with different dimensionality.

TRY THESE
—7—

In each of the following cases, tell whether or not the indicated process is possible:
1.   57 feet + 2 miles =
2.   30 acres − 140 square meters =
3.   25 meters + 12 kilograms =
4.   12 ounces − 4 cubic centimeters =
5.   17 liters + 2 kilograms =

New Dimensionalities

While you cannot compare, add, or subtract quantities unless their dimensionalities agree, you can always multiply or divide any quantity by any other. Just as squaring a linear dimension creates a new label and a new dimension, multiplying or dividing labeled quantities creates a new label with its own dimensionality. The newly created label is formed by treating labels just as though they are algebraic quantities.

Dividing one label by another creates a label expressed as a ratio.

What, for example, is the linear density of a wire if a piece 550 yards long weighs 210 pounds?

$$d = \frac{w}{L} = \frac{210 \text{ lb}}{550 \text{ yd}} = 0.38 \text{ lb/yd}$$

The answer is read "zero point three eight pound per yard." The label of a ratio is always written as a fraction.

EXAMPLE 1.7   What is the average speed of a car that travels 240 miles in 6.2 hours?

SOLUTION

Average speed is distance divided by time:

$$\frac{240 \text{ mi}}{6.2 \text{ hours}} = 39 \text{ mi/hour}$$

Multiplication also produces new labels. Your public utility, for example, decides how much to charge you for the electric energy you use by multiplying your power consumption by the length of time you have it turned on. Suppose you burn a 100-watt bulb for 6 hours. How much energy do you use? Solution:

$$(100 \text{ watts})(6 \text{ hours}) = 600 \text{ watt} \cdot \text{hours}$$

The answer is read as "six hundred watt-hours." You would be charged the same amount if you burned a 60-watt bulb for 10 hours, or a 200-watt bulb for 3 hours.

CORE CONCEPT

Labeled quantities can be multiplied and divided, and the labels follow the ordinary rules of algebra.

TRY THESE
—8—

1. Divide 125 pounds by 30 square inches.
2. Multiply 20 feet by 16 pounds.
3. It takes 100 centimeters to make a meter. Take the cube of 100 centimeters to find the number of cubic centimeters in a cubic meter.

Le Système Internationale d'Unités

The name of this system is in French because the International Bureau of Standards has its headquarters in Paris, where the metric system was invented. The *Système Internationale* (SI), used throughout this book, is a special simplification of the metric system.

Traditional systems of measurement have evolved through the centuries and are different in every country. The only thing they have in common is that they make no sense at all. No one in his or her right mind would sit down and invent the English system, in which there are 437.5 grains to the ounce, 16 ounces to the pound, 14 pounds to the stone, 8 stone to a hundredweight, and 20 hundredweight to a long ton.

In the metric system, which *was* invented, things are much simpler. All of the multipliers are powers of 10. There are 100 centimeters to a meter, 1 000 cubic centimeters to a liter, 1 000 000 micrograms to a gram, and so on.

Even simpler is the SI, which uses no multipliers at all. Some are made available to those who feel they need them, but they are not necessary. The system is extremely simple because, as long as you stick to it, it is never necessary to convert units.

In the SI, there are exactly seven fundamental units. Dozens of other units are derived by multiplication and division of these seven, but without introducing any numbers at all. These are the fundamental units:

the meter (m) for distance

the kilogram (kg) for mass

the second (s) for time

the kelvin (K) for temperature

the ampere (A) for electric current

the candela (cd) for luminous intensity

the mole (mol) for number of particles

Here are a few examples of derived SI units:

the square meter (m^2) for area

cubic meter (m^3) for volume

the kilogram per cubic meter (kg/m^3) for mass density

the coulomb (C) or ampere-second (A·s) for charge

the newton (N) or kilogram-meter per second squared ($kg \cdot m/s^2$) for force

The system allows certain multiples and submultiples to be used also. They are limited to powers of 10^3. Here are the ones we will use:

nano $= 10^{-9}$; example: 1 nanometer (nm) $= 10^{-9}$ m

micro $= 10^{-6}$; example: 1 microcoulomb (μC) $= 10^{-6}$ C

milli $= 10^{-3}$; example: 1 milliampere (mA) $= 10^{-3}$ A

kilo $= 10^{3}$; example: 1 kilowatt (kW) $= 10^{3}$ W

mega $= 10^{6}$; example: 1 megajoule (MJ) $= 10^{6}$ J

giga $= 10^{9}$; example: 1 gigavolt (GV) $= 10^{9}$ V

You will find that there are some metric units that always refuse to fit this pattern, but they are used all the time anyway from pure force of long habit:

1 centimeter (cm) $= 10^{-2}$ m

1 liter (L) $= 10^{3}$ cm^3

While this system is simple, it is not always convenient. Some of the units are too large or too small for practical use, and we still like the old, familiar ones. Don't ask your grocer for 0.00091 cubic meters of milk; tell him you want a quart. Soon enough you will be getting a slightly larger bottle when you ask for a liter.

CORE CONCEPT
The SI is a convenient system of units for scientific work in which each dimensionality is expressed in only one unit.

TRY THIS
—9—

The rate of flow of water in a pipe can be expressed as the amount of mass per unit of time. What would be the correct SI unit for this rate?

REVIEW EXERCISES FOR CHAPTER 1

Fill-In's. For each of the following, fill in the missing word or phrase.

1. The testing of a physical theory requires scientists to have a good understanding of _____.

2. If the label of a measurement is kilograms, the correct unit for the limit of uncertainty is _____.

3. The accuracy of a measurement is best expressed in terms of the _____ of uncertainty.

4. The nature of the thing being measured places a limit on the possible _____ of the measurement.

5. Zeroes at the _____ of a number are never significant.

6. Final zeroes may not be significant if there is no _____.

7. When performing _____ or _____ on data, the number of significant digits in the answer depends on the position of the estimated digits in the data.

8. In scientific notation, order of magnitude is indicated by powers of _____.

9. Nonsignificant zeroes are never needed if a number is expressed in _____.

10. In multiplying numbers in scientific notation, _____ the exponent of 10.

11. Two quantities cannot be compared unless they have the same _____.

12. In doing arithmetic, quantities with different dimensionality can be _____ or _____.

13. A given number of gallons, converted into SI units, would be expressed in _____.

14. New labels are created by the processes of _____ and _____.

15. The SI unit of time is the _____.

Multiple Choice.

1. A chemist weighs a sample and finds it to be 236.50 grams. In this measurement, the estimated digit is the
    (1) 2   (2) 3   (3) 6   (4) 5   (5) 0.

2. A table is measured and found to be 220.10 centimeters long. The number of significant digits in this reading is
    (1) 1   (2) 2   (3) 3   (4) 4   (5) 5.

3. You walk 15 miles in 4 hours and divide the distance by the time to find your average speed. It is
    (1) 3 miles per hour
    (2) 3.0 miles per hour
    (3) 3.75 miles per hour
    (4) 4 miles per hour
    (5) 4.0 miles per hour.

4. You have a board 10 feet long and cut 2 inches off one end. The length of board remaining is
    (1) 9 feet   (2) 9.8 feet   (3) 9.83 feet   (4) 10.0 feet   (5) 10 feet.

5. The height of *any* high school student is
    (1) 2 meters
    (2) 2.0 meters
    (3) 2.00 meters
    (4) approximately 2.00 meters
    (5) not capable of being expressed as a single number.

6. A school bell is controlled by a clock that is accurate to the nearest second. The length of the period is
    (1) 40 minutes
    (2) 40.0 minutes
    (3) 40.00 minutes
    (4) 40.000 minutes
    (5) 40.0000 minutes.

7. Which of the following is an impossible process?
    (1) 3 cubic feet + 7 liters
    (2) 3 miles $\times$ 6 kilograms per hour
    (3) 4 quarts – 3 square feet
    (4) 20 miles per hour + 35 feet per second
    (5) 16 kilograms $\div$ 25 seconds.

8. In the Système Internationale, the unit of area is the
    (1) acre
    (2) meter
    (3) square meter
    (4) cubic meter
    (5) liter.

9. If a, b, and c are all measurements of length, which of the following is mathematically impossible?
    (1) $4\pi a^2 bc$ (2) $b^2 - 3ac$ (3) $b^3 + ac$ (4) $ac^2 - 2b^2c$ (5) $abc + 3b^3$.

10. The fundamental unit of mass in the SI is the
    (1) kilogram   (2) gram   (3) newton   (4) ounce   (5) centimeter.

11. You are 180 centimeters tall and have a mass of 60 kilograms. Which of the following statements is correct?
    (1) Your height is 3 times your mass.
    (2) The sum of your height and your mass is 240.
    (3) Each centimeter of your height has a mass of 1 kilogram.
    (4) The product of your height and your mass is 10 800.
    (5) The ratio of your height to your mass is 3 centimeters per kilogram.

12. $47 + 1\ 216 + 29.30 =$
    (1) 1 300   (2) 1 290   (3) 1 292   (4) 1 292.3   (5) 1 292.30.

13. $(4.0 \times 10^6) + (6.5 \times 10^5) =$
    (1) 10.5×10^6 (4) 4.7×10^6
    (2) 10.5×10^5 (5) 4.7×10^5.
    (3) 4.65×10^6

14. If a volume measurement in m^3 is divided by an area measurement in m^2, the answer will be expressed in
    (1) m (2) m^2 (3) m^3 (4) m^4 (5) m^5.

Problems.

1. What is the percent of uncertainty of a measurement given as 35 ± 0.2 grams?

2. What is the limit of uncertainty of a measurement given as 655 kilometers $\pm$ 3%?

3. How many significant digits are there in a measurement given as 0.0360 kilograms?

4. $(231 + 45.7 + 6.195) \div 0.056 = ?$

5. $(1.90 \times 66.1) - 122 = ?$

6. $(1.55 \times 10^4) \div 2.8 \times 10^6 = ?$

7. $(6.50 \times 10^{-7}) \times (2.950 \times 10^3) = ?$

8. $(4.775 \times 10^4) + (8.02 \times 10^3) = ?$

9. (3.1 meters) $\times$ (2.7 meters) $\times$ (8.95 meters) = ?

10. (1.07 kilograms) $\div$ (3.88 seconds) = ?

11. $(7.0 \times 10^4$ meters per second) $\div (2.251 \times 10^2$ seconds) = ?

12. $(8.22 \times 10^{-12}$ square meters) $\times (2.2 \times 10^{-3}$ meters per kilogram) = ?

Notions of Motion

WHAT YOU WILL LEARN

It is the job of physics to explain why things happen. But the first task is to describe *what* happens. Here we are going to use algebra to write mathematical descriptions of the way things move.

SECTIONS IN THIS CHAPTER

- How Fast, How Far?
- Vectors: Which Way?
- Velocity Vectors
- A Vector in Parts
- Acceleration
- Speeding Up Steadily
- Falling Freely
- More Free Falls
- Trajectories
- Going in Circles
- It's Relative

How Fast, How Far?

We must begin by defining our variables carefully. You no doubt already have some idea of what is meant by *speed*. Suppose that, as you start on a trip, you note that your watch reads 4:10 and the odometer of your car says 45 350 miles. When you arrive at your destination, your watch says 5:40 and the odometer reads 45 422 miles. Now you do the following calculation:

distance traveled: 45 422 − 45 350 = 72 miles
time elapsed: 5:40 − 4:10 = 1:30 = 1.50 hours
speed = 72 miles/1.50 hours = 48 miles per hour

Now, you do not know how fast you were going at any one time. On the highway, you were doing 55, or perhaps more if no one was watching; when you stopped for a red light, your speed was zero. What you have calculated is the *average* speed for the whole trip.

Your calculation was correct, because it agrees with the standard kinematic definition of average speed, which is

$$v_{av} = \frac{\Delta s}{\Delta t}$$

(Equation 2.1)

Let's decode this equation:

v_{av} stands for average speed
s stands for distance
t stands for time
Δ is the Greek capital letter delta

The symbol Δ calls for some explanation. It can be translated as "change in" or "difference of." Thus Δs stands for the difference between the original odometer reading and the final reading. Therefore Δs is the distance traveled during this particular trip. Similarly, Δt is the time elapsed during the trip. The equation says that average speed is the distance traveled divided by the elapsed time. The unit of speed in the Système Internationale (SI) is the meter per second (m/s), which is a little over 2 miles per hour.

EXAMPLE 2.1 If it takes you 45 seconds to run 340 meters, what is your average speed?

SOLUTION

According to Equation 2.1,

$$v_{av} = \frac{\Delta s}{\Delta t} = \frac{340 \text{ m}}{45 \text{ s}} = 7.6 \text{ m/s}$$

EXAMPLE 2.2 How long does it take a skier to travel 650 meters, going 18.5 meters per second?

SOLUTION

Equation 2.1 can be transformed into

$$\Delta t = \frac{\Delta s}{v_{av}} = \frac{650 \text{ m}}{18.5 \text{ m/s}} = 35 \text{ s}$$

Note how the units behave:

$$\frac{\text{m}}{\text{m/s}} = \cancel{\text{m}}\left(\frac{\text{s}}{\cancel{\text{m}}}\right) = \text{s}$$

EXAMPLE 2.3

At 9:00 A.M. you start on a walk. At 10:30 A.M. you are 4.6 miles from your starting point. In SI units, what was your average speed? (There are about 1 600 meters to a mile.)

SOLUTION

Distance traveled, Δs, is

$$4.6 \text{ mi} \times \frac{1\ 600 \text{ m}}{1 \text{ mi}} = 7\ 360 \text{ m}$$

Elapsed time, Δt, is

$$1.5 \text{ hours} \times \frac{3\ 600 \text{ s}}{1 \text{ hour}} = 5\ 400 \text{ s}$$

From Equation 2.1,

$$v_{\text{av}} = \frac{\Delta s}{\Delta t} = \frac{7\ 360 \text{ m}}{5\ 400 \text{ s}} = 1.36 \text{ m/s}$$

Your speedometer does not indicate averages; it tells you how fast you are going *now*. This is your *instantaneous speed*. If the speedometer reading does not change during the trip, you are said to be going at *constant speed*. In this case the constant value of the speed will be equal to your average speed for the whole trip.

CORE CONCEPT

Average speed is distance traveled divided by the elapsed time:

$$v_{\text{av}} = \frac{\Delta s}{\Delta t}$$

TRY THIS
—1—

What is the average speed of a runner who finishes the 1 500-meter race in 3 minutes 30 seconds?

Vectors: Which Way?

Your speedometer gives you very incomplete information. You cannot rely on it to tell you where you will end your trip, because it says nothing about *which way* you are going. The quantity that tells you that is called *velocity*.

Velocity is one of the many physical quantities that have direction. Quantities of this kind are called *vectors*. In contrast, quantities that have no directional property, such as time, temperature, energy, and volume, are called *scalars*. A scalar quantity is completely specified when you give its magnitude and units. A vector quantity is incomplete unless its direction is also stated. Speed, as distinguished from velocity, is a scalar.

In printing algebraic expressions, it is sometimes important to distinguish vector from scalar quantities. Symbols for scalars are written in lightface italics—for example, *t* for time. In expressions in which direction is important, vectors are printed in boldface roman type, as **v** for velocity. If the magnitude, but not the direction of a vector is significant, italic type may be used. Thus, speed, the magnitude of velocity, will be represented as *v*. Abbreviations for units are always in lightface roman; 35 V means 35 volts.

The most basic of all vectors, the quantity that defines the properties of a vector, is *displacement*. Displacement is simply distance and direction from some starting point or origin of coordinates. Suppose you start from your home and walk 2 miles. Where are you? Well, which way did you go? If you walked straight north, your displacement is 2 miles north.

Vectors can be added, but they do not obey the same rules of mathematics as scalars. If you walk 2 miles north and then 3 miles east, you have walked 5 miles. But your displacement from your starting point, as shown in Figure 2.1, is 3.6 miles at 56° east of north. This quantity is called the *vector sum* of the other two vectors.

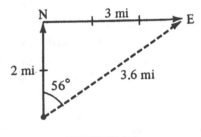

FIGURE 2.1

Look at the vectors labeled **A** and **B** in Figure 2.2. They might represent displacements, velocities, electric fields, or any other physical quantity that has direction. The vector **A** points up and a little to the right. The vector **B** is longer, indicating that its magnitude is greater than that of **A**. Also, its direction is different—down and to the right.

To perform a *vector addition,* just imagine that these two vectors represent displacements. You start at the tail end of **A** and walk the distance and direction that this vector represents. Then, from your new position, you walk the distance and direction specified by **B**. Your total displacement, the vector sum of **A** and **B**, can be found by drawing the vector **B** with its tail at the head of **A**. Then, an arrow starting at the tail of **A** (your original starting point) and drawn to the head of **B** (your endpoint) is the vector sum of **A** and **B**, written as **A** + **B**. When the vectors represent quantities other than displacements, the length of each vector must be proportional to the magnitude of the quantity. Thus a vector 2 cm long would represent an electrical attraction twice as strong as a vector 1 cm long, for example.

You might have elected to add the two vectors in the opposite order. The last diagram in the figure represents the vector sum **B** + **A**, and it gives the same sum as **A** + **B**. Maybe it looks different because it is in a different configuration, but if you examine it carefully, you will see that it is the same length as **A** + **B** and points in the same direction. Therefore it is the same vector.

When vectors are at right angles to each other, they can be added very easily by means of the simple trigonometry rules.

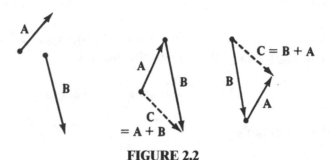

FIGURE 2.2

 EXAMPLE 2.4 You walk 7 miles south and then 3 miles west. What is your displacement from your starting point?

SOLUTION

The sum of the two displacements is represented in the diagram of Figure 2.3. The vector pointing south is 7 units long and the one pointing west is 3 units long. The magnitude of the vector sum is found by the theorem of Pythagoras:

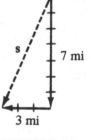

7 mi

3 mi

FIGURE 2.3

$$(7 \text{ mi})^2 + (3 \text{ mi})^2 = s^2$$
$$s = 7.6 \text{ mi}$$

The direction is found by using the tangent function:

$$\frac{3 \text{ mi}}{7 \text{ mi}} = \tan\theta$$
$$\theta = 23°$$

The displacement is $s = 7.6$ mi 23° W of S.

EXAMPLE 2.5 A child is playing with a toy car on the floor of a train that is moving eastward. While the train travels 12.0 meters, the child pushes the car 2.6 meters northward on the floor of the train. What is the resulting displacement of the car?

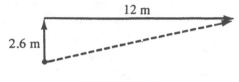

FIGURE 2.4

SOLUTION

The two displacements can be added in either sequence. Figure 2.4 shows the vector addition. The magnitude of the total displacement is

$$s = \sqrt{(2.6 \text{ m})^2 + (12.0 \text{ m})^2} = 12.3 \text{ m}$$

The angle has a tangent of

$$\frac{12.0 \text{ m}}{2.6 \text{ m}} = 4.615$$

so the direction of the vector sum is 78° E of N.

CORE CONCEPT

Vectors are quantities with both magnitude and direction, and they can be added geometrically.

TRY THIS
—2—

Find the vector sum of 3.5 meters east and 5.8 meters south. Give both magnitude and direction of the sum.

Velocity Vectors

Sometimes you have two different velocities at the same time. Your total velocity is then the vector sum of the separate velocities. It is found by vector addition, using the same rules as in adding displacement vectors.

Suppose, for example, a boat can travel 4.0 meters per second in still water. It is in a river that flows southward at 5.5 meters per second, as shown in Figure 2.5. If the boat heads eastward, directly across the river, what are the direction and magnitude of its total velocity?

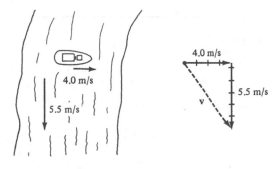

FIGURE 2.5

The vector diagram in Figure 2.5 shows how the two velocity vectors can be added. You can add them in either sequence; in the diagram, the order shown is

boat's velocity + river's velocity = total velocity

Since these velocities are vectors, they must be added by the rules of vector mathematics, that is, as shown in the vector diagram. We get the answer by solving the triangle. The magnitude of the total velocity of the boat is the hypotenuse of the triangle. Using the theorem of Pythagoras, we obtain

$$v_{total}{}^2 = (4.0 \text{ m/s})^2 + (5.5 \text{ m/s})^2$$

from which

$$v_{total} = 6.8 \text{ m/s}$$

The boat will obviously not travel directly eastward, but at some angle south of east. To find the angle, use the tangent function:

$$\tan\theta = \frac{5.5 \text{ m/s}}{4.0 \text{ m/s}}$$

which gives $\theta = 54°$. The total velocity of the boat is therefore $\mathbf{v} = 6.8$ meters per second 54° south of east.

EXAMPLE 2.6 An airplane is headed directly east at 340 miles per hour when the wind is from the south at 45 miles per hour. What is its velocity with respect to the ground?

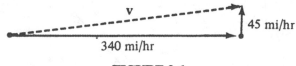

FIGURE 2.6

SOLUTION

The vector diagram for adding these two velocities is shown in Figure 2.6. The hypotenuse of a right triangle is found this way:

$$v^2 = \left(340 \frac{\text{mi}}{\text{hour}}\right)^2 + \left(45 \frac{\text{mi}}{\text{hour}}\right)^2$$

and the angle is

$$\tan\theta = \frac{45 \text{ mi/hour}}{340 \text{ mi/hour}}$$
$$\theta = 7.5°$$

Total velocity is **v** = 343 mi/hour at 7.5° N of E.

 EXAMPLE 2.7 You are exercising by running around the deck of an ocean liner traveling west at 18.5 meters per second. What is your total velocity if you are running athwartships, going south at 7.3 meters per second?

SOLUTION

Since the two velocities are perpendicular to each other, they add by the theorem of Pythagoras:

$$v = \sqrt{(18.5 \text{ m/s})^2 + (7.3 \text{ m/s})^2} = 19.9 \text{ m/s}$$

The direction west of south is the angle whose tangent is the ratio of the west velocity to the south velocity:

$$\tan\theta = \frac{18.5 \text{ m/s}}{7.3 \text{ m/s}}$$
$$\theta = 68° \text{ W of S}$$

 EXAMPLE 2.8 You want to drive your motorboat directly across a stream that flows at 3.5 meters per second, and you know that your boat can do 4.6 meters per second in still water. Find (a) the angle upstream at which you must point the boat; (b) the resulting speed of the boat in the cross-stream direction.

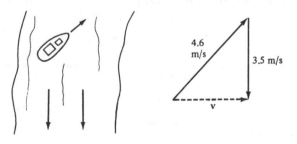

FIGURE 2.7

SOLUTION

As shown in Figure 2.7, the velocity of the boat is now the hypotenuse of a right triangle and the resulting velocity is its short leg.

(a) To find the angle, note that we know the hypotenuse of the triangle and the side opposite the angle we are looking for. Then

$$\sin\theta = \frac{\text{opp}}{\text{hyp}} = \frac{3.5 \text{ m/s}}{4.6 \text{ m/s}}$$

so

$$\theta = 50°$$

(b) To find the resulting speed:

$$v^2 + (3.5 \text{ m/s})^2 = (4.6 \text{ m/s})^2$$
$$v^2 = (4.6 \text{ m/s})^2 - (3.5 \text{ m/s})^2$$
$$v = 3.0 \text{ m/s}$$

CORE CONCEPT

Velocities are vectors and can be added by the same mathematical rules that govern displacement vectors.

TRY THIS
—3—

In an airplane that goes 620 miles per hour, the pilot wants to fly directly north when the wind is blowing at 110 miles per hour from the east. In what direction must he head the airplane?

A Vector in Parts

In some cases, a vector tells us more than we really want to know. If you get stopped for going 75 miles per hour in a 55 miles per hour zone, the cop is not concerned with whether the direction of your motion is east or north. The only part of the vector that is of any interest is its magnitude.

In other cases, we want to know only what effect the vector has in a particular direction. For example, if you want to know how your time zone changes as you travel, your displacement northward is of no significance; you need to know only how far east or west you have come. And if you are concerned with the height of the sun in the sky at noon, it is only your north-south displacement that counts. Typically, a problem might be stated this way: If I travel 350 miles in a direction 30° west of north, how much is my displacement in the northward direction? We want to ignore any movement along an east-west axis. The quantity we are looking for is the *northward component* of the displacement vector.

It is easily found, as shown in Figure 2.8. From the definition of the cosine, we can write

$$\cos 30° = \frac{s_N}{350 \text{ mi}}$$

where s_N is the northward component of the displacement vector. Its value is

$$s_N = (350 \text{ mi})(\cos 30°) = 303 \text{ mi}$$

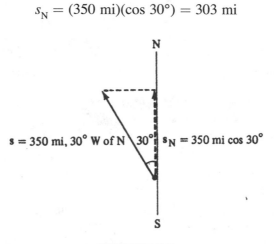

FIGURE 2.8

If you are worried about the time of sunrise, you want the component of your displacement vector on an east-west axis. The principle is the same—see Figure 2.9. This time the displacement vector is making an angle of 60° with the axis, so the equation is

$$s_W = (350 \text{ mi})(\cos 60°) = 175 \text{ mi}$$

This process works on any kind of vector, including velocity vectors. We can write a perfectly general rule for finding the component of any vector on any axis:

$$P_x = \mathbf{P} \cos \theta_x \qquad \text{(Equation 2.2)}$$

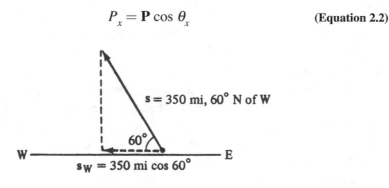

FIGURE 2.9

where **P** is a vector (as indicated by boldface type), P_x is its component on the x axis, and θ_x is the angle between the vector and the axis.

**EXAMPLE
2.9** You are driving along at 20 meters per second on a road that heads at an angle of 25° W of N. At what rate are you going (a) north; (b) west?

SOLUTION

(a) Since your velocity makes an angle of 25° with the north axis, your velocity northward is

$$v_N = (20 \text{ m/s})(\cos 25°) = 18 \text{ m/s}$$

(b) The angle that your velocity makes with the west axis is $(90° - 25°) = 65°$. Therefore your westward velocity is

$$v_W = (20 \text{ m/s})(\cos 65°) = 8.5 \text{ m/s}$$

It should be noted that $\cos (90° - 25°) = \sin 25°$. In general, the component of any vector on an axis perpendicular to the axis given is the vector times the sine of the angle. Thus the west component is

$$v_W = (20 \text{ m/s})(\sin 25°) = 8.5 \text{ m/s}$$

**EXAMPLE
2.10** An airplane descending to the runway at 250 meters per second is going at an angle of 22° with the horizontal. Find (a) its horizontal velocity; (b) its rate of descent.

SOLUTION

(a) Since the airplane makes an angle of 22° with the horizontal, the horizontal component of its velocity is

$$v_{horiz} = (250 \text{ m/s})(\cos 22°) = 232 \text{ m/s}$$

(b) The rate of descent is its vertical velocity, at right angles to the horizontal, so

$$v_{vert} = (250 \text{ m/s})(\sin 22°) = 94 \text{ m/s}$$

CORE CONCEPT

The component of a vector on any axis is found by multiplying the vector by the cosine of the angle that the vector makes with the axis:

$$P_x = \mathbf{P} \cos \theta_x$$

You are going 40 miles per hour up a hill that makes an angle of 22° with the horizontal. How fast are you going (a) horizontally; (b) vertically?

Acceleration

Things get more complicated when you change your velocity. You do it in your car by stepping on the gas or the brakes, or by turning the steering wheel. Remember, velocity is a vector, so a change in direction is a change in velocity. Whenever velocity is changing—in either magnitude or direction—the motion is said to be *accelerated*.

Let's take the simplest case first. Suppose you are trying out your new car. You are satisfied to find that you can get from 0 to 60 miles per hour in 9.2 seconds. What you have just measured is the *acceleration* of the car.

Acceleration is the rate at which velocity changes. You can find the average acceleration of the car this way:

$$\text{acceleration} = \frac{60 \text{ mi/hour} - 0}{9.2 \text{ sec}} = 6.5 \text{ mi/hour/sec}$$

The average acceleration is 6.5 miles per hour per second. In other words, the speed increases an average of 6.5 miles per hour in each second of the trip.

This leads to an algebraic definition of average acceleration:

$$\mathbf{a}_{av} = \frac{\Delta \mathbf{v}}{\Delta t} \qquad \textbf{(Equation 2.3)}$$

In words: average acceleration is the change in velocity divided by the length of time required to make that change.

Let's take a specific case. Suppose a car is traveling at 12 meters per second and increases its speed uniformly to 30 meters per second, taking 15 seconds to do so. "Uniformly" means that the acceleration of the car is constant, so it must be equal to the average acceleration during that time. The question is, What is the value of that acceleration?

First of all, $\Delta v = 18$ meters per second, representing an increase from 12 meters per second to 30 meters per second. Therefore

$$a = \frac{\Delta v}{\Delta t} = \frac{18 \text{ m/s}}{15 \text{ s}} = 1.2 \text{ m/s}^2$$

Note that the SI unit of acceleration is the meter per second per second, or meter per second squared. In each second, the speed of the car increased by 1.2 meters per second.

See the sample problems for other examples of the use of this relationship.

EXAMPLE 2.11 What is the acceleration of a rocket ship in outer space that takes 5.0 seconds to increase its speed from 1 240 meters per second to 1 300 meters per second?

SOLUTION

The change in its speed is

$$\Delta v = 1\ 300\ \text{m/s} - 1\ 240\ \text{m/s} = 60\ \text{m/s}$$

Then

$$a = \frac{\Delta v}{\Delta t} = \frac{60\ \text{m/s}}{5.0\ \text{s}} = 12\ \text{m/s}^2$$

EXAMPLE 2.12 How much does the speed of a car increase if it accelerates uniformly at 2.5 meters per second squared for 5 seconds?

SOLUTION

Equation 2.3 gives the definition of average acceleration:

$$a_{\text{av}} = \frac{\Delta v}{\Delta t}$$

Since the acceleration is constant, it is equal to a_{av}. Multiply the equation through by Δt to get

$$\Delta v = a\ \Delta t$$

$$\Delta v = \left(2.5\frac{\text{m}}{\text{s}^2}\right)(5\ \text{s}) = 12.5\ \text{m/s}$$

EXAMPLE 2.13 A car is going 8.0 meters per second on an access road into a highway, and then accelerates at 1.8 meters per second squared for 7.2 seconds. How fast is it then going?

SOLUTION

Equation 2.3 can be transformed to

$$\Delta v = a\Delta t = \left(1.8\frac{\text{m}}{\text{s}^2}\right)(7.2\ \text{s}) = 13.0\ \text{m/s}$$

This is the *increase* in the speed of the car. Since it started at 8.0 meters per second, its new speed is

$$8.0 + 13.0 = 21\ \text{m/s}$$

CORE CONCEPT

Acceleration is the rate of change of velocity:

$$\mathbf{a}_{av} = \frac{\Delta \mathbf{v}}{\Delta t}$$

TRY THIS
—5—

In outer space a rocket ship is traveling at the enormous speed of 2 800 meters per second. What is its acceleration if it increases its speed uniformly and is going 2 840 meters per second after 25 seconds?

Speeding up Steadily

When motion is accelerated, velocity varies as a function of time. Like any function, this one can be made visual by representing it as a graph. Such a graph is shown in Figure 2.10.

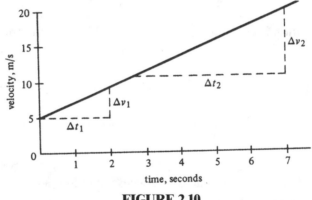

FIGURE 2.10

What does this graph tell you? First of all, note that at time 0 the moving object is going 5 meters per second. Time 0 is purely arbitrary; it is the moment when we start our stopwatch. Two seconds later, the velocity has increased to 8 meters per second, and it keeps on increasing for the entire 7 seconds plotted in the graph. We do not know what it was doing before we started our clock, or after the first 7 seconds.

The graph can tell us the acceleration of the object. From Equation 2.3,

$$a = \frac{\Delta v}{\Delta t}$$

In the first 2 seconds, $\Delta t = 2$ seconds and $\Delta v = 8$ m/s – 5 m/s = 3 m/s. The acceleration is therefore 1.5 m/s^2.

It is clear from the figure that some other time interval would give the same value of acceleration. From simple geometry,

$$\frac{\Delta v_1}{\Delta t_1} = \frac{\Delta v_2}{\Delta t_2}$$

The acceleration is the same regardless of which time interval we elect in order to calculate it. This is true because the graph is a straight line.

We can do a lot of physics with a simple form of motion, which can be handled algebraically with ease. This is motion that is uniformly accelerated, such as shown in the straight line of Figure 2.10. Let's get a useful set of equations relating distance traveled (s), time elapsed (t), acceleration (a), and the initial and final speeds (v_i and v_f).

When acceleration is constant, it is equal to the average acceleration, so Equation 2.3 applies. Then the increase in speed is just at. To get the final speed, just add this to the initial speed:

$$v_f = v_i + at \qquad \textbf{(Equation 2.4a)}$$

While the object is speeding up, its average speed is just the average of the initial and final speeds. Equation 2.1 tells us we get the distance traveled by multiplying the average speed by the time:

$$s = \left(\frac{v_i + v_f}{2}\right)t$$

Now we get a useful equation by substituting into this equation the value of v_f from Equation 2.4a:

$$s = v_i t + \frac{1}{2}at^2 \qquad \textbf{(Equation 2.4b)}$$

The first term is the distance that would be traveled at the original speed; the second term is the additional distance due to the acceleration.

Combining these equations produces another useful form, relating the speeds to the acceleration and distance:

$$v_f^2 = v_i^2 + 2as \qquad \textbf{(Equation 2.4c)}$$

EXAMPLE 2.14

How far does a car travel if it is going 4.0 meters per second and accelerates at 3.5 meters per second squared for 5.0 seconds?

SOLUTION

$$s = v_i t + \frac{1}{2}at^2$$
$$s = (4.0 \text{ m/s})(5.0 \text{ s}) + \frac{1}{2}(3.5 \text{ m/s}^2)(5.0 \text{ s})^2$$
$$s = 64 \text{ m}$$

What is the acceleration of a rocket that speeds up uniformly from rest and travels 650 meters in the first 12 seconds?

SOLUTION

In Equation 2.4b, $v_i = 0$; solving for a gives

$$a = \frac{2s}{t^2} = \frac{2(650 \text{ m})}{(12 \text{ s})^2}$$
$$a = 9.0 \text{ m/s}^2$$

If a car is going at 12 meters per second, how long will it take it to reach a speed of 26 meters per second if it accelerates at 2.2 meters per second squared?

SOLUTION

Solving Equation 2.4a for t gives

$$t = \frac{v_f - v_i}{a} = \frac{26 \text{ m/s} - 12 \text{ m/s}}{2.2 \text{ m/s}^2}$$
$$t = 6.4 \text{ s}$$

How far does a car travel in speeding up from 4.5 meters per second to 22 meters per second if its acceleration is 3.6 meters per second squared?

SOLUTION

Solving Equation 2.4c for s gives

$$s = \frac{v_f^2 - v_i^2}{2a} = \frac{(22 \text{ m/s})^2 - (4.5 \text{ m/s})^2}{2 \,(3.5 \text{ m/s}^2)}$$
$$s = 66 \text{ m}$$

A car going 19 meters per second jams on its brakes, slowing down to 8.4 meters per second and leaving skid marks 24.6 meters long. What was its acceleration?

SOLUTION

Slowing down is also a kind of acceleration; it will have a negative value. From Equation 2.4c, the acceleration is

$$a = \frac{v_f^2 - v_i^2}{2s} = \frac{(8.4 \text{ m/s})^2 - (19 \text{ m/s})^2}{2 \,(24.6 \text{ m})}$$
$$a = -5.9 \text{ m/s}^2$$

CORE CONCEPT

Problems involving objects that accelerate uniformly, starting or ending at rest, can be solved with a set of three simple equations.

TRY THIS
—6—

What is the acceleration of a rocket-driven sled that travels 360 meters in 8.3 seconds, starting at 22 meters per second?

Falling Freely

When something is dropped, its velocity starts at zero and increases. It is accelerated. It is a remarkable fact that in a vacuum all objects—rocks, books, feathers, raindrops, dust particles—have the same acceleration. If dropped together, all will accelerate uniformly and strike the ground at the same time. This rule does not hold in everyday experience only because the resistance of the air tends to hold things back—and it holds feathers more effectively than rocks.

The acceleration of an object in free fall (in a vacuum) does not depend on the nature of the object. It depends only on where the object happens to be. The acceleration is uniform, and its value is the *acceleration due to gravity,* represented by the letter g. Near the earth's surface, g is about 9.8 meters per second squared, or 32 feet per second squared.

The value is far different on other planets—3.3 meters per second squared on Mars, 25.6 meters per second squared on Jupiter; and when our astronauts walked on the moon, anything they dropped accelerated downward at a mere 1.67 meters per second squared. Even on earth, the acceleration due to gravity varies a little. It is about 9.78 meters per second squared at the equator, increasing to 9.83 meters per second squared at the poles. It also decreases with altitude, dropping from 9.80 meters per second squared at the surface (at latitude 40°) to 9.79 meters per second squared at altitude 10 miles.

Whatever the local value of g, it can be used in Equations 2.4b to find out how long it takes something to hit the ground or how fast it will be going when it hits. Just remember that for an object in free fall, $a = g$. Sample Problems 2.19 and 2.20 show how to proceed. A word of caution, however: If the object is very light, or if it has a lot of flat surface, or if it falls a great distance, air resistance becomes significant, and use of these simple equations will not be appropriate.

EXAMPLE 2.19 A high-wire artist missteps and falls 9.2 meters to the ground. What is the acrobat's speed on landing?

SOLUTION

The motion is uniformly accelerated and starts at rest, so $v_i = 0$ and from Equation 2.4c,

$$v = \sqrt{2as} = \sqrt{2\left(9.8\,\frac{m}{s^2}\right)(9.2\ m)} = 13\ m/s$$

 EXAMPLE 2.20 A cat falls out of a tree, dropping 16 meters to the ground. How long is the cat in the air?

SOLUTION

Since the cat started at rest and accelerated uniformly, $v_i = 0$, so, from Equation 2.4b,

$$s = \frac{1}{2}gt^2$$

Multiply through by $2/g$ and take the square root of both sides to get

$$t = \sqrt{\frac{2s}{g}}$$

$$t = \sqrt{\frac{2\,(16\ m)}{9.8\ m/s^2}} = 1.8\ s$$

 EXAMPLE 2.21 You throw a ball straight up into the air at a speed of 22.0 meters per second. If someone catches it 3.6 seconds later, how fast is it going?

SOLUTION

You have to be careful of the signs in this problem. The upward direction is taken as positive, so the initial speed is $+22.0$ m/s, and $a = g = -9.8$ m/s^2. We can find the final speed from Equation 2.4a:

$$v_f = v_i + at = 22.0\ m/s + (-9.8\ m/s^2)(3.6\ s)$$
$$v_f = -13\ m/s$$

The ball was caught on the way down.

CORE CONCEPT

An object in free fall accelerates uniformly, and the value of the acceleration due to gravity depends only on the object's location in space.

You drop a rock from a bridge, and it hits the water 2.3 seconds later. Find (a) the height of the bridge; (b) the velocity of the rock when it hits.

More Free Falls

To put a rock into free fall, you do not necessarily just drop it. You can also throw it— up, down, or sideways. Whichever you do, the rock is in free fall once it leaves your hand, and it must therefore accelerate downward at the rate of the acceleration due to gravity—9.8 meters per second squared.

When you throw the rock, it is moving upward, slowing down as it rises. *Any change in velocity, speeding up, slowing down or changing direction, is an acceleration.* When an object is slowing down, its acceleration is in the opposite direction to its velocity. The rock, rising and slowing down, is accelerated downward by the earth's gravity.

Suppose, for example, you throw the rock straight upward and it leaves your hand going 25 meters per second. Since its acceleration is downward, opposite to its velocity, it must slow down. After 1 second, it is going (25 meters per second – 9.8 meters per second) = 15.2 meters per second. It loses another 9.8 meters per second of velocity during the next second. Eventually, it comes to rest and starts downward. Then it speeds up, since its acceleration is now in the same direction as its velocity.

Look at the graph of Figure 2.11 to see how this accelerated motion takes place. At time 0, the velocity is 25 meters per second. When 2.6 seconds have passed, the velocity has dropped to zero—the rock has come to rest at the top of its flight. From that point on, the velocity is negative, meaning that the rock is now on the way down. It stops at the point from which it was thrown, going 25 meters per second. But through the entire flight, the slope of the graph—the acceleration—has not changed. Even when the rock is at the top of its flight, at rest, its velocity has not stopped changing, at the rate of 9.8 meters per second squared.

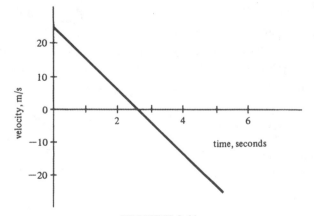

FIGURE 2.11

EXAMPLE 2.22

You throw a baseball straight up, and it leaves your hand at 15 meters per second. What is its velocity 2.0 seconds later?

SOLUTION

Equation 2.4a provides the answer:

$$v_f = v_i + at = 15 \text{ m/s} + (-9.8 \text{ m/s}^2)(2.0 \text{ s})$$
$$v_f = -4.6 \text{ m/s}$$

The ball is on the way down.

CORE CONCEPT

The acceleration due to gravity is the same for all objects in free fall, regardless of the direction of their velocities.

TRY THIS
—8—

A tennis ball is struck with a racket, firing it straight upward at 22 meters per second. After how much time will it be falling at 15 meters per second?

Trajectories

If a shell is to be fired from a cannon, it would be useful to be able to predict where it will land. The path of a projectile in free fall is called its *trajectory*.

The first thing to understand about the trajectory of a projectile is that, neglecting air resistance, the object is in free fall. That means that its motion is affected only by gravity; it accelerates downward at –9.8 meters per second squared. This rule applies whether it is moving up, down, or sideways.

Suppose you stand on a roof and throw a ball horizontally. You have given it no vertical motion at all. Vertical velocity will be supplied by gravity at the rate of 9.8 meters per second squared, starting at zero. No matter how hard you throw it, the ball will reach the ground in exactly the same length of time as if you just dropped it—as long as the initial velocity is strictly horizontal.

Now, how about the horizontal motion? Nothing is acting on the ball except gravity, and that affects only the vertical motion. Whatever speed you give the ball, it will keep on flying horizontally at that speed until gravity brings it to the ground. The net result of these two motions is shown in Figure 2.12. The horizontal velocity is constant, and the vertical velocity is accelerated.

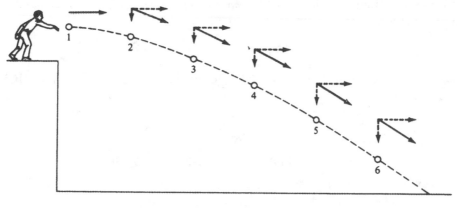

FIGURE 2.12

| EXAMPLE 2.23 | A ball is thrown horizontally at 25 meters per second from a roof that is 15 meters high. How far does it travel before hitting the ground? |

SOLUTION

The first step is to find how long the ball is in the air. This depends only on its vertical velocity, which starts at zero and increases at the rate of 9.8 m/s². From Equation 2.4b, with $v_i = 0$,

$$t = \sqrt{\frac{2s}{g}} = \sqrt{\frac{2\,(15\text{ m})}{9.8\text{ m/s}^2}}$$
$$t = 1.75 \text{ s}$$

It travels horizontally at its initial speed for this length of time, so

$$s = vt = (25\text{ m/s})(1.75\text{ s}) = 44\text{ m}$$

Now suppose that ball is not thrown horizontally but at an angle upward. This will surely carry it further. We still have to deal with the two components of its velocity separately. The first step is to separate the initial velocity into its vertical and horizontal parts. Then follow pretty much the same precedure as in the previous problem:

| EXAMPLE 2.24 | The ball is now thrown with the same speed at an angle of 35° upward. How far does it travel? |

SOLUTION

The horizontal component of its velocity, which does not change, is (from Equation 2.2)

$$v_{\text{hor}} = (25\text{ m/s}) \cos 35° = 20.5\text{ m/s}$$

The angle with the vertical is $(90° - 35°) = 55°$, so the vertical component of its initial velocity is

$$v_{vert} = (25 \text{ m/s}) \cos 55° = 14.3 \text{ m/s}$$

Equation 2.4a tells us how long it takes for the ball to reach its maximum height, where $v_f = 0$:

$$t = \frac{-v_i}{a} = \frac{-14.3 \text{ m/s}}{-9.8 \text{ m/s}^2} = 1.46 \text{ s}$$

Now we have to know how high the ball rises. From Equation 2.4b:

$$s = \frac{1}{2} at^2 = \frac{1}{2}(9.8 \text{ m/s}^2)(1.46 \text{ s})^2 = 10.4 \text{ m}$$

and it has to fall to the ground a distance of 10.4 m + 15 m, so

$$t = \sqrt{\frac{2s}{a}} = \sqrt{\frac{2(25.4 \text{ m})}{9.8 \text{ m/s}^2}} = 2.28 \text{ s}$$

So the total time of flight, up and down, is 1.46 s + 2.28 s = 3.74 s. During all that time, it has kept up its constant horizontal speed. Therefore, the horizontal distance it travels is

$$s = vt = (20.5 \text{ m/s})(3.74 \text{ s}) = 77 \text{ m}$$

After all that work, you might expect that you now know how far the ball will travel. You don't unless you are living in a vacuum. Air resistance will surely reduce that distance considerably. At least you know that it cannot be more than 77 meters.

CORE CONCEPT

Trajectories can be calculated by dealing separately with the accelerated vertical motion and the steady horizontal velocity.

TRY THIS
—9—

A slingshot fires a stone horizontally from a tower 32 meters high, and it lands 135 meters from the foot of the tower. What was the speed with which it left the slingshot?

Going in Circles

When an object is traveling in a circular path at constant speed, its direction is constantly changing. Therefore this is an example of accelerated motion.

The motorcyclist in Figure 2.13 is going around a circular track of radius r, and his speed is v. The direction of his velocity is tangential; at the moment, he is at the northernmost point of the track and traveling eastward. Let T stand for the *period* of his motion, that is, the length of time it takes him to go around once. Since the distance around the track is $2\pi r$, his speed is

$$v = \frac{2\pi r}{T}$$

(Equation 2.5a)

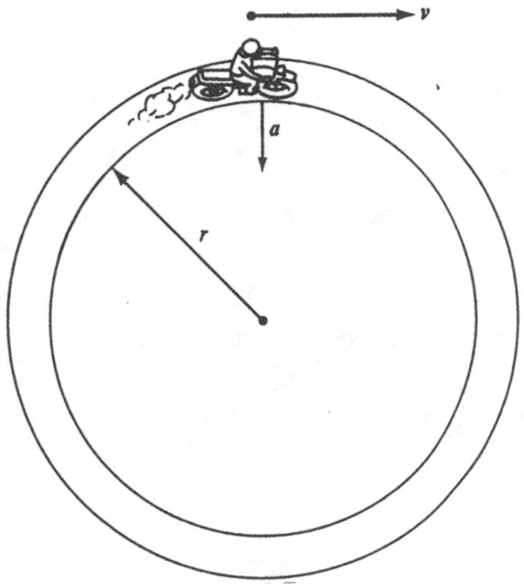

FIGURE 2.13

What is the period of the motion of a runner going 9.2 meters per second on a circular track whose radius is 22 meters?

SOLUTION

From Equation 2.5a,

$$T = \frac{2\pi r}{v} = \frac{2\pi\ (22\ \text{m})}{9.2\ \text{m/s}} = 15.0\ \text{s}$$

As the motorcyclist travels, he is constantly turning to the right, and his speed is constant. This means that his acceleration is to the right and perpendicular to his motion, always toward the center of the circle. The acceleration is said to be *centripetal,* which simply means toward the center of a circle.

The magnitude of centripetal acceleration can be calculated from this formula:

$$a_c = \frac{v^2}{r}$$

(Equation 2.5b)

See the sample problems for examples.

EXAMPLE 2.26 What is the acceleration of a motorcycle going 28 meters per second on a circular track whose radius is 140 meters?

SOLUTION

From Equation 2.5b,

$$a_c = \frac{v^2}{r} = \frac{(28 \text{ m/s})^2}{140 \text{ m}} = 5.6 \text{ m/s}^2$$

EXAMPLE 2.27 To dry clothes in a rotary dryer, the clothes should be accelerated toward the center of the tub at $15g$. If the tub has a diameter of 80 centimeters, how many revolutions per second must it make?

SOLUTION

First, find the necessary velocity of the rim of the tub. The radius of the tub is 0.40 m, and $15g$ means 15 times the acceleration due to gravity. From Equation 2.5b,

$$v = \sqrt{ar} = \sqrt{(15 \times 9.8 \text{ m/s}^2)(0.40 \text{ m})} = 7.7 \text{ m/s}$$

From Equation 2.5a,

$$T = \frac{2\pi r}{v} = \frac{2\pi \, (0.40 \text{ m})}{7.7 \text{ m/s}} = 0.33 \text{ s}$$

Since the tub goes around once every 1/3 second, it must be making 3 revolutions per second.

CORE CONCEPT

When acceleration has constant magnitude and is kept always perpendicular to velocity, motion is in a circle at constant speed and the centripetal acceleration can be calculated:

$$a_c = \frac{v^2}{r}$$

TRY THIS
—10—

The motorcyclist of Figure 2.13 is on a track whose radius is 150 meters, and he makes one complete circuit every 30 seconds. Find the direction and magnitude of his (a) velocity and (b) acceleration when he is at the southernmost point of the track.

It's Relative

You are in a train going 40 miles an hour, playing a game of catch. You throw a ball forward to your friend, at 30 miles an hour. How fast is the ball going?

Well, that depends. To someone watching from the station, the ball is going 70 miles an hour. Your friend, however, does not have to cope with such a fast ball. To him, it is traveling 30 miles an hour, whether the train is moving or not. Whether the ball is going at 30 or 70 miles an hour depends on whether the frame of reference is attached to the train or to the station.

Now suppose the object in question is not a ball but a toy rocket, which accelerates from you to your friend. In the frame of reference attached to the train, the forward velocity of the rocket when your friend catches it is $\frac{1}{2}at^2$. The viewer from the station sees it as $v_i + \frac{1}{2}at^2$. The two viewers disagree on the forward velocity, but they agree on the amount of acceleration. If the train is going at constant speed velocity is *relative*, but the acceleration is *invariant*, which simply means that it is independent of the frame of reference.

Now suppose the ball is dropped out of the window. What is its path as it falls? The rules governing the trajectory of a falling object are invariant. When the person inside the train drops the ball, the ball is initially at rest. From that frame of reference, the ball falls straight down. The person in the station sees the ball in horizontal motion when it drops, so the path is curved (actually, it is a parabola—see "Trajectories" page 38). While the laws governing the motion of the ball are invariant, the actual path of the ball is relativistic.

What is the true path of the ball? Must we take into account the rotation of the earth and its revolution around the sun? We would find a different path from each frame of reference. All these paths are true, each in its own frame of reference.

A guiding principle of twentieth-century physics is the rule that *the laws of nature are equally true in every frame of reference*. This *relativity principle* is now considered to be a fundamental criterion in the decision as to whether a statement or an equation is in fact a general law of nature.

A simple equation can transform the value of a velocity from one frame of reference to another. If **v** and **v′** are the velocities in two different frames of reference and **u′** is the relative velocity of the two frames, then the *Galilean velocity transformation* is

$$\mathbf{v'} = \mathbf{v} + \mathbf{u} \qquad \text{(Equation 2.6)}$$

Note that it is a vector equation.

EXAMPLE 2.28   Suppose you throw that ball across the width of the train. What is its speed in a frame of reference attached to the station?

SOLUTION

Add **v** and **u** vectorially. They are at right angles, so

$$\mathbf{v'} = \sqrt{\mathbf{v}^2 + \mathbf{u}^2} = \sqrt{(30 \text{ mi/hr})^2 + (40 \text{ mi/hr})^2} = 50 \text{ mi/hr}$$

CORE CONCEPT

Measurements of physical quantities differ according to frame of reference, but the laws of nature apply in all frames.

TRY THIS
—11—

An elevator is moving upward at a speed of 3.0 meters per second. A passenger in the elevator drops a watch. Find the velocity and acceleration of the watch in a frame of reference attached to (a) the elevator; and (b) the building.

REVIEW EXERCISES FOR CHAPTER 2

Fill-In's. For each of the following, fill in the missing word or phrase.

1. Distance traveled divided by elapsed time gives _____.

2. The SI unit of velocity is the _____.

3. A quantity that has magnitude but no direction is called a(n) _____.

4. A(n) _____ has magnitude and direction.

5. Depending on their relative directions, the vector sum of two vectors 6 meters and 2 meters long must be somewhere between _____ meters and _____ meters long.

6. If an airplane is flying in a direction 30° west of north, the northward component of its velocity is the velocity times the _____ of 30°.

7. The rate of change of velocity is called _____.

8. The SI unit of acceleration is the _____.

9. If a car starts at rest and accelerates uniformly, the distance it travels is proportional to the _____ of the time it travels.

10. All objects in free fall at a given place have the same _____.

11. If a car is going northward and the driver jams on its brakes, the direction of its acceleration is _____.

12. If a car is going northward and starts to turn left, the direction of its acceleration is _____.

13. For a baseball hit on a fly to center field, the methods of calculation of this chapter assume that only the _____ component of its velocity will change.

14. The assumptions used in Question 13 will give the wrong answer because they neglect to take _____ into account.

15. When an object is going in a circular path at constant speed, the direction of its acceleration is _____.

16. The acceleration of the moon is toward the _____.

17. There are _____ radians in a circle.

Multiple Choice.

1. If a body is moving at constant speed in a circular path, its
   (1) velocity is constant and its acceleration is zero
   (2) velocity and acceleration are both changing direction only
   (3) velocity and acceleration are both increasing
   (4) velocity is constant and acceleration is changing direction
   (5) velocity and acceleration are both constant.

2. Which of the following quantities is not fully specified unless its direction is given?
   (1) time   (2) velocity   (3) temperature   (4) mass   (5) speed

3. A flowerpot dropped from a window and fell for 3.3 seconds to the ground. How high was the window?
   (1) 16 meters (2) 32 meters   (3) 50 meters   (4) 100 meters   (5) 2.1 meters

4. A rock is thrown straight up and reaches a height of 12 meters before starting to fall. When it is at rest at the top of its path, its acceleration is
   (1) 0   (4) 11 meters per second squared
   (2) 1.2 meters per second squared   (5) 20 meters per second squared.
   (3) 9.8 meters per second squared

5. If a velocity of 3 meters per second is added to another of 5 meters per second, the sum is
   (1) 2 meters per second
   (2) 4 meters per second
   (3) anything over 3 meters per second
   (4) 8 meters per second
   (5) between 2 meters per second and 8 meters per second.

6. A graph is plotted showing the velocity of a car as a function of time. If the graph is a straight line, it means that
   (1) the car started at rest
   (2) acceleration was constant
   (3) acceleration was increasing
   (4) velocity was constant
   (5) velocity was increasing.

7. The acceleration of an object will be 9.8 meters per second squared if the object is falling freely
   (1) near the surface of the earth
   (2) anywhere
   (3) traveling straight down
   (4) traveling upward
   (5) anywhere within the earth's gravitational pull.

8. If a car is traveling north on a straight road and its brakes are applied, it will
   (1) have no acceleration
   (2) accelerate to the south
   (3) accelerate to the north
   (4) accelerate either east or west
   (5) maintain a constant acceleration.

9. An artificial satellite is circling the globe at the equator, going eastward at constant speed. Its acceleration is
   (1) zero
   (2) eastward
   (3) northward
   (4) downward
   (5) upward.

10. A tennis ball is struck into a high lob. As it travels, it will have a constant
    (1) horizontal velocity
    (2) vertical velocity
    (3) horizontal acceleration
    (4) net velocity
    (5) centripetal acceleration.

Problems.

1. A bicycle averages 4.5 meters per second while traveling for 10 minutes. How far does it travel?

2. What is the average speed of a car that travels 4.6×10^4 meters in 1 hour? Give your answer in SI units.

3. You are on an ocean liner that is going eastward at 12.0 meters per second, and you run southward at 3.6 meters per second. Find the magnitude and direction of your resulting velocity.

4. A pilot wants to fly a plane directly eastward when the wind is from the north at 55 miles per hour. If the air speed is 230 miles per hour, in what direction must the plane be headed?

5. You walk 2.0 miles north, then 4.5 miles east, then 6.2 miles south. What is your displacement from your starting point?

6. You drive a car 45 miles in a direction north 30° W. How much farther west are you?

7. A sailor's compass says that the ship is traveling N 55° W, and the ship's sextant says that at the end of 6.0 hours the ship is 35 miles farther north. How fast is he going?

8. What is the acceleration of a car that speeds up from 12 meters per second to 30 meters per second in 15 seconds?

9. If a car can accelerate at 3.2 meters per second squared, how long will it take to speed up from 15 meters per second to 22 meters per second?

10. How far does a motorcycle travel if it starts at rest and is going 22 meters per second after 15 seconds?

11. What is the acceleration of a car that gets to a speed of 18 meters per second from rest while traveling 240 meters?

12. A ball is dropped from a window 24 meters high. How long will it take to reach the ground?

13. An arrow is fired straight up, leaving the bow at 15 meters per second. If air resistance is negligible, how high will the arrow rise?

14. A firefighter drops from a window into a net. If the window is 34 meters above the net, at what speed does the firefighter hit the net?

15. A trained acrobat can safely land on the ground at speeds up to 15 meters per second. What is the greatest height from which the acrobat can fall?

16. An elevator descending at 4.4 meters per second is accelerated upward at 1.5 meters per second squared for 2.0 seconds. What is its velocity at the end of that time?

17. A toy train is traveling around a circular track 2.0 meters in radius, and it makes a complete circuit every 4.5 seconds. Find (a) its velocity; (b) its acceleration.

18. A carousel is considered safe if no rider is accelerated at more than 3.0 meters per second squared. What is the greatest permissible speed of a rider at the outer edge of a carousel that is 6.5 meters in radius?

19. A car will skid if its acceleration as it makes a turn is more than 3.5 meters per second squared. If the car is traveling at 20 meters per second, what is the radius of the smallest circle it can travel in without skidding?

20. In the spin cycle of a washing machine, the clothes must be accelerated at 75 meters per second squared to squeeze the water out of them. If the radius of the basket is 30 centimeters, how many revolutions must it make per minute?

21. A baseball is thrown horizontally from a window that is 22 meters high. If the initial speed of the ball is 18 meters per second, find (a) how long the ball takes to reach the ground; (b) how far from the building it lands.

22. A train going north at 22 meters per second passes a second train going south at 14 meters per second. A child in the second train fires a paper airplane eastward at 8.0 meters per second. What is the speed of the aiplane in the frame of reference of the first train?

23. A thief snatches a purse and runs due west, going 6.0 meters per second. A policeman, 15 meters to the east, sees the event and gives chase. If the officer is a good sprinter, going at 8.5 meters per second, how far does he have to run to catch the thief?

24. In a test race, an automobile must complete a 10-kilometer run in 4 minutes. If the car goes at 36 meters per second for the first half of the track, what must its speed be for the second half?

25. A bullet is fired from a rifle, emerging from the muzzle at 340 meters per second. It strikes a sandbag some distance away, having lost 10 percent of its velocity due to air resistance. If it penetrates the sandbag to a depth of 12.0 centimeters, how long did it take for the bullet to come to rest in the sandbag?

26. A car going at 24 meters per second passes a motorcycle at rest. As it passes, the motorcycle starts up, accelerating at 3.2 meters per second squared. If the

motorcycle can keep up that acceleration, how long will it take for it to catch the car?

27. You are driving along at a steady speed of 26 meters per second when you suddenly see a child in the road, 150 meters in front of you. Your brakes can produce an acceleration of −2.5 meters per second squared, but it takes time to get your foot from the gas pedal to the brake pedal. How much time do you have if you are to avoid hitting the child?

28. A pitcher throws his fastball horizontally at 42.1 meters per second. How far does it drop before crossing the plate, 18.3 meters away?

29. A rock is thrown horizontally from a cliff overlooking the ocean. If the cliff is 45 meters high and the rock is thrown at 18 meters per second, what is the angle at which it strikes the ocean?

30. What is the centripetal acceleration of a point on the earth's equator? The radius of the earth is 6 400 kilometers.

31. A race car on a circular track with circumference 1 200 meters speeds up from rest with an acceleration of 1.2 meters per second squared. After how much time will its linear and centripetal accelerations be equal in magnitude?

Forces:
Push and Pull

WHAT YOU WILL LEARN

Why do we need a different kind of bat to use on a Ping-Pong ball than the one we use on a baseball? This section deals with the factors that determine how things move.

SECTIONS IN THIS CHAPTER

- What Is a Force?
- Mass and Weight
- Friction Forces
- Some Other Forces
- Action and Reaction

- Balanced Forces
- Components of a Force
- Equilibrium with Several Forces
- The Inclined Plane
- Making Things Turn

What Is a Force?

Every now and then, the whole structure of physical theory has to be reorganized. This happens when the theory comes into conflict with observations of the real world. Then someone with an analytical mind looks at the theory and says, "That can't be right!"

It happened in the sixteenth century, when the Italian professor Galileo examined the accepted answer to the question of what makes things move. The Greek philosopher Aristotle had given the answer centuries before: things keep moving, he said, as long as there are forces acting on them. When the horse stops pulling, the wagon stops moving. The harder the horse pulls, the faster the wagon goes. The spear was a problem. What keeps it going after it leaves the soldier's hand? Aristotle said that a current of air displaced by the spear and coming back behind it keeps pushing the spear forward.

Galileo looked at this theory and said, "That can't be right." The theory could not account for the endless motion of the planets through airless space. By adopting a whole new set of assumptions, Galileo was able to calculate, for the first time, the flight of a cannonball.

The trouble, said Galileo, is that the problem has not been correctly stated. We do not have to explain why something keeps on moving, once started. Anything in motion will keep on moving in a straight line forever, unless something is done to it. The thing we have to explain is the *change* in velocity, and the "something" we have to apply to cause the change is called a *force*.

This is surely a much more reasonable proposition today than it was in Aristotle's time. When you take your foot off the gas pedal, your car does not suddenly come to a stop. It coasts on, only gradually losing its velocity. If you want the car to stop, you have to do something to it. That is what your brakes are for: to exert a force that decreases the car's velocity.

A spacecraft illustrates this point vividly. When a planetary explorer is launched, its rockets burn out quickly as it enters outer space. From that moment on, it simply coasts. *Voyager* coasted through the solar system for years, studying the surfaces of Mars, Jupiter, Saturn, Neptune, and Pluto before it wandered out of the solar system. Nothing pushed it all that distance. While it was exploring, signals from Earth could fire its control rockets to change its velocity, making it speed up, slow down, or change direction. These firings produced the forces that accelerated the spacecraft.

The weight of an object depends, in part, on its mass. An elephant has more mass than a mouse, and that makes the elephant weigh more. Mass and weight are so closely interconnected that students tend to confuse them. One reason for the confusion is that the English system unit, the pound, is a unit of force, or weight, but is often carelessly used as a unit of mass. It is essential that you learn the difference.

Let's invent a parable to eliminate the confusion. Imagine you are holding two dinner parties to celebrate the first colonization of Mars. One is on Earth, and the other is on Mars, and 100 guests are at each party. Both parties have the same menu, involving 30 pounds of potato salad. The potato salad weighs 30 pounds on Earth but only 10 pounds on Mars; gravity is much weaker on Mars. However, both batches of potato salad will still feed the same 100 people; both have the same mass. The *weight* of the potato salad is different. However, the *amount*, its mass, is the same.

One of the nice features of the SI is that this confusion does not exist. The potato salad weighs 133 newtons on Earth and only 40 newtons on Mars; its mass is 14 kilograms on Earth, on Mars, or anywhere else. Weight and other forces are measured in newtons; mass is measured in kilograms.

The weight of anything depends on its mass, but there is more to it. Weight depends on where the object is in space. As you climb a mountain or fly high above the earth in an airplane, your weight gets smaller. At the surface of the earth, the earth's gravity pulls on a 60-kilogram rocket with a force of 588 newtons. If the rocket rises to a height of 2 600 kilometers, it weighs half as much, only 294 newtons.

We can think of the earth as being surrounded by a region of space in which any object feels a pull downward, toward the center of the earth. The force of that pull depends on where in that space the object is located. That space is called the earth's *gravitational field*. At any given point in that space, the field has a certain strength. The strength of the field diminishes as you go further from the earth, approaching zero in the outer reaches of space. Since the gravitational field strength has direction, always toward the center of the earth, it is a vector quantity. The weight of any object is found by multiplying the gravitational field strength, at the object's location, by the mass of the object.

$$\mathbf{w} = m\mathbf{g} \qquad\qquad \text{(Equation 3.1)}$$

One of the simplifying features of the SI is that the force unit is defined so as to make the earth's gravitational field strength equal to the acceleration due to gravity. Thus, at the surface of the earth, where the acceleration due to gravity is 9.8 meters per second squared, the gravitational field strength is 9.8 newtons per kilogram. In the SI, $N/kg = m/s^2$. In using Equation 3.1, when the mass is in kilograms and the g is in meters per second squared, the weight comes out in newtons.

When an object has no forces acting on it, like that *Voyager* space capsule, it is said to be in a state of *equilibrium*. In this state, its velocity does not change. If it is moving, it continues to move at constant speed in a straight line. If it is not moving, it stays right where it is. An object with more than one force acting on it can also be in equilibrium, provided that all the forces act so as to cancel each other out. In a tug-of-war, the rope may remain at rest because the two teams are equally matched. As they pull with equal force in opposite directions, the two forces they exert on the rope cancel each other. The direction at which a force is applied matters, and this shows that force is a vector quantity.

The spring scale of Figure 3.1 is a device for measuring force. The pull on the shackle is measured in force units: pounds or (in SI) *newtons*. A newton (abbreviated as N) is a fairly small unit of force; it takes 4.45 newtons to equal one pound.

CORE CONCEPT

A force is a vector quantity that can change the velocity of an object. If no force acts, or if all forces cancel each other, there is no change in velocity and the object is in equilibrium.

TRY THIS
—1—

What is the evidence that force is a vector quantity?

FIGURE 3.1

Mass and Weight

Probably the first law of physics that everyone learns is this: If you drop something, it falls. Since its velocity keeps on changing as it falls, there must be a force acting on it all the while. That is the force we call *gravity*.

You can measure gravity by balancing it off with a spring scale. When you hang something on a spring scale and read the scale in pounds or newtons, you usually call the force of gravity acting on the object by a special name. You call it the *weight* of the object.

EXAMPLE 3.1

If a hammer has a mass of 2.5 kilograms, how much does it weigh (a) on earth; (b) on Mars?

SOLUTION

(a) Equation 3.1 tells us that

$$w = mg$$

and we know that g on earth is 9.8 m/s². Therefore

$$w = (2.5 \text{ kg})\left(9.8\frac{\text{m}}{\text{s}^2}\right)$$

$$w = 24.5\frac{\text{kg} \cdot \text{m}}{\text{s}^2} = 24.5 \text{ N}$$

(b) On Mars:

$$w = (2.5 \text{ kg})\left(3.3\frac{\text{m}}{\text{s}^2}\right) = 8.3 \text{ N}$$

EXAMPLE 3.2

A 62-kilogram astronaut lands on a strange planet. He drops his phasor, and finds that it falls the 3.5 m from his doorway to the ground in 1.6 s. Find (a) the acceleration due to gravity; (b) the astronaut's weight.

SOLUTION

(a) From Equation 2.4b,

$$a = \frac{2s}{t^2} = \frac{2 (3.5 \text{ m})}{(1.6 \text{ s})^2} = 2.73 \text{ m/s}^2$$

(b)

$$w = mg = (62 \text{ kg})\left(2.73\frac{m}{\text{s}^2}\right) = 170\frac{\text{kg} \cdot \text{m}}{\text{s}^2} = 170 \text{ N}$$

One of the features of the SI system is that it generates many different units of measure. The newton is a derived unit. *Kilogram*, *second*, and *meter* are fundamental units of the system, but combinations of them generate many different units of measure. This is a matter of convenience. Weight is mass times acceleration, but it would be cumbersome to write kg · m/s^2 every time we wanted to refer to a weight, so we decide to give the weight unit a special name, and call it a newton.

CORE CONCEPT

Weight, the force of gravity, is the product of mass and the acceleration due to gravity:

$$\mathbf{w} = m\mathbf{g}$$

TRY THIS
—2—

What is the mass of a girl who weighs 340 newtons on earth?

Friction

Consider the brick resting on a table top and being pulled with a spring scale, as in Figure 3.2. At first, the brick does not move; it stays in equilibrium. This means that the force being exerted by the spring scale must be balanced by some other force, pulling in the opposite direction. This force is called *static friction*. It results from the interaction between the surface of the brick and the surface on which it rests.

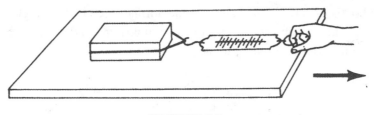

FIGURE 3.2

As the spring scale pulls harder, the static friction also increases. At some point, the brick starts to move; the brick has then gone out of the equilibrium state. The force that is just sufficient to break out of the static condition is known as the *starting friction*.

Once the brick is in motion, it can be kept in motion with a much smaller force than the starting friction. If it is pulled along at constant speed, it is in equilibrium. Therefore the force exerted by the spring scale must be just equal in magnitude and opposite in direction to a force that holds it back. The force that holds it back, opposing the motion, is called *sliding friction*. As the brick moves along at constant speed, the spring scale registers the magnitude of the sliding friction.

Sliding friction is a force that is generated whenever two surfaces rub against each other. It always acts in such a direction as to oppose the motion. The amount of sliding friction depends very critically on the kind of surfaces. Rubber, for example, is a high-friction material, which is why it is so useful for automobile tires and tennis shoes. On the other hand, if you want to make a closet door slide easily, make the skids out of a low-friction material such as nylon. Also, make the surface smooth; sliding friction is greater when the surface is rough.

At ordinary speeds, sliding friction does not depend on speed. Also, it does not depend on the amount of surface area in contact. If the brick in Figure 3.2 were placed on end, the amount of surface in contact with the table top would be much less, but the force on each unit area would be proportionally greater. The result would be no net change in the friction. Sliding friction depends on the total *normal force* pushing the two surfaces together; that is, on the force perpendicular to the surface. The frictional force, F_{fr}, is proportional to the normal force, F_{nor}:

$$F_{fr} = \mu F_{nor} \qquad \text{(Equation 3.2)}$$

The proportionality constant μ (the Greek letter mu) is called the *coefficient of friction*. It depends only on the nature of the two surfaces.

EXAMPLE 3.3

A cardboard carton weighing 165 newtons is resting on a marble floor. If the coefficient of friction between cardboard and smooth marble is 0.20, how much force would it take to keep the box sliding along at constant speed?

SOLUTION

Since the box is on a horizontal surface, the normal force is the weight of the box. If the box is to slide at constant speed, it is in equilibrium, so the force needed to push it is equal to the friction holding it back. Then

$$F_{fr} = \mu F_{nor} = (0.20)(165 \text{ N}) = 33 \text{ N}$$

EXAMPLE 3.4

A girl is pressing a 1.8-kilogram book against a vertical wall, as shown in Figure 3.3. If the coefficient of friction between the book and the wall is 0.35, how hard would she have to push in order to allow the book to slide down at constant speed?

FIGURE 3.3

SOLUTION

If the book is to slide down at constant speed, it is in equilibrium. Then the frictional force holding it back must be equal to the gravitational force (weight) pulling it down. Its weight is

$$w = mg = (1.8 \text{ kg})(9.8 \text{ m/s}^2) = 17.6 \text{ kg} \cdot \text{m/s}^2 = 17.6 \text{ N}$$

The force the girl exerts is the normal force, which is

$$F_{\text{nor}} = \frac{F_{\text{fr}}}{\mu} = \frac{17.6 \text{ N}}{0.35} = 50 \text{ N}$$

CORE CONCEPT

Friction is a retarding force that acts whenever two surfaces move against each other. Sliding friction is independent of speed and surface area and is proportional to the normal force.

TRY THIS
—3—

What is the coefficient of friction between a 65-kilogram chair and a floor if the chair can be pushed at constant speed with a force of 220 newtons?

Some Other Forces

A wide variety of forces act to make things move, to slow them down, or just to distort their shapes. In the MKS system, all forces are measured in newtons. Here are a few:

Elastic recoil   The special property of solids, as opposed to liquids and gases, is that they resist changing their shape. When you push on a solid, it pushes back. Look at the plank in Figure 3.4. The plank exerts an upward force against the hand because the hand is changing the shape of the board. The floor pushes you up when your weight pushes down on it, because your weight changes its shape. A feather resting on your desk changes the shape of the desktop—but not much.

Tension   This is a special form of elastic recoil, resulting from stretching something. In a tug-of-war, the rope is under great tension, so it pulls back on the teams at the opposite ends. The spring scale of Figure 3.1 is actually a tension-measuring device. The harder you pull on both ends of the spring, the more it stretches, and the harder it pulls back.

Compression   This is the opposite of tension. Pushing on a window pole shortens the pole—a little—and the pole pushes back. Actually, the bending board of Figure 3.4 exerts its recoil because its upper surface is compressed and its lower surface is tensed.

Buoyancy   Anything submerged in a liquid or a gas experiences an upward force. A rock under water feels lighter than the same rock in the air. If buoyancy is greater than weight, as for a cork under water or a helium-filled balloon in the air, the object rises.

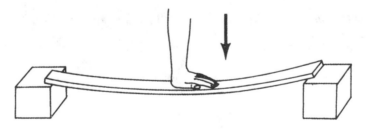

FIGURE 3.4

If you are under water, you can rise to the top until part of you emerges into the air. Then, buoyancy and gravity are equal, and you float.

Viscous drag   (That's VIS-kus, not vicious!) Anything moving through a liquid or gas feels a retarding, friction-like force. That is why a boat needs an engine, why you cannot swim fast enough to get into the Olympics, why automobiles are streamlined, and why a parachute is advisable if you plan to fall out of an airplane.

Magnetism There is a lot more to magnetism than little red horseshoes picking up nails. This topic is covered in Chapter 11.

CORE CONCEPT

There are many other kinds of force, including buoyancy, which acts upward on anything submerged in a liquid or a gas.

TRY THIS
—4—

An object submerged in water has a mass of 6 kilograms, and the buoyant force acting on it is 75 newtons. Will the object rise or fall in the water?

TRY THIS
—5—

Figure 3.5 shows a boat that has sunk to the bottom of a lake and is being dragged out by a man pulling on its painter. There is also a diagram indicating eight directions of space. From these eight, select the one that indicates the direction of each of the following forces acting on the boat: (a) friction; (b) viscous drag; (c) buoyancy; (d) weight; (e) elastic recoil of the rope; (f) elastic recoil of the lake bottom.

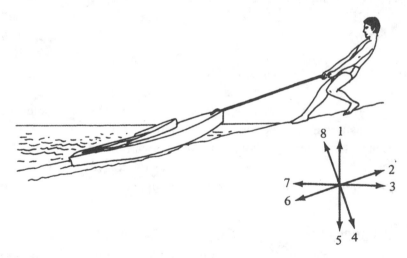

FIGURE 3.5

Action and Reaction

The batter steps up to the plate and takes a healthy swing, sending the ball into left field. The bat has exerted a large force on the ball, changing both the magnitude and the direction of its velocity. But the ball has also exerted a force on the bat, slowing it down. The batter feels this when the bat hits the ball. Sometimes this reaction force breaks the bat.

Next time up, he strikes out. He has taken exactly the same swing but exerts no force on anything. (Air doesn't count.) The batter has discovered that it is impossible to exert a force unless there is something there to push back. Forces exist *only* in pairs. When object *A* exerts a force on object *B,* then *B* must exert a force on *A.* The two forces are sometimes called *action* and *reaction,* although which is which is often rather arbitrary. The two parts of the interaction are equal in magnitude, are opposite in direction, and act on different objects.

Each of the six forces acting on the boat of Figure 3.5 is half of an interaction pair. Friction of the lake bottom pulls the boat in direction 6; the boat pulls the sand of the lake bottom in direction 2. Elastic recoil of the lake bottom supports the boat; elastic recoil of the boat depresses the lake bottom. The viscous drag of the water pulls the boat in direction 6, and the boat pulls some water along with it, in direction 2. Tension in the rope pulls the boat in direction 2; the boat stretches the rope in direction 6.

The weight of the boat is also one-half of an interaction. It is the gravitational influence of the whole earth that pulls the boat down. If the law of interaction is valid, the gravity of the boat must be pulling the earth up. This effect cannot be detected in the context under discussion, but that does not invalidate the law. If we calculated the effect the force has on the earth, we would find that it is far too small to detect.

The law of action and reaction leads to an apparent paradox, if you are not careful how it is applied. The horse pulls on the wagon. If the force of the wagon pulling the horse the other way is the same, as the law insists, how can the horse and wagon get started?

The error in the reasoning is this: If you want to know whether the horse gets moving, you have to consider the forces acting *on the horse.* The force acting on the wagon has nothing to do with the question. The horse starts up because the force he exerts with his hooves is larger than the force of the wagon pulling him back. And the wagon starts up because the force of the horse pulling it forward is larger than the frictional forces holding it back. To know how something moves, consider the forces acting *on it.* The action and reaction forces *never* act on the same object.

CORE CONCEPT
Forces exist only in pairs, equal in magnitude and opposite in direction, acting on different objects.

TRY THIS
—6—

When you jam on the brakes, friction with the pavement brings your car to a screeching halt. What is the other half of this interaction?

Balanced Forces

If a single force acts on an object, the velocity of the thing must change. If two or more forces act, however, their effects may eliminate each other. This is the condition of equilibrium, in which there is no net force and the velocity does not change. We saw such a condition in the case in which gravity is pulling an object down and the spring scale, used for weighing it, is pulling it upward.

An object in equilibrium may or may not be at rest. A parachutist, descending at constant speed, is in equilibrium. His weight is just balanced by the viscous drag on the parachute—which is why he put it on in the first place. A heavier parachutist falls a little faster; his speed increases until the viscous drag just balances his weight.

Balancing the vertical forces is not enough to produce equilibrium. An airplane traveling at constant speed, as in Figure 3.6, is in equilibrium under the influence of four forces, two vertical and two horizontal. Vertical: gravity (down) is just balanced by the lift produced by the flow of air across the wing. Horizontal: viscous drag is just balanced by the thrust of the engines. Both the vertical and the horizontal velocities are constant.

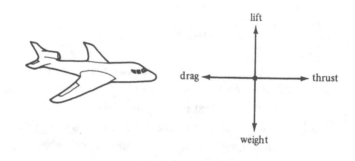

FIGURE 3.6

The brick of Figure 3.2, resting on a tabletop and being pulled along at constant speed, is another example. Vertical: the downward force of gravity is balanced by the upward force of the elastic recoil of the tabletop. Horizontal: the tension in the spring scale, pulling to the right, is balanced by the friction pulling it to the left, opposite to the direction of motion.

CORE CONCEPT

If an object is in equilibrium—at rest or moving at constant speed in a straight line—the total force acting on it in any direction is exactly equal in magnitude to the force in the opposite direction.

The mass of the brick in Figure 3.2 is 1.6 kilograms, and the spring scale reads 4 newtons. Find (a) the force of friction on the brick; (b) the elastic recoil of the tabletop.

Components of a Force

The crate of Figure 3.7 is being dragged along the floor by means of a rope, which is not horizontal. The rope makes an angle θ to the floor.

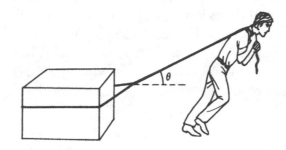

FIGURE 3.7

The tension in the rope, acting on the crate, does two things to it. First, it drags the crate across the floor. Second, it tends to lift the crate off the floor. The smaller the angle θ, the larger the effective force that is dragging the crate, and the smaller the effective force that is lifting it. When $\theta = 0$, the entire force is dragging and there is no lifting at all. Conversely, when $\theta = 90°$, the entire force is lifting the crate.

How can you find out how much force is being used to drag the crate? Force is a vector, and it obeys the same mathematical rules as velocity vectors and displacement vectors, which you used in Chapter 2. The dragging force is the component of the tension in the rope acting parallel to the floor, that is, the horizontal component. According to Equation 2.2, it is found as follows:

$$F_{horiz} = F \cos \theta$$

Suppose, for example, that the tension in the rope is 250 newtons and the angle the rope makes with the ground is 25°. Then the dragging force, as shown in Figure 3.8, is

$$F_{horiz} = (250 \text{ N})(\cos 25°) = 227 \text{ N}$$

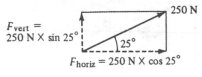

FIGURE 3.8

The vertical component is (250 N)(cos 65°); but remember, cos 65° = sin 25°. Therefore:

$$F_{vert} = (250 \text{ N})(\sin 25°) = 106 \text{ N}$$

This is not enough to get the crate off the floor, but it relieves the floor of some of the weight.

How much is the friction in this situation? If the crate is moving at constant speed, the force pulling the crate to the right must equal the friction holding it back. Since the force pulling to the right is the horizontal component of the tension in the rope, the friction must be the same—227 newtons.

And how much is the elastic recoil? The upward forces must equal the downward forces. The only downward force is the weight of the crate, say, 500 newtons. The *total* upward force must then be 500 newtons. But 106 newtons of this is provided by the upward component of the tension in the rope. The rest—394 newtons—is the elastic recoil of the floor.

EXAMPLE 3.5   Using a window pole that makes an angle of 23° with the window, you push up on the pole with a force of 85 newtons to close the window. Find (a) the effective force that is pushing the window up; (b) the force pushing the window against its sash.

SOLUTION
(a) Since the force makes an angle of 23° with the vertical, its vertical component is (85 N)(cos 23°) = 78 N.
(b) The horizontal component is (85 N)(sin 23°) = 33 N.

CORE CONCEPT
The effect that a force has in any direction can be found by calculating its component in that direction.

TRY THIS
—8—

You push a lawn mower with a force of 160 newtons, exerted directly along its shaft. The shaft makes an angle of 30° with the ground. Find (a) how much force is moving the lawn mower; (b) how much force is pushing the lawn mower into the ground.

Equilibrium with Several Forces

For the purpose of making a complete analysis of the forces acting on an object, a *vector diagram* is a useful device.

Figure 3.9 is a vector diagram showing the forces on a brick being dragged along a tabletop. Four forces act: gravity (weight), friction, elastic recoil of the tabletop, and tension in the cord. Each force is represented by a vector, drawn at the correct angle and with its length proportional to the force. For the purpose of analysis, all vectors are represented by components along a pair of axes perpendicular to each other. In this case, we elect to use horizontal and vertical axes, since three of the forces are already on these axes. To do the analysis, we have to resolve the tension vector into its components on the vertical and horizontal axes. Then we can write two equations: one says that there is no net vertical force, and the other says that there is no net horizontal force.

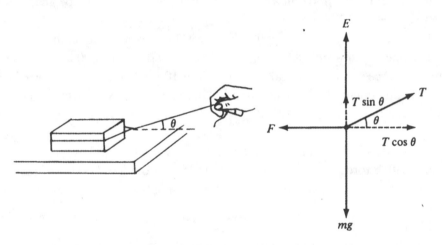

FIGURE 3.9

EXAMPLE 3.6 The 2.5-kilogram brick of Figure 3.9 is being pulled by a cord that makes an angle of 20° with the horizontal and has 7.0 newtons of tension in it. Find (a) the force of friction; (b) the elastic recoil of the tabletop.

SOLUTION

(a) On the horizontal axis, the friction must be equal to the horizontal component of the tension, so

$$F = (7.0 \text{ N})(\cos 20°)$$
$$F = 6.6 \text{ N}$$

(b) On the vertical axis, the downward force (weight) must equal the sum of upward forces, so

$$mg = E + T \sin \theta$$
$$E = mg - T \sin \theta$$

$$E = (2.5 \text{ kg})\left(9.8 \frac{\text{m}}{\text{s}^2}\right) - (7.0 \text{ N})(\sin 20°)$$

$$E = 24.5 \text{ N} - 2.4 \text{ N} = 22 \text{ N}$$

Another example is the lawn mower. In this case, there are four forces acting: gravity, friction, the force along the shaft, and the elastic recoil of the ground. The force exerted by the poor fellow doing the work is resolved into vertical and horizontal components.

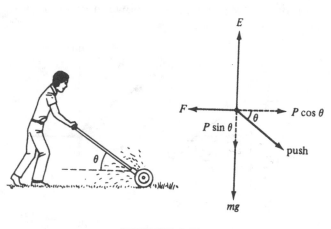

FIGURE 3.10

EXAMPLE 3.7 The lawn mower of Figure 3.10 has a mass of 22 kilograms, and is being pushed at constant velocity against a frictional resistance of 150 newtons, with the shaft making an angle of 30°. Find (a) the compression in the shaft; and (b) the elastic recoil of the ground.

SOLUTION

(a) The compression in the shaft is the push (P) force, and its horizontal component must equal the frictional resistance (F):

$$F = P \cos \theta$$

$$P = \frac{F}{\cos \theta} = \frac{150 \text{ N}}{\cos 30°}$$

$$P = 173 \text{ N}$$

(b) Now we can use the value of the push force to find the elastic recoil:

$$E = mg + P \sin \theta$$

$$E = (22 \text{ kg})\left(9.8 \frac{m}{s^2}\right) + (173 \text{ N})(\sin 30°)$$

$$E = 216 \text{ N} + 87 \text{ N} = 300 \text{ N}$$

Here is one more example: What is the tension in the rope holding up the sign in Figure 3.11? This can be found by analyzing the forces acting *on the bar*. The rope pulls the bar up and to the left; the weight of the sign pulls it straight down; the elastic recoil of the wall pushes it to the right. If the weight of the bar itself is too small to worry about, the solution to the problem is as shown in Example 3.8.

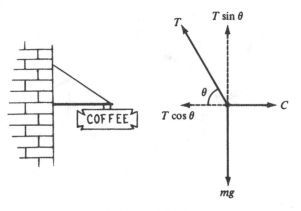

FIGURE 3.11

EXAMPLE 3.8   If the sign of Figure 3.11 weighs 240 newtons and the angle the rope makes with the bar is 55°, how much is the tension in the rope?

SOLUTION

One equation is enough to get this answer. On the vertical axis:

$$mg = T \sin \theta$$

$$T = \frac{mg}{\sin \theta} = \frac{240 \text{ N}}{\sin 55°}$$

$$T = 290 \text{ N}$$

CORE CONCEPT

To calculate the forces acting on an object, first analyze them into components along two axes at right angles to each other.

A helium-filled balloon weighs 25 newtons, and is acted on by four forces, as shown in Figure 3.12. These forces are its weight, the buoyant force of the air, the wind pushing the balloon to the right, and the tension in the rope that is holding it down. The rope makes an angle of 20° with the vertical, and the tension in it is 16 newtons. Find the buoyancy and the force exerted by the wind.

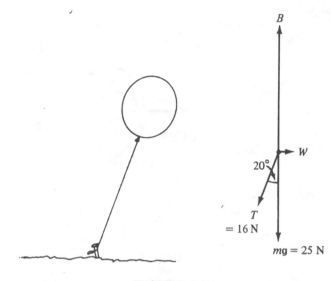

FIGURE 3.12

The Inclined Plane

A wagon rolls downhill, propelled only by its own weight. But gravity pulls straight down, not at an angle downhill. What makes the wagon go at an angle downhill is a *component* of its weight, a part of its weight acting downhill, parallel to the surface the wagon rests on.

A component of a force can act in any direction, not just vertically or horizontally. On the inclined plane, the weight of the wagon has two different effects: it acts *parallel* to the surface of the hill, pushing the wagon downhill; and it acts perpendicular (or *normal*) to the surface, pushing the wagon into the surface. As the hill gets steeper, the parallel component becomes larger and the normal component decreases.

Figure 3.13 shows the wagon and a vector diagram of the components of its weight. Geometrically, the components are found by dropping perpendiculars from the end of the vector to the two axes, one parallel to the surface and the other normal. The angle marked θ in the vector diagram is equal to the slope of the hill. This can easily be shown by a little geometry; both angles θ are complements of the two angles marked ϕ. Sample Problem 3.1 shows how the two components of the weight are calculated.

EXAMPLE 3.9

The wagon of Figure 3.13 weighs 40 pounds, and the angle that the hill makes with the horizontal is 35°. Find (a) the force pushing the wagon downhill; (b) the force pushing the wagon into the surface.

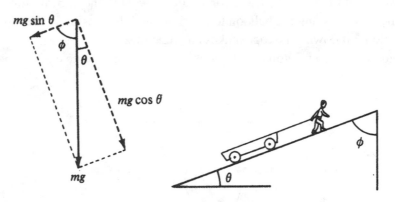

FIGURE 3.13

SOLUTION

(a) Parallel (downhill) component of weight:

$$F_P = w \sin \theta$$
$$F_P = (40 \text{ lb})(\sin 35°)$$
$$F_P = 23 \text{ lb}$$

(b) Normal component of weight:

$$F_N = w \cos \theta$$
$$F_N = (40 \text{ lb})(\cos 35°)$$
$$F_N = 33 \text{ lb}$$

EXAMPLE 3.10

A well-oiled, frictionless wagon with a mass of 75 kilograms is pulled uphill, using a force of only 110 newtons. What is the angle that the hill makes with the horizontal?

SOLUTION

As before, the component of weight parallel to the plane is the weight times the sine of the slope of the plane. The weight of the wagon (Equation 3.1) is

$$mg = (75 \text{ kg})(9.8 \text{ m/s}^2) = 735 \text{ N}$$

Then

$$\sin \theta = \frac{110 \text{ N}}{735 \text{ N}} \quad \text{and } \theta = 8.6°$$

EXAMPLE 3.11 A plank will break if a force of 350 newtons is applied to its center. What is the largest weight it can support if it is tilted to an angle of 35°?

SOLUTION

At the breaking point, the normal component of the weight is 350 N, which is equal to the weight times the cosine of the angle. Then

$$F_N = w \cos \theta$$

so

$$w = \frac{F_N}{\cos \theta} = \frac{350 \text{ N}}{\cos 35°} = 430 \text{ N}$$

When the wagon is resting on the surface, the elastic recoil of the surface is just enough to cancel the normal component of the wagon's weight. If the wagon is to stay in equilibrium, you have to pull on it, uphill, to prevent it from running away. If there is no friction, the uphill force needed is the same whether the wagon is standing still, or going either uphill or downhill at constant speed.

The situation is different if the wagon is moving and there is friction. If the wagon is going uphill, you have to pull harder, because the friction is working against you, holding it back. The total force you need to keep the wagon going is then equal to the parallel component of the weight plus the friction. On the other hand, if you are lowering the wagon down the hill, holding the rope to keep it from running away from you, friction is acting uphill, helping you to hold the wagon back. Then the force you must exert is the parallel component of the weight *minus* the friction.

CORE CONCEPT

When an object is on an inclined plane, it is propelled down the plane by a force equal to $w \sin \theta$, and a force $w \cos \theta$ pushes it into the surface.

TRY THIS
—10—

1. The 120-kilogram wagon of Figure 3.13 is being pulled up a 20° slope. If there is no friction, find (a) the tension in the rope; (b) the elastic recoil of the surface.
2. A 60-kilogram crate is resting on a ramp that slopes at 32°. A rope is attached to it that pulls, uphill, just hard enough to keep the crate sliding downhill at constant speed. If the tension in the rope is 240 newtons, what is the coefficient of sliding friction between the crate and the ramp?

Making Things Turn

A force, properly applied, can make something rotate. Roughly speaking, a *force* makes things move, and a *torque* makes things rotate. Consider the wrench being used to tighten a nut in Figure 3.14. To get the nut tight, you want to use a large force (*F*) and to exert that force at the largest distance (*r*) from the point of rotation. You might even rig up an extension of the wrench to increase the value of *r*. The *torque* you apply to the nut is the product of the force and its distance from the axis. Torque is the product of a force and a distance, so it is measured in newton-meters (N·m).

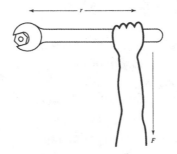

FIGURE 3.14

In Figure 3.15, you cannot get your arm into the best position to exert a force on the wrench. The force you are exerting is not perpendicular to the wrench. The resulting torque is less than you might like. In fact if you exert the force parallel to the wrench, you would get no torque at all. It is only the component of force perpendicular to the wrench that produces a torque. That component is the force on the wrench times the sine of the angle that the force makes with the radius.

$$\tau = r F \sin \theta \qquad\qquad \textbf{(Equation 3.3)}$$

The torque is at maximum when the full force is perpendicular to the radius, and the torque is zero if the force is parallel to the radius.

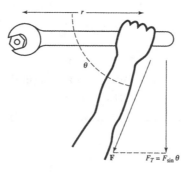

FIGURE 3.15

The radius of the wheel of fortune in Figure 3.16 is 1.2 meters, and the operator applies a force of 45 newtons tangentially to get it spinning. What torque has he supplied?

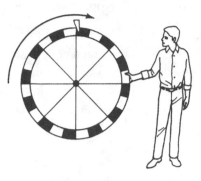

FIGURE 3.16

SOLUTION

$$\tau = rF \sin 90°$$
$$\tau = (45 \text{ N})(1.2 \text{ m})$$
$$\tau = 54 \text{ N·m}$$

  A 32-kg child sits on a seesaw. If she is 2.2 meters from the pivot, what is the torque that her weight exerts, making the seesaw rotate around the pivot?

SOLUTION

Her weight is a vertical force, perpendicular to her distance from the pivot, so

$$\tau = rF = mgr = (32 \text{ kg})(9.8 \text{ m/s}^2)(2.2 \text{ m}) = 690 \text{ N·m}$$

  The 52-kilogram boy in Figure 3.17 is standing on the pedal of his bicycle at a moment when the crank arm, which is 54 centimeters long, makes an angle of 30° with the ground. How much torque is turning the pedal arm?

FIGURE 3.17

SOLUTION

The radius vector is half the crank arm, or 0.27 meters. The force is vertically downward, at an angle of 60° with the radius vector. Therefore, the torque is

$$\tau = rF \cos \theta = (0.27 \text{ m})(52 \text{ kg})(9.8 \text{ N/kg}) \sin 60° = 120 \text{ N·m}$$

Torques may balance each other in such a way that they do not produce rotation. In that situation the object is said to be in rotational equilibrium. In Figure 3.18, three people are sitting on a seesaw. Their weights produce downward forces. All radius vectors are horizontal; one side can be designated positive and the other negative. With vertical forces and horizontal radius vectors, sin $\theta = 1$ and can be left out.

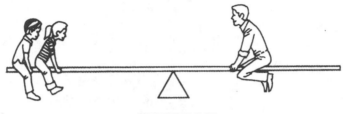

FIGURE 3.18

 The seesaw of Figure 3.18, balanced in the middle, is 4.0 meters long. Georgie weighs 240 newtons and is at one end; Sally weighs 170 newtons and is 1.0 m from Georgie. If Daddy weighs 620 newtons, how far from the pivot must he sit?

SOLUTION

Set the clockwise torques equal to the counterclockwise torques:

$$(240 \text{ N})(2.0 \text{ m}) + (170 \text{ N})(3.0 \text{ m}) = (620 \text{ N})r$$

from which $r = 1.6$ m. Daddy has to sit 1.6 m on the other side of the pivot.

Here's one more example. A pile of books rests on a plank that is supported on either end by a chair, as in Figure 3.19. The books are not in the center. How much of the weight of the books is supported by each of the chairs on which the plank rests? You solve this by imagining that the right-hand chair is pushing up on the plank, making it try to turn counterclockwise around the left-hand chair. The counterclockwise torque produced by the elastic recoil of the right-hand chair must be canceled by the clockwise torque, which comes from the weight of the books.

EXAMPLE 3.16 The mass of the books of Figure 3.19 is 35 kilograms; the plank is 3.7 meters long, and the books are centered 1.2 meters from the left-hand chair. What is the elastic recoil of each chair?

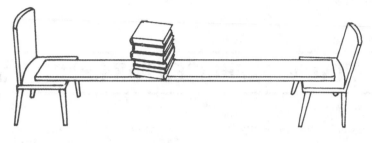

FIGURE 3.19

SOLUTION

Consider the left-hand chair to be a pivot around which the whole plank is trying to turn. Then the weight of the books is producing a clockwise torque around that pivot, and the elastic recoil of the other chair is producing a counterclockwise torque. These two torques produce rotational equilibrium, so they must be equal to each other. Then

$$(mgr)_{books} = (Er)_{\text{right-hand chair}}$$

$$(35 \text{ kg})\left(9.8\frac{\text{m}}{\text{s}^2}\right)(1.2 \text{ m}) = (3.7 \text{ m})E_{right}$$

$$412 \text{ N·m} = (3.7 \text{ m}) \, E_{right}$$

$$E_{right} = 111 \text{ N}$$

To get the elastic recoil of the left-hand chair, note that the books are also in translational equilibrium, so the total upward force must equal the total downward force:

$$mg = E_{right} + E_{left}$$

$$(35 \text{ kg})\left(9.8\frac{\text{m}}{\text{s}^2}\right) = 111 \text{ N} + E_{left}$$

$$E_{left} = 230 \text{ N}$$

CORE CONCEPT

An object is in rotational equilibrium when the counterclockwise torques acting on it are equal to the clockwise torques.

TRY THIS
—11—

A horizontal plank weighing 80 newtons and 3.5 meters long rests with one end on a rock and the other on a scale. A woman is standing on the plank, 1.0 meter from the rock, and the scale reads 160 newtons. How much does the woman weigh?

REVIEW EXERCISES FOR CHAPTER 3

Fill-In's. For each of the following, fill in the missing word or phrase.

1. Application of a force changes the _____ of an object.

2. The direction of the force of sliding friction is always _____ to that of velocity.

3. An object moving at constant speed is in a state of _____.

4. The SI unit of force is the _____.

5. It takes about 4.45 newtons to equal a _____.

6. The force that retards a solid moving through a liquid is called _____.

7. The force of gravity acting on an object is called the object's _____.

8. The amount of material in an object is called the object's _____.

9. Mass times acceleration due to gravity equals _____.

10. One kilogram-meter per second squared = 1 _____.

11. When the shape of a solid is distorted, it produces a force called _____.

12. When a rope is stretched, the _____ in it results in an elastic recoil force at each end.

13. A solid submerged in a liquid experiences an upward force called _____.

14. If you stand next to a wall and push it northward with a force of 30 pounds, it will push you with a force of _____ pounds.

15. When you stand on a floor, your weight is balanced by the _____ force of the floor.

16. If a 50-pound girl on roller skates stands on a hill where the slope is 30°, the gravitational force pushing her down the hill is _____.

17. To find the component of a force on any axis, multiply the force by the _____ of the angle the force makes with the axis.

18. If it takes a force of 200 newtons to move a wagon up a frictionless hill at constant speed, the force needed to let the wagon roll downhill at constant speed is _____.

Multiple Choice.

1. If an object is in a state of equilibrium,
   (1) it is at rest
   (2) it is in motion at constant velocity
   (3) it is in free fall
   (4) it is in accelerated motion
   (5) it may be more than one of the above.

2. The direction of the force of elastic recoil depends on the direction of
   (1) gravity   (4) acceleration
   (2) velocity   (5) friction.
   (3) applied force

3. A fish is swimming upward at an angle of 30° with the horizontal. The direction of the force of gravity acting on it is
   (1) upward   (4) at an angle upward
   (2) downward   (5) at an angle downward.
   (3) horizontal

4. On the moon, $g = 1.6$ meters per second squared. If an astronaut has a mass of 100 kilograms on earth, his mass on the moon will be
   (1) 100 kilograms   (4) 160 kilograms
   (2) 980 newtons   (5) 63 kilograms.
   (3) 160 newtons

5. A 50-newton bowling ball is placed on a surface inclined at 20° to the horizontal. The force propelling it down the slope is
   (1) 50 newtons   (4) (50 newtons) (sin 20°)
   (2) (50 newtons)/(cos 20°)   (5) (50 newtons) (cos 20°)
   (3) (50 newtons)/(sin 20°)

6. In the SI, weight is measured in
   (1) grams   (4) pounds
   (2) kilograms   (5) more than one of the above.
   (3) newtons

7. If a boat is moving along at constant speed, it may be assumed that
    (1) a net force is pushing it forward
    (2) the sum of only vertical forces is zero
    (3) the buoyant force is greater than gravity
    (4) the sum of only horizontal forces is zero
    (5) the sum of all forces is zero.

8. The force of sliding friction depends on the coefficient of friction and the
    (1) weight of the object   (4) force perpendicular to the surface
    (2) slope of the surface   (5) kind of surface.
    (3) mass of the object

9. Daddy is pulling a child along on a sled by means of a rope. The reaction force to the force he exerts on the rope is
    (1) the friction of the sled against the snow
    (2) the rope pulling on the sled
    (3) the rope pulling on Daddy
    (4) the sled pulling on the rope
    (5) Daddy's feet pushing on the snow.

10. The sled of Question 9 is being pulled uphill. The direction of the friction acting on the sled is
    (1) uphill   (4) straight up, vertically
    (2) downhill   (5) horizontal.
    (3) straight down, vertically

Problems.

1. What is the mass of a dog that weighs 75 newtons?

2. An astronaut with all her equipment has a mass of 95 kilograms. How much will she weigh on the moon, where the acceleration due to gravity is 1.67 meters per second squared?

3. A rope is attached to a 35-kilogram rock to lift the rock from the bottom of the lake. If the buoyancy on the rock is 50 newtons and the viscous drag is 25 newtons, how much is the tension in the rope?

4. How much is the viscous drag acting on a rocket-driven sled that is going at constant speed against a frictional force of 22 000 newtons when the thrust of the engine is 31 000 newtons?

5. A sled is being pulled along a horizontal road at constant speed by means of a rope that makes an angle of 25° with the horizontal. If the friction between the sled and the snow is 85 newtons, how much is the tension in the rope?

6. On a camping trip, you stretch a rope between two trees and hang your knapsack from the middle of it, to keep it safe from bears. The mass of the knapsack is 32 kilograms, and each half of the rope makes an angle of 40° with the horizontal. Find (a) the amount of weight supported by each half of the rope; (b) the tension in the rope.

7. A 35-kilogram child is on a swing supported by two ropes. A baby-sitter is holding the swing back so that the ropes make an angle of 25° with the vertical. How much is the tension in each rope?

8. In a sign supported as shown in Figure 3.11, the tension in the rope is 350 newtons. How much does the sign weigh if the angle between the rope and the wall is 40°?

9. A 20-kilogram pile of books is resting on a plank tilted so that it makes an angle of 20° with the ground. How much force do the books exert normally against the plank?

10. A force of 40 pounds is needed to push a wagon up a 35° slope. How much does the wagon weigh? (Neglect friction.)

11. A 7.0-kilogram bowling ball is placed on a ramp sloped at 15°. Find (a) the force that propels the ball down the ramp; (b) the force that compresses the surface of the ramp.

12. How much force is needed to keep a 62-kilogram crate moving across a floor at constant speed if the coefficient of friction between the floor and the crate is 0.18?

13. A heavy trunk rests on the floor. George pulls at it with a rope, using a force of 85 newtons. Janet pulls with another rope at right angles to George's, with a force of 62 newtons. How hard must Henry pull on another rope to put the trunk into equilibrium?

14. A 45-kilogram table can be pushed along the floor at constant speed with a force of 170 newtons. What is the coefficient of friction between the table and the floor?

15. A heavy picture, with a mass of 18 kilograms, is 45 centimeters wide. A wire 62 centimeters long is stretched across its back. If the picture is suspended in the center of the wire by a nail in the wall, find (a) the force on the nail; (b) the tension in the wire.

16. On a strange planet, a 65-kilogram astronaut steps on a scale and finds that the scale reads 240 newtons. If a survival pack is dropped a distance of 8.5 meters to the ground, how fast will it be going when it lands?

17. A crowbar 2.0 meters long is used to pry a rock out of the ground, by pivoting the crowbar 0.5 m from the rock and pushing down on the other end. The end is pushed with a force of 250 newtons. Find (a) the torque being applied to the crowbar by the person doing the job; (b) the force being applied to the rock.

18. A light scaffold is supported by a rope at each end, and a painter is standing 1.0 meter from the left end. If the painter weighs 600 newtons and the scaffold is 2.5 meters long, how much is the tension in each of the ropes?

19. A winch, consisting of a crank that turns a shaft with a rope wrapped around it, is being used to lift cargo into a boat. The shaft has a diameter of 3.0 centimeters, and the crank is 45 centimeters long. If a load of cargo weighs 1 200 newtons, how much force is needed to lift it?

20. You have a rope attached to a cart that weighs 450 newtons, and you are lowering the cart down a 25° slope. If the friction is 75 newtons, how hard do you have to pull on the rope to prevent the cart from running away from you?

21. A plank 4.0 meters long has its left end resting on a rock and the other supported by a rope, in a horizontal position. If the greatest tension the rope can stand without breaking is 350 newtons, how far from the rock can a 55-kilogram girl walk out on the plank before the rope breaks?

22. A 45-kilogram crate is on a ramp inclined at 38° to the horizontal, and the coefficient of friction between the crate and the ramp is 0.26. Find (a) the force that would be needed to push the crate uphill if there were no friction; (b) the normal force that the crate exerts on the ramp; (c) the force of friction; (d) the actual force that would be needed to move the crate up the ramp.

23. A rope is attached to tree A at a height of 2.0 meters and to tree B, 16 meters away, at a height of 5.0 meters. When a 28-kilogram bag of goodies is attached to the rope 4.5 meters from tree A, segment A of the rope (attached to tree A) is horizontal. Find the tension in each of the two ropes.

24. An instrument weighing 85 newtons is being towed under water, by a boat. The line, attached at the boat's waterline, is 17.0 meters long, and the instrument is 5.6 meters deep. If the buoyancy on the instrument is 22 newtons, how much is the viscous drag on it?

Making Things Move

WHAT YOU WILL LEARN

Forces change velocities, and this section deals with the calculations that you need to determine just how velocities change in various circumstances.

SECTIONS IN THIS CHAPTER

- How to Change a Velocity
- Inertia
- Multiple Forces
- Momentum
- Momentum Endures
- Making Circles
- Rules of Spinning
- The Moon in Orbit
- Gravity Is Everywhere
- Weightlessness

How to Change a Velocity

When one of those space exploration machines is traveling in outer space, it keeps on going in a straight line at constant speed, unless otherwise instructed. The instructions sent by radio take the form, "Fire rocket c for t seconds." The purpose of the firing is to change the velocity of the spacecraft—to speed it up, slow it down, or change its direction.

How much does the velocity change? That depends on which rocket fires, and for how long. Rockets vary according to the amount of force they exert on the spacecraft. A powerful rocket fired for a short burst might produce the same amount of change in velocity as a weak rocket fired gently for a long time. The amount by which the velocity changes depends on the product of the force exerted by the rocket and the length of time for which it is exerted. This product is called the *impulse*, represented by **P** in the definition

$$\mathbf{P} = \mathbf{F} \, \Delta t \qquad \text{(Equation 4.1a)}$$

In the SI, impulse is expressed in newton-seconds (N · s). The amount that the velocity of an object changes depends directly on the amount of impulse applied to it. An impulse of 500 newton-seconds, for example, could be produced by a force of 100 newtons operating for 5 seconds, or by 500 newtons operating for 1 second. In either case, it produces twice as much change in the velocity of a given object as would an impulse of 250 newton-seconds. This relationship is expressed in the proportionality

$$\Delta \mathbf{v} \propto \mathbf{F} \, \Delta t \qquad \text{(Equation 4.1b)}$$

EXAMPLE 4.1   To slow down a car, a braking force of 1 200 newtons is applied for 10 seconds. How much force would be needed to produce the same change in velocity in 6 seconds?

SOLUTION

To produce the same change in velocity, the same impulse is needed. Thus

$$F_1 \, \Delta t_1 = F_2 \, \Delta t_2$$
$$(1\ 200\ \text{N})(10\ \text{s}) = F_2(6\ \text{s})$$
$$F_2 = 200\ \text{N}$$

EXAMPLE 4.2   A frictionless wagon is pushed, from rest, with a force of 60 newtons for 14 seconds. If it then strikes a wall and comes to rest in 0.15 second, how much average force does the wall exert on it?

SOLUTION

The impulse that puts the wagon in motion must be the same as the impulse that stops it and in the opposite direction:

$$(60\ \text{N})(14\ \text{s}) = F_2(0.15\ \text{s})$$
$$F_2 = 5\ 600\ \text{N}$$

EXAMPLE 4.3   A spacecraft has two rocket engines, one producing a thrust of 300 newtons and the other 750 newtons. Firing the smaller engine for 10 s speeds the ship up from 80 meters per second to 95 meters per second, and the large engine is then fired for 12 seconds. How fast is the ship then going?

SOLUTION

The two impulses are

$$\text{impulse}_1 = F\,\Delta t = (300\text{ N})(10\text{ s}) = 3\,000\text{ N}\cdot\text{s}$$
$$\text{impulse}_2 = F\,\Delta t = (750\text{ N})(12\text{ s}) = 9\,000\text{ N}\cdot\text{s}$$

So the second impulse must produce three times as much change in velocity as the first. The first change was 95 m/s – 80 m/s = 15 m/s, so the second impulse increased the velocity by 45 meters per second, from 95 meters per second to 140 meters per second.

CORE CONCEPT

Change in velocity of an object is proportional to the product of force and the time during which the force acts:

$$\Delta\mathbf{v} \propto \mathbf{F}\,\Delta t$$

TRY THIS
—1—

A spacecraft is going 350 meters per second; a retro rocket that provides a force of 520 newtons for 10 seconds slows it down to 300 meters per second. Then another retro rocket fires, with a force of 130 newtons for 4 seconds. How fast is the craft then going?

Inertia

How much change in velocity does a given force produce? An impulse of 50 newton-seconds would send a billiard ball moving rapidly across the table but would barely nudge a bowling ball. Every object has a tendency to resist changes in its velocity, a tendency that we call the *inertia* of the object. A bowling ball has a lot more inertia than a billiard ball.

Equation 4-1b tells us that $\mathbf{F} \propto \Delta\mathbf{v}/\Delta t$. In other words, the acceleration of an object is proportional to the force applied to it. However, the acceleration also depends on the inertia of the object; the more inertia, the less acceleration. The concept of inertia can be made quantitative by using it as the constant of proportionality between force and acceleration:

$$\mathbf{F} = i\mathbf{a}$$

When an object is in free fall, this equation becomes F = ig, where g is the acceleration due to gravity. Since this quantity is universally proportional to the gravitational field strength, we define the units such that g stands for both the acceleration due to gravity and the gravitational field strength. The weight of an object is the force of gravity acting on it; Equation 3.1 tell us that this force is equal to mg. So

$$F = mg = ig$$

When we define the units such that acceleration due to gravity is equal to gravitation field strength, it follows that inertia equals mass. Now we can write the definition of inertia in its more usual form. Newton's law of inertia, one of the most fundamental equations in all of physics, is

$$\mathbf{F} = m\,\mathbf{a} \qquad\qquad \textbf{(Equation 4.2)}$$

It is this choice of units that makes $\text{m/s}^2 = \text{N/kg}$.

EXAMPLE 4.4 What force is needed to accelerate a 60-kilogram wagon from rest to 5.0 meters per second in 2.0 seconds?

SOLUTION

First, find the acceleration from Equation 2.3:

$$a = \frac{\Delta v}{\Delta t} = \frac{5.0 \text{ m/s}}{2.0 \text{ s}} = 2.5 \text{ m/s}^2$$

Then apply Equation 4.2:

$$F = ma = (60 \text{ kg})(2.5 \text{ m/s}^2) = 150 \text{ kg} \cdot \text{m/s}^2 = 150 \text{ N}$$

EXAMPLE 4.5 A frictionless wagon going at 2.5 meters per second is pushed with a force of 380 N, and its speed increases to 6.2 meters per second in 4.0 seconds. What is its mass?

SOLUTION

The acceleration of the wagon is

$$a = \frac{\Delta v}{\Delta t} = \frac{6.2 \text{ m/s} - 2.5 \text{ m/s}}{4.0 \text{ s}} = 0.93 \text{ m/s}^2$$

From Equation 4.2,

$$m = \frac{F}{a} = \frac{380 \text{ N}}{0.93 \text{ m/s}^2} = 410 \text{ kg}$$

EXAMPLE 4.6 What acceleration would be given to a 7.5-kilogram bowling ball being swung with a propelling force of 120 newtons?

SOLUTION

From Equation 4.2,

$$a = \frac{F}{m} = \frac{120 \text{ N}}{7.5 \text{ kg}} = 16 \text{ m/s}^2$$

CORE CONCEPT

Force equals mass times acceleration ($\mathbf{F} = m\mathbf{a}$).

TRY THIS
—2—

What is the mass of a frictionless sled that will be accelerated at 3.0 meters per second squared by a force of 130 newtons?

Multiple Forces

Spaceships in outer space and things falling freely have one thing in common: Each has only a single force acting on it. We rarely deal with this sort of situation. When we push something, there are other forces acting as well. The $\mathbf{F}$ in Equation 4.2 is not any one of these forces; it is the vector sum of all of them. If they all add up to zero, then the object will not accelerate; it will be in equilibrium. $\mathbf{F}$ is the *net force* on the object.

Consider, for example, a 350-kilogram rocket about to take off vertically from its launching pad. You decide that you would like it to zoom upward with a substantial acceleration—say, $a = 8$ meters per second squared. So you apply Equation 4.2 to this acceleration and conclude that you need an engine with a thrust of $ma = 2\,800$ newtons. If you design your rocket engine that way, you will be severely disappointed. Your rocket weighs $mg = 3\,430$ newtons; 2 800 newtons will never get it off the ground. The thrust you need is 2 800 newtons *more than* the weight of the rocket. Example 4.7 shows how the needed thrust must be calculated.

EXAMPLE 4.7   What thrust is needed to fire a 350-kilogram rocket straight up with an acceleration of 8.0 meters per second squared?

SOLUTION

The net force needed to produce this acceleration is

$$F = ma = (350 \text{ kg})(8.0 \text{ m/s}^2) = 2\,800 \text{ N}$$

However, there is a downward force acting on it as well, equal to

$$w = mg = (350 \text{ kg})(9.8 \text{ m/s}^2) = 3\,430 \text{ N}$$

The net force is the thrust minus the weight, or

$$\text{thrust} - 3\,430 \text{ N} = 2\,800 \text{ N}$$

so the rocket engine must product a thrust of 6 230 N.

Here's another example: The child and sled of Figure 4.1 have a mass of 40 kilograms. The acceleration of the sled is produced by the net horizontal force. The horizontal component of the tension in the tow rope pulls the sled to the right, while friction is pulling it to the left. The net force is the difference—and that is what accelerates the sled.

FIGURE 4.1

EXAMPLE 4.8   What is the acceleration of the child on the sled, with combined mass 40 kilograms, if the friction is 60 newtons and the rope is being pulled with a force of 170 newtons at an angle of 35° with the ground?

SOLUTION

The horizontal component of the force pulling the sled forward is (from Equation 2.2)

$$T_{horiz} = T \cos \theta = (170 \text{ N})(\cos 35°) = 139 \text{ N}$$

To get the net force, subtract the force to the right from the frictional force to the left:

$$F = 139 \text{ N} - 60 \text{ N} = 79 \text{ N} = 79 \text{ kg} \cdot \text{m/s}^2$$

From Equation 4.2,

$$a = \frac{F}{m} = \frac{79 \text{ kg} \cdot \text{m/s}^2}{40 \text{ kg}} = 2.0 \text{ m/s}^2$$

EXAMPLE 4.9   A 2.0-kilogram weather balloon is released and begins to rise against 6.5 newtons of viscous drag. If its buoyancy is 32 newtons, what is its acceleration?

SOLUTION

First, we have to find the net force on it. Its weight is $mg = (2.0 \text{ kg})(9.8 \text{ m/s}^2) = 19.6 \text{ N}$. Since it is rising, the viscous drag is another downward force, so the total downward force is 19.6 N + 6.5 N = 26.1 N. Since the buoyancy pushing it upward is 32 N, the net upward force is 32 N – 26.1 N = 5.9 N. Now that we know the net force, we can find the acceleration:

$$a = \frac{F}{m} = \frac{5.9 \text{ N}}{2.0 \text{ kg}} = 3.0 \text{ m/s}^2$$

CORE CONCEPT

Acceleration must be calculated from the vector sum of all forces acting on the object.

TRY THIS
—3—

What is the acceleration of a 1 200-kilogram boat if its motor produces 8 500 newtons of forward thrust and the viscous drag is 6 200 newtons?

Momentum

Which does more damage in striking a tree, a Cadillac or a Toyota? If both hit at the same speed, the Cadillac, with its larger mass, will surely exert more force on the tree. But suppose the Toyota is going much faster than the Cadillac. Now is there any way to figure it out?

Well, maybe we can. We might start by rewriting Equation 4.2 like this:

$$F = m\frac{\Delta v}{\Delta t}$$

Now multiply both sides by Δt:

$$F \Delta t = m \Delta v$$

and we recognize the left-hand side of this equation as the impulse that stops the car. Remember, as you learned in Chapter 3, that the force the tree exerts on the car has to be the same in magnitude as the force the car exerts on the tree. Both are damaged in the collision. And the amount of damage depends on the product of the mass of the car and the speed at which it was going before the tree abruptly brought it to rest.

The quantity *mv*—mass times velocity—is called *momentum*. In bringing the car to rest, the tree suddenly reduced the momentum of the car to zero. And the amount by which the momentum changed is equal to the impulse applied to the car:

$$\mathbf{F} \Delta t = \Delta(m\mathbf{v}) \qquad \text{(Equation 4.3)}$$

Or: The impulse applied to anything is equal to the amount by which its momentum changes.

This equation says nothing that we have not already used in Equation 4.2, but it is often more convenient. In some problems, it saves us the trouble of calculating the acceleration as a separate step.

 EXAMPLE 4.10 What braking force is needed to bring a 2 200-kilogram car going 18 meters per second to rest in 6.0 seconds?

SOLUTION

Using Equation 4.3, we have

$$F(6.0 \text{ s}) = (2\,200 \text{ kg})(18 \text{ m/s}) - 0$$

which says that the momentum is changing from mv to 0. This gives $F = 6\,600$ N.

EXAMPLE 4.11 A 650-kilogram rocket is to be speeded up from 440 meters per second to 520 meters per second in outer space. If the thrust of the engine is 1 200 newtons, for how long must the engine be fired?

SOLUTION

The change in the momentum of the rocket is $(650 \text{ kg})(520 \text{ m/s})$ $- (650 \text{ kg})(440 \text{ m/s}) = 52\,000 \text{ kg} \cdot \text{m/s}$. This must be equal to the impulse, so

$$(1\,200 \text{ N})(\Delta t) = 52\,000 \text{ kg·m/s}$$
$$\Delta t = 43 \text{ s}$$

There are some situations in which it is either difficult or impossible to use Equation 4.2, but Equation 4.3 comes to the rescue. This happens when the mass of the object is changing. In that case, even with a constant force, the acceleration is changing, and the problem becomes complex. Here is a case in point: A rocket ship traveling in outer space speeds up by turning on its engines. How much does its velocity change? You would have to know the thrust of the engine (F) and the period of time for which the engine burned (Δt). But if the fuel consumed is any substantial part of the mass of the rocket, the mass of the rocket is also changing. The impulse gives you the change in the momentum of the rocket ship, and you can use this to determine the final speed only if you know how much fuel was consumed.

EXAMPLE 4.12 A rocket ship in outer space has a mass of 650 kilograms, including 120 kilograms of fuel, and it is going 140 meters per second. Burning all the fuel produces a thrust of 1 200 newtons for 25 seconds. How fast is the ship then going?

SOLUTION

The impulse applied to the ship is

$$P = F\,\Delta t = (1\,200 \text{ N})(25 \text{ s}) = 30\,000 \text{ N·s}$$

and this must equal the increase in the momentum of the ship. Its original momentum is

$$mv = (650 \text{ kg}) (140 \text{ m/s}) = 91\,000 \text{ kg} \cdot \text{m/s}$$

Note that, since $N = \text{kg·m/s}^2$, the impulse and the momentum are expressed in the same units and can be added. The final momentum is the initial momentum plus the impulse, or

$$mv_{\text{final}} = 91\,000 \text{ N·s} + 30\,000 \text{ N·s}$$

Since the fuel has been burned up, the final mass is only 530 kg, so

$$v_{\text{final}}(530 \text{ kg}) = 121\,000 \text{ kg·m/s}$$
$$v_{\text{final}} = 228 \text{ m/s}$$

A study of Equation 4.3 will explain why trapeze artists keep a net under them as they work. If they fall, they will have a certain definite momentum when they hit the ground. They must lose all this momentum; the impulse applied to their bodies will be $F \Delta t$. The purpose of the net is to increase the value of Δt, the time it takes them to come to rest. The hard, unyielding ground brings them to rest within a fraction of a second. The net spreads out the impulse over a much longer period of time, so that the force is proportionally smaller.

CORE CONCEPT

Impulse equals change in momentum:

$$\mathbf{F} \Delta t = \Delta(m \mathbf{v})$$

TRY THIS
—4—

A pitcher throws a ball whose mass is 0.30 kilogram, bringing it from rest to a speed of 35 meters per second. If the motion of his arm while holding the ball lasts 0.50 second, find (a) the momentum of the ball; (b) the impulse the pitcher applied to it; and (c) the force the pitcher used in throwing it.

Momentum Endures

The greatest value of the concept of momentum is in the calculation of what happens when two or more objects interact. It turns out that we can often find out the results of a collision, for example, without knowing anything at all about the forces involved or how long they persist.

The explanation is this: Think of two objects, A and B, colliding. Object A exerts a force F on object B, and B exerts a force $-F$ on A. The negative sign indicates that the two forces are in opposite directions. They have equal magnitudes. Since the duration of the impact, Δt, is the same for both, the impulse that B exerts on A is the negative of the impulse A exerts on B.

Now, impulse is equal to change in momentum. It follows that, in the collision, the change in the momentum of A is the negative of the change in the momentum of B. The sum of the two changes is thus zero. In other words, during the collision, the total momentum does not change! The increase in the momentum of one object is exactly equal to the decrease in the momentum of the other. This gives us the very fundamental and important law of nature called the *law of conservation of momentum: In an isolated system, the total momentum does not change.*

You should think of using this rule whenever you have to deal with an interaction between two objects. For example, a little girl is standing on a wagon, and jumps off to the rear of it. To do so, she has to kick the wagon so that it moves forward. Before she jumped, the total momentum of the system was zero, so (if friction is small enough to ignore) it must continue to be zero afterward. Her momentum backward must equal the wagon's momentum forward.

EXAMPLE 4.13

The 35-kilogram girl is standing on a 20-kilogram wagon and jumps off, giving the wagon a kick that sends it off at 3.8 meters per second. How fast is the girl moving?

SOLUTION

Total momentum is initially zero, and must remain zero. Therefore, the momentum acquired by the wagon in one direction equals the momentum acquired by the girl in the other:

$$(mv)_{\text{girl}} = (mv)_{\text{wagon}}$$
$$(35 \text{ kg}) (v_{\text{girl}}) = (20 \text{ kg}) (3.8 \text{ m/s})$$
$$v_{\text{girl}} = 2.2 \text{ m/s}$$

Now suppose the girl is running and jumps into the wagon. If you know how fast she is going, you can figure out how fast she and the wagon will be traveling together when she is in it. Before she hit the wagon, her momentum was the total momentum of the system, and that value does not change.

EXAMPLE 4.14

The same 35-kilogram girl is now running along at 5.2 meters per second and jumps into the 20-kilogram wagon. How fast is the wagon moving with the girl in it?

SOLUTION

At first the girl has all the momentum, and it must be equal to the total momentum of the girl and the wagon after she lands in it. So

$$(mv)_{\text{before}} = (mv)_{\text{after}}$$

$$(35 \text{ kg})(5.2 \text{ m/s}) = (35 \text{ kg} + 20 \text{ kg})v_{\text{after}}$$

$$3.3 \text{ m/s} = v_{\text{after}}$$

In dealing with problems of this sort, be careful to take the direction of the momentum into account; it is a vector. In one dimension, it is enough to call one direction negative and the other positive. Thus, if a ball hits a wall perpendicularly, going 20 meters per second, and bounces off at 15 meters per second, the change in its velocity is

$$20 \text{ meters per second} - (-15 \text{ m/s}) = 35 \text{ meters per second}$$

EXAMPLE 4.15

A 1.2-kilogram basketball traveling at 7.5 meters per second hits the back of a 12-kilogram wagon and bounces off at 3.8 meters per second, sending the wagon off in the original direction of travel of the ball. How fast is the wagon going?

SOLUTION

The momentum of the ball before it hit has to equal the sum of the momenta of the two objects after the collision. Let's call the original direction of travel of the ball positive. Then the equation becomes

$$(1.2 \text{ kg}) (7.5 \text{ m/s}) = (1.2 \text{ kg}) (-3.8 \text{ m/s}) + (12 \text{ kg}) (v)$$
$$1.1 \text{ m/s} = v$$

CORE CONCEPT

In a closed system, the total momentum does not change.

TRY THIS
—5—

A 2 000-kilogram limousine going east at 22 meters per second strikes a 1 200-kilogram sports car going west at 30 meters per second. How fast are the two cars going together after the collision?

Making Circles

As we saw in Chapter 2, a change in the direction of motion is an acceleration, even if the speed remains constant. And when an object is traveling in a circular path, its direction is constantly changing. Its acceleration is centripetal—toward the center of the circle. The only way to produce a centripetal acceleration is to apply a centripetal force.

Consider, for example, a ball attached to the end of a string and whirled around overhead in a horizontal circle, as shown in Figure 4.2. If you let go of the string the ball will keep going in the direction in which it is traveling; it will fly off on a path that is tangent to the circle. To keep it going in the circular path, the string must constantly exert a force on the ball. The force is at all times perpendicular to the velocity, so it changes only the direction of the velocity, not the speed.

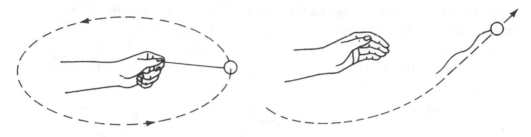

FIGURE 4.2

This, then, is the condition necessary for an object to travel at constant speed in a circular path: There must be a force acting on it, a *centripetal force* that is kept perpendicular to the velocity as the direction of the velocity changes. The force will be directed toward the center of the circle. Let's look at some examples.

When a clothes dryer spins around, the walls of the tub exert a centripetal force on the clothes, making them travel in a circular path. The water can pass through the holes in the tub, so there is no effective centripetal force on it, and it flies off through the holes.

When a car travels in a circular path, the centripetal force is supplied by the friction of the road against the front tires. If the road is banked, the elastic recoil of the road helps by pushing the car toward the center of the circle.

When a satellite is in circular orbit above the earth, the earth's gravity is constantly pulling it toward the center of the earth, perpendicular to the satellite's velocity. Gravity is, in this case, a centripetal force.

How much is the centripetal force? Equation 4.2 tells us that the force that produces an acceleration is ma, and this law is not repealed when the force is centripetal. And we learned, in Equation 2.5b, that centripetal acceleration is equal to v^2/r —the square of its velocity divided by the radius of the circle in which it is traveling. Using these two relationships, we find that centripetal force is given by

$$F_c = \frac{mv^2}{r}$$
(**Equation 4.4**)

EXAMPLE 4.16

The ball at the end of the string of Figure 4.2 has a mass of 150 grams. It is traveling at 4.0 meters per second and the string is 65 centimeters long. How great is the tension in the string?

SOLUTION

If we are to get the answer in newtons, we have to solve the problem in SI units; 150 g = 0.150 kg; 65 cm = 0.65 m. First, we need the acceleration of the ball, from Equation 2.5b:

$$a_c = \frac{v^2}{r} = \frac{(4.0 \text{ m/s})^2}{0.65 \text{ m}} = 24.6 \text{ m/s}^2$$

To get the force, use Equation 4.2:

$$F = ma = (0.150 \text{ kg})(24.6 \text{ m/s}^2) = 3.7 \text{ N}$$

 EXAMPLE 4.17 In outer space, a 1 200-kilogram spacecraft is traveling at 60 meters per second. In order to change its direction, a rocket is fired off to one side, in a direction perpendicular to the velocity. If the thrust of the rocket is 3 820 newtons, what is the radius of the circle in which the spacecraft travels?

SOLUTION

From Equation 4.4,

$$r = \frac{mv^2}{F} = \frac{(1\ 200 \text{ kg})(60 \text{ m/s})^2}{3\ 820 \text{ N}} = 1\ 130 \text{ m}$$

EXAMPLE 4.18 A 1.5-kilogram rock is tied to a string 40 centimeters long and spun around in a horizontal circle. If the string will break when the tension in it exceeds 250 newtons, what is the greatest speed the rock can have without breaking the string?

SOLUTION

From Equation 4.4,

$$v = \sqrt{\frac{Fr}{m}} = \sqrt{\frac{(250 \text{ N})(0.40 \text{ m})}{1.5 \text{ kg}}} = 8.2 \text{ m/s}$$

CORE CONCEPT

The force that keeps an object going in a circular path at constant speed is equal to the mass of the object times its centripetal acceleration.

$$F_c = \frac{mv^2}{r}$$

A bicycle and rider, with a combined mass of 55 kilograms, are traveling at 6.0 meters per second. If they are going in a circular path whose radius is 30 meters, how much is the force acting on the tires of the bicycle?

Rules of Spinning

Consider the case of a bicycle wheel mounted vertically. If you want to make it spin, you have to apply a force to it. But it makes a big difference where the force is applied and in what direction.

Your first impulse, which is quite correct, is to apply the force on the rim of the wheel and to exert it along a tangent to the rim. Torque is what makes things spin, so it is this combination of location and direction of the force that gives you the most torque from the force you applied. Pushing the wheel radially, whether outward or inward, produces no torque no matter how hard you push. Only a component of force perpendicular to the radius produces torque. And the further out from the center of rotation you apply the force, the more torque you get for the force you apply. This is according to the definition of torque, given in Equation 3.3.

Just as in linear systems, applying a force produces momentum; in rotating systems, applying a torque produces *angular momentum*. The simplest case to consider is one in which the mass of the object is concentrated at a given radial distance from the center of rotation, as the ball spinning at the end of a string, shown in Figure 4.3. Its angular momentum is simply its mass times its speed times its radial distance from the center: angular momentum = *mvr*. Just as a change in linear momentum is equal to force times time (Equation 4.3), change in angular momentum is equal to torque times time.

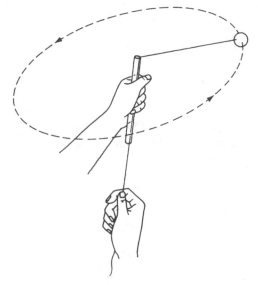

FIGURE 4.3

What happens if you pull the string in Figure 4.3 to draw the ball closer in? You are not applying any torque to the ball, since the force exerted on it by the string is radial. The angular momentum *mvr* cannot change. As the radius gets smaller, the velocity must increase. The ball speeds up.

The rule works in more complex systems as well. The ice skater in Figure 4.4 starts his spin with arms outstretched. When he draws his arms into the center of his body, he decreases the radius of their spin, and his whole body rotates faster. The calculations are difficult because every little part of his arm starts out with a different radius of spin, but the principle still holds.

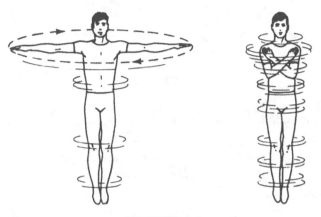

FIGURE 4.4

This principle is incorporated into one of the most fundamental laws of mechanics, known as the law of conservation of angular momentum. *In a closed system, the total angular momentum remains constant.* If you jump off a merry-go-round, it spins a little faster.

CORE CONCEPT
Angular momentum is conserved in all interactions.

The Moon in Orbit

Suppose you could build a tower several miles high and mount a horizontal cannon at its top, as shown in Figure 4.5. The cannonballs will fall to earth, always accelerating downward in free fall. The greater the speed, the farther a ball will travel before it strikes. The fastest ball in the picture will never hit the ground at all. It keeps falling toward the earth, but it never lands; the earth curves away from it as it falls. It is in orbit.

FIGURE 4.5

Objects in orbit around the earth—artificial satellites, Spacelabs, the moon—are in free fall. They must obey the usual rule of free fall: their acceleration depends only on where they are. A satellite orbiting the earth at low altitude has a downward acceleration somewhat less than 9.8 meters per second squared.

The earth's influence on objects out in space does not go on indefinitely. The gravitational field gets weaker with greater distance from the earth becoming vanishingly small at a great distance. The relationship between the field strength and the distance from the earth can be deduced by studying the orbit of the moon.

The moon orbits the earth in a nearly circular path, with a period of 27.3 days. From this information, we can calculate its centripetal acceleration; this must be equal to the gravitational field at the distance of the moon's orbit.

Combining Equations 2.5a and 2.5b tells us that centripetal acceleration is equal to $4\pi^2r/T^2$. Let's put this into SI units. The moon is 3.82×10^8 kilometers away and the period of its revolution is

$$(27.3 \text{ days})\left(\frac{24 \text{ hours}}{\text{day}}\right)\left(\frac{60 \text{ minutes}}{\text{hour}}\right)\left(\frac{60 \text{ seconds}}{\text{minute}}\right) = 2.36 \times 10^6 \text{ s}$$

Now let's find its centripetal acceleration.

$$a_c = \frac{4\pi^2r}{T^2} = \frac{4\pi^2 \, (3.82 \times 10^8 \text{ km})}{(2.36 \times 10^6 \text{ s})^2} = 2.71 \times 10^{-3} \text{ m/s}^2$$

Remember, now, the moon is in free fall since the only force acting on it is the earth's gravity. Clearly, the acceleration due to gravity at the moon's location is not 9.8 meters per second squared but only 0.0027 meters per second squared. This important constant, g, is substantially lower at the moon's altitude than at the surface of the earth.

There must be some mathematical rule that tells us how gravity varies as the distance from the earth increases. To get this rule, let's first find the ratio of the two known values of g.

$$\frac{\text{acceleration of moon}}{g \text{ at surface of earth}} = \frac{2.7 \times 10^{-3} \text{ m/s}^2}{9.8 \text{ m/s}^2} = \frac{1}{3\ 600}$$

Now let's get the ratio of the two distances from the center of the earth.

$$\frac{\text{distance of moon}}{\text{radius of earth}} = \frac{243\ 000 \text{ mi}}{4\ 000 \text{ mi}} = 60$$

The moon is 60 times as far away and experiences 1/3 600 the gravitational field. The ratio of the two values of g (1:3 600) is the inverse square of the ratio of the distances (60:1). This relationship between the ratios holds for any distance from the center of the earth. With this information, you can find the value of g, the acceleration of any object in free fall at that distance, whether it is going up, is going down, or is in orbit. It is useful to express this rule as the ratio between g at any distance r and the field and distance at the surface, g_0 and r_0. Remember that the distance must be measured from the *center* of the earth or whatever other celestial body is involved.

$$\frac{g}{g_0} = \left(\frac{r_0}{r}\right)^2 \qquad \textbf{(Equation 4.5)}$$

EXAMPLE 4.19 What is the acceleration of a piece of space junk when it is at an altitude of 22 000 km?

SOLUTION

Note that the problem does not specify which way the object is traveling. It will have the same acceleration whether it is going up, down, or sideways in orbit, so long as it is in free fall.

Since the r in Equation 4.8 is distance from the center of the earth, the first thing we need is the correct value of r for the object. This is its altitude plus the radius of the earth: 22 000 km + 6 400 km = 28 400 km.

Equation 4.5 can be applied:

$$\frac{g}{g_0} = \left(\frac{r_0}{r}\right)^2$$

where g and g_0 are values of acceleration due to gravity and r and r_0 are corresponding radial distances from the center of the earth. Since we know the value of g_0 at the surface (6 400 km from the center), we may write

$$\frac{9.8 \text{ m/s}^2}{g} = \left(\frac{28\ 400 \text{ km}}{6\ 400 \text{ km}}\right)^2$$

from which $g = 0.50 \text{ m/s}^2$.

We do not build towers to put satellites in orbit, but we get the satellites up there. They are lifted by rockets. Once the desired altitude has been reached, the last-stage rocket is pointed horizontally and fired, placing the satellite into a horizontal path. Its acceleration is down, perpendicular to its velocity. If the velocity is exactly right, the direction of the satellite will change just enough as it travels to keep the velocity always perpendicular to the acceleration. The path will then be a circle.

EXAMPLE 4.20

What is the acceleration due to gravity at the surface of a planet with a radius of 1.6×10^6 meters if a satellite near the surface makes one circular orbit every 1 700 seconds?

SOLUTION

The acceleration due to gravity must be the centripetal acceleration of the satellite. We can find its speed from Equation 2.5a:

$$v = \frac{2\pi r}{T} = \frac{2\pi (1.6 \times 10^6 \text{ m})}{1\ 700 \text{ s}} = 5\ 910 \text{ m/s}$$

Now we can find the acceleration:

$$a = \frac{v^2}{r} = \frac{(5\ 910 \text{ m/s})^2}{1.6 \times 10^6 \text{ m}} = 22 \text{ m/s}^2$$

EXAMPLE 4.21

An artificial satellite is to be put into orbit at an altitude of 15 000 kilometers. What speed must it be given to make it go into a circular orbit?

SOLUTION

First, find the gravitational field into which this satellite is being launched. The radius of the earth, r_0, is 6 400 km. The distance of the satellite from the center of the earth is $r = 15\ 000 + 6\ 400 = 21\ 400$ km. Then, from Equation 4.5,

$$g = g_0 \left(\frac{r_0}{r}\right)^2 = (9.8 \text{ m/s}^2)\left(\frac{6\ 400 \text{ km}}{21\ 400 \text{ km}}\right)^2 = 0.88 \text{ m/s}^2$$

For a circular orbit, the gravitational field must be equal to the centripetal acceleration of the satellite. From Equation 2.5b,

$$v = \sqrt{ar} = \sqrt{(0.88 \text{ m/s}^2)(2.14 \times 10^7 \text{ m})} = 4\,300 \text{ m/s}$$

CORE CONCEPT

An object in orbit is in free fall with centripetal acceleration due to gravity.

TRY THIS
—7—

How fast was the Lunar Orbiter traveling when it was in orbit around the moon, near its surface? The radius of the moon is 3.5×10^6 meters, and the acceleration due to gravity at its surface is 1.67 meters per second squared.

Gravity Is Everywhere

What determines the amount of gravitational force the earth exerts on the moon, or on anything else? From the physics you already know, we can figure this out.

We saw in Chapter 3 that the weight of anything, which is the force of gravity acting on it, is equal to mg. The force of the earth's gravity acting on an object on earth, then, is proportional to the mass of the object:

$$F_{\text{grav}} \propto m_{\text{o}}$$

We also know that every force is one-half of an interaction pair; if the earth is pulling on any object, the object must be pulling on the earth with an equal force in the opposite direction. This force the moon exerts on the earth, for example, produces a detectable wobble in the motion of the earth. The only reason it is so small is that the earth has such a large mass that the force cannot produce much acceleration.

How much force does an object on earth exert on the earth? If the force the earth exerts on the object is proportional to the mass of the object, it is reasonable to suppose that the force the object exerts on the earth is proportional to the mass of the earth:

$$F_{\text{grav}} \propto m_{\text{e}}$$

Since this force is the same in both directions, it must be proportional to *both* the mass of the moon and the mass of the earth:

$$F_{\text{grav}} \propto (m_{\text{o}})(m_{\text{e}})$$

We also know that this force varies inversely as the square of the distance from the earth to the object, so

$$F_{grav} \propto \frac{m_o m_e}{r^2}$$

We must assume that the earth and the moon obey the same laws of nature as all other objects in the universe. This relationship, as applied to everything, is called the *law of universal gravitation*. It states that any two objects in the universe attract each other with a force proportional to the product of their masses and inversely proportional to the square of the distance between them. Newton was the first to prove that this law could explain the motions of objects in the solar system.

By inserting a constant, whose value depends only on the choice of units, we can write the law algebraically this way:

$$F_{grav} = \frac{G m_1 m_2}{r^2} \qquad \textbf{(Equation 4.6)}$$

G is the same for any pair of objects at all; it is a universal constant whose value in the SI is

$$G = 6.67 \times 10^{-11} \ \text{N·m}^2/\text{kg}^2$$

This is an extremely small number. What it tells us is that the gravitational attraction between ordinary objects is usually too small to be detected. Gravity becomes a significant force only when one of the objects is very large—like the earth, for example. Example 4.22 shows how this law can be applied in determining the force between two ordinary objects.

EXAMPLE 4.22   What is the gravitational attraction between a 70-kilogram boy and a 60-kilogram girl who are 3 meters apart?

SOLUTION

Apply Equation 4.6, the law of universal gravitation:

$$F_{grav} = \frac{G m_1 m_2}{r^2}$$

$$F_{grav} = \frac{(6.67 \times 10^{-11} \ \text{N·m}^2/\text{kg}^2)(70 \ \text{kg})(60 \ \text{kg})}{(3 \ \text{m})^2}$$

$$F_{grav} = 3 \times 10^{-8} \ \text{N}$$

which is barely sufficient to move two fruit flies toward each other.

CORE CONCEPT

All objects attract each other with a force proportional to the masses of both and inversely proportional to the distance between them:

$$F_{\text{grav}} = \frac{Gm_1m_2}{r^2}$$

TRY THIS
—8—

What is the force of attraction between two asteroids 4.0×10^5 meters apart if their masses are 5 700 kilograms and 14 000 kilograms, respectively?

Weightlessness

When astronauts are in an orbiting spacecraft, gravity seems to have disappeared. They, and everything in the craft, float freely around the cabin. Why?

You might notice the same phenomenon if you are ever unlucky enough to be trapped inside a falling elevator. Of course, it would have to be a frictionless elevator so that it would fall freely, accelerating at 9.8 meters per second squared. Then, if you should drop a pencil, it would fall along with you, appearing to stand still in space.

If someone is watching this whole event from outside the elevator, he would see what is happening somewhat differently. He would say, "Your pencil is falling. It is obvious, however, that it has the same acceleration as the elevator, so it never gets any closer to the floor of the elevator. You have that acceleration also, just like everything else in the elevator. You only *think* gravity has disappeared."

What will a scale read if you step on it while the elevator is falling? If you step on a scale in your own bathroom, you are in equilibrium. The scale pushes up on you with a force equal to your weight, and you push down on the scale with an equal force (action equals reaction). In the falling elevator, you fall with an acceleration *g* because the only force acting on you is gravity. There is no upward force at all. The scale reads zero.

This condition of weightlessness exists for all measurements made inside the elevator. It occurs because the elevator is in free fall. Remember, however, that when a satellite is in orbit, it also is in free fall, accelerated centripetally by the force of gravity. And as long as the spacecraft is in free fall, everything in it has the same acceleration, and everything appears weightless to anyone inside.

Acceleration of a vehicle you are in by some force other than gravity produces a gravity-like effect. When you accelerate your car forward, you feel yourself pushed back; when you turn to the right, you feel yourself pushed to the left. If your car turns in a circle, you feel yourself pushed to the outside of the circle, as in Figure 4.6, and you say that you feel a centrifugal force. Anyone standing on the road outside looks at you and says, "Nonsense. You only think there is a force pushing you outward. Actually, your body is trying to travel in a straight line while the body of the car is pushing you toward the center of the circular path."

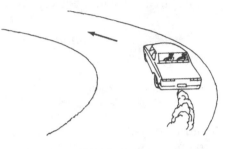

FIGURE 4.6

CORE CONCEPT

Acceleration of a vehicle changes the apparent gravity inside it, and when the vehicle is in free fall there are no gravitational effects inside.

TRY THIS
—9—

You are in a spacecraft in outer space somewhere, and everything inside it is weightless. Suddenly the things around you start to fall. What could you conclude?

REVIEW EXERCISES FOR CHAPTER 4

Fill-In's. For each of the following, fill in the missing word or phrase.

1. Impulse is the product of force and _____.

2. The change in the velocity of an object is proportional to the_____ applied to it.

3. The ratio of net force applied to an object to the acceleration it produces is the _____ of the object.

4. A kilogram-meter per second squared is called a(n)_____.

5. If there are several forces on an object, its acceleration depends on its mass and the _____ force.

6. Momentum is the product of mass and _____.

7. The change in the momentum of an object is equal to the _____ applied to it.

8. In any interaction between two or more isolated objects, the total _____ does not change.

9. An object will move in a circular path at constant speed if a force is applied to it in a direction that is kept _____ to its velocity.

10. The force that accelerates an object into a circular path is called a(n) _____ force.

11. The force of _____ is the centripetal force on the moon.

12. The force of gravity between two objects is directly proportional to the product of their _____.

13. The force of gravity between two objects is inversely proportional to the square of the _____.

14. The acceleration due to gravity at the surface of a planet depends on the _____ and the _____ of the planet.

15. There is no perceptible gravity within an orbiting spacecraft because the craft is in a state of _____.

16. To place a satellite in circular orbit, it must be lifted to the right height and given the correct _____.

17. The acceleration of a satellite in orbit depends only on its _____.

18. The centripetal force acting on a satellite is produced by _____.

19. When a dancer in a pirouette flings her arms out to slow her revolution, there is no change in her _____.

20. A 0.50-kilogram basketball going 10 meters per second strikes a wall and rebounds at 8 meters per second. The change in its momentum is _____ kilogram-meters per second.

21. A 200-gram croquet ball is going north at 5 meters per second, and it is struck by another ball going east, applying an impulse to the first ball of 1 newton-second. After the impact, the first ball is traveling to the _____.

Multiple Choice.

1. A 500-kilogram rocket is fired straight up from the earth, the engines providing 7 500 newtons of thrust. Its acceleration is
   (1) 4.5 meters per second squared
   (2) 5.2 meters per second squared
   (3) 9.8 meters per second squared
   (4) 15 meters per second squared
   (5) 20.2 meters per second squared.

2. The gravitational force on an object in motion is centripetal and constant in magnitude whenever the object is
   (1) in motion   (4) in any orbit
   (2) in any kind of free fall   (5) in circular orbit only.
   (3) in outer space

3. The gravitational field at the surface of Mars is 3.3 meters per second squared and its radius is 3 400 kilometers. When the Viking spacecraft was 3 400 kilometers above the surface of Mars, its acceleration was
   (1) 14 meters per second squared
   (2) 6.6 meters per second squared
   (3) 3.3 meters per second squared
   (4) 1.65 meters per second squared
   (5) 0.83 meters per second squared.

4. In outer space, a 250-kilogram rocket is to be speeded up from 60 meters per second to 75 meters per second in 5.0 seconds. The thrust needed is
   (1) 17 newtons   (4) 3 200 newtons
   (2) 750 newtons   (5) 3 750 newtons.
   (3) 3 000 newtons

5. A force of 40 newtons keeps a rock going in a horizontal circle by means of a string 2.0 meters long. If the rock is moving at 10 meters per second, its mass is
   (1) 0.8 kilogram   (4) 8 kilograms
   (2) 2 kilograms   (5) 16 kilograms.
   (3) 1.6 kilograms

6. Two stars of masses m and $5m$ are 3 000 parsecs apart. If the gravitational force on the large one is F, the force on the small star is
    (1) $F/25$ (2) $F/5$ (3) F (4) $5F$ (5) $25F$.

7. A 500-gram sinker is dropped into the water. After falling several centimeters, it stops gaining speed. At that point, its acceleration is
    (1) zero and constant
    (2) zero and increasing
    (3) zero and decreasing
    (4) 9.8 meters per second squared and constant
    (5) 9.8 meters per second squared and decreasing.

8. An astronaut in an orbiting spacecraft is weightless because
    (1) he is beyond the influence of the earth's gravity
    (2) the spacecraft is in free fall
    (3) the walls of the craft protect him from the earth's gravity
    (4) there is no air resistance in that region
    (5) there is no inertia in outer space.

9. A billiard ball and a Ping-Pong ball are at rest on a pool table. Equal impulses are applied to them with a cue stick. Then
    (1) they will have the same momentum but different velocities
    (2) they will have the same velocity but different momenta
    (3) the forces applied were equal
    (4) they will have the same acceleration
    (5) they apply equal reaction forces to the cue stick.

10. A moving object can come to rest only if it
    (1) has a frictional force acting on it
    (2) has no net force acting on it
    (3) is completely isolated
    (4) is in a gravitational field
    (5) applies an impulse to something else.

11. A land mine explodes and pieces of it fly out in all directions. After the explosion, the vector sum will be zero for all of the particles'
    (1) velocities   (4) masses
    (2) momenta   (5) weights.
    (3) accelerations

Problems.

1. If a jet engine provides a thrust of 45 000 newtons, how long must it fire to produce 1 million newton-seconds of impulse?

2. A rocket-driven sled speeds up from 40 meters per second to 55 meters per second in 5.0 seconds, using an engine that produces 3 500 newtons of thrust. How much thrust would be needed to get the same increase in speed in 2.0 seconds?

3. What force is needed to speed up a frictionless 60-kilogram cart from 4.0 meters per second to 6.5 meters per second in 3.0 seconds?

4. A 1 400-kilogram car strikes a telephone pole and comes to rest, driving the front bumper 40 centimeters into the engine. This means that it continued to travel 40 centimeters while the retarding force was being applied. If it was going 25 meters per second before it hit, find (a) its average acceleration while coming to rest; and (b) the average force that produced the damage.

5. What force must the brakes and tires apply to a 2 800-kilogram truck going 30 meters per second to bring it to rest in 8.0 seconds?

6. A 680-kilogram rocket is to be lifted off the surface of the moon, where $g = 1.67$ meters per second squared. What force is needed to give it an upward acceleration of 2.0 meters per second squared?

7. How much is the frictional force acting on a 75-kilogram cart if a push of 220 newtons gives it an acceleration of 2.0 meters per second squared?

8. A 35-kilogram girl on roller skates, standing still, throws a 6-kilogram medicine ball forward at 3.5 meters per second. How much is her recoil velocity (the backward speed she acquires as a result of the throw)?

9. What is the recoil velocity of a 7.5-kilogram rifle if it fires an 8.0-gram bullet with a muzzle velocity of 640 meters per second?

10. A 750-kilogram rocket, at rest in outer space, propels itself forward by ejecting 45 kilograms of hot gas, which leaves the nozzle at 85 meters per second. Find (a) the change in the momentum of the fuel; and (b) the final speed of the rocket ship.

11. A rocket engine ejects 60 kilograms of hot gas at a speed of 95 meters per second in 20 seconds. Find (a) the change in the momentum of the hot gas; and (b) the thrust of the engine.

12. Two hockey players make a head-on collision and cling to each other. One has a mass of 62 kilograms and is going 3.5 meters per second; the other has a mass of 53 kilograms and is going 5.0 meters per second. How fast are they going together after the collision?

13. What is the radius of the circle in which a 240-kilogram motorcycle and rider are traveling if they are going 35 meters per second and the force exerted on the tires perpendicular to the velocity is 6 200 newtons?

14. A 45-kilogram woman is on a rotating platform in an amusement park, 2.5 meters from the center of rotation. If she is going 4.5 meters per second, what frictional force between her body and the platform is needed to keep her from flying off at a tangent?

15. What is the acceleration of a meteor passing by the earth at an altitude of 55 000 kilometers?

16. How much is the gravitational force that keeps an artificial satellite of mass 3 500 kilograms in orbit around the earth at an altitude of 4 200 kilometers?

17. Determine the gravitational attraction between two asteroids separated by 22 000 meters if their masses are 450 000 kilograms and 700 000 kilograms, respectively.

18. A 64-kilogram water-skier is being towed along at a constant speed of 6.0 meters per second against a viscous drag of 240 newtons. If the tension in the tow rope is increased to 510 newtons how long will it take for the skier to get to a speed of 10.0 meters per second?

19. The radius of the earth is 6 400 kilometers. What is its mass? (Hint: How much does a kilogram weigh at the surface?)

20. When a satellite is in the synchronous orbit above the equator, it stays in one place with reference to the earth by making each revolution in just the same time as it takes the earth to rotate once. What is the altitude of the synchronous orbit?

21. An astronaut in a spacecraft goes into circular orbit above a strange planet that has a radius of 2 100 kilometers. The craft is going at 1 200 meters per second at an altitude of 6 500 kilometers. The astronaut's life-support pack has a mass of 60 kilograms. What will the life-support pack weigh on the surface of the planet?

Energy: They Don't Make It Anymore

WHAT YOU WILL LEARN

Energy takes many different forms, and one kind can be converted into another. However, the total amount does not change. Conservation of energy is one of the most important basic laws of physics.

Doing Work

You have to lift a heavy trunk onto a platform 2 meters high, and the darned thing weighs 600 newtons, too heavy to lift. So you set up a ramp, inclined at 30° and roll the thing up, using a (conveniently weightless and frictionless) dolly. The force you have to exert is (600 newtons)(sin 30°) = 300 newtons. You have cut the load in half.

However, there is a loss. You have to push it further than 2 meters. The length of the ramp, in fact, is (2 meters)/(sin 30°) = 4 meters. You are pushing only half as hard, but you have to do it over twice the distance.

Will it help to change the angle? No. If you use a smaller angle, you reduce the force still further and increase the distance by the same ratio. If you try some other system, such as rigging a pulley, you will find the same result. The thing you cannot reduce is the product of force and distance. This product is called *work*, represented in equations as W. In the SI, the product of force and distance is measured in newton-meters (N·m). This is such an important quantity that it is given a name of its own. A newton-meter is called a *joule* (J).

EXAMPLE 5.1   Using a force of 150 newtons, you push a sofa to the other side of a room, a distance of 3.1 meters. Then you decide you do not like it there, so you push it back. How much work have you done?

SOLUTION

In the first move, the work you did is (150 N)(3.1 m) = 465 J. It takes just as much work to move it back, so the total is 930 joules.

There is a lesson in that sample problem. You wind up with the sofa just where it was; its net displacement is zero. The work you did is not zero. Displacement is a vector, but work is scalar.

Now suppose you decided to move that sofa by attaching a rope to it and dragging it across the floor. The rope makes an angle of, say, 25° with the ground, as in Figure 5.1. Now you will have to use a larger force since only the horizontal component of the force moves the sofa. The only part of a force that does work is the component in the direction of motion.

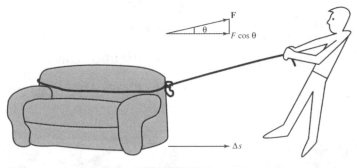

FIGURE 5.1

EXAMPLE 5.2   If the tension in the rope of Figure 5.1 is 165 newtons, how much work is done in moving the sofa to its new position?

SOLUTION

The horizontal component of the tension in the rope is (165 N)(cos 25°) = 150 N. As you might expect, the amount of work is the same whether you push it directly or use the rope.

Let θ be the angle between the direction of the force vector and the direction of the displacement vector. Then this is the formal definition of work:

$$W = F\Delta s \cos\theta \qquad \text{(Equation 5.1)}$$

Now suppose you decided to carry the sofa. Once you get it off the floor, how much work do you do in carrying it horizontally? You are exerting a vertical, upward force to keep it off the floor and carrying it horizontally. Force and displacement are at right angles. Since cosine 90° is zero, you do no work. Clearly, the physical definition of work differs from its customary meaning. Yet the physical definition is meaningful. To move that sofa horizontally, put it on your frictionless dolly. You will have give it a little push to get it moving, but then it will coast all the way to the other side of the room. It takes a little work to get it moving, but no work at all to keep it moving horizontally as long as there is no friction.

CORE CONCEPT

Work is the product of distance moved and the component of force in the direction of motion.

TRY THIS
—1—

Determine the amount of work done in each of the following cases: (1) to keep a door closed in a hurricane, you push it with a force of 300 newtons for 30 seconds; (2) a sled is pulled a horizontal distance of 20 meters with a force of 55 newtons, using a rope that makes an angle of 20° with the horizontal; (3) a 2.0-kilogram sweater is spinning in a dryer tub with a radius of 20 centimeters, going at 22 meters per second for 30 seconds.

What Is Energy?

Some people have a lot of energy when they get up in the morning. Carbohydrates are high-energy foods. Oil is the main source of the energy that keeps industry and cars going. The center in a football formation has to put a lot of energy into his attack.

The word energy has a different meaning in each of those four sentences. As used in physics, the word has a very precise meaning, although it is a little difficult to define because energy takes many different forms. We can approach a definition by noting the relationship between energy and work.

Doing work on any sort of system is sure to change the system in some way. The change, whatever it may be, is called an increase in the *energy* of the system. *Doing work on any system increases its energy by an amount equal to the work done.* Conversely, when the system does work, its energy decreases by that amount. By this definition, it is clear that the units of energy are the same as those of work: foot-pounds, for example, or joules. Energy is such a widely used concept that many other units are in use as well.

It takes work to raise a drop hammer, so its height is one measure of energy, but there are many others. Any observable and measurable change may serve as a measure of energy. If you do a lot of work in sandpapering a plank, for example, you must be adding energy to the system, so you look around to see how that energy can be measured. The plank is smoother—so perhaps smoothness is a measure of energy. The plank is also hotter, and you might well wonder whether there is some way of using the temperature as a measure of the work that has been done on the plank.

If you push a cart—doing work on it—you increase its energy, by definition. If that cart were on a horizontal, frictionless surface, it would keep going faster as long as you kept pushing it. It would surely seem that its velocity is some sort of measure of its energy.

And suppose you grab the ends of a spring and stretch it. Surely this takes a force, moving through a distance, so work is being done. Unquestionably, then, the length of the spring at any given moment is a measure of the energy in it.

Suppose you have to pump up a bicycle tire, doing work by pushing the air into a smaller space at higher pressure. Isn't this acceptable evidence that the pressure of a gas can be used as a measure of energy if only we know how to go about it?

Now take a look at the two bar magnets in Figure 5.2. At (a), they are shown close together; at (b) they are farther apart. The two N poles are facing each other. Knowing that two N poles repel each other, can you tell whether (a) or (b) is the high-energy state? All you have to do is ask yourself whether you would have to do work on the system to change it from state (a) to state (b), or vice versa. Since the magnets repel each other, work would have to be done to push them closer together, so (a) is the high-energy state. You can measure the energy of the system by the distance between the two N poles.

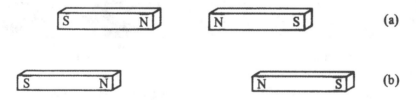

FIGURE 5.2

So we have many different ways of measuring energy. We classify the various forms of energy according to the way in which they can be measured. The energy of the drop hammer, measured by its height, is called *gravitational potential energy*. If we measure energy by means of the velocity at which the thing is going, we call it *kinetic* energy. Energy measured in terms of temperature is called *internal* energy. Energy stored in a stretched spring—or in any other deformed solid—is called *deformational* energy. And so on.

Energy has no direction. If you do a lot of work in pushing a sofa across the floor, and then decide to return it to its old position, you have done work both times. The system has not returned to its original state, because friction warmed up the floor and scraped the finish twice as much in the two moves as in one. Energy is scalar.

CORE CONCEPT

Energy is a scalar quantity, measured in joules, that increases in any system on which work is done.

TRY THIS
—2—

It is known that the opposite electric charges in a pair attract each other. Does the energy of the system increase or decrease as the charges are moved closer together?

Energy and Gravity

Since it takes work to lift a drop hammer, or anything else, any object always has more energy when it is at a higher elevation. The energy that an object has because of its position in a gravitational field is called gravitational potential energy.

A useful definition of any form of energy must include some way of calculating the amount of energy. When something is lifted, the increase in energy is equal to the work done in lifting it. Remember that work is

$$W = F \, \Delta s \cos \theta$$

Gravity pulls things down; if you are holding something or lifting it, the force you must exert is upward. It is equal to the weight of the object. If you lift something straight up $\theta = 0$ and $\cos \theta = 1$. Therefore the work done is simply the weight of the object times the increase in elevation (Δh):

$$\Delta E_{grav} = mg \, \Delta h$$

Suppose you carry something horizontally. Then the force you exert is still upward, but the displacement is horizontal. This means that $\theta = 90°$ and $\cos \theta = 0$. No matter how hard it is to do this, no work is done, and the energy does not increase.

If something is moved at some angle, as in carrying it up a ramp or a staircase, it is only the vertical component of the displacement that counts. The increase in height measures the gravitational potential energy, regardless of any horizontal motion.

Is it possible to state, at any given time, the actual amount of gravitational potential energy in an object? Yes, provided that you decide, arbitrarily, where $h = 0$. You need some reference level from which elevations can be measured. In any problem, put this reference level wherever it is convenient. Then an expression can be given for the actual amount of gravitational energy stored in a system:

$$E_{grav} = mgh$$

(Equation 5.2)

EXAMPLE 5.3

You lift a 2.0-kilogram basketball from the floor, where its gravitational energy is zero. You raise it 2.4 meters, carry it horizontally to a window, and drop it out. It falls 12.0 meters. What is its gravitational potential energy when it hits the ground?

SOLUTION

Only the vertical distance counts. In lifting it, you raised the basketball's energy; carrying it horizontally made no change; when it fell to the ground, its energy dropped. Since it dropped below the floor, its energy decreased more than the increase when you raised it. On the ground, it is 9.6 meters below the zero level. Therefore

$$E_{grav} = mgh = (2.0 \text{ kg})(9.8 \text{ m/s}^2)(-9.6 \text{ m}) = -188 \text{ J}$$

Two points should be noted in this example: first, gravitational potential energy can be negative; it depends on where you decide to place the zero level. Second, the unit called a joule is equal to 1 kilogram-meter squared per second squared, which is the same as 1 newton-meter.

CORE CONCEPT

Gravitational potential energy, the work done in raising the elevation of an object, is equal to mgh.

TRY THIS
—3—

If a 280-kilogram drop hammer was lifted with 12 000 joules of work, how high is it above the zero level?

Energy of Motion

A cart moving on a horizontal, frictionless surface will keep going at constant speed forever. To speed it up, you have to push it through a distance—do work on it. Then it follows that its speed must be a measure of its energy. The energy an object has because of its speed is called *kinetic energy*.

How can you determine the kinetic energy of a moving object? It must be equal to the work that was done in making it move. If an object is at rest and is pushed through a distance s with a force F, the work done on it is equal to the kinetic energy it acquires, so, from Equation 5.1,

$$E_{kin} = Fs$$

The force produces an acceleration; since it is the only force acting, $F = ma$ (Equation 4.2), so

$$E_{kin} = mas$$

The acceleration is uniform and starts at rest, so Equation 2.4c tells us that

$$as = \frac{1}{2}v^2$$

and substituting this into the equation above gives the standard formula for kinetic energy:

$$E_{kin} = \frac{1}{2}mv^2 \qquad \textbf{(Equation 5.3)}$$

Therefore, we can determine the kinetic energy of anything if we know only its mass and velocity.

EXAMPLE 5.4 What is the kinetic energy of a 6.0-kilogram medicine ball traveling at 4.5 meters per second?

SOLUTION

Use Equation 5.3:

$$E_{kin} = \frac{1}{2}mv^2$$

$$E_{kin} = \frac{1}{2}\ (6.0\ \text{kg})\left(4.5\frac{\text{m}}{\text{s}}\right)^2$$

$$E_{kin} = 60.75\frac{\text{kg}\cdot\text{m}^2}{\text{s}^2}$$

$$E_{kin} = 61\ \text{J}$$

EXAMPLE 5.5 In pushing a frictionless cart from rest on a horizontal surface, 2 400 joules of work is done. If the mass of the cart is 65 kilograms, how fast is it then going?

SOLUTION

The work done is the amount of kinetic energy it gains, starting with none. From Equation 5.3,

$$v = \sqrt{\frac{2E_{kin}}{m}} = \sqrt{\frac{2(2\ 400\ \text{J})}{65\ \text{kg}}} = 8.6\ \text{m/s}$$

EXAMPLE 5.6 What is the mass of a baseball that has 110 joules of energy when it is going 22 meters per second?

SOLUTION

From Equation 5.3,

$$m = \frac{2E_{kin}}{v^2} = \frac{2(110\ J)}{(22\ m/s)^2} = 0.45\ kg$$

Like all energy, kinetic energy is a scalar. It is true that v is a vector, but any vector squared is a scalar, equal to the square of the magnitude of the vector. Kinetic energy does not change when the direction of motion changes. Consider, for example, a ball attached to a string and whirled around in a horizontal circle. A (centripetal) force is constantly being exerted to keep it in the circle. Does this change its kinetic energy?

It does not. Remember the definition of work (Equation 5.1)?

$$W = \mathbf{F}\, \Delta s\, \cos \theta$$

In this case, the force is radially inward, while the velocity is tangential. Since the force is perpendicular to the displacement, q = 90°. The cosine of 90° = 0, so you are doing no work on the ball. Its energy does not change, even though the direction of its velocity is constantly changing.

CORE CONCEPT

Kinetic energy $= \frac{1}{2}mv^2$; it is the energy anything has because of its motion.

TRY THIS
—4—

What is the kinetic energy of a 1 200-kilogram car going 22 meters per second?

Energy Transformed

Let's take another look at that drop hammer. Work was done on it to lift it, thereby increasing its (gravitational potential) energy. When the hammer was dropped onto a piling, it did no work until it actually struck the piling. Where was all that energy just before the hammer hit the piling?

Well, the hammer was moving fast at that moment. In fact, from Equation 2.4c, we can easily see that its speed was

$$v = \sqrt{2gh}$$

where h is the height from which it was dropped. From this equation

$$\frac{1}{2}v^2 = gh$$

and multiplying by m gives

$$\frac{1}{2}mv^2 = mgh$$

In other words, the kinetic energy it has just before it hits the piling is equal to the gravitational potential energy it lost while falling through a distance h. Its total energy has not changed.

If you calculate the kinetic energy gained and the gravitational potential energy lost when a cart rolls down a frictionless inclined plane, you will get the same result. In the absence of friction, the total energy does not change. This result will hold for a hill of any shape at all.

Once we know this, certain problems become solvable. Consider, for example, the roller coaster cart of Figure 5.3. It is pushed a short distance on a level track and then released, so that it rolls downhill, uphill, back down, and so on. We would like to know how fast it is going at point P. An impossible problem, right? You have to know the slope of the hill at all points so you can calculate the acceleration everywhere, add the velocities, and so on.

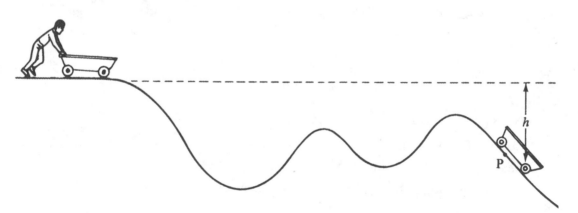

FIGURE 5.3

Wrong. It's easy. The nice thing about energy is that it is a scalar, so you do not have to worry about directions. Also, once you know that its total value never changes, you can deal with the initial state and the final state and forget about everything that happens in between.

To take the simplest case first, let's consider a 50-kilogram cart starting at rest and rolling, frictionless, down a slope 20 meters long and 10 meters high. Its kinetic energy at the top is 0, and its gravitational potential energy, from Equation 5.2, is mgh

= (50 kg)(9.8 m/s²)(10 m) = 4 900 J. The force pushing it downhill, as we saw in Chapter 4, is its weight times h/l= (50 kg)(9.8 m/s²)(10 m/20 m) = 245 N. This produces an acceleration, according to Equation 4.2, of

$$a = \frac{F}{m} = \frac{245 \text{ N}}{50 \text{ kg}} = 4.9 \text{ m/s}^2$$

Now we can find out how fast it is going at the bottom, using Equation 2.4c

$$v = \sqrt{2as} = \sqrt{2(4.9 \text{ m/s}^2)(20 \text{ m})} = 14 \text{ m/s}$$

So its kinetic energy at the bottom is

$$E_{kin} = \frac{1}{2}mv^2 = \frac{1}{2}(50 \text{ kg})(14 \text{ m/s})^2 = 4 900 \text{ J}$$

This is exactly equal to the gravitational potential energy it had before it started down the hill!

In a frictionless system like this, the amount of kinetic energy the object gains is equal to the potential energy it loses. It works just as well if the hill has curves in it. And it works if the cart is in motion before it starts down the hill. Using this principle, we can easily find out how fast the cart is going at any point on the roller coaster hill of Figure 5.3. Its kinetic energy at any point, such as P, is equal to the amount of kinetic energy it had originally, plus the amount of gravitational potential energy it lost while rolling downhill.

EXAMPLE 5.7 Starting at rest, the cart of Figure 5.3 glides frictionlessly to point P, which is 4.5 meters below the top of the hill. How fast is it going at P?

SOLUTION

Considering its gravitational energy to have dropped to zero at P, its gravitational energy when it is at the top of the hill is $(mgh)_{start}$. Its kinetic energy starts at zero and increases to $(\frac{1}{2}mv^2)_{end}$. Since its total energy does not change,

$$(mgh)_{start} = \left(\frac{1}{2}mv^2\right)_{end}$$

The m's drop out, and the equation becomes

$$v = \sqrt{2gh} = \sqrt{2(9.8 \text{ m/s}^2)(4.5 \text{ m})} = 9.4 \text{ m/s}$$

EXAMPLE 5.8 The same cart, going 9.4 meters per second at point P, continues to the bottom of the path and climbs up to a point Q, 3.0 meters higher than P. How fast is it then going?

SOLUTION

The total energy does not change, so

$$(E_{grav})_P + (E_{kin})_P = (E_{grav})_Q + (E_{kin})_Q$$

For convenience, we may consider P to be the level of zero gravitational energy. The equation then becomes

$$\left(\frac{1}{2}mv^2\right)_P = (mgh)_Q + \left(\frac{1}{2}mv^2\right)_Q$$

Again, the m's drop out, and we have

$$\frac{1}{2}(9.4 \text{ m/s})^2 = (9.8 \text{ m/s}^2)(3.0 \text{ m}) + \frac{1}{2}v_Q{}^2$$

$$\frac{1}{2}v_Q{}^2 = 44.2 \text{ m}^2/\text{s}^2 - 29.4 \text{ m}^2/\text{s}^2$$

$$v_Q = \sqrt{2(14.8 \text{ J})} = 5.4 \text{ m/s}$$

EXAMPLE 5.9

The cart has a mass of 650 kilograms and is pushed on level ground for 8.0 meters with a force of 520 newtons and then released at the lip of the hill. How fast is it moving at point P, which is 3.0 meters below the lip? (Neglect friction.)

SOLUTION

The cart has some kinetic energy when the pushing stops, equal to the work done in pushing it:

$$W = F\,\Delta s = (520 \text{ N})(8.0 \text{ m}) = 4\,160 \text{ J}$$

When it is at point P, it has lost gravitational potential energy equal to

$$E_{grav} = mgh = (650 \text{ kg})\left(9.8\frac{\text{m}}{\text{s}^2}\right)(3.0 \text{ m})$$

$$E_{grav} = 19\,110 \text{ J}$$

All this was converted to kinetic energy, which is added to its original kinetic energy, giving a total of 23 270 joules. Therefore,

$$E_{kin} = \frac{1}{2}mv^2$$

$$\frac{1}{2}(650 \text{ kg})v^2 = 23\,270 \text{ J}$$

$$v = 8.5 \text{ m/s}$$

In practice, of course, the cart will be going slower than that because a real roller coaster is not frictionless. But you can be quite certain that it cannot be going faster than the speed you calculated.

CORE CONCEPT

When an object is moving in a frictionless system, the sum of its kinetic and potential energies does not change.

TRY THIS
—5—

A child in a wagon, combined mass 30 kilograms, is going 6.0 meters per second on level ground. Then the wagon coasts up a hill. Find (a) its kinetic energy at the bottom of the hill; and (b) how high it rises on the hill before coming to rest. Neglect friction.

Energy in Space

Suppose you had the job of figuring out how much work a rocket engine has to do to lift an artificial satellite into outer space. This work must be equal to the increase in its gravitational energy, so you look at the formula and start to calculate the value of mgh.

When you get to the g, you are in trouble. When the rocket takes off, it is in a place where $g = 9.8$ m/s^2. But it is rising into regions where g is getting smaller. What value should you use? It seems that the expression mgh is suitable only in cases where h is so small that g remains practically constant.

When dealing with outer-space distances, you need a different expression for gravitational potential energy. In arriving at such an expression, it is most convenient to adopt the convention that, when two objects are infinitely far apart, there is no gravitational potential energy. They do not influence each other, and it requires no work to move them around—at least as far as their effect on each other is concerned.

As the two objects come closer to each other, their gravitational potential energy decreases. This is consistent with the usual use of the mgh formula. But if the energy was zero at infinite separation, and less than that at closer distances, then it must always be negative at any real distance! The formula, in fact, is this:

$$E_{\text{grav}} = \frac{-Gm_1m_2}{r} \qquad \textbf{(Equation 5.4)}$$

where G is the universal gravitation constant, the m's are the masses of the two objects, and r is the distance between their centers.

EXAMPLE 5.10

How much is the gravitational energy of a 720-kilogram meteoroid when it is at an altitude of 2 200 kilometers? (See Appendix 2 for constants.)

SOLUTION

Apply Equation 5.4

$$E_{grav} = \frac{-Gm_1m_2}{r}$$

$$E_{grav} = \frac{-(6.67 \times 10^{-11} \text{ N·m}^2/\text{kg}^2)(6.0 \times 10^{24}\text{kg})(720 \text{ kg})}{8.6 \times 10^6 \text{ m}}$$

Note that the value of r must be measured in meters to the center of the earth. The answer comes out to be -3.4×10^{10} J.

EXAMPLE 5.11

How much work must be done on a 5 800-kilogram satellite to lift it into orbit at an altitude of 2 600 kilometers?

SOLUTION

The work that has to be done is the difference between the satellite's gravitational potential energy when it is on earth and its energy when it is at the high altitude. First, calculate the energy when it is on earth; there, r is the radius of the earth, 6 400 km:

$$E_{grav} = \frac{-Gm_1m_2}{r}$$

$$E_{grav} = \frac{-(6.67 \times 10^{-11} \text{ N·m}^2/\text{kg}^2)(6.0 \times 10^{24}\text{kg})(5\ 800 \text{ kg})}{6.4 \times 10^6 \text{ m}}$$

$$E_{grav} = -3.63 \times 10^{11} \text{ J}$$

When it is in orbit, r is its altitude plus the radius of the earth, or 9.0×10^6 m. Repeating the calculation with this value of r gives

$$E_{grav} = --2.58 \times 10^{11} \text{ J}$$

which is an increase of 1.05×10^{11} J.

CORE CONCEPT

For calculating gravitational potential energy over large distances, use a formula that gives a negative value, rising to zero at infinite separation:

$$E_{grav} = \frac{-Gm_1m_2}{r}$$

If a 60-kilogram meteor falls to earth from a great distance, how much gravitational potential energy does it lose?

Escape from Earth

According to the formula for gravitational attraction, the force does not drop to zero at any finite distance. Yet it is possible to fire a rocket straight up, carrying a space probe, in such a way that it will coast away into outer space and never return.

The reason is that it is coasting continually into regions of ever-weakening gravity. Its acceleration is always toward the earth, slowing it down. But the acceleration is getting smaller and smaller as it moves away from the earth. If it is going fast enough when it is flung into space, it will still have some velocity left when the acceleration due to the earth's gravity is too small to make any difference.

To allow the space probe to escape, it has to be given a certain minimum speed when it is thrown upward. That speed is known as its *escape velocity*. Its value can be calculated by careful consideration of the energy changes of the space probe as it rises.

When anything rises in free fall—after its rockets have stopped firing—it is losing kinetic energy and gaining gravitational potential energy. The sum of the two does not change. If it is to escape, it needs enough total energy so that it will not come to rest until it is infinitely far away. At that point, its total energy is zero. Then its total energy is great enough so that, if you fire it straight up, it will keep going forever. The space probe (or any projectile) needs enough kinetic energy so that its total energy will be zero:

$$\frac{1}{2}mv^2 + \frac{-Gm_1m_2}{r} = 0$$

where m and m_1 are both the mass of the projectile. Note that you can divide the equation through by the mass of the projectile and solve for v. If air drag is neglected, the equation becomes

$$v_{\text{escape}} = \sqrt{\frac{2Gm_2}{r}}$$

where m_2 is the mass of the earth and r is its radius. The mass of the projectile is not in the equation; all objects have the same escape velocity, which turns out to be about 11 000 m/s. This is about 20 times the speed of a rifle bullet.

EXAMPLE 5.12 What is the escape velocity from the surface of the earth?

SOLUTION

The object needs enough kinetic energy so that the total energy will be at least zero. The text shows that this condition exists when

$$v > \sqrt{\frac{2Gm_2}{r}}$$

$$v > \sqrt{\frac{2\left(6.67 \times 10^{-11}\ \text{N} \cdot \text{m}^2/\text{kg}^2\right)\left(6.0 \times 10^{24}\ \text{kg}\right)}{6.4 \times 10^6\ \text{m}}}$$

$$v > 11\ 200\ \text{m/s}$$

In practice, we often put spaceships into orbit before sending them off into outer space. Their escape velocity is considerably smaller, since they are leaving from a point where r is larger.

CORE CONCEPT

Escape velocity, the same for all objects, is the velocity needed to bring the total energy to zero.

TRY THIS
—7—

What is the escape velocity from a point in space at an altitude equal to the radius of the earth?

You Can't Get Energy for Nothing

If you shove a crate across a floor and back again, you have done considerable work. The only outcome, however, is a rise in temperature—the crate and the floor get warmer. Since, by definition, work creates energy, it follows that a rise in temperature is a measure of an increase in energy. Something has happened inside the material of the crate and the wood of the floor, something that is detectable by the warmth produced. The kind of energy that is measured by a thermometer is called *internal energy*.

At first glance, it might seem that you have created some energy. Not so. There has been another change. It happened inside you; muscles contracted, propelled by chemical changes inside your cells. There was a decrease in the energy inside your body, equal to the increase in energy of the crate and the floor. The work you did was no more than a means of transferring energy from one system to another. It is a good general rule that any increase in one form of energy is accompanied by an equal decrease in some other form.

With this principle, it is possible to calculate the amount of internal energy produced in a real mechanical system.

A boy in a wagon, combined mass 55 kilograms, coasts on level ground, slowing down from 4.5 meters per second to 2.2 meters per second. How much internal energy is created?

SOLUTION

Since the ground is level, there is no change in gravitational energy; all the lost kinetic energy is converted to internal energy. Therefore,

$$E_{int} = \left(\frac{1}{2}mv^2\right)_{start} - \left(\frac{1}{2}mv^2\right)_{end}$$

$$E_{int} = \frac{1}{2}(55 \text{ kg})(4.5 \text{ m/s})^2 - \frac{1}{2}(55 \text{ kg})(2.2 \text{ m/s})^2$$

$$E_{int} = 420 \text{ J}$$

EXAMPLE 5.14

A 1 550-kilogram car going 20 meters per second coasts uphill, coming to rest after rising a vertical distance of 11.0 meters. How much internal energy is produced?

SOLUTION

The kinetic energy of the car at the bottom of the hill is

$$E_{kin} = \frac{1}{2}mv^2 = \frac{1}{2}(1\ 550 \text{ kg})(20 \text{ m/s})^2$$

$$E_{kin} = 3.10 \times 10^5 \text{ J}$$

As it rises up the hill, the amount of gravitational potential energy it gains is

$$E_{grav} = mgh = (1\ 550 \text{ kg})\left(9.8\ \frac{\text{m}}{\text{s}^2}\right)(11.0 \text{ m})$$

$$E_{grav} = 1.67 \times 10^5 \text{ J}$$

Since all the kinetic energy was lost as the car went up the hill, the rest of it must have been converted into internal energy. This energy comes to 1.43×10^5 J.

Once it is recognized that temperature is a measure of energy, it becomes necessary to enlarge the definition of energy. To make a teakettle warm you do not have to do work on it. All you have to do is put it on a fire. It seems, then, that energy can be transferred from one system to another by means other than work. One such means is called heat; another is electromagnetic waves.

There are many forms of energy in addition to internal, gravitational, and kinetic. Work done on or heat added to a system can change a shape, a pressure, a phase, or a chemical composition. Each such change can be used to measure the energy added to the system. Once the many forms of energy can be measured, regularities are found. No exception has ever been detected to the rule that any increase in one form of energy is matched by a corresponding decrease. This has led to the statement known as the *first law of thermodynamics,* or the law of conservation of energy: *In any interaction, the total amount of energy does not change.* If a stick of dynamite explodes, the chemical energy stored in the dynamite is exactly equal to the energy of the heat, violent motion, sound, and light produced in the explosion and the remaining chemical energy in the gases produced in the explosion.

CORE CONCEPT

In any system, energy may change from one form to another, but the total remains constant.

TRY THIS
—8—

How much internal energy is produced in the brakes, tires, and roadbed when a 2 200-kilogram car going at 26 meters per second applies the brakes and slows down to 11 meters per second?

Energy of a Spring

When a child bounces up and down on a pogo stick, he is doing an exercise in energy transformation. As he hits the ground and comes to rest, all his gravitational and kinetic energy disappear. A lot of it comes back, however, when he bounces up again. The energy has been stored in the spring. This is no different from what happens when he bounces a ball. The ball is elastic; it is deformed while in contact with the ground. Deformation absorbs energy, and some or all of this energy becomes internal. If the object is elastic, it can store its *deformation energy* and return some of it in some other form.

   In the case of a linear spring, the energy stored can be calculated with a simple formula. When a spring is stretched, the tension in it is proportional to the increase in its length:

$$F_{spr} = k \, \Delta l$$

As you stretch a spring to increase its length by Δl, the force you apply increases from 0 to F_{spr}. Since the force increases linearly, the average force applied during this stretching process is one-half of this. Multiplying this average force by the distance through which it moves (Δl) gives the amount of work done in stretching the spring by this amount:

$$E_{\text{spr}} = \frac{1}{2} k \Delta l^2 \qquad \textbf{(Equation 5.5)}$$

which is, of course, the energy stored in the spring.

EXAMPLE 5.15   A pogo stick is to be designed so that a 22-kilogram girl, dropping from a height of 0.50 meters, will compress the spring by 12 centimeters. What must be the constant of the spring?

SOLUTION

Assuming that all of the girl's gravitational energy will be stored in the spring, we have

$$mgh = \frac{1}{2} k \Delta l^2$$

She has fallen a total distance of 0.50 m plus the 0.12 m that the spring compressed. Thus

$$k = \frac{2mgh}{\Delta l^2} = \frac{2(22 \text{ kg})(9.8 \text{ m/s}^2)(0.62 \text{ m})}{(0.12 \text{ m})^2} = 18\ 600 \text{ N/m}$$

Many factors other than spring constant go into the proper design of a spring. The spring of the pogo stick, for example, must be long enough so that it does not collapse completely when the girl lands on it. Also, never assume that all the energy put into the spring can be delivered to a load. If you put a Ping-Pong ball on the spring of that pogo stick and expected to calculate how high it would rise when the spring was released, you would surely be disappointed. There is no way such a heavy spring could deliver a lot of energy to a Ping-Pong ball. Most of the spring's energy would be converted into internal energy inside the spring. If a spring is to deliver the largest possible part of the energy stored in it, the spring has to be matched to the job it is expected to do.

CORE CONCEPT

Energy is stored in the deformation of elastic objects: for a linear spring, there is a simple formula for the amount.

TRY THIS
—9—

A 2.0-kilogram rock is attached to the lower end of a vertical spring 12 centimeters long that has a constant of 150 newtons per meter. If the rock is dropped, how far does it fall before being brought to rest by the stretching of the spring? (Hint: The distance the rock drops is equal to the amount the spring stretches.)

Power

A piano mover can rig a set of pulleys that will make it possible to lift a 3 000-newton weight with only 600 newtons of force. There is a trade-off, of course, since there is no way to get more work out of a system than is put into it. The mover will have to pull five times as much rope through the system as the height of the window the piano is going into.

The mover must decide in advance what kind of pulley system to use. The system must be designed so as to limit the force that the mover must exert to some lower value congenial to the worker. Theoretically, it is possible to use enough pulleys to make the input force very small indeed.

However, there is another consideration. There is not much point in making the force very small if you have to keep exerting that force for a much longer period of time. The system has to provide an appropriate input force, but it must also adjust the rate at which the work is done. If the work is stretched out over a long period of time, there is a lot of waste; much of the work will be done in just lifting the worker's arms.

The rate at which work is done is called *power*. Remember that work done is just one example of a transfer of energy, and we will meet many others. More generally, power is the rate at which energy is converted. Here is its definition:

$$P = \frac{\Delta E}{\Delta t}$$

(Equation 5.6)

The unit of power is the joule per second, called a *watt (W)*. This is a very small unit, and the kilowatt (kW) is often used.

EXAMPLE 5.16   How much power must a motor have to operate a pump that raises 1 500 kilograms of water every minute to a distance of 12 meters?

SOLUTION

Work is a kind of energy conversion. The work done per unit time is

$$P = \frac{\Delta E}{\Delta t} = \frac{(1\,500\ \text{kg})(9.8\ \text{N/kg})(12\ \text{m})}{60\ \text{s}} = 2\,940\ \text{J/s} = 2.9\ \text{kW}$$

If you are planning to do this, you had better get at least a 4-kilowatt motor, because there is going to be a lot of loss due to friction and viscosity.

CORE CONCEPT

Power is the rate of transforming energy by any means, including by doing work.

TRY THIS
—10—

What power is used by a 40-kilogram child who runs upstairs, going a vertical distance of 25 meters in 30 seconds?

Energy and Mass

In 1905, Albert Einstein made a profound change in the basic concepts of physics. His Special Theory of Relativity made physicists discard certain assumptions that were held without question ever since the days of Aristotle.

The new theory, based on both theoretical analysis and the results of certain experiments, started with the idea that the speed of light in a vacuum is invariant. This means that it does not transform from one frame of reference to another. The Galilean relativity principle, Equation 2.6 does not apply to light. This speed, $c = 3.0 \times 10^8$ m/s, is the same no matter what frame of reference it is measured from.

Think what this means. Imagine you are in a car going at half the speed of light (do not try this) and you turn on your headlights. If you measure the speed of the light, you will find it to be c. Now suppose someone is standing at the side of the road and measuring the speed of the light in your beam. According to the Galilean velocity transformation it ought to be $1.5c$. The relativity theory states that it is not; the speed of light is invariant. The Galilean velocity transformation, a bedrock concept in classical physics, does not apply to light.

Logical conclusions following from this assumption indicate that other parameters, thought to be invariant, are not. Length, mass, and time all transform from one frame of reference to another. The person on the side of the road will find you to be skinnier and heavier, although you, at rest in your frame of reference, will not find any difference. The outside observer will even find that your watch is keeping different time from his.

The speed of light is the ultimate speed. No material object, no matter how hard pushed, can never reach this speed. Now suppose an object is pushed with a constant force to make it go faster; the work done on it increases its energy. Its velocity increases, but as it approaches the speed of light, the gain in velocity gets less and less. The work on it must increase its energy, but its velocity cannot go up proportionally. Instead, its mass increases.

Since doing work on this thing increases its mass, it must follow that mass is a measure of energy.

The mass-energy corollary of the theory of relativity is that what we call mass and what we call energy are simply two different ways of measuring the same thing. Every kind of energy has mass, and all mass is a form of energy. The relationship between the two methods of measurement is given by the famous equation

$$E = mc^2 \qquad \text{(Equation 5.7)}$$

in which E is any kind of energy, m is the corresponding mass, and c is the speed of light, 3.00×10^8 meters per second.

EXAMPLE 5.17

In a nuclear power plant, uranium atoms spilt, and the mass of the resulting smaller nuclei is about 2 percent less than the mass of the uranium nucleus. If a 10-kilogram fuel rod undergoes fission, converting the missing mass into electrical energy, how much electrical energy is produced?

SOLUTION

The missing mass is 0.2 kg, and the equivalent amount of electrical energy is

$$E = mc^2 = (0.2 \text{ kg})(3.0 \times 10^8 \text{ m/s})^2 = 1.8 \times 10^{16} \text{ J}$$

The usual unit of electrical energy is the kilowatt-hour (kWh), 1 000 watts for 1 hour. Its joule equivalent is

$$1 \text{ kWh} = (1\,000 \text{ W})(3\,600 \text{ s}) = 3.6 \times 10^6 \text{ J}$$

Converting units from joule to kWh, the electrical energy produced is 5.0×10^9 kWh. A house may use about 2 000 kWh every month, so the energy of this bit of mass would keep the house supplied with electricity for 200 thousand years.

This theory revises the laws of conservation of mass and energy, uniting them into a single law. When you burn the gasoline, the energy it releases causes a corresponding reduction in its mass. The car in motion has more mass than it had at rest, the extra mass being its kinetic energy. The hot engine and the hot exhaust gases also have extra mass, which they lose as they cool off. Thus, when all the internal energy has been dissipated and the car is once more at rest, the whole system has less mass than it had before the gasoline was burned. The chemist's law of conservation of mass is defective because it has always failed to take into account the mass of the energy converted.

How is it that, in a century of quantitative chemistry, no one detected this discrepancy? The reason is that c^2 is an extremely large number. The energy changes in a chemical reaction are not great enough to produce a mass discrepancy that is detectable. Even with the extraordinary accuracy of today's chemists, they can safely ignore any mass changes and continue to do business with the old rules.

A truck with a mass of 22 000 kilograms goes from rest to 27 meters per second. How much does its mass increase?

SOLUTION

The increase in mass corresponds to the increase in kinetic energy, which is

$$E_{\text{kin}} = \frac{1}{2}mv^2 = \frac{1}{2}(22\ 000\ \text{kg})(27\ \text{m/s})^2 = 8.0 \times 10^6\ \text{J}$$

To get the amount of mass corresponding to this much energy, use Equation 5.12:

$$m = \frac{E}{c^2} = \frac{8.0 \times 10^6\ \text{J}}{\left(3.0 \times 10^8\ \text{m/s}\right)^2} = 8.9 \times 10^{-11}\ \text{kg}$$

EXAMPLE 5.19

In burning a gallon of gasoline, a half-billion joules (5.0×10^8 J) are converted into internal energy. When all the exhaust products have cooled off, how much less is their mass than the mass of the original gasoline?

SOLUTION

From Equation 5.11,

$$m = \frac{E}{c^2} = \frac{5.0 \times 10^8\ \text{kg·m}^2/\text{s}^2}{\left(3.0 \times 10^8\ \text{m/s}\right)^2}$$

$$m = 6 \times 10^{-9}\ \text{kg}$$

CORE CONCEPT

Mass and energy are two aspects of the same thing, and their measurements are related by the square of the speed of light:

$$E = mc^2$$

TRY THIS
—11—

If a 1 400-kilogram car is going 30 meters per second, how much is the increase in its mass due to its kinetic energy?

You Can't Break Even

The *first law of thermodynamics* tells us that in any interaction, the total amount of energy does not change. If there is just as much energy as there ever was, why do we have to keep drilling for oil? The reason is found in another law, which says you can't even break even. There are always *nonconservative* forces that steal some energy and convert it into a useless form.

If we could invent a frictionless roller coaster, it would not need a motor to run it. The car, at the end of its run, would return all the way up to its starting point. While it traveled, its gravitational potential energy would increase and decrease repeatedly, but there would be no net loss. Gravity is an example of a *conservative* force. This means that, when an object moves around under the influence of gravity, none of its energy is converted into a useless form. The object can return to its starting point without putting any additional energy into it. The moon has been making nearly the same orbit, under the influence of gravity only, for millions of years.

In solving problems, it may be convenient to assume that only conservative forces are acting. This assumption often gives us good approximations to the real world, but something is missing. There is always friction, viscous drag, deformation, or some other kind of force that converts kinetic or potential energy into a useless form. These are called *nonconservative* forces. Nonconservative forces convert useful energy into useless, low-temperature internal energy.

Internal energy can be highly useful; we use a lot of it when we burn coal to produce electricity. But it is useful only at high temperatures. Hot steam can turn a turbine that runs an electric generator. If you were to take that steam and mix it with cold water, you would still have the same amount of internal energy you started with, but it would be spread thin, through a large mass of material, and the temperature would be low. You cannot turn a turbine with warm water.

A steam engine wastes a good deal of energy. It must be cooled, usually by running lots of cold water through it. The cooling water emerges warm and is discharged into a river or the ocean. The best steam engines are only about 48 percent efficient; 52 percent of the internal energy given up by the steam leaves the engine in the form of useless warm water. According to the *second law of themodynamics,* there is no way to convert internal energy into other forms with 100 percent efficiency. In practice, you cannot come anywhere near 100 percent.

What happens to the other 48 percent of the energy, the useful part? It is used to drive an electric generator—which is also a little less than 100 percent efficient. So another part of the total becomes useless, low-temperature internal energy, because of the friction in the generator and the heat generated by the current flowing in its wires. What is left goes out into transformers and transmission lines, where some more is lost as heat. Finally it reaches your home.

The process continues at the point where the electric energy is used. You plug in a toaster, thereby converting some of the electricity to internal energy—100 percent! If you run a mixer, it creates kinetic energy—and more internal energy. Finally, all that kinetic energy does nothing but warm up the batter. In the end every bit of the energy that was in the steam in the form of usable, high-temperature internal energy is dissipated into the environment as internal energy once again. But this time, it is at low temperature—spread thin and useless.

And that is why we have to keep looking for oil. The first law of thermodynamics tells us that the total amount of energy does not change; the second law informs us that every time energy is converted from one form to another, part of it becomes useless, low-temperature internal energy.

CORE CONCEPT

Every energy conversion results in degradation of part of the energy into a useless form.

TRY THIS
—12—

You drive your car up a hill and then come back and stop at the place where you started. Describe the ultimate fate of all the energy in the gasoline you burned.

REVIEW EXERCISES FOR CHAPTER 5

Fill-In's. For each of the following, fill in the missing word or phrase.

1. The energy of a system is always increased when _____ is done on the system.

2. If two objects attract each other, their potential energy _____ when they are moved farther apart.

3. The energy of an object as measured by its motion is called _____ energy.

4. If the speed of an object doubles, its kinetic energy is multiplied by a factor of _____.

5. Since energy does not depend on direction, energy is a _____ quantity.

6. As an object falls freely, it loses _____ energy and gains an equal amount of _____ energy.

7. Neglecting friction, when an object slides down a slope of any shape, its _____ energy remains constant.

8. The formula *mgh* for gravitational energy is an approximation useful whenever the change in _____ is small.

9. Increase in temperature indicates an increase in _____ energy.

10. According to the first law of thermodynamics, the total amount of _____ in a closed system does not change.

11. A fire converts _____ energy to internal energy.

12. According to relativity theory, energy and _____ are different aspects of the same thing.

13. The mass corresponding to any energy can be found by dividing the energy by the square of_____.

14. As energy is used, all of it eventually turns into useless _____ energy.

15. When the gravitational potential energy of an object on earth is calculated with respect to the universe, the largest value it can have is _____ .

Multiple Choice.

1. Which of the following properties is not a measure of energy?
   (1) temperature
   (2) surface area
   (3) altitude
   (4) mass
   (5) velocity.

2. A clock is on a vertical wall. As it runs, the kinetic energy of its second hand
   (1) remains constant
   (2) alternately increases and decreases
   (3) increases steadily
   (4) changes in direction only
   (5) changes in both magnitude and direction.

3. A bricklayer carries a load that weighs 200 newtons up a ramp 10 meters long that is inclined at an angle of 30°. The amount of work he does on the load is
   (1) 20 joules
   (2) 200 joules
   (3) 1 000 joules
   (4) 2 000 joules
   (5) 5 000 joules.

4. A car of mass *m* traveling at speed *v* on a horizontal road applies its brakes and comes to rest in time *t* while traveling a distance *d*. The amount of internal energy created is
   (1) 0   (2) *mv*   (3) *mv/d*   (4) *mvt*   (5) $\frac{1}{2}mv^2$

5. A joule is equivalent to one
   (1) newton-square meter
   (2) newton-meter
   (3) kilogram-meter per second squared
   (4) newton-square meter per second squared
   (5) newton-second.

6. A car is going downhill and its brakes are applied, bringing it to rest. Its gravitational energy was
   (1) converted to kinetic energy
   (2) unchanged
   (3) converted to internal energy
   (4) completely destroyed
   (5) converted to work.

7. A rock is thrown straight up into the air. While it rises and falls, its kinetic energy
   (1) increases then decreases
   (2) decreases then increases
   (3) increases steadily
   (4) remains constant
   (5) changes direction only.

8. If a car is running along on a level road at constant speed, the energy suppied by its engine is converted to
   (1) internal energy only
   (2) kinetic energy only
   (3) potential energy only
   (4) internal and kinetic energy
   (5) potential and kinetic energy.

9. A bicyclist starts at rest at the top of a hill and pedals furiously on the way down. At the bottom, the kinetic energy of the bicycle and rider will be equal to
   (1) work done plus lost potential energy
   (2) lost potential energy
   (3) lost potential and kinetic energy
   (4) work done plus kinetic energy
   (5) work done plus kinetic energy plus potential energy.

10. It is possible to get more work out of a system by using a proper combination of
    (1) ramps and pulleys
    (2) gears
    (3) levers
    (4) towing ropes
    (5) none of the above.

Problems.

1. Using appropriate machinery, you lift a load by pulling 22 meters of rope with a force of 150 newtons. How much does the energy of the whole system increase?

2. A frictionless 380-kilogram cart is pushed on a horizontal track for 15 meters, using a force of 75 newtons. Find (a) the kinetic energy of the cart; and (b) its velocity.

3. A 20-kilogram boulder falls off a cliff and strikes the ground going 10.0 meters per second. How much work does it do on the ground while coming to rest?

4. A pendulum, swinging back and forth, rises at the end of its swing to a position 15 centimeters higher than its lowest point. How fast is it going at the lowest point?

5. A 300-gram toy rocket is fired straight upward, its engine providing a thrust of 6.5 newtons. The rocket engine burns out at an altitude of 25 meters. Find (a) the work done by the rocket engine; (b) the gravitational potential energy of the rocket when it comes to rest at its highest position; and (c) the height above the ground at which the rocket comes to rest.

6. A 160-gram golf ball is dropped from a height of 5.0 meters and bounces back up, reaching a height of 4.2 meters on the rebound. Find (a) the amount of internal energy produced in the impact; and (b) the speed of the ball as it comes off the ground.

7. A force of 250 newtons is used to push a 30-kilogram wagon up a hill, a distance of 12 meters, bringing it to a point 6.0 meters higher than its starting point. If 10 percent of the work done is used in overcoming friction, find (a) the amount of work done; (b) the increase in the gravitational potential energy of the wagon; (c) the increase in the internal energy of the system; (d) the increase in the kinetic energy of the wagon; and (e) the speed of the wagon at the top.

8. The brakes are applied in a 1 200-kilogram car and it slows down from 22 meters per second to 15 meters per second. How much internal energy is created in the brakes and tires?

9. How much is the gravitational potential energy of a 1 800-kilogram elevator when it is 15 meters above its lowest level?

10. If 20 000 joules of work is done in pumping water up to a height of 12 meters, how much water is pumped?

11. A 1.5-kilogram rock traveling at 4.0 meters per second strikes a stationary 2.5-kilogram lump of clay, and the two bodies then move together. Find (a) the speed of the combination after the impact; and (b) the amount of internal energy produced in the collision. (Hint: Momentum is conserved and some of the kinetic energy is converted to internal energy.)

12. It takes an average force of 60 newtons to pull the string of a bow a distance of 35 centimeters. How much energy is stored in the bow?

13. What mass of fuel is used in a nuclear power plant that produces nuclear energy at the rate of 10^9 watts during a whole year?

14. How much energy is there in the mass of a grain of sugar, which has a mass of about 10^{-4} gram?

15. An electric hoist must lift 500 kilograms of material to the top of a building under construction, 75 meters high, using a 20-kilowatt motor. Operating through an appropriate pulley system, how long does it take?

16. A rocket engine is to lift a 250-kilogram space probe so that it will coast all the way out of the solar system. (a) How much energy must the rocket supply? (b) If the thrust of the engine is 640 000 newtons, at what altitude will the engine stop firing?

17. A 60-kilogram meteor falls from outer space to the moon, which has a radius of 1.73×10^6 meters and a mass of 7.48×10^{22} kilograms. How much work does it do on the surface of the moon when it strikes?

18. Using the data of Problem 17, determine the escape velocity from the surface of the moon.

19. How much work must be done on a spring with a constant of 75 newtons per meter to increase its length from 15.0 centimeters to 31.5 centimeters?

20. The spring of a dart gun has a constant of 28 newtons per meter. When a 12-gram dart is forced into the gun, the length of the spring is decreased from 12.0 centimeters to 5.5 centimeters. If the release of the spring transfers half its energy to the dart, how fast is the dart traveling when it leaves the gun?

Things That Flow

WHAT YOU WILL LEARN

To understand the behavior of liquids and gases, we need a special set of physical laws. Many of them depend on an understanding of the quantity called *pressure*.

Phases of Matter

To the ancient Greeks, the universe was composed of four elements: fire, earth, water, and air. To us, these "elements" can be used to define everything that composes the universe as we know it: fire stands for energy, and the others represent matter in its three phases—solid, liquid, and gas.

The objects of our world are in the solid phase. Each has its own special size and shape, its characteristic mass, weight, volume, and so on.

Liquids and gases are fluids, things that flow. They have no shape of their own, but will assume the shape of whatever container they are in. While a given sample of water, air, carbon dioxide, or milk has a definite, measurable mass, it does not have the solid's capacity to resist changes in its shape by exerting an elastic recoil force.

A given sample of a liquid has a definite volume, whereas a sample of a gas does not. If you pour a pint of coffee from a pot into three cups, it is still a pint. The coffee settles into the cup, and its volume is defined by the cup and by the upper, horizontal surface. It is this surface that distinguishes a liquid from a gas.

It is quite possible to pour a heavy, visible gas like chlorine, for example, from one container into another. It will not stay in the container, however, unless you put a cap on it. A gas has no upper surface and will spread out indefinitely. The volume of a gas is the volume of the container it is in.

CORE CONCEPT

A solid has a definite shape and a definite volume; a liquid has only a definite volume; a gas has neither.

Try This
—1—

Tell whether each of the following properties is a characteristic of a given sample of a solid, of a liquid, or of a gas: (a) mass; (b) volume; (c) weight; (d) elasticity; (e) surface; (f) shape; (g) density; (h) color; (i) length.

Density

Which is heavier, milk or alcohol?

The word "heavy," unfortunately, has at least three different meanings. If you say that a rock is too heavy to lift, you are talking about the *weight* of the rock. When "heavy" is used in that sense, there is no answer to the question about the milk and the alcohol. A gallon of either weighs more than a drop of the other.

The question as asked is clearly intended to refer to a general property of the kind of substance—milk or alcohol. It is not a question about any particular sample, so it cannot be asking about the weight. It is a question not about weight, but about *density,* which is a property of the kind of material, regardless of the size of the sample.

Density tells you to what extent the mass of a substance is concentrated. Lead is a high-density material because a great deal of mass is concentrated into a small volume. Air, on the other hand, has a very low density.

It is easy to compare the densities of two liquids, or of a solid and liquid. If something is less dense than the liquid, it will float; if more dense, it will sink. Most wood is less dense than water. A toothpick will float, and so will a thousand-pound pine log. Both have the same density.

The exact definition of density is the ratio of mass to volume:

$$D = \frac{m}{V}$$

(Equation 6.1)

In the SI, the unit of density is the kilogram per cubic meter (kg/m^3), but this is rarely used. The commonly used unit is the gram per cubic centimeter (g/cm^3). This is a very convenient unit because most ordinary liquids and solids have densities that are given by numbers of reasonable size in this unit. The density of water is just 1 g/cm^3, varying a little as the temperature changes. Gases have much lower densities.

EXAMPLE 6.1 What is the density of a liquid if 75 cubic centimeters of it has a mass of 93 grams?

SOLUTION
From Equation 6.1,

$$D = \frac{m}{V}$$

$$D = \frac{93 \text{ g}}{75 \text{ cm}^3}$$

$$D = 1.24 \text{ g/cm}^3$$

EXAMPLE 6.2 A "heavy" oil has a density of 0.91 grams per cubic centimeter. What is the mass of 2.5 liters of this oil? (A liter is 1 000 cubic centimeters.)

SOLUTION
From Equation 6.1

$$m = DV = \left(0.91\frac{\text{g}}{\text{cm}^3}\right)(2\ 500 \text{ cm}^3) = 2\ 300 \text{ g, or 2.3 kg}$$

Beware of the third meaning of "heavy." A heavy oil does not have a high density. In fact, it will float on water. "Heavy" in this sense refers to gooeyness, or *viscosity*.

EXAMPLE 6.3 How big a bottle is needed to hold 5.0 kilograms of mercury, which has a density of 13.6 grams per cubic centimeter?

SOLUTION
From Equation 6.1

$$V = \frac{m}{D} = \frac{5\ 000 \text{ g}}{13.6 \text{ g/cm}^3} = 370 \text{ cm}^3$$

CORE CONCEPT

Density is mass per unit volume:

$$D = \frac{m}{V}$$

TRY THIS
—2—

What is the volume of 650 grams of ether, which has a density of 0.62 grams per cubic centimeter?

Pressure

The effect that a force has on a surface depends on how the force is applied. A man in spiked shoes dents the turf or the floor that he stands on. Put the same man in ordinary shoes and he does no damage. Yet the force he applies to the floor, his weight, is the same in both cases.

The difference is that the spikes concentrate all the force into a small area. The spikes do not change the total force, but they greatly increase the *force per unit area*. This quantity, the force per unit area, is called *pressure*. This is how it is defined mathematically:

$$p = \frac{F}{A} \qquad \textbf{(Equation 6.2)}$$

The bottom of a size 10 shoe is about 11 inches long and 4 inches wide, so the total area of two such shoes is 2 (11 in) (4 in) $= 88$ in^2. In a pair of size 10 shoes, the pressure a 180-pound man exerts on the ground is

$$p = \frac{F}{A} = \frac{180 \text{ lb}}{88 \text{ in}^2} = 2.0 \text{ lb/in}^2,$$

which is read as "two point zero pounds per square inch." If the man walks in snow, this pressure is enough to compress the snow, and he sinks into it. He can put on skis, which have a much larger area, to reduce the pressure he exerts on the snow.

There are many units of pressure in use, and they can be confusing. A convenient unit, one that can serve as a reference for the others, is the *atmosphere* (atm), defined as the average pressure exerted by the air at sea level. It is about 14.7 lb/in^2. The SI unit is the newton per square meter (N/m^2), which is called a *pascal* (Pa). This is a very small unit, and the *kilopascal* (kPa), equal to 1 000 Pa, is coming into use. It takes about 100 kPa to equal 1 atmosphere. Meteorologists have been using the *millibar,* which is one-tenth of a kPa.

What pressure is exerted on the snow by a 180-pound skier if his skis are 6 feet long and 5 inches wide?

SOLUTION

The area of the skis is

$$2 \times (72 \text{ in}) \times (5 \text{ in}) = 720 \text{ in}^2$$

To get the pressure, we have Equation 6.2

$$p = \frac{F}{A}$$

$$p = \frac{180 \text{ lb}}{720 \text{ in}^2}$$

$$p = 0.25 \text{ lb/in}^2$$

A boy and a sled have a combined mass of 38 kilograms. The runners of the sled are 1.60 meters long and 1.2 centimeters wide. Find the pressure exerted on the snow in (a) pascals; (b) atmospheres.

SOLUTION

(a) The force exerted on the snow is the weight of the boy on the sled:

$$mg = (38 \text{ kg})(9.8 \text{ m/s}^2) = 372 \text{ N}$$

The area of contact is $2(1.60 \text{ m})(0.012 \text{ m}) = 0.0384 \text{ m}^2$. Then

$$p = \frac{F}{A} = \frac{372 \text{ N}}{0.0384 \text{ m}^2} = 9\ 700 \text{ Pa}$$

(b) Converting the units, with $100\ 000 \text{ Pa} = 1$ atm, we have

$$(9\ 700 \text{ Pa})\left(\frac{\text{atm}}{10^5 \text{ Pa}}\right) = 0.097 \text{ atm}$$

EXAMPLE 6.6

Soft snow can be compressed by about 3 000 pascals of pressure. What is the smallest area that a pair of snowshoes must have if they will enable a 70-kilogram person to walk over the snow without sinking in?

SOLUTION

The force on the snow is the person's weight $= mg = (70 \text{ kg})(9.8 \text{ m/s}^2) = 686$ N. Then, from Equation 6.2,

$$A = \frac{F}{p} = \frac{686 \text{ N}}{3\ 000 \text{ N/m}^2} = 0.23 \text{ m}^2$$

CORE CONCEPT

Pressure is force per unit area:

$$p = \frac{F}{A}$$

Try This

—3—

What is the pressure exerted on the ground by a camera and tripod weighing 50 newtons if each leg of the tripod has a tip whose area of contact with the ground is 0.2 square centimeter?

Atmospheric Pressure

A liquid or a gas exerts pressure against any object immersed in it. This is called *hydrostatic* pressure. Your eardrums are sensitive to changes in the hydrostatic pressure on them, and you feel these changes when you dive deep into the water or when you rise suddenly in an elevator or an airplane.

We all live at the bottom of a sea of air, and this air is always exerting pressure on everything. The classic method of measuring this pressure—still widely used—is with a *mercury barometer* (Figure 6.1). It is an easy thing to make. The method is as follows (see warning below): Take a glass tube about a meter long and fill it with mercury. Hold your finger over the open end and push it into an open bowl of mercury. When you remove your finger, the level of mercury in the tube will drop, leaving a vacuum in the space above the mercury. The mercury in the tube is supported by the pressure of the air on the open surface of the mercury in the bowl, transmitted through the liquid. The height of the mercury level in the tube, above that in the bowl, is a measure of the pressure of the air. *(Don't try this; mercury vapor is poisonous, and special precautions must be taken!)*

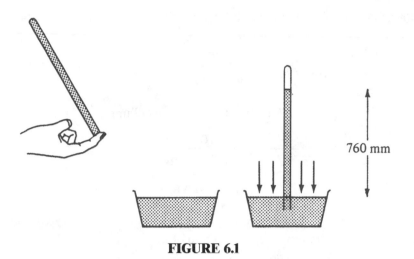

FIGURE 6.1

The first barometer of this sort used water in the tube, but this called for a column over 34 feet high. The change to mercury made it possible to use a shorter tube, since mercury has more than 13 times the density of water.

Let's figure out how high the mercury column will stand in the barometer when the air pressure outside is 1 standard atmosphere, or 101.3 kilopascals. Equilibrium is reached when the pressure at the bottom of the tube, caused by the weight of all the mercury above it, is equal to the atmospheric pressure holding up the column. Therefore, we want to know what length of mercury column produces a pressure of 101.3 kilopascals at the bottom.

The volume of the mercury is the length of the mercury column *(h)* times its cross-sectional area *(A)*:

$$V = hA$$

From Equation 6.1, the mass of the mercury is

$$m = VD = hAD$$

and from Equation 3.1, its weight is

$$w = mg = hADg$$

The force that the weight of the mercury exerts at the bottom is just its weight: $F = w$. Then, from Equation 6.2,

$$p = \frac{F}{A} = \frac{w}{A} = \frac{hADg}{A}$$

so, finally,

$$p = hDg$$ **(Equation 6.3)**

To find the height of the mercury column, we had better get the density into SI units first:

$$D = \left(13.6\frac{g}{cm^3}\right)\left(\frac{100\ cm}{m}\right)^3\left(\frac{kg}{1\ 000\ g}\right) = 13\ 600\ kg/m^3$$

Now we can solve Equation 6.3 for the height of the column:

$$h = \frac{p}{Dg} = \frac{101\ 300\ N/m^2}{\left(13\ 600\ kg/\ m^3\right)\left(9.8\ m/s^2\right)}$$

$$h = 0.760\ m$$

Pressures are often expressed in terms of the height of the mercury column they will support, and 1 standard atmosphere is 760 mm of mercury.

 EXAMPLE 6.7   How tall would a barometer tube have to be if it were filled with alcohol, density 0.87 grams per cubic centimeter?

SOLUTION
The density of alcohol, in SI units, is 870 kg/m³. Then, from Equation 6.3,

$$h = \frac{p}{Dg} = \frac{101\ 300\ N/m^2}{\left(870\ kg/m^3\right)\left(9.8\ m/s^2\right)} = 11.9\ m$$

EXAMPLE 6.8   In a partially evacuated chamber, the barometer stands at 210 mm. How much is the pressure in the chamber in (a) atmospheres; (b) pascals?

SOLUTION
(a) Since it takes 760 millimeters of mercury to make 1 atmosphere, the pressure is

$$\left(\frac{210\ mm}{760\ mm}\right)atm = 0.28\ atm$$

(b) One atmosphere is 101.3 kilopascals, so the pressure is

$$(0.28)(101.3\ kPa) = 28\ kPa, \text{ or } 28\ 000\ Pa$$

EXAMPLE 6.9   If a wall is 3 meters high and 7 meters long, how much force does the atmosphere exert on each side of it?

SOLUTION
The area is $(7\ m) \times (3\ m) = 21\ m^2$. Then, from Equation 6.3,

$$F = pA = \left(101\ 300\ \frac{N}{m^2}\right)\left(21\ m^2\right) = 2\ 100\ 000\ N$$

CORE CONCEPT

The atmosphere exerts pressure which can be measured in terms of the height of the mercury column it will support in a barometer:

$$p = hDg$$

TRY THIS
—4—

You have a table whose top measures 1 meter by 2 meters. What is the total force exerted by the atmosphere on the top surface of the table?

Diving Deep

All divers know that the pressure on them increases as they go deeper into the water, and it can become dangerous if they go too deep. This pressure is created by the weight of all that water above the diver, just as the pressure at the bottom of a barometer tube is produced by the weight of the mercury.

Notice that in deriving the expression for the pressure at the bottom of a mercury column, Equation 6.3, we found that the width of the column did not matter. The pressure at a depth of 1 foot in Lake Erie is the same as that 1 foot down from the surface of the water in your bathtub. In both cases, you can find the pressure by applying Equation 6.3.

The shape of the bathtub does not matter, either. This can be shown by the well-known Pascal's vases of Figure 6.2. In the narrow tube, the wide tube, the conical tube, and the squiggly tube, the water level at the top is the same. This can only be so because the pressure is the same at the bottom of all the tubes.

Hydrostatic pressure at any point is exerted equally in all directions. Imagine a thin slice of water, set horizontal at some arbitrary depth. It is not going anywhere, so the force pushing it up from the bottom must be equal to the force pushing it down from above. You could do the same mental trick with any slice of water at all, no matter how it is oriented. If an object is immersed in water, the force on any surface will always be perpendicular to the surface. Hydrostatic pressure has no direction (it is a scalar), and the direction of the force it exerts on a surface depends only on the orientation of the surface.

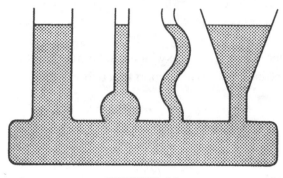

FIGURE 6.2

Equation 6.3 gives the pressure at any depth, in any liquid. In using it, you must be sure to add the pressure at the surface, if there is any. In the barometer, the top surface of the mercury faces a vacuum, so no such correction is needed. But the pressure on a diver's eardrums is 1 atmosphere before he jumps into the water, and it increases from there as he dives deeper. Therefore, after calculating hDg, you have to add 1 atmosphere, or 101.3 kilopascals, or 14.7 pounds per square inch, or 760 millimeters of mercury, or any other measure of the air pressure at sea level.

 What is the pressure at a depth of 23 meters in water?

SOLUTION

First, we need the density of water in SI units. As before, this is 1 000 times as much as the numerical value in grams per cubic centimeter. Therefore, $D = 1\,000$ kilograms per cubic meter. Now, from Equation 6.3,

$$p = hDg = (23 \text{ m})\left(1\,000\,\frac{\text{kg}}{\text{m}^3}\right)\left(9.8\,\frac{\text{m}}{\text{s}^2}\right) = 225\,000 \text{ Pa}$$

This is 225 kilopascals. To get the actual pressure, we have to add the atmospheric pressure at the surface, which is 101 kilopascals. Therefore, $p = 330$ kilopascals.

EXAMPLE 6.11 Through what vertical distance does a diver move to increase the pressure on herself by 1 atmosphere?

SOLUTION

From Equation 6.3,

$$h = \frac{p}{Dg} = \frac{101\,000 \text{ N/m}^2}{\left(1\,000\,\frac{\text{kg}}{\text{m}^3}\right)\left(9.8\,\frac{\text{m}}{\text{s}^2}\right)} = 10.3 \text{ m}$$

A diver should know that each 10 meters of descent increases the pressure on the body by about 1 atmosphere. A healthy diver can tolerate a pressure of 10 atmospheres, but must be extremely careful to come up slowly. A sudden reduction in pressure can be fatal, as it causes nitrogen dissolved in the blood to form bubbles that block the circulation.

CORE CONCEPT

Hydrostatic pressure in a liquid is directly proportional to the depth of the liquid:

$$p = hDg$$

TRY THIS
—5—

A large tank holds salad oil, density 0.9×10^3 kilograms per cubic meter, to a depth of 6 meters. What is the pressure at the bottom of the tank?

Buoyancy

When you take a deep dive, why does the water push you upward?

Buoyancy is caused by a difference in hydrostatic pressure. The water pushes against your body on all sides. But it pushes harder on the lower surface because that is deeper in the water. Since the water is pushing upward from below and downward from above, the upward force is greater than the downward force. The difference between these two forces is the buoyancy.

Let's make it quantitative. Suppose you have a can with a side spout, as shown in Figure 6.3, and you fill the can with water up to the spout. Then you take a rectangular block of some kind and immerse it in the water. The block will displace an amount of water equal to its own volume, and that much water will flow out of the spout.

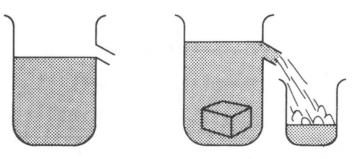

FIGURE 6.3

The pressure against the top surface of the block is p_{top}, and the pressure against the bottom surface (upward) is p_{bottom}. Letting h stand for the depth of the water at top and bottom, and D for the density of the water, we can write, from Equation 6.3,

$$p_{top} = h_{top}Dg$$
$$p_{bottom} = h_{bottom}Dg$$

If the block has length l and breadth b, the area of both top and bottom is bl. Since the force is pA (Equation 6.2),

$$F_{top} = (blh)_{top}Dg \quad \text{and} \quad F_{bottom} = (blh)_{bottom}Dg$$

Buoyancy is the difference between these two forces:

$$B = F_{bottom} - F_{top} = bl(h_{bottom} - h_{top})Dg$$

But $(h_{bottom} - h_{top})$ is the thickness of the block, so $bl(h_{bottom} - h_{top})$ is its volume—and also the volume of the water it displaces. Therefore

$$B = VDg$$

Furthermore, VD is the mass of the displaced water (Equation 6.1) and mg is its weight (Equation 3.1). Therefore, *the force of buoyancy acting on a submerged object is equal to the weight of the fluid displaced.* This is Archimedes' principle, known since ancient Greek times. Example 6.12 shows this principle in operation.

EXAMPLE 6.12

A steel box 40 centimeters long, 22 centimeters high, and 35 centimeters wide is submerged in water. How large is the force of buoyancy exerted on it by the water?

SOLUTION

The volume of the box is

$$(40 \text{ cm}) \times (22 \text{ cm}) \times (35 \text{ cm}) = 30\ 800 \text{ cm}^3$$

and this is the volume of water that it displaces, since it is completely submerged. To get the mass of the water, we use Equation 6.2:

$$m = DV$$
$$m = \left(1\frac{\text{g}}{\text{cm}^3}\right)(30\ 800 \text{ cm}^3)$$
$$m = (30\ 800 \text{ g})\left(\frac{\text{kg}}{1\ 000 \text{ g}}\right) = 30.8 \text{ kg}$$

The buoyancy on the box is the weight of this much water. Using Equation 3.1, we have

$$w = mg$$

$$w = (30.8 \text{ kg})\left(9.8\,\frac{\text{N}}{\text{kg}}\right)$$

$$w = 302 \text{ N}$$

EXAMPLE 6.13

A rock with a mass of 22 kilograms and a volume of 0.018 cubic meters is attached to a spring scale, calibrated in newtons, and immersed in water. What does the spring scale read?

SOLUTION

The scale must read the difference between the weight of the rock and its buoyancy. Its weight is

$$mg = (22 \text{ kg})\,(9.8 \text{ m/s}^2) = 216 \text{ N}$$

Its buoyancy is the weight of the 0.018 m^3 of water that it displaces, $w = mg$; but $m = DV$, so

$$w = DVg = \left(\frac{1\,000 \text{ kg}}{\text{m}^3}\right)(0.018 \text{ m}^3)(9.8 \text{ m/s}^2) = 176 \text{ N}$$

Therefore, the scale reads

$$216 \text{ N} - 176 \text{ N} = 40 \text{ N}$$

A geologist in the field makes use of this principle to determine the density of a rock. All he needs is a spring scale calibrated in grams—mass units. This kind of scale will tell the mass of an object, provided that it is used where the acceleration due to gravity is 9.8 meters per second squared. On Mars, it will tell you lies.

Here is how the method works: Hang the rock on the spring scale to find its mass. Then submerge the rock in water and read the scale again. The difference between the two readings gives the mass of the water displaced. Since the density of water is 1 gram per cubic centimeter, the mass in grams is numerically equal to the volume in cubic centimeters—which is also the volume of the rock. Dividing the mass of the rock by its volume gives its density. (See Example 6.14.)

EXAMPLE 6.14

A rock is suspended from a gram-calibrated spring scale, which reads 155 grams when the rock is in air and 83 grams when it is in water. What is the density of the rock?

SOLUTION

Since the buoyancy equals the weight of water displaced, the mass of displaced water must be (155 g − 83 g) = 72 g. The density of water is 1 g/cm^3, so the volume of displaced water is 72 cm^3; this must also be the volume of the rock. Then, from Equation 6.1,

$$D = \frac{m}{V}$$

$$D = \frac{155 \text{ g}}{72 \text{ cm}^3} = 2.2 \text{ g/cm}^3$$

CORE CONCEPT

When an object is immersed in a fluid, the force of buoyancy on it is equal to the weight of the displaced fluid.

TRY THIS
—6—

An unknown fluid is poured into an overflow can like that of Figure 6.3, up to the spout. A block of aluminum is suspended from a scale and lowered into the fluid. If the scale reads 25 grams when the block is in air and 15 grams when the block is in water, and the amount of fluid that flows out of the spout is 8 cubic centimeters, what is the density of the fluid?

It Floats!

If you push a log under water and release it, it will rise to the top and float there. Clearly, this implies that, when the log is submerged, the force of buoyancy acting on it is greater than its weight.

When the log is submerged, it displaces its own volume of water. But that volume of water weighs more than the log does, so the buoyancy is larger than the weight of the log. This is true only if the density of the log is less than the density of the water. Generally, any object will rise in a fluid if the fluid has a greater density than the object.

How high will the log rise? It gets to the surface and sticks out a little. Now, the volume of water it displaces is equal to the volume of only that part of the log that is below the water surface, as in Figure 6.4. The log sits happily at the surface because the buoyancy force on it is equal to its weight. And since the buoyancy is equal to the weight of displaced water, it follows that the log sinks into the water just deeply enough to displace its own weight of water.

FIGURE 6.4

EXAMPLE 6.15

An overflow can is filled up to the spout. Then a chunk of wood is floated on the water, and 330 cubic centimeters of water flows out of the spout. If the wood is then pushed down until it is completely submerged, an additional 75 cubic centimeters of water flows out. What is the density of the wood?

SOLUTION

The volume of the wood is the amount of water it displaces when completely submerged: $330 \text{ cm}^3 + 75 \text{ cm}^3 = 405 \text{ cm}^3$. Its mass is the same as the mass of water it displaces when floating, which is 330 g. Therefore, its density is

$$\frac{m}{V} = \frac{330 \text{ g}}{405 \text{ cm}^3} = 0.81 \text{ g/cm}^3$$

EXAMPLE 6.16

If a log floats with 40 percent of its volume above the surface of the water, what is its density?

SOLUTION

Since the force of buoyancy on the log is equal to its weight, the weight of the log is equal to the weight of the displaced water. But this much water takes up only 60 percent as much volume as the whole log does, since only 60 percent of the log is underwater. Therefore the density of the log must be 60 percent of that of the water, or 0.60 grams per cubic centimeter.

A steel ship floats, too, even though the density of steel is more than 7 times that of water. The reason is that the ship is not mainly steel; it is mainly air. When the enclosed air is taken into account, the *average* density of the ship is much less than 1 gram per cubic centimeter.

A balloon filled with helium or with hot air will rise for the same reason that the log rose to the top of the water. But it cannot rise to the top of the atmosphere because there is no top. Remember, however, that the density of the air decreases with altitude. The balloon rises until its average density is equal to the density of the air outside. If the balloonist wants to go higher, she increases the volume of the balloon by adding more helium or hot air to it. This increases its weight, to be sure; helium has mass. But it increases the buoyancy more, because the additional displaced air weighs more than the added helium.

CORE CONCEPT

A floating object displaces its own weight of fluid.

TRY THIS
—7—

An ice cube 2 centimeters on a side floats in water with 0.3 centimeter of ice projecting above the surface of the water. What is the density of the ice?

The Skin of a Liquid

Try this: Take a clean razor blade and lay it gently on the surface of a quiet pot of clean water. With a little care, you can make it rest on the surface of the water.

Is the razor blade floating, the way a cork floats? Certainly not; the density of steel is many times that of water. If you look carefully at the blade, you will see that no part of it is underwater, as shown in Figure 6.5. The blade is actually resting on the surface. It even depresses the surface a little, as though it were resting on a slightly elastic membrane. If you push it through the surface, it will sink.

FIGURE 6.5

This elastic property of the surface of a liquid is called *surface tension*. Surface tension is responsible for many of the familiar properties of liquids. For example, a thin stream of water flowing out of a tap will break up into droplets because the surface tension makes the water contract into the smallest possible space. Also, it is the surface tension of water that prevents water from penetrating into tiny spaces, such as the pores in dirty clothing. Soap and other detergents promote penetration because they reduce surface tension. You can't float a razor blade on soapy water.

Many insects make use of the surface tension of water. A water strider actually rests on the surface film (Figure 6.6). A mosquito larva lives under water, with its breathing apparatus projecting through the surface. Its body is supported by the tension of the surface film acting on its snorkel.

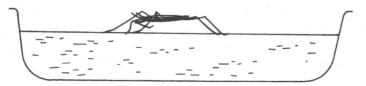

FIGURE 6.6

CORE CONCEPT

The surface of a liquid acts like an elastic membrane.

TRY THIS
—8—

How would you go about drowning mosquito larvae?

Drag Again

As we saw in Chapter 3, when an object moves through a fluid, a retarding force, called viscous drag, is acting on it. If you are planning to move a boat through the water, or a train through the air, it is important to design your vehicle so as to keep the drag as low as possible. On the other hand, if you fall out of an airplane, you want as much drag as you can get to slow down your fall—hence the parachute.

The amount of viscous drag depends on three factors: (1) the nature of the fluid; (2) the speed of the motion; and (3) the size and shape of the object.

Every fluid has a characteristic *viscosity*. You could compare the viscosities of different fluids by dropping a steel ball into them and measuring the rate at which it falls in each. In air, it will be retarded little, and it will accelerate at very nearly the acceleration due to gravity in a vacuum. Water is more viscous; the ball will not speed up as much. If you drop the ball into a bucket of molasses, it will soon reach a slow, constant speed, and will take a long time to get to the bottom of the bucket. Molasses is a highly viscous fluid.

Viscous drag increases with speed. As the steel ball falls through the fluid, accelerating all the while, it will eventually be going fast enough so that the viscous drag on it is equal to its weight. Then it will be in equilibrium and will continue to fall at constant speed. This speed is called its *terminal velocity*.

If a man falls out of an airplane at a fairly low altitude, he is not in free fall. Figure 6.7 shows what happens. When the man leaves the plane, he is moving slowly, so air resistance is small, and he starts to accelerate at 9.8 meters per second squared. Note that the slope of the graph, at the beginning, is 9.8 meters per second squared. However, the curve soon begins to level off. The rate at which the man's speed increases is getting smaller, since the increasing air resistance prevents him from accelerating so rapidly. Eventually, at a speed of about 50 meters per second, the speed stops increasing altogether. The air resistance is now equal to the man's weight, and he continues to fall at constant speed, in equilibrium.

Sky divers, jumping out at very high altitudes, have more complex motions. When they deplane at altitudes of a couple of miles, they often assume the "delta position"—falling head first with a stiff body and arms at the sides. This position produces minimum drag. After about 15 seconds, the diver has fallen about 800 meters and is going at a terminal velocity of 100 meters per second. He or she does not dare open the parachute at that speed because the sudden increase in drag could produce a severely injuring jerk on the shroud lines. As the diver falls into denser air, the drag increases, and is increased still further by flattening the body and spreading the arms and legs. The terminal velocity decreases to about 50 meters per second. Near the ground, the parachute is opened, increasing the drag still more and reducing the terminal velocity to 15 meters per second—still enough to produce quite a jolt on landing.

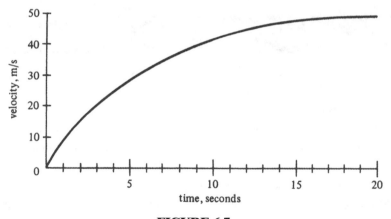

FIGURE 6.7

The viscous drag on the sky diver depends on the shape of his or her body. In the delta position, the air flows around the body smoothly, like the air around the airfoil surface in Figure 6.12. This flow is said to be *laminar,* that is, in layers. Laminar flow produces the minimum of viscous drag. If the surface of the object is more irregular, so that the air forms swirls and eddies, the drag is greatly increased. The flow is then said to be *turbulent*.

The engineer who designs a boat to travel through water, or a car or plane to go through the air, makes the shape smooth and rounded, so that the air flow will be laminar. This is especially important if the vehicle is to travel at high speed. A vehicle designed to produce laminar flow as it races through the air or the water is said to be *streamlined*.

CORE CONCEPT
Viscous drag is least when the flow of fluid is laminar, and greatest when it is turbulent.

TRY THIS
—9—

In terms of fluid flow, explain the utility of the shape of (a) a parachute; (b) a fish.

Pipelines

When water flows through a smooth pipe, the flow is laminar and obeys certain rules. Some of them are obvious, but others are rather surprising.

Let's look at what happens when the pipe gradually narrows, as in Figure 6.8. In the picture, one section of the water has been shown in three different positions. It flows at *A* through a wide pipe; then it passes into *B*, where the pipe is getting narrower; finally, at *C*, it is in a narrow pipe of uniform width.

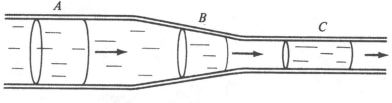

FIGURE 6.8

First of all, notice that this particular piece of water occupies a much longer section of pipe at *C* than it does at *A*. Also, it has to take the same length of time to travel its own length in either pipe. Therefore, it must be going faster at *C* than at *A*. You have probably made use of the fact that the water in a pipe has to speed up as it goes through a constriction; did you ever partly cover the end of a hose (Figure 6.9) to make the water flow out faster?

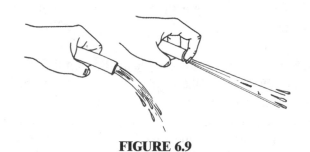

FIGURE 6.9

A second rule about flowing water is more unexpected. The section of water at *A* (Figure 6.8) is flowing along at constant speed, so it must be in a state of equilibrium. Therefore, the force pushing it to the left is equal to the force pushing it to the right, just as a cart moves along at constant speed when all the forces on it balance out. This means that, in the wide pipe, the pressure is constant throughout.

Now look at what happens when the water enters section *B*, where the pipe is gradually constricted. In that region, it is speeding up—after all, it is going faster at *C* than at *A*. This means that the force pushing it to the right is greater than the force pushing it to the left. It follows that *the pressure at A must be greater than the pressure at C!*

The rule we have just found is called *Bernoulli's theorem;* it states that in a flowing liquid or gas, the pressure is least where the speed is greatest. That this is true is easily demonstrated with a device like that of Figure 6.10. The water in the vertical tubes stands highest in the widest part of the pipe, showing that the pressure is greatest where the water flows slowest.

There are a number of practical applications of Bernoulli's principle. The venturi meter is a device for measuring the rate of flow of a fluid through a pipe; the principle is shown in Figure 6.10. The higher the fluid stands in the vertical tube, the slower the liquid is flowing.

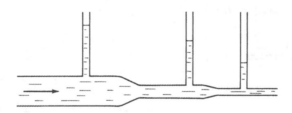

FIGURE 6.10

CORE CONCEPT

In a liquid or gas in which the flow is laminar, the pressure is least where the speed of flow is greatest.

TRY THIS
—10—

Sometimes a part of an artery in the body becomes swollen and widens. Explain why this is dangerous.

The familiar atomizer or spray gun also uses the Bernoulli principle (Figure 6.11). The flow of air across the mouth of the vertical pipe is constricted. This reduces the air pressure at that point, making it less than the atmospheric pressure on the open surface of the liquid. Thus the fluid rises up into the vertical tube, just as it does when you use your mouth to reduce the pressure at the top of a drinking straw.

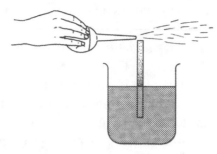

FIGURE 6.11

What Makes It Fly?

Long ago, nature designed a structure that makes use of Bernoulli's principle to produce lift, enabling birds to fly. It is no coincidence that the cross-sectional shape of an airplane wing is much like that of a bird's wing. Both operate on the basis of the same law of physics.

The principle on which the wing functions is illustrated in Figure 6.12. As the wing—of bird or airplane—passes through the air, there is a laminar flow of the air across both its upper and lower surfaces. But the upper surface has a curve to it. Air, hugging the surface as it flows, must go farther across the upper surface than the lower one. Since the upper and lower streams come together behind the wing, without turbulence, they must be making the trip across the wing surface in equal times. The upper stream must be going faster, and, according to Bernoulli, is therefore under less pressure than the air flowing across the flat lower surface. The reduced pressure on the upper surface means that the upward force on the wing is greater than the downward force.

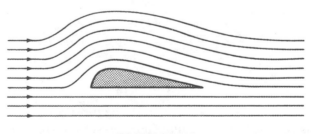

FIGURE 6.12

The amount of lift depends on how fast the air flows across the surface, that is, on how fast the plane is flying. Lift increases with the square of velocity. A plane designed for high speed needs a lot less wing than a slow-flying sport plane. There is a trade-off; the military fighter plane with small wings has to get up a lot of speed to take off, so it needs a powerful engine and a long runway. One compensation is that the small wing produces less drag.

Lift also depends on the *angle of attack* of the wing. This is the angle that the lower wing surface presents to the air flow. In Figure 6.13, the angle of attack has been increased from the position of Figure 6.12, as it would be when the plane is taking off or landing. With the larger angle, there is some loss of laminar flow across the upper surface, as you can see by the turbulence in the picture. This increases drag considerably. But it also increases lift, which is why the pilot tilts the plane in this way. The extra lift comes about because the lower wing surface is now interfering with the flow of air. When the air flow on the lower surface slows down, its pressure increases.

FIGURE 6.13

There is a limit to how much extra lift the pilot can get this way. If he tilts the plane too much, the air flow across the upper surface breaks away from the surface and becomes turbulent, as in Figure 6.14. Lift disappears. In this condition, the plane is said to *stall*. The pilot can recover from a stall if he has enough space and control to speed up sufficiently to get back his lift at a smaller angle of attack.

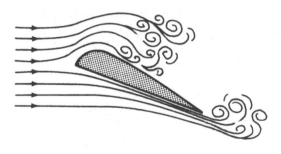

FIGURE 6.14

When a plane is coming in for a landing, it has to slow down to well below its stall speed at the usual angle of attack. That is why the wing flaps are lowered as the plane approaches the runway, giving the wing cross section the configuration shown in Figure 6.15.

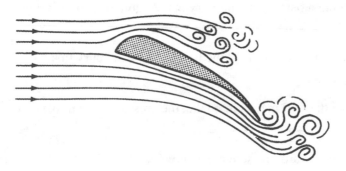

FIGURE 6.15

CORE CONCEPT

Lift is produced by the reduced air pressure in the fast-flowing air across the upper surface of the wing.

TRY THIS
—11—

Explain the two important ways in which lowering the wing flaps helps prepare an airplane for landing.

REVIEW EXERCISES FOR CHAPTER 6

Fill-In's. For each of the following, fill in the missing word or phrase.

1. A fluid with no upper surface is called a(n) _____.

2. Every solid is characterized by a definite _____.

3. The ratio of mass to volume is called _____.

4. A rock will sink in water because its _____ is more than that of water.

5. Pressure is the ratio of force applied to the _____ over which it is applied.

6. The pressure a woman exerts on the ground increases as the heels of her shoes are made _____.

7. In making a barometer, mercury is the liquid of choice because of its high _____.

8. In a standard atmosphere, the height of the mercury in a barometer is _____.

9. At a depth of 10 meters in a lake, the pressure is 2 atmospheres. At 20 meters, the pressure is _____.

10. If the downward pressure at some point is 2 200 millibars, the upward pressure is _____.

11. Difference of pressure on top and bottom of a submerged object produces the force known as _____.

12. The buoyancy on an object is equal to the _____ of the fluid it displaces.

13. When an object is floating, the _____ of the displaced fluid is equal to that of the floating object.

14. An object will float on a fluid whenever its _____ is less than that of the fluid.

15. The elastic property that tends to hold a liquid together is called _____.

16. When an object is moving through a fluid, the amount of viscous drag depends, in part, on the _____ of the fluid.

17. Viscous drag increases as the _____ of the object increases.

18. At the terminal velocity of an object that is falling, the viscous drag is equal to the object's _____.

19. Viscous drag is least when the flow of fluid around the object is _____.

20. As a fluid passes into a narrower section of pipe, its speed _____.

21. As the fluid in a pipe speeds up, its pressure _____.

22. Lift on an airplane wing results from the fact that the air is flowing _____ across the upper surface than the lower.

23. Both lift and drag increase as the _____ of an airplane increases.

24. Both lift and drag increase as the _____ of the wing into the air increases.

25. When a fluid flows from a narrow section of a pipe into a wider section, the rate of flow of the liquid _____.

26. When a fluid flows from a narrow section of a pipe into a wider section, the velocity of flow _____.

27. When a fluid flows from a narrow section of a pipe into a wider section, the pressure _____.

Multiple Choice.

1. A given quantity of a substance occupies a definite volume if the substance is a
   (1) gas only
   (2) liquid only
   (3) solid only
   (4) liquid or solid
   (5) liquid, solid, or gas.

2. Lead is said to be a "heavy" substance because it has a large
   (1) density
   (2) weight
   (3) mass
   (4) pressure
   (5) volume.

3. The function of snowshoes is to
   (1) increase weight
   (2) reduce pressure
   (3) increase pressure
   (4) reduce area
   (5) reduce weight.

4. The height of the column in a mercury barometer is independent of the
   (1) atmospheric pressure
   (2) density of mercury
   (3) diameter of the tube
   (4) altitude at which the measurement is made
   (5) weather conditions.

5. The pressure in a liquid is independent of
   (1) the density of the liquid
   (2) depth in the liquid
   (3) pressure at the surface
   (4) gravitational field
   (5) shape of the container.

6. One kilopascal is equal to
   (1) 1 000 newton-meters
   (2) 1 000 kilograms per meter squared
   (3) 1 000 newtons per meter
   (4) 1 000 kilogram-square meters
   (5) 1 000 newtons per square meter.

7. A cube of wood 5 centimeters on a side is floating in water. The mass of the cube must be
   (1) 125 grams
   (2) at least 125 grams
   (3) less than 125 grams
   (4) 25 grams
   (5) not determinable from the information given.

8. Gases do not exhibit the property of
   (1) mass
   (2) weight
   (3) surface tension
   (4) pressure
   (5) volume.

9. A 500-newton sky diver is floating downward, attached to a 100-newton parachute. The viscous drag is
   (1) 500 newtons
   (2) 600 newtons
   (3) at least 500 newtons
   (4) at least 600 newtons
   (5) between 500 and 600 newtons.

10. An airfoil stalls when the
    (1) flow across its surface is laminar
    (2) pressure in the upper surface is less than that on the lower surface
    (3) flow on the lower surface becomes turbulent
    (4) flow on the upper surface becomes turbulent
    (5) air flow is too small to maintain the pressure difference.

Problems.

1. To determine the density of a liquid, an empty graduated cylinder is placed on a balance, and its mass is found to be 120 grams. The liquid is then added, up to the 100 cubic centimeter mark, and the cylinder placed back on the balance. If the balance now reads 265 grams, what is the density of the liquid?

2. What is the mass of 1 liter (= 1 000 grams per cubic centimeter) of alcohol, whose density is 0.83 grams per cubic centimeter?

3. A new plastic is to be made, to be fabricated into little balls that will sink in any liquid whose density is less than 1.20 grams per cubic centimeter. What is the volume needed in a container to package 500 grams of this plastic?

4. What pressure is exerted on the ground by a 2 500-pound buffalo if each hoof has an area of 4 square inches?

5. What is the total force, in pounds, exerted by the atmosphere on one side of a sheet of paper that measures $8\frac{1}{2} \times 11$ inches?

6. How many standard atmospheres are represented by a pressure of 62 pounds per square inch?

7. What is the pressure, in atmospheres, on the ears of a diver who is 65 meters deep in the water?

8. A dam is to be made to hold back an artificial lake 110 meters deep. Find the pressure that the bottom of the dam must withstand, in pascals (1 pascal = 1 newton per square meter). Note that the density of water, in SI units, is 10 kilograms per cubic meter.

9. What is the pressure at the bottom of a tank of salad oil, density 880 kilograms per cubic meter, if the tank is 7.5 meters deep? Express your answer in atmospheres; 1 atmosphere = 10^5 pascals.

10. What is the force of buoyancy, in newtons, acting on a rock that has a volume of 650 cubic centimeters and is immersed in water?

11. An anchor is suspended from a spring scale that reads 350 newtons, and the reading drops to 245 newtons when the anchor is lowered into the water. What is the volume of the anchor?

12. To find the density of a rock, a geologist suspends it from a spring scale calibrated in mass units—grams. The scale reads 45 grams when the rock is in air, and 32 grams when it is in water. What is the density of the rock?

13. A cubical block of wood, 3.0 centimeters on a side, is floated in an overflow can, and it displaces 18 cubic centimeters of water. Find the block's (a) mass; (b) density.

14. A rectangular plank is 6.5 centimeters thick; it floats in water with 1.5 centimeters projecting above the surface. What is its density?

15. An iceberg floats with 86 percent of its volume under water. What is the density of ice?

Heat: Useful and Wasted

WHAT YOU WILL LEARN

What is the form of energy we call heat, and how does it interact with matter?

Heat and Temperature

As we saw in Chapter 5, we have good reason to believe that, when something gets warmer, its energy content goes up. We called that kind of energy *internal* or *thermal energy*. With confidence in the law of conservation of energy, we think there ought to be some way to measure internal energy. When friction makes the temperature rise, the increase in internal energy ought to be equal to the loss of mechanical energy.

The first step is to quantify the statement that some things are warmer than others. We need a device that will measure the degree of warmness or coldness of something—a thermometer. Touching the thing with your hand can give you a rough idea of how hot it is, but your sense of temperature is highly unreliable, and you surely would not want to touch things that are very hot or very cold.

Fortunately, there are a lot of measureable properties of things that change when the objects heat up. A liquid or a solid expands when heated. The pressure in a gas increases. Electrical resistance changes. Any of these properties may be useful in defining what we mean by temperature.

What do we mean by the statement that two objects are at the same temperature? Rigorously defined, it means this: If the two objects are placed together, and neither one of them undergoes any change, then they are said to be at the same temperature.

If the objects are not at the same temperature, the hot one gets cooler and the cold one gets warmer, until both of them stop changing. Then, they are at the same temperature. No further change takes place, and the objects are said to be in a state of *thermal equilibrium.*

Since we start with the assumption that an object has more internal energy in it if it is warmer, this means that the cool object has gained internal energy and the warm one has lost some. Since energy is conserved, this implies that some energy has passed from the warm object to the cool one. Energy that flows from a warmer object to a cooler one because of the difference in temperature is called *heat.*

There is a strong analogy between heat and work. Both are modes of transferring energy from one system to another. When you push a crate uphill, the energy of the crate goes up and your energy goes down. Work is a means of transferring mechanical energy. When a hot rock is dropped into cold water, the internal energy of the rock goes down and the internal energy of the water goes up. The energy transferred is called heat.

CORE CONCEPT

Heat is energy transferred because of a temperature difference; no heat is transferred spontaneously unless there is a temperature difference.

TRY THIS
—1—

To measure your body temperature, you insert a fever thermometer into your mouth. How do you know when thermal equilibrium has been reached?

Traveling Heat

For one object to warm another, actual contact is not necessary. The sun keeps us all warm.

There are three distinctly different mechanisms by which heat can be transferred. Each occurs under a particular set of conditions. The only thing they all have in common is the fundamental definition of heat: The transfer of energy takes place, in all cases, from something warmer to something colder.

If one end of a hot poker is placed in the fire and left there for any length of time, the other end gets hot. Heat is transferred through the rod, from the hot end to the cold end. This method of heat transfer is called *conduction*. It is a flow of energy through a material. Metals—particularly silver, copper, and aluminum—are good conductors of heat, while most nonmetals conduct poorly. The copper bottom of a frying pan conducts the heat in all directions, spreading it out evenly over the bottom. The handle is made of wood or plastic—poor conductors. These materials, when used for the purpose of preventing conduction of heat, are called *insulators*.

Water is a very poor conductor of heat. If you place a pan of water under the broiler of your stove, it is not difficult to get it boiling at the surface while it remains ice cold at the bottom. To heat the water uniformly throughout, you have to place the flame *under* the pot. This warms the water at the bottom, making it expand. Its density drops, and it therefore rises above the surrounding cool water. The result is a circulation that continuously brings colder water near the flame, until the entire potful gets to the boiling point. This process is called *convection*.

Heat transfer by convection occurs only under certain special conditions. You need a fluid, for one thing, so that the warm material can rise through the colder material. And the fluid must either be heated at the bottom or cooled at the top. The radiator in your room is placed near the floor, while the cooling element in your refrigerator is at the top of the box. We have a lot of devices for preventing transfer of heat by convection. They do this by preventing air from circulating. This is why we fill the wall spaces of our homes with foam, why we wrap ourselves in layers of wool, why we make our picnic refrigerator of styrofoam. Air is a very poor conductor, and if we can trap it, it will not be able to transfer heat by convection, either.

Heat comes to us from the sun through the vast vacuum of empty space. It travels in the form of rays, similar to light rays but invisible for the most part. Every warm object—including you—gives off rays of this sort. Cool objects absorb them, so the net transfer of heat, as always, is from warmer to colder objects. Heat being thus transferred by *radiation* travels in straight lines, like light. That is why you can sit in front of an open fire with your face toasting gently and your back growing icicles.

The efficiency with which objects emit or absorb radiation depends on their color. A black object is black because it does not reflect any of the visible light that falls on it. It does not reflect the heat radiation either. Any radiation that falls on a cold black surface passes right in, warming up the interior of the object. Radiation passes through a black surface in the other direction just as easily, and a hot black object cools down more quickly than a white one. That is why the cooling fins of an engine are painted black. A white surface absorbs or radiates less effectively, and a silvery surface is worst of all.

Any device that is designed to prevent transfer of heat must deal with all three kinds of transfer. Air is a very poor conductor, but it flows easily and thus is an excellent convector. A woolen blanket keeps you warm because it traps air between the fibers, preventing it from circulating. The fibrous padding that insulates a house uses the same principle and also includes a sheet of aluminum foil to stop radiation. The most effective device for keeping hot things hot and cold things cold is the Dewar flask (more popularly known by its trade name, Thermos bottle). It consists of two bottles, one inside the other, with a vacuum in between. Both bottles are silvered to prevent radiation.

CORE CONCEPT

Heat travels through metals by conduction; upward through fluids by convection; through empty space by radiation.

TRY THIS
—2—

Explain why a vacuum bottle is silvered, and why the space between its double walls is evacuated.

What's a Thermometer?

To make the concept of coldness-versus-hotness quantitative, we need some sort of device that changes observably when it is warmed up or cooled down. One such device is the mercury–glass thermometer, illustrated in Figure 7.1. It consists of a glass tube with a very fine bore running throughout its length. At the bottom is a reservoir containing a quantity of liquid—often, but not always, mercury. Warming the bulb causes the liquid and the glass to expand, but the liquid expands much more than the glass does. The extra volume of liquid has no place to go but up into the bore of the tube. If the bore is thin enough, a small amount of extra volume will rise a good way into it.

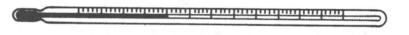

FIGURE 7.1

Now all we have to do is mark a scale on the stem of our thermometer. Any sort of scale will do, provided only that it is clearly defined so that everyone knows what the numbers mean. We have to find a couple of fixed points that can be given arbitrary values and then simply divide up the stem into equal spaces.

The most convenient fixed points, points that can be carefully defined and used with any thermometer, are the transition temperatures between different phases of a substance—between liquid and solid or between liquid and gas. What happens, for example, if you put some ice into warm water in a completely insulated container? Since the ice is colder than the water, heat passes from the water into the ice. This makes some ice melt; it also makes the water cool down. This continues until the ice and the water are in thermal equilibrium, and then no further change takes place. The crucial point is that there is only one temperature at which liquid water and ice can exist together in equilibrium, and we can use that point to define a temperature scale.

That temperature is known as the *melting point* of ice. Melting point is not a good expression because the ice does not necessarily melt there. Equilibrium point would be better.

Another fixed point can be found by heating water until it boils. At that temperature, both liquid water and water vapor can exist in equilibrium. This *boiling point,* however, is not always the same; it depends on the pressure of the vapor.

Now we can calibrate our thermometer—that is, fix the scale. Put the bulb into an ice–water mixture and wait for the mercury to stop falling. Then the thermometer will be in thermal equilibrium with the ice–water mixture, and therefore at the same temperature. Mark the mercury level 0 degrees Celsius. Then put the thermometer in boiling water at a pressure of 1 atmosphere, wait for equilibrium again, and mark it 100 degrees Celsius. Now divide the space between the two fixed points into 100 equal spaces, and the thermometer is finished. If you wish, you can extend the scale some distance below zero and above 100 degrees Celsius, using the same spacing for the degree marks.

This procedure defines the Celsius temperature scale, which is universally used in laboratories and is in common use in most of the world. In the United States, we still cling to the old-fashioned Fahrenheit scale, in which the melting point of ice is 32 degrees Fahrenheit and the boiling point of water is 212 degrees Fahrenheit. A comparison of the two scales is shown in Figure 7.2.

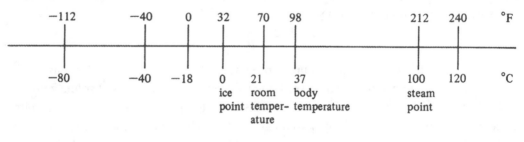

FIGURE 7.2

CORE CONCEPT
The Celsius scale defines 100 degrees between the melting point and the boiling point of water.

TRY THIS
—3—

On your mercury–glass thermometer, the distance from the melting point to the boiling point of water is 15.0 centimeters. What is the temperature of a liquid if the thermometer comes to equilibrium with it when the mercury stands 11.5 centimeters above the melting-point mark?

A Better Temperature Scale

A bucket of water is at a temperature of 10 degrees Celsius. If its temperature doubles, what will the temperature be? Careful, now, the answer is *not* 20 degrees Celsius!

Let's see why not. Suppose I decide to do my calculation using the more familiar Fahrenheit scale. I find that 10 degrees Celsius = 50 degrees Fahrenheit and double it to get 100 degrees Fahrenheit, which is 38 degrees Celsius, not 20 degrees Celsius. What happened? The problem is that the two scales have different zeroes. On either scale, the zero point is not a real zero but a value selected arbitrarily, for convenience. That is why we have negative temperatures.

What we need is a temperature scale with a real zero—a scale that starts at the lowest possible temperature. Only then will it be possible to answer the question, How much is double a given temperature? The SI temperature scale, the Kelvin scale, has a real zero. On this scale, temperature is measured in kelvins (K), and 0 kelvins is the lowest possible temperature, also called *absolute zero*. For this reason, temperatures in kelvins can be multiplied and divided. Twice 20 kelvins is 40 kelvins, and one-third of 300 kelvins is 100 kelvins.

A fairly good way of defining the Kelvin scale depends on the fact that, when an enclosed gas is heated, its pressure rises. This is the principle of the constant-volume hydrogen thermometer. Fill a rigid can with hydrogen gas at low pressure, and attach a pressure gauge. Then the Kelvin scale can be defined by the relation

$$T \propto p$$

which says that the Kelvin temperature is proportional to the pressure in the can.

The Kelvin scale needs only one fixed point since the other point is absolute zero. The fixed point is the *triple point* of water. This is the temperature at which water in its three phases—ice, liquid, and vapor—can all exist together in equilibrium. To make our Kelvin thermometer, place the pressure gauge into water at its triple point. Mark that point on a graph, as in Figure 7.3. Then draw a straight line through that point. Label the point where the line meets the pressure axis "zero kelvins." We could use any convenient number for the temperature of the triple point, and the graph is now a calibration curve for our constant-volume hydrogen thermometer. If the pressure in the thermometer drops to half, the temperature is half of the triple point temperature.

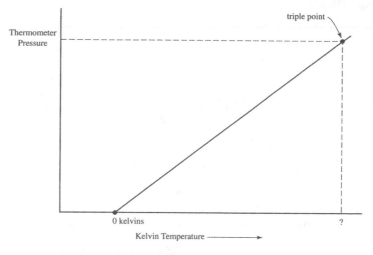

FIGURE 7.3

Now, what shall we call the triple point temperature? It would be convenient to make it easy to translate from the Kelvin scale to the Celsius scale and vice versa. The way to provide this convenience is to label the triple point temperature 273.15 kelvins. This choice of fixed point establishes the number of kelvins between the normal melting and boiling point of water as exactly 100, so the kelvin and the degree Celsius are the same size. The normal melting point of ice is 0.01 kelvin above the triple point; ice melts at 273.16 kelvins and boils at 373.16 kelvins. With this system, a temperature increase from 20 degrees Celsius to 50 degrees Celsius is an increase of 30 kelvins. Temperature *differences* are numerically the same in the two systems; actual temperatures are not, but can be converted from one system to the other by this equation:

$$T_K = T_C + 273.15 \text{ K}$$ **(Equation 7.1)**

EXAMPLE 7.1

The gauge of Figure 7.4 (calibrated to read absolute pressure) reads 120 kilopascals when the thermometer is in water at its triple point and 175 kilopascals when it is in the other beaker. What is the temperature of the liquid in the other beaker?

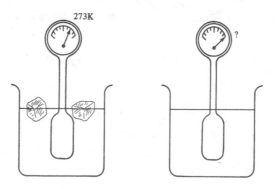

FIGURE 7.4

SOLUTION

Since $T \propto p$, T/p *is* a constant, so

$$\frac{T_1}{p_1} = \frac{T_2}{p_2}$$

$$\frac{273 \text{ K}}{120 \text{ kPa}} = \frac{T}{175 \text{ kPa}}$$

$$T = 398 \text{ K}$$

EXAMPLE 7.2

How much, in degrees Celsius, is twice 10 degrees Celsius?

SOLUTION

Temperature ratios must be calculated in kelvins, not degrees Celsius.

$$T_K = T_C + 273 \text{ K} = 10°\text{C} + 273 \text{ K} = 283 \text{ K}$$

Twice this is 566 kelvins; to put it into degrees Celsius:

$$566 \text{ K} = T_C + 273 \text{ K}$$

$$T = 293°\text{C}$$

CORE CONCEPT

The pressure of an ideal gas at constant volume is proportional to its Kelvin temperature, which is equal to the Celsius temperature plus 273.15 K:

$$T_K \propto p; \ T_K = T_C + 273.15 \text{ K}$$

TRY THIS
—4—

A gas in a bottle is at a temperature of 40 degrees Celsius. If it is cooled down until its pressure drops to half, what is its Celsius temperature?

Temperature and Internal Energy

If a rise in temperature indicates an increase in internal energy, we might suspect that the change in temperature is proportional to the energy change:

$$\Delta H \propto \Delta T$$

However, there is another factor. It takes a lot more energy to heat a gallon of water to the boiling point than to heat a pint through the same change in temperature. When the temperature rises by a given amount, the energy added must also be proportional to the amount of material that was heated up. Therefore,

$$\Delta H \propto m \ \Delta T$$

And one more point. We have no reason to believe that all materials respond in the same way when heat is added to them. On the same flame, some materials will get hot a lot faster than others. We can make an equation out of this relationship by introducing a constant that is characteristic of the material whose thermal energy is being changed:

$$\Delta H = cm \, \Delta T \qquad \text{(Equation 7.2)}$$

Now, how can we find out the value of that constant? One way is to add a measured amount of energy to something and then determine its temperature change. This was done in the last century in an elegant series of experiments by James Prescott Joule. In one experiment, he allowed a falling weight to spin a set of paddles that were in a carefully insulated container of water. He did it enough times to get a measurable increase in the temperature of the water. The loss in the potential energy of the falling weights was the internal energy added to the water, the ΔH in Equation 7.2. In a whole series of such experiments, he found a constant relationship between the increase in the internal energy of the water and the decrease in the mechanical energy that heated the water. Modern experiments have confirmed the results and perfected the measurements, giving, in modern units, the value of the constant c for water as 4.187 joules per gram per kelvin.

(Gram? Why are we suddenly using a unit that does not belong to the SI system, but is just a thousandth of a kilogram? If we did this work with SI units, the numbers would become very unwieldy. You can adjust.)

Equation 7.2 has been found to work well for small changes in temperature. The constant c is a characteristic of the kind of substance whose temperature is changing, known as the *specific heat* of the substance. Values of different materials are determined experimentally; see the next section for the method. Once the value of specific heat is known, it is possible to calculate energy changes.

EXAMPLE 7.3

The specific heat of aluminum is 0.90 joules per gram-kelvin. How much heat does it take to raise the temperature of a 350-gram aluminum pot from 20 degrees Celsius to 85 degrees Celsius?

SOLUTION

The temperature increase is 65 K, and

$$\Delta H = cm \, \Delta T = \left(0.90 \, \frac{\text{J}}{\text{g} \cdot \text{K}}\right)(350 \text{ g})(65 \text{ K}) = 20\ 000 \text{ J}$$

EXAMPLE 7.4

What is the specific heat of a metal if the addition of 15 000 joules of heat raises the temperature of a 620-gram sample of it from 15 degrees Celsius to 90 degrees Celsius?

SOLUTION

$$c = \frac{H}{m\Delta T} = \frac{15\ 000\ \text{J}}{(620\ \text{g})(75\ \text{K})} = 0.32\ \text{J/g}\cdot\text{K}$$

EXAMPLE 7.5

To find the specific heat of a metal, you take 500 grams of shot and drop it through a distance of 2.0 meters, so that it is warmed up by the impact with the ground. After you have done this 1 000 times, the temperature of the shot has risen from 10 degrees Celsius to 38 degrees Celsius. What is the specific heat of the metal?

SOLUTION

Looking at Equation 7.2, you have to know the mass of the sample (500 g), the increase in its temperature (28 K), and the amount of added energy that produced this increase. The energy was added by lifting the shot 1 000 times. From Equation 5.2:

$$\Delta E = (1\ 000)(0.5\ \text{kg})\left(9.8\frac{\text{m}}{\text{s}^2}\right)(2.0\ \text{m})$$

$$\Delta E = \Delta H = 9\ 800\ \text{J}$$

Now we can apply Equation 7.2:

$$\Delta H = cm\,\Delta T$$
$$9\ 800\ \text{J} = c(500\ \text{g})(28\ \text{K})$$

which gives $c = 0.70\ \text{J/g}\cdot\text{K}$

CORE CONCEPT

Change in the internal energy of an object, the heat added or removed, is proportional to the mass and temperature change and depends on the kind of material:

$$\Delta H = cm\,\Delta T$$

TRY THIS
—5—

How much heat must be added to a 250-gram aluminum pan to raise its temperature from 20 degrees Celsius to 85 degrees Celsius? The specific heat of aluminum is 0.90 joules per gram-kelvin.

The Method of Mixtures

The principle of conservation of energy tells us that the total energy of a system must remain constant. If a system is isolated in such a way that no energy comes in or goes out, the energy lost by one part of the system must be equal to the energy gained by other parts. With two parts of a system at different temperatures, the internal energy lost by one part is equal to the internal energy gained by the other part. The energy transferred is called *heat*.

Using this principle, we can investigate various properties of materials. It is often most convenient to use water as one part of the system. Experiment has shown that the specific heat of water is very high, about 4.19 joules per gram-kelvin. If water is placed in contact with something else, in a well-insulated system, the internal energy lost by the water must be equal to the internal energy gained by the other material (or vice versa):

$$(mc\Delta T)_{\text{water}} = (mc\Delta T)_{\text{other material}} \qquad \textbf{(Equation 7.3)}$$

There are six variables in this equation. If any five of them are known, the other one can be found. The following four sample problems show how the method of mixtures can be used to calculate masses, specific heats, and temperatures.

(In the past, much of this work was done using a special unit of energy, the calorie. This was defined originally by the specific heat of water, so that 1 calorie = 4.19 joules. Today, the calorie is defined in terms of the joule; 1 calorie = 4.18605 joules. The "Calorie" used by nutritionists is really what a chemist calls a kilocalorie, which equals 1 000 calories. The calorie is not an SI unit, and we will not make use of it.)

In Example 7.6, we determine an unknown mass. Hot lead, at a known temperature, is added to cold water with a thermometer in it. When equilibrium is reached, both the lead and the water are at the same temperature, and we can easily find ΔT for both of them. Knowing the mass of the water and the two specific heats, we can solve for the unknown value of the mass of the lead.

In Example 7.7, the unknown is the initial temperature of the iron ingot; everything else is known. In Example 7.8, we are using the method of mixtures to find the specific heat of an unknown material. In each case, only one of the six variables is not known.

Example 7.9 introduces an additional complication. We do not know the final temperature, so we cannot put the ΔT's into the equation. The equation must be written with the final temperature T *(not ΔT)* as the unknown. Since the water cools down from 98 degrees Celsius to T, its $\Delta T = 98$ degrees Celsius $- T$. And the aluminum warms up from 20 degrees Celsius to T, so its $\Delta T = T - 20$ degrees Celsius.

EXAMPLE 7.6

A handful of lead shot (specific heat 0.130 joules per gram-kelvin) is heated to 95 degrees Celsius and dumped into 230 grams of cold water at a temperature of 18 degrees Celsius, The mixture comes to thermal equilibrium at 32 degrees Celsius. What is the mass of the lead?

SOLUTION

While the lead cools down from 95°C to 32°C, the water warms up from 18°C to 32°C. The heat lost by the lead must be equal to the heat gained by the water:

$$(cm\Delta T)_{\text{lead}} = (cm\Delta T)_{\text{water}}$$

$$\left(0.130\,\frac{J}{g\cdot K}\right)(m)(63\ K) = \left(4.19\,\frac{J}{g\cdot K}\right)(230\ g)(14\ K)$$

which gives $m = 1\,650$ g.

EXAMPLE 7.7

To determine the temperature of a hot 1.8-kilogram iron ingot, it is dropped into 500 grams of water at 20 degrees Celsius, and thermal equilibrium is reached at 89 degrees Celsius. If the specific heat of iron is known to be 0.48 joules per gram-kelvin, what was the initial temperature of the iron?

SOLUTION

Heat lost by the iron equals heat gained by the water:

$$\left(0.48\,\frac{J}{g\cdot K}\right)(1\,800\ g)(\Delta T) = \left(4.19\,\frac{J}{g\cdot K}\right)(500\ g)(69\ K)$$

so $\Delta T = 167$ K. This is the amount that the temperature of the iron ingot dropped, so it started at 89°C + 167 K = 260°C.

EXAMPLE 7.8

To determine the specific heat of a metal, 650 grams of the metal are heated to 100 degrees Celsius and dropped into 240 grams of water at 17 degrees Celsius. If the mixture reaches equilibrium at 38 degrees Celsius, what is the specific heat of the metal?

SOLUTION

The heat lost by the metal is equal to the heat gained by the water; both are equal to $cm\,\Delta T$. Therefore:

$$\left(4.19\,\frac{J}{g\cdot K}\right)(240\ g)(21\ K) = c_{\text{metal}}(650\ g)(62\ K)$$

which gives $c_{\text{metal}} = 0.52$ J/g·K.

EXAMPLE 7.9

What equilibrium temperature will be reached if you pour 450 grams of hot water at 98 degrees Celsius into a 600-gram aluminum pot at 20 degrees Celsius? The specific heat of aluminum is 0.92 joules per gram-kelvin.

SOLUTION

We can write an equation stating that the heat lost by the water is equal to the heat gained by the aluminum:

$$(cm\Delta T)_{\text{water}} = (cm\Delta T)_{\text{aluminum}}$$

There is a difficulty. Since we do not know the final temperature, we cannot know ΔT. All we can do is put in the final temperature as our unknown; call it T. Then

$$\Delta T_{\text{water}} = 98°C - T, \quad \text{and} \quad \Delta T_{\text{aluminum}} = T - 20°C$$

Our equation then becomes

$$\left(4.19 \frac{J}{g \cdot K}\right)(450 \text{ g})(98°C - T) = \left(0.92 \frac{J}{g \cdot K}\right)(600 \text{ g})(T - 20°C)$$

This simplifies to

$$185\,000°C - 1\,890T = 552T - 11\,040°C$$

So $T = 80°C$.

CORE CONCEPT

The method of mixtures can be used to find many properties of a system.

TRY THIS
—6—

How much water at 20 degrees Celsius should you use to cool a 350-gram iron pot from 195 degrees Celsius to 75 degrees Celsius? The specific heat of iron is 0.48 joules per gram-kelvin.

Melting and Boiling

As we have seen, if you put a large amount of ice in warm water, heat will flow from the water to the ice until thermal equilibrium is reached at 0 degrees Celsius. The water cools down, but the ice does not warm up.

Our energy equation is in trouble. What happened to the heat that the water lost? As usual, we examine the situation to find out what change has taken place in the system. Easy: some ice has melted. This implies that, at a temperature of 0 degrees Celsius, water contains more internal energy than an equal mass of ice. In absorbing heat at

0 degrees Celsius, ice does not change temperature. Instead, it changes phase, from solid to liquid.

To investigate this phenomenon we might start with, say, 100 grams of very cold ice at –40 degrees Celsius. We will set it up in some sort of closed system and arrange to keep adding heat to it at a steady rate. Figure 7.5 is a graph of temperature as a function of time, showing what will happen. In the first minute, the ice warms up to 0 degrees Celsius. Then it starts to melt, and there is a mixture of ice and water at 0 degrees Celsius for the next 4 minutes. When all the ice has melted, the water begins to warm up, reaching 100 degrees Celsius in the next 5 minutes. Then it starts to boil, and as the water turns to steam, we will have a mixture of water and steam. The graph does not show it, but it would take 27 minutes for all the water to vaporize and the steam to begin to get hotter than 100 degrees Celsius.

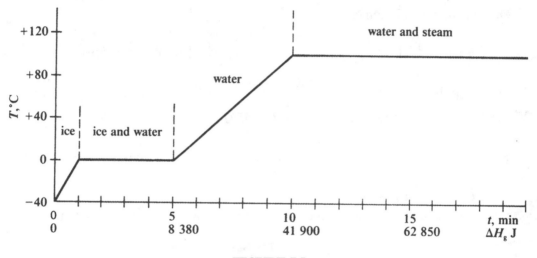

FIGURE 7.5

Looking at the graph, we can calibrate it in terms of the rate at which heat is being added. It takes 5 minutes to raise the temperature of the water from 0 degrees Celsius to 100 degrees Celsius; Equation 7.2 tells us that this temperature increase for 100 grams of water calls for the addition of 41 900 joules. Therefore, we must be adding 8 380 joules of heat every minute. Now we can add another scale to the graph on the horizontal axis, showing the amount of heat that has been added to the system.

Between 1 and 5 minutes, we added 33 500 joules to our material. There was no temperature change; the only effect of this added heat was to melt the ice. Apparently, it takes 335 joules to melt each gram of ice, at 0 degrees Celsius. This quantity is called the *latent heat of fusion* of ice; it is 335 joules per gram. If you want to find out how much energy it takes to melt any given amount of ice, use this equation:

$$\Delta H = mL_F \qquad \text{(Equation 7.4)}$$

where L_F = latent heat of fusion.

EXAMPLE 7.10   How much heat is needed to melt 300 grams of ice at 0 degrees Celsius?

SOLUTION

From Equation 7.4,

$$H = mL_F = (300 \text{ g})\left(335\frac{\text{J}}{\text{g}}\right) = 101\ 000 \text{ J}$$

EXAMPLE 7.11   You put ice cubes into 350 grams of tap water at 18 degrees Celsius. How much ice melts?

SOLUTION

Heat lost by the water in coming to 0 degrees Celsius equals heat gained by the ice in melting:

$$(cm\Delta T)\text{water} = (mL_F)\text{ice}$$

$$\left(4.19\frac{\text{J}}{\text{g}\cdot\text{K}}\right)(350 \text{ g})(18 \text{ K}) = m\left(335\frac{\text{J}}{\text{g}}\right)$$

So $m = 79$ g.

EXAMPLE 7.12   You take 250 grams of ice from the refrigerator at –8 degrees Celsius. How much heat does it absorb in warming up to room temperature at 22 degrees Celsius?

SOLUTION

First, the ice has to warm up to 0 degrees Celsius; the specific heat of ice is 2.3 joules per gram-kelvin, so

$$\Delta H = cm\Delta T = \left(2.3\frac{\text{J}}{\text{g}\cdot\text{K}}\right)(250 \text{ g})(8 \text{ K})$$

$$\Delta H = 4\ 600 \text{ J}$$

Next, the ice has to melt, absorbing its heat of fusion:

$$\Delta H = mL_F = (250 \text{ g})\left(355\frac{\text{J}}{\text{g}}\right)$$

$$\Delta H = 88\ 800 \text{ J}$$

Finally, it must warm up to 22°C. Now it is water, so

$$\Delta H = cm\Delta T = \left(4.19\frac{\text{J}}{\text{g}\cdot\text{K}}\right)(250 \text{ g})(22 \text{ K})$$

$$\Delta H = 23\ 000 \text{ J}$$

The total heat absorbed is the sum of these three quantities, or 116 000 J = 116 kJ.

Ice is an excellent way to cool things down because it does not warm up as it absorbs heat. It will keep on taking heat from its surroundings until the things around it have reached the melting point of the ice, 0 degrees Celsius.

The process also works in reverse. To convert water to ice, you have to remove 335 joules for each gram that is to be frozen. That is what happens in the freezing compartment of your refrigerator. Your freezer, however, does not stop there; it continues removing heat from the ice until its temperature drops to about –5 degrees Celsius. To find out how much heat has to be removed, you would use Equation 7.2, with a value of $c = 2.30$ joules per gram-kelvin.

Figure 7.5 shows that something similar happens at the second phase change, from water to steam at 100 degrees Celsius. Again, there is no increase in temperature as long as there is still some liquid water present. The *latent heat of vaporization* of water is very high: 2 260 joules per gram. You can use Equation 7.4 to calculate heat changes in *any* phase change; for this one, the L_v stands for latent heat of vaporization.

EXAMPLE 7.13

How much heat must be added to 180 grams of water at 20 degrees Celsius to convert it into steam at 100 degrees Celsius?

SOLUTION

First, you have to heat the water to 100°C:

$$\Delta H = cm\Delta T = \left(4.19\frac{J}{g \cdot K}\right)(180 \text{ g})(80 \text{ K}) = 60\ 300 \text{ J}$$

Then you have to vaporize it:

$$\Delta H = mL_v = (180 \text{ g})(2\ 260 \text{ J/g}) = 407\ 000 \text{ J}$$

for a total of 467 000 J, or 470 k J.

EXAMPLE 7.14

What final temperature is reached if 30 grams of steam at 100 degrees Celsius is bubbled into 200 grams of water at 12 degrees Celsius? All the steam condenses in the water.

SOLUTION

The steam must condense, giving up its latent heat of vaporization. Then it is water; now it cools down to the final temperature T, giving up some more heat. All this released heat goes into warming up the water:

$$(30 \text{ g})\left(2\ 260\frac{J}{g}\right) + \left(4.19\frac{J}{g \cdot K}\right)(30 \text{ g})(100°C - T) = \left(4.19\frac{J}{g \cdot K}\right)(200 \text{ g})(T - 12°C)$$

The label J·g/g appears in every term and can be divided out. Multiplying the whole equation through by K and clearing parentheses gives

$$67\ 800\ \text{K} + 12\ 570°\text{C} - 126T = 838T - 10\ 060°\text{C}$$
$$964T = 22\ 630°\text{C} + 67\ 800\ \text{K}$$
$$T = 23°\text{C} + 70\ \text{K} = 93°\text{C}$$

It is the extremely high value of the latent heat of vaporization of water that makes steam heat practical. A boiler adds 2 260 joules of heat every time it vaporizes a gram of water. In a radiator, the steam condenses back to liquid water, releasing all that latent heat to warm the room.

CORE CONCEPT

Change of phase is accompanied by a change in internal energy with no change in temperature:

$$\Delta H = mL$$

TRY THIS
—7—

A 600-gram iron ball (specific heat 0.48 joules per gram-kelvin) at 240 degrees Celsius is dropped onto a very large cake of ice. How much ice melts?

Gases Expand

Consider the cylinder containing a sample of gas shown in Figure 7.6. The piston sits on top of the gas, adding its weight to the pressure of the atmosphere pushing down on the gas. Since the piston is free to move, the pressure on the gas will not change. Now we light a fire under the cylinder.

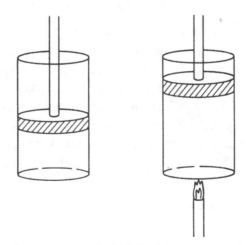

FIGURE 7.6

Like nearly anything else, a gas expands when heated. Gases are different from other materials, however, in a very important way. At constant pressure, the volume of a gas is approximately proportional to its Kelvin temperature:

$$V \propto T_K$$

Thus, if pressure is kept constant, the volume of a gas will shrink to half if its Kelvin temperature is halved.

EXAMPLE 7.15

The cylinder of Figure 7.6 contains 350 cubic centimeters of air at a temperature of 20 degrees Celsius. What will the volume of the cylinder be if the air is heated to 130 degrees Celsius with no change in pressure?

SOLUTION

First of all, all temperatures must be in kelvins. The starting temperature, from Equation 7.4, is 20°C + 273 K = 293 K; the final temperature is 130°C + 273 K = 403 K.

Since $V \propto T$, V/T is constant, so

$$\frac{350 \text{ cm}^3}{293 \text{ K}} = \frac{V_{final}}{403 \text{ K}}$$

and $V_{final} = 480 \text{ cm}^3$.

Volume is proportional to temperature only if the pressure is kept constant. According to the definition of the Kelvin scale (see "A Better Temperature Scale"), the pressure in a gas at constant volume is proportional to its temperature. When a gas is heated, it may expand, or its pressure may increase, or some of each may occur. We need an equation that says: At constant pressure, volume is proportional to temperature, and at constant volume, pressure is proportional to temperature. Here it is:

$$pV \propto T_K \qquad \text{(Equation 7.5)}$$

This equation is known as the *ideal gas law,* and Examples 7.16 and 7.17 show how it can be applied. The law is very simple and easy to use. Unfortunately, it is only an approximation. An ideal gas is one that obeys this law, but no real gas fits it exactly. The fit is best when the gas is at high temperature and low pressure. Cooling a gas down toward zero kelvin will not make it shrink to nothing; it will freeze long before that. And high pressure may make it liquefy at even higher temperatures. Furthermore, the ideal gas law becomes an ever-poorer approximation as temperature drops and pressure rises. Still, it is a very good, useful approximation at ordinary temperatures and pressures.

EXAMPLE
7.16

A cylinder contains 380 cubic centimeters of gas at 1.5 atmospheres of pressure and 20 degrees Celsius. If it is expanded to a volume of 120 cubic centimeters and its pressure goes up to 4.6 atmospheres, what will its temperature be?

SOLUTION

From Equation 7.5,

$$\frac{p_1 V_1}{T_1} = \frac{p_2 V_2}{T_2}$$

WARNING! All temperatures must be in kelvins! So

$$\frac{(1.5 \text{ atm})(380 \text{ cm}^3)}{293 \text{ K}} = \frac{(4.6 \text{ atm})(120 \text{ cm}^3)}{T_2}$$

and $T_2 = 284 \text{ K} = 11°C$.

EXAMPLE
7.17

At sea level, 440 liters of helium are put into a balloon with the temperature at 28 degrees Celsius. If the balloon rises to a point where the pressure is 35 kilopascals and the temperature is –12 degrees Celsius, what volume does the helium occupy?

SOLUTION

The pressure at sea level is about 100 kilopascals, and the temperature (from Equation 7.1) is 301 kelvins. At high altitude, the temperature is 261 kelvins. Therefore, from Equation 7.5,

$$\frac{(100 \text{ kPa})(440 \text{ L})}{301 \text{ K}} = \frac{(35 \text{ kPa})(V)}{261 \text{ K}}$$

which gives $V = 1\,100$ liters.

In applying the ideal gas law, there are three precautions you must be sure to observe: (1) Use only Kelvin temperatures. (2) The law works only as long as you do not add or remove any gas from the sample; you cannot use it to calculate temperatures and pressures in a bicycle tire when you blow it up. (3) Pressure must be *absolute* pressure. Many pressure gauges read gauge pressure—the difference between the pressure inside and the pressure outside. If your automobile tire is flat, your pressure gauge will read zero, even though the pressure in it is surely 1 atmosphere. Gauge pressure is 1 atmosphere less than absolute pressure.

EXAMPLE 7.18

A 600-cubic centimeter sample of gas is in a cylinder under a gauge pressure of 1.8 atmosphere. What will the pressure gauge read if the sample is compressed to 180 cm^3 at constant temperature?

SOLUTION

With the temperature constant, pV is constant, so

$$p_1V_1 = p_2V_2$$

But the pressure in the sample starts out at 2.8 atm, 1 atm higher than the gauge reads. Therefore

$$(2.8 \text{ atm})(600 \text{ cm cm}^3) = p_2(180 \text{ cm}^3)$$

giving an absolute pressure $p_2 = 9.3$ atm. The gauge will read 1 atm less than this, or 8.3 atm.

CORE CONCEPT

For an ideal gas, Kelvin temperature is proportional to the product of pressure and volume:

$$pV \propto T_K$$

TRY THIS
—8—

A cylinder contains 300 cubic centimeters of gas at room temperature of 20 degrees Celsius and 1 atmosphere of pressure—that is, it is open to the outside air. If it is sealed and plunged into boiling water, while a piston compresses the gas into a volume of 100 cubic centimeters, what will a pressure gauge indicate for the pressure of the gas in the cylinder?

The Size of the Sample

We can write the ideal gas law, Equation 7.5, this way:

$$\frac{pV}{T_K} = \text{constant}$$

where the constant depends, somehow, on the size of the sample of gas. If we add some gas, or remove some, the value of pV/T_K changes. Let's take a look at that constant.

Experimental chemistry showed long ago that, if you take equal volumes of gas at the same temperature and pressure, the two samples will contain the same number of molecules, regardless of what kind of gas they are. (This is called *Avogadro's law.*) In other words, the value of the constant in the equation above depends *only* on the number of molecules in the sample. Then we can write the ideal gas law this way:

$$pV = NkT_K \qquad \text{(Equation 7.6)}$$

where N is the number of molecules in the sample and k is a universal constant of nature. It does not depend on the kind of gas in the sample; for a sample with a given number of molecules, pressure is determined only by the volume and temperature. The constant k is called the *Boltzmann constant,* and its value is

$$k = 1.38 \times 10^{-23} \frac{J}{\text{molecule} \cdot K}$$

It is often messy to apply this law, because it will come out right only if all quantities are expressed in SI units: newtons per square meter (pascals) for pressure, cubic meters for volume, and kelvins for temperature.

EXAMPLE 7.19   How many molecules (mcl) are there in 2.0 liters of gas at 75 degrees Celsius if the absolute pressure is 3.0 atmosphere?

SOLUTION

We need SI units throughout. The temperature is 75°C + 273 K = 348 K. Since there are 10^3 liters in 1 m^3, the volume is 2.0×10^{-3} m^3. And 1 atm is very nearly 1.0×10^5 Pa, so the pressure is 3.0×10^5 Pa. Now solve Equation 7.6 for N:

$$N = \frac{pV}{kT_K}$$

$$N = \frac{\left(3.0 \times 10^5 \text{ Pa}\right)\left(2.0 \times 10^{-3} m^3\right)}{\left(1.38 \times 10^{-23} J / \text{mcl} \cdot K\right)\left(348 \text{ K}\right)}$$

$$N = 1.25 \times 10^{23} \text{ molecules}$$

EXAMPLE 7.20   A 1.0-liter bottle at 20 degrees Celsius contains 1.5×10^{22} molecules of oxygen. What is the pressure of the gas?

SOLUTION

From Equation 7.6,

$$p = \frac{NkT}{V} = \frac{\left(1.5 \times 10^{22} \text{mcl}\right)\left(1.38 \times 10^{-23} J/\text{mcl} \cdot K\right)\left(293 \text{ K}\right)}{1.0 \times 10^{-3} m^3}$$

$$p = 6.1 \times 10^4 \text{ Pa}$$

CORE CONCEPT

For an ideal gas, the quantity pV/T depends only on the number of molecules in the sample:

$$pV = NkT_K$$

TRY THIS
—9—

What pressure is exerted by a sample of hydrogen containing 5.0×10^{22} molecules in a 0.50-liter bottle at 30 degrees Celsius?

The Why of the Gas Law

Theoretical physicists puzzled for years over the gas law, looking for a way to explain it. One model has shown itself to be extremely successful, both intuitively and mathematically. It is called the *kinetic molecular theory of gases*.

The model assumes the following: (1) A gas consists of a very large number of extremely small molecules, with great empty spaces between them; (2) the molecules are moving at random in all directions; (3) they interact with each other only through perfectly elastic collisions, which means that the total kinetic energy is conserved through the collision; (4) they exert pressure on the walls of their container by making elastic collisions with them.

This model accounts nicely for the relationship between the pressure and the number of molecules in the sample. Surely, if you double the number of molecules, there will be twice as many collisions with the walls, so the pressure will double. Then

$$p \propto N$$

How does volume affect the pressure? If you double the volume, every molecule will have twice as far to travel between collisions with the wall and so will collide half as often, so the pressure on the walls will drop to half. Therefore,

$$p \propto \frac{1}{V}$$

The speed of the molecules influences the pressure. The faster any molecule is moving, the more often it will hit any given wall; as it bounces from wall to wall, it returns to any given wall twice as often if it is going twice as fast. The molecules are traveling at various speeds, and the pressure is proportional to the average speed:

$$p \propto v_{av}$$

And one more. A faster-moving molecule hits the wall harder than a slower one; a molecule with a larger mass hits it harder than one with less mass. It can be proved from this that the impulse the molecules exert against the wall is proportional to their momenta (that's the plural of momentum). Again, we need the average, so

$$p \propto (mv)_{av}$$

Now we can combine all these proportionalities:

$$p \propto (N)\left(\frac{1}{V}\right)(v_{av})(mv)_{av}$$

Detailed mathematical analysis, allowing for the fact that the molecules move in three dimensions, tells us that the proportionality constant in this relationship is $\frac{1}{3}$. Then, finally, the equation becomes

$$p = \frac{Nmv_{av}^2}{3V}$$

which can be written as

$$pV = \frac{2}{3}N\left(\frac{mv_{av}^2}{2}\right)$$

Now, the average kinetic energy of the molecules is $(mv_{av}^2)/2$, so our equation becomes

$$pV = \frac{2}{3}N\overline{E}_{kin}$$

where the bar over the E means "average."

Comparing this equation with Equation 7.6 suddenly gives us a whole new insight into the meaning of temperature. We find that

$$\frac{2}{3}\overline{E}_{kin} = kT$$

or, in its more usual form,

$$\overline{E}_{kin} = \frac{3}{2}kT \qquad \text{(Equation 7.7)}$$

which says that temperature depends *only* on the average kinetic energy of the molecules.

EXAMPLE 7.21 (a) At what temperature is the average kinetic energy of the molecules of oxygen equal to 6.8×10^{-22} joules? (b) What is the temperature of hydrogen molecules with the same average kinetic energy?

SOLUTION

(a) From Equation 7.7,

$$T = \frac{2}{3}\frac{\overline{E}_{kin}}{k} = \frac{2(6.8 \times 10^{-22} \text{ J})}{3(1.38 \times 10^{-23} \text{ J/mcl} \cdot \text{K})} = 33 \text{ K}$$

(b) The same.

Does this relationship apply only to gases? By no means. Consider what happens when these gas molecules strike the walls of their container. They must surely set the molecules of the walls into motion. When equilibrium is reached, the molecules of the walls will be at the same temperature as the gas molecules—and they will also have the same average kinetic energy. Of course, they will not be bouncing around, like gas molecules; they will simply be oscillating around some central position.

EXAMPLE 7.22 What is the average kinetic energy of the molecules of *anything* at a room temperature of 300 kelvins?

SOLUTION

$$\overline{E}_{kin} = \frac{3}{2}\left(1.38 \times 10^{-23}\,\frac{\text{J}}{\text{mcl} \cdot \text{K}}\right)(300 \text{ K})$$
$$\overline{E}_{kin} = 6.2 \times 10^{-21} \text{ J/mcl}$$

CORE CONCEPT

Temperature is the average random kinetic energy of molecules:

$$\overline{E}_{kin} = \frac{3}{2}kT$$

TRY THIS
—10—

What is the temperature of a gas if the average random kinetic energy of its molecules is 3.5×10^{-21} joules?

Heat Engines

Many devices involve the interaction between work done and heat exchanged. A refrigerator or a heat pump does work to transfer heat up to a higher temperature level. An automobile engine uses heat to produce the work that moves the car. All such devices are firmly bound by the law of conservation of energy. Any heat added to the system has only two places to go: it can do work or it can add to the internal energy of the system, or some combination of both:

$$\Delta H = W + \Delta E_{int}$$

A heat engine, such as a steam engine or a gasoline engine, operates in a cyclic manner. It repeats the same process over and over, each time returning to the same state at which the process began. This means that there is no net change in its internal energy during a cycle, so $\Delta H = W$.

Now, what is this quantity called ΔH? In a steam turbine, heat is supplied in the form of hot steam. After the steam is used, it passes though a cooling system, where it discharges its heat, probably into a river. With the system working steadily, its internal energy remains constant. The work the engine does must be equal to the difference between the heat that goes into it and the heat that it discharges to its environment:

$$W = H_{in} - H_{out}$$

Anyone who runs a heat engine of any kind wants to get the most bang for the buck—the most work output for the amount of heat that is put into the engine. The *thermodynamic efficiency* (ε, the Greek letter epsilon) of an engine is the ratio of work output to heat input:

$$\varepsilon = W/H_i$$

Substituting the value of work from the previous equation:

$$\varepsilon = \frac{H_i - H_0}{H_i} = 1 - \frac{H_0}{H_i}$$

EXAMPLE 7.23 An engine uses 5 000 kilojoules of heat and discharges 1 400 kilojoules. What is its efficiency?

SOLUTION

$$\varepsilon = 1 - \frac{H_0}{H_i} = 1 - \frac{1\ 800\ \text{kJ}}{5\ 000\ \text{kJ}} = 64\%$$

The system operates between two heat reservoirs, a high-temperature reservoir at the input end and a low-temperature reservoir at the output end. It can be shown that, in an ideal case in which frictional and other losses are neglected, the efficiency is absolutely limited by the temperatures of the input and output heat reservoirs. It is the ratio of these temperatures that places the limit, according to the equation

$$\varepsilon = 1 - \frac{T_0}{T_i}$$

(Equation 7.8)

 EXAMPLE 7.24   A steam turbine operates on superheated steam at a temperature of 180 degrees Celsius, and its exhaust is cooled by a flow of river water at 12 degrees Celsius. What is its greatest possible efficiency?

SOLUTION

Any ratio of temperatures must be expressed in kelvins; add 273 kelvins to the Celsius temperatures:

$$\varepsilon = 1 - \frac{285 \text{ K}}{453 \text{ K}} = 0.37 \text{ , which is 37\%}$$

EXAMPLE 7.25   What is the maximum work the turbine of Example 7.24 can perform in a single cycle if it absorbs 1 800 kilojoules from the steam?

SOLUTION

The efficiency rating says that only 37% of the heat input is converted to work; $(0.37)(1\ 800 \text{ kJ}) = 670 \text{ kJ}$.

The figures obtained in these two sample problems are maximum. Under ideal conditions, a real turbine would not do that well.

One unavoidable consequence of Equation 7.8 is that the only way to get 100 percent efficiency would be to discharge the heat into a heat reservoir of 0 kelvins. This, of course, is impossible.

CORE CONCEPT

Heat engines convert heat input into work, with efficiency limited by the input and output temperatures.

TRY THIS
—11—

The highest efficiency of a gasoline engine is about 25 percent. If the engine is air cooled by the outside air, which is at 20 degrees Celsius, what is the temperature of the engine?

REVIEW EXERCISES FOR CHAPTER 7

Fill-In's. For each of the following, fill in the missing word or phrase.

1. If two objects are placed together and neither one undergoes any change, they have the same _____.

2. Energy traveling spontaneously from a warm object to a cold one is called _____.

3. No heat will pass from one object to another in contact with it if the objects are in a state of _____.

4. Through _____ , heat travels best by radiation.

5. Through_____ , heat travels best by conduction.

6. Convection occurs in fluids that are _____ at the bottom.

7. A mercury–glass thermometer works on the principle that, when heated, mercury _____ more than glass.

8. The definition of 0 degrees Celsius is the equilibrium temperature between water and _____.

9. For a given object, the change in its temperature is proportional to the _____ added or removed.

10. The heat needed to raise the temperature of a gram of a substance by 1 degree Celsius is called the _____ of the substance.

11. The specific heat of _____ is 4.19 joules per gram-kelvin.

12. Latent heat is the energy per unit mass absorbed or released in the process of changing _____.

13. It takes _____ joules to convert a gram of water at 100 degrees Celsius into steam at 100 degrees Celsius.

14. At constant volume, the _____ of an ideal gas is proportional to its Kelvin temperature.

15. To find Kelvin temperature, add _____ to Celsius temperature.

16. The SI temperature unit, equivalent to the Celsius degree, is the _____.

17. Gauge pressure is _____ less than absolute pressure.

18. For an ideal gas, the product of _____ and _____ is proportional to the Kelvin temperature.

19. The ideal gas law is a good approximation at _____ pressure and _____ temperature.

20. The value of pV/T_K, for any sample of gas, depends only on the _____ of the sample.

21. The average kinetic energy of molecules is a measure of _____.

22. In an engine, the difference between the heat put in and the heat discharged will appear as _____.

23. The ratio of input to output temperatures of an engine determines the engine's _____.

Multiple Choice.

1. A Thermos bottle is silvered to
   (1) prevent radiation   (4) reduce convection
   (2) promote conduction   (5) eliminate conduction.
   (3) increase heat loss

2. On a sunny day, a road surface is probably hotter than a cold sidewalk because the road surface is
   (1) thicker   (2) smoother   (3) blacker   (4) oilier   (5) wider.

3. On a typical summer day, the air temperature might be about
   (1) 25 kelvins   (4) 250 kelvins
   (2) 75 kelvins   (5) 300 kelvins.
   (3) 150 kelvins

4. Ice melts at exactly 0 degrees Celsius because
   (1) that is the highest temperature it can reach
   (2) that is the definition of zero degrees Celsius
   (3) water freezes at that temperature
   (4) that is when molecular motion stops
   (5) it cannot expand at higher temperatures.

5. What temperature is half as much as the boiling point of water at a pressure of 1 atmosphere?
   - (1) 50 degrees Celsius
   - (2) −87 degrees Celsius
   - (3) 50 kelvins
   - (4) 273 kelvins
   - (5) −50 degrees Celsius.

6. Fifty grams of hot aluminum shot at 100 degrees Celsius are dropped into 100 grams of water at 0 degrees Celsius. The equilibrium temperature will be about
   - (1) 10 degrees Celsius
   - (2) 30 degrees Celsius
   - (3) 50 degrees Celsius
   - (4) 70 degrees Celsiuis
   - (5) 90 degrees Celsius.

7. The specific heat of a material is a measure of the relationship between the heat added to it and
   - (1) its melting point
   - (2) its temperature
   - (3) its internal energy
   - (4) changes in its temperature
   - (5) changes in its internal energy.

8. Knowledge of the latent heat of fusion of a substance is useful only when
   - (1) the substance is at its melting point
   - (2) its temperature is increasing
   - (3) its temperature is decreasing
   - (4) the substance is at its boiling point
   - (5) the substance is at its triple point.

9. A gas is heated, and it is found that its pressure did not change. This is possible if
   - (1) additional gas is added
   - (2) the volume of gas increased
   - (3) some gas was removed
   - (4) its internal energy increased
   - (5) its internal energy decreased.

10. There are no negative Kelvin temperatures because
    - (1) the fixed point of the scale is the triple point of water
    - (2) the size of the kelvin is the same as the degree Celsius
    - (3) the scale has a natural zero
    - (4) no one has been able to devise a thermometer that will read negative values
    - (5) the triple point of water is nearly the same as the melting point of ice.

Problems.

1. In a mercury–glass thermometer, the length of the mercury column is 3.5 centimeters when the bulb is in ice water, 19.0 centimeters when it is in boiling water, and 12.2 centimeters when it is in warm water. What is the temperature of the warm water?

2. A lead bullet (specific heat = 0.130 joules per gram-kelvin) going 460 meters per second strikes a tree. If all the heat generated remains in the bullet, how much does its temperature rise?

3. If a waterfall is 110 meters high, how much warmer is the water at the bottom than at the top?

4. How much heat does it take to heat a 580-gram aluminum pot (specific heat = 0.92 joules per gram-kelvin) from 20 degrees Celsius to 180 degrees Celsius?

5. To measure the specific heat of an unknown metal, 760 grams of it is heated to 95 degrees Celsius and dumped into 300 grams of cold water at 5 degrees Celsius. If the mixture reaches equilibrium at 28 degrees Celsius, what is the specific heat of the metal?

6. To heat 220 grams of water from 20 degrees Celsius to 50 degrees Celsius, an iron ball (specific heat = 0.48 joules per gram-kelvin) at 210 degrees Celsius is dropped into the water. What mass of iron is needed?

7. How much heat is needed to convert 200 grams of ice at 0 degrees Celsius to steam at 100 degrees Celsius?

8. A batch of hot copper shot with a mass of 350 grams (specific heat = 0.38 joules per gram-kelvin) is dropped onto a cake of ice at 0 degrees Celsius, and 75 grams of ice melts. What was the temperature of the copper?

9. What equilibrium temperature will be reached if 300 grams of water at 15 degrees Celsius is poured into a hot 580-gram copper vessel at 145 degrees Celsius. (Specific heat of copper = 0.38 joules per gram-kelvin).

10. A constant-volume hydrogen thermometer registers an absolute pressure of 95 kilopascals when in ice water and 62 kilopascals in a freezing mixture. What is the temperature of the freezing mixture in (a) kelvins; and (b) degrees Celsius?

11. If a constant-volume hydrogen thermometer registers 1,400 millibars in ice water, what will it read in boiling water?

12. Atmospheric air is admitted to a cylinder with a volume of 2.5 liters and sealed in. What will the gauge pressure be if a piston then compresses the air into a volume of 1.0 liter at constant temperature?

13. At the beginning of the compression stroke of a diesel engine, the cylinder contains 600 cubic centimeters of air at atmospheric pressure and a temperature of 22 degrees Celsius. At the end of the stroke, the air has been compressed to 50 cubic centimeters and the absolute pressure is 40 atmospheres. What is the temperature?

14. A liter of helium under a gauge pressure of 2 atmospheres and a temperature of 22 degrees Celsius is heated until both the pressure and the volume have doubled. What is its temperature?

15. How many molecules of air are there in a 1-liter flask at 20 degrees Celsius if the pressure in it is 0.3 atmosphere? (1 atmosphere $= 1 \times 10^5$ newtons per square meter.)

16. A cylinder contains 500 cubic centimeters of air at 1 atmosphere pressure and 20 degrees Celsius. Air is pumped in to double the mass of air in the cylinder, and the piston is pushed in until the volume is reduced to 200 cubic centimeters. If the temperature is kept constant, what is the pressure in the cylinder?

17. What is the average kinetic energy of the molecules of boiling water?

18. In a mixture of hydrogen (molecular mass $= 2$ daltons) and oxygen (molecular mass 32 daltons), what is the ratio of the speed of the hydrogen molecules to the speed of the oxygen molecules?

19. An engine has an efficiency of 25 percent. In each cycle, it discharges 20 000 joules into the coolant. How much work does it do?

20. A steam engine does 540 joules of work in each cycle; its efficiency is 30 percent. How much heat is it absorbing per cycle?

21. In an operating gasoline engine, the surface is at 240 degrees Celsius, and it is cooled to the outside air at 20 degrees Celsius. If each cycle absorbs 2 400 joules of heat, how much work does it do?

22. A gas turbine is operated by burning gas at a temperature of 175 degrees Celsius. What must the temperature of the cooling water be to achieve an efficiency of 70 percent?

Making Waves

WHAT YOU WILL LEARN

How energy moves in the form of waves, and how this is manifested in the phenomenon of sound.

Good Vibrations

Lots of things move back and forth over and over again: a child on a swing, the pendulum of a clock, the plucked string of a guitar, a metronome, an athlete on a trampoline, a child on a pogo stick. Many such motions are good approximations of a special kind of oscillation called *simple harmonic motion*.

In Figure 8.1, the knob moves in a circular path at constant speed; its shadow moves back and forth with simple harmonic motion. Simple harmonic motion is the component along one axis of circular motion.

FIGURE 8.1

The *period* (T) of a simple harmonic motion is the time it takes for the motion to complete one full cycle. This would have to be the same for the knob and its shadow. The *amplitude* (A) of a simple harmonic motion is the greatest distance the shadow moves from its central position; this must be the same as the radius of the circle in which the knob travels.

The velocity of the shadow at various points can be found as a component of the velocity of the knob. Figure 8.2 shows the path of the knob (○) as it travels in a circle. The path of the shadow (●), moving back and forth, is shown on the next page. Four positions of the knob are labeled. At (1) the knob is moving to the right, and its shadow follows at the same speed. At (2) the knob is moving at an angle downward; the horizontal component of its velocity is the velocity of the shadow. At (3) the shadow again copies the horizontal component of the knob's velocity. At (4) the knob is moving upward and its velocity has no horizontal component; the shadow is at rest, reversing its direction.

A real object moving with simple harmonic motion has the motion of the shadow of Figure 8.2. Its parameters can be calculated by endowing it with an imaginary object moving in a circle, like the knob in the figure. The imaginary object has the same period as the oscillating object; its radius is the amplitude of the simple harmonic motion of the real object.

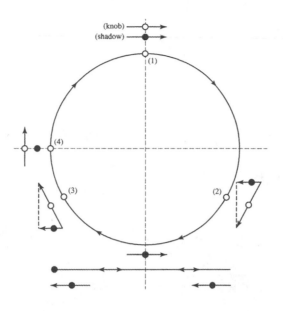

FIGURE 8.2. Velocities of Knob and Shadow

EXAMPLE 8.1

A pendulum swings back and forth in simple harmonic motion, a distance of 50 centimeters end to end. If it is going at 3.5 meters per second as it passes the center of its swing, what is its period?

SOLUTION

The reference circle has a radius of 0.25 meter, and an object in it is moving at 3.5 meters per second.

$$T = \frac{2\pi r}{v} = \frac{2\pi(0.25 \text{ m})}{3.5 \text{ m/s}} = 0.45 \text{ s}$$

Now let's look at the accelerations of the shadow, shown in Figure 8.3. The acceleration of the knob is centripetal; that is, it is always toward the center of circle, as you can see in the figure. At (1), the acceleration of the knob has no horizontal component, so the shadow has zero acceleration. This is the position at which its velocity has reached a maximum and starts to slow down. At (2) the centripetal acceleration of the knob has a component pointing to the left and the shadow is speeding up as it moves toward the center. At (3), the shadow is moving toward the left, and is accelerated to the right; it is slowing down. The acceleration of the shadow is always toward the center of its motion; as it recedes from the center it slows down, and as it approaches the center it speeds up. At (4), it is reversing direction, and its acceleration is a maximum, equal to the centripetal acceleration of the knob.

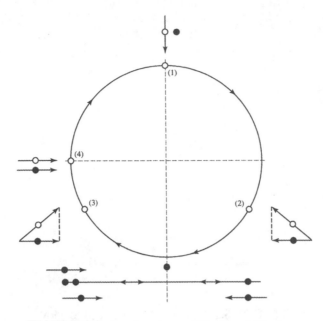

FIGURE 8.3. Accelerations of Knob and Shadow

Now let's look at a real physical system—an object hanging at the end of a spring, Figure 8.4. The upward force on the object is the tension in the spring; the downward force is the weight of the object. In that position, the object is in equilibrium.

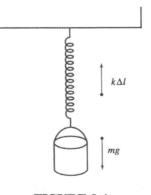

$k\Delta l$

mg

FIGURE 8.4

Now suppose you pull the weight down further and then release it. This stretches the spring further, so the upward force on the object is larger than the weight. It accelerates upward. After it passes the equilibrium point, the spring is not stretched as much, so the upward force is less than the weight; acceleration is downward. The force, and therefore the acceleration, is always toward the equilibrium point. The motion is simple harmonic.

Another well-known case is a pendulum (Figure 8.5). Two forces act on the bob: the tension in the supporting string and the weight. The vector addition diagram in Figure

8.5 shows that the sum of these forces acts to push the bob toward the equilibrium position. The motion is simple harmonic. Analysis shows that, provided the angle is small, the period of a pendulum is given by

$$T = 2\pi\sqrt{\frac{l}{g}}$$

(Equation 8.1)

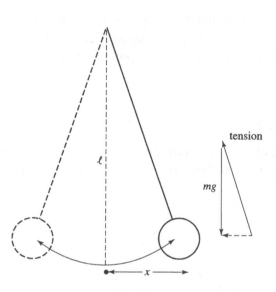

FIGURE 8.5

where l is the length of the string and g is what it always is. The weight of the bob does not matter as long as it is a lot more than the weight of the string.

CORE CONCEPT

Simple harmonic motion, as of a pendulum or a weight suspended on a string, is a projection of circular motion.

TRY THIS
—1—

How long must a pendulum be to mark off seconds?

Calculating SHM

The condition for an object to be in simple harmonic motion (SHM) is that its acceleration is always proportional to its displacement from the equilibrium position and toward that position. If x is its displacement, then the condition for simple harmonic motion is

$$a \propto -x$$

Now we can find an expression for the period of any simple harmonic motion. The period must be the same as that of the reference circle to which we compare simple harmonic motion. Equations 2.5a and 2.5b combine to give an expression for centripetal acceleration:

$$a = \frac{4\pi^2 r}{T^2}$$

When the oscillating object is at it furthest displacement, $x = r$, and the acceleration of the object is the same as that of the centripetal acceleration of the reference object. Then

$$a = \frac{4\pi^2 x}{T^2}$$

so $$T = 2\pi\sqrt{\frac{x}{a}}$$ **(Equation 8.2)**

With this equation, we can find the period of any simple harmonic motion. All we need is the ratio of displacement to acceleration.

Let's look at the object hanging on a spring in Figure 8.4 to see whether it qualifies. First of all, when it is in equilibrium, its weight must be equal in magnitude to the upward force exerted by the spring:

$$mg = k\,\Delta l$$

where k is the spring constant. Now suppose we pull it down a distance x and then release it. The net force on it (which $= ma$) must be

$$ma = k(\Delta l + x) - mg = k\Delta l + kx - mg$$

Since $k\,\Delta l = mg$, this reduces to

$$a = \frac{k}{m}x$$

so the motion is simple harmonic. Furthermore, this equation tells us all we need to know to find the period of the motion.

EXAMPLE 8.2
A kilogram mass standard suspended from a spring with a constant of 120 newtons per meter bobs up and down. What is the period of its oscillation?

SOLUTION

The ratio x/a is equal to m/k. Therefore, from Equation 8.2:

$$T = 2\pi\sqrt{\frac{m}{k}} = 2\pi\sqrt{\frac{1\ \text{kg}}{120\ \text{N/m}}} = 0.57\ \text{s}$$

The pendulum is another example. Comparing the vector diagram in Figure 8.5 with the physical dimensions of the pendulum reveals that $ma/mg \approx x/l$ where l is the length of the pendulum string. The approximation holds well as long as the angular displacement is small. Therefore, $x/a = l/g$.

EXAMPLE 8.3
A pendulum is used on a strange planet to determine the acceleration due to gravity. If the string is 85 centimeters long and the pendulum oscillates with a period of 1.6 seconds, what is the acceleration due to gravity?

SOLUTION

Since $x/a = l/g$, Equation 8.2 becomes

$$g = \frac{4\pi^2 l}{T^2} = \frac{4\pi^2\,(0.85\ \text{m})}{(1.6\ \text{s})^2} = 13\ \text{m/s}^2$$

CORE CONCEPT

In simple harmonic motion, the ratio of displacement to acceleration is constant and determines the period.

TRY THIS
—2—

To measure the mass of an object in outer space, the object is attached to a flexible rod. The rod is clamped at one end. The other end, with the mass attached, is displaced by 5.0 centimeters, using a force of 40 newtons. When the end is released, it oscillates with simple harmonic motion with a period of 1.5 seconds. What is the mass of the object?

What Is a Wave?

As you sit on the shore and watch the waves roll in, you can see them pounding against the rocks and roiling the sand. They carry a lot of energy, transporting it from far out at sea. Yet a boat 100 yards from shore is not carried in with the waves. It just bobs up and down. Waves keep coming in to shore, but the water merely moves up and down, or back and forth. The waves deliver energy, not water, to the shore.

That disturbance on the surface of the ocean is what the physicist calls a *periodic wave*. It repeats the same pattern endlessly. The world is full of periodic waves, although most of them are invisible. Sound waves, light waves, and electromagnetic waves surround us moment by moment. All are cyclical disturbances that have much in common with each other.

That boat in the continuous wave of the water is moving up and down in regular, repeated cycles. It is *oscillating*. Many things oscillate: a pendulum, a guitar string, the air column in a trumpet, a child on a pogo stick. If an oscillation is simple harmonic, it can be described mathematically by giving its amplitude and its period.

It is often convenient to express the time property of an oscillation in terms of *frequency* instead of period. Frequency is the number of cycles completed per unit time. Thus, if a short pendulum swings back and forth with a period of one-third of a second, its frequency is 3 cycles per second. The frequency is the reciprocal of the period:

$$f = \frac{1}{T}$$

(Equation 8.3)

The cycle per second is such an important unit that it is given a special name: a *hertz*, abbreviated as Hz. Some vibrations are extremely fast and call for multiples: kilohertz (kHz = 1 000 Hz) and megahertz (MHz = 10^6 Hz).

EXAMPLE 8.4   One end of a meter bar is clamped to a table top, and the other end is struck. It oscillates so as to make one complete cycle every fifth of a second. What is its frequency?

SOLUTION

$$f = \frac{1}{T} = \frac{1}{0.2 \text{ s}} = 5 \text{ cycles per second} = 5 \text{ Hz}$$

Consider the two identical pendulums of Figure 8.6. They have the same period, but pendulum A is swinging a little behind pendulum B. B has reached its equilibrium position, but A will get there a little later. The two oscillations are different in *phase*. One way of expressing this difference is in terms of fractions of a cycle: A is swinging about $\frac{1}{8}$ cycle behind B. If 360° represents a full cycle, this can be expressed by saying that pendulum A lags behind pendulum B by 45°.

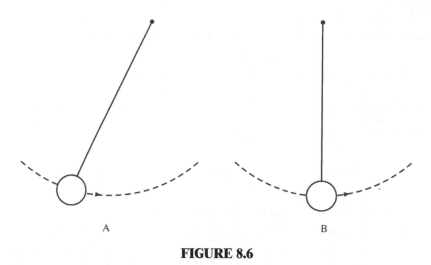

FIGURE 8.6

EXAMPLE 8.5   A strong breeze makes two slats on a venetian blind oscillate with a frequency of 8.0 hertz. If one slat reaches its maximum displacement 0.050 seconds after the other, what is the phase difference, in degrees, between the oscillations?

SOLUTION

The period of the oscillation is

$$T = \frac{1}{f} = \frac{1}{8.0 \text{ Hz}} = 0.125 \text{ s}$$

The time delay, in fractions of a full cycle, is thus

$$\frac{0.050 \text{ s}}{0.125 \text{ s}} \times 360° = 144°$$

What happens if there are two boats in the water, bobbing up and down in the periodic wave? Both oscillations will have the same amplitude and frequency, but will differ in phase, like the two pendulums. A wave consists of a whole series of oscillations, all having the same frequency but differing in phase. For a simple model of a wave, consider the rope in Figure 8 .7. You grasp one end and shake it up and down. If there is a marker on the rope, some distance from your hand, it will oscillate up and down. If there are two markers, they will have the same frequency, but will differ in phase. The farther apart the two markers are, the greater the phase difference. It is not hard to find two positions that are completely out of phase; that is, 180 degrees apart in phase. When one zigs, the other zags.

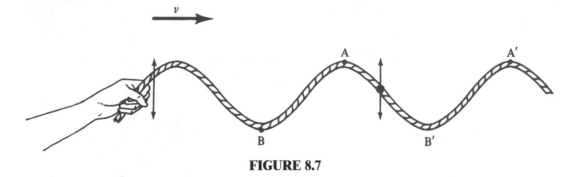

FIGURE 8.7

The result of all these oscillations is that alternating highs and lows *(crests* and *troughs)* seem to follow each other down the rope. They travel at a definite speed, the wave velocity. Each point on the rope pulls on the point beyond, transferring energy as the wave travels. The energy will be delivered at the other end of the rope. The greater the amplitude of the wave, the more energy it carries. In an ideal rope, all the energy put in at one end comes out the other end. Of course, in a real rope, some energy is converted to internal energy as the rope moves, and the amplitude of the wave decreases as the wave travels. This process of energy loss is called *damping.*

EXAMPLE 8.6

What is the frequency of a wave if two points 90 degrees out of phase reach their peaks 0.10 second apart?

SOLUTION

The time difference of 0.10 s represents a phase difference of 90°/360°, or just one-fourth of a cycle. Therefore, a full cycle takes 0.40 s, which is the period of the wave. Then, by Equation 8.3,

$$f = \frac{1}{T} = \frac{1}{0.40 \text{ s}} = 2.5 \text{ Hz}$$

CORE CONCEPT

Waves are described in terms of period, frequency, amplitude, phase, energy, and damping:

$$f = \frac{1}{T}$$

TRY THIS
—3—

What is the phase difference between two points if they reach their peak displacements 0.05 second apart when the frequency is 7.5 hertz?

Traveling Waves

Note the two points marked A and A' in Figure 8.7. At the moment the picture was made, both were at the peak of a crest. Therefore, they must be in phase with each other. Similarly, points B and B' are in phase, since both are at the bottom of a trough at the same moment. The distance between A and A', or between B and B', or between the points in any other pair in phase, with no other such points in between them, is called the *wavelength* of the wave.

As the crest travels from A to A', point A' goes through one full cycle, from crest to trough and back to crest again. This gives us an important relationship between period and wavelength: Since the wave travels one wavelength (λ, the Greek letter lambda) in the time of one period, the speed of the wave (v) is given as

$$v = \frac{\lambda}{T}$$

This can be put into its more familiar form by applying Equation 8.3:

$$v = f \lambda \qquad \text{(Equation 8.4)}$$

EXAMPLE 8.7

What will be the wavelength of the wave in a rope if its speed in that rope is 12 meters per second and the end of the rope is shaken with a frequency of 6.0 hertz?

SOLUTION

From Equation 8.4,

$$\lambda = \frac{v}{f} = \frac{12 \text{ m/s}}{6.0/\text{s}} = 2.0 \text{ m}$$

Note that Hz = 1/s.

EXAMPLE 8.8

What is the speed of the wave in a rope if two points out of phase by 180 degrees are 35 centimeters apart and the frequency is 5.0 hertz?

SOLUTION

Two points out of phase by 180 degrees are half a wavelength apart, so the wavelength is 70 centimeters. Then, from Equation 8.4,

$$v = f \lambda = (5.0 \text{ Hz})(0.70 \text{ m}) = 3.5 \text{ m/s}$$

In the wave we have been looking at so far, the particles of the rope are moving at right angles to the rope, while the wave travels along the rope. Any wave in which the particle motion is perpendicular to the direction of travel of the wave is called a *transverse* wave. The wave in the rope is transverse whether the rope is shaken up and down, from side to side, or in any other direction perpendicular to the direction in which the wave is traveling.

Figure 8.8 illustrates another kind of wave. The medium in which the wave travels is a long, stretched spring. The hand that makes the waves is moving back and forth, parallel to the length of the spring. When the hand moves to the right, it creates a *compression;* moving to the left, it makes a *rarefaction.* The separate particles of the spring vibrate back and forth, parallel to the velocity of the wave, as the compressions and rarefactions follow each other down the spring. This is called a *longitudinal* wave.

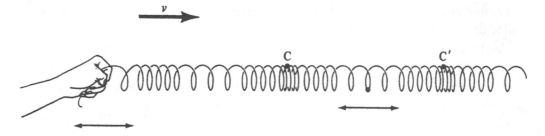

FIGURE 8.8

In a longitudinal wave, two points of greatest compression will be in phase with each other, and the distance between them is a wavelength. C and C′ are a wavelength apart.

In either kind of wave, the velocity of the wave depends primarily on the nature of the medium—the rope or the spring. In some kinds of waves, velocity increases a little as the frequency rises, but it is usually a small effect. Also, the amplitude of the wave may affect the speed slightly, but usually not much. Velocity depends on the kind of medium, and frequency depends on the nature of the disturbance; together they determine the wavelength according to Equation 8.4.

CORE CONCEPT

In any wave, transverse or longitudinal, velocity depends primarily on the nature of the medium and is equal to the product of frequency and wavelength:

$$v = f\,\lambda$$

TRY THIS
—4—

What is the velocity of a wave if the frequency is 6.0 hertz and two points 90 degrees out of phase with each other are 25 centimeters apart?

Sound Waves

Sound travels through space as a wave, transmitted by vibrations of air molecules. To see how this works, look at Figure 8.9, which shows successive stages in the vibration of a tuning fork.

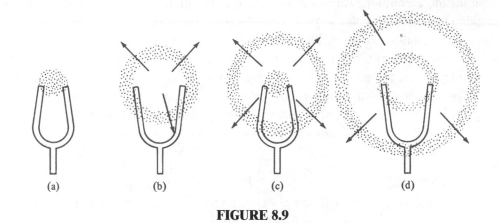

(a)   (b)   (c)   (d)

FIGURE 8.9

At (a) the two prongs of the fork have moved closer to each other, creating a region of compressed air between them. Since the pressure here is higher than it is outside this region, the air in the compression zone expands outward. Meanwhile, the tuning fork prongs move farther apart, as at (b). This creates a low-pressure region, a rarefaction, between the prongs. The rarefaction zone follows the compression zone out from the tuning fork—and the process repeats itself as long as the fork continues to vibrate. Thus, the compression and rarefaction zones follow each other out, spreading radially.

When a wave consists of zones of compression and rarefaction, we know that the particles are vibrating parallel to the direction in which the wave travels. In this case, the direction of travel at any point is directly away from the source—radially outward. The wave is longitudinal.

The speed with which a sound wave travels in air depends only on the temperature. If the piccolo and the bassoon are playing together in the orchestra pit, the sounds will arrive at your ear together, no matter how far you are from the stage; frequency does not affect the wave velocity. But waves do travel faster in warmer air. The speed at any temperature can be found from this equation:

$$v_{air} = 331\frac{m}{s} + \left(0.6\frac{m}{s}\right)T_C \qquad \textbf{(Equation 8.5)}$$

where T_C is the Celsius temperature.

EXAMPLE 8.9 How fast will a wave travel in air at a temperature of 15 degrees Celsius?

SOLUTION
From Equation 8.5,

$$v_{air} = 331\frac{m}{s} + \left(0.6\frac{m}{s}\right)(15)$$

$$v_{air} = 340 \text{ m/s}$$

Unlike the wave in the rope or the spring, the amplitude of a sound wave diminishes drastically as the wave travels. This occurs because the wave is spreading out in three-dimensional space instead of traveling in a single line along a one-dimensional medium. Each crest is a sphere of compression, growing as it travels. Such a spherical crest or trough, or any other sphere of points in phase with each other, is called a *wave front*. The area of this wave front increases as the square of the radius, but the total energy in it (neglecting damping) does not change. Thus, the amplitude of the wave must drop off at a greater distance from the source.

CORE CONCEPT

A sound wave is a longitudinal wave spreading through three-dimensional space at a speed that depends only on the air temperature:

$$v_{\text{air}} = 331\frac{\text{m}}{\text{s}} + \left(0.6\frac{\text{m}}{\text{s}}\right)T_C$$

TRY THIS
—5—

With the temperature at 24 degrees Celsius, a tuning fork vibrates at 320 hertz. What is the wavelength of the sound wave it produces?

Music and Noise

What you hear depends on the properties of the sound wave that enters your ear. A loud sound represents a high-amplitude wave, one with lots of energy.

Musicians use a letter code to describe the *pitch* of a musical sound. The pitch of a sound represents the frequency of the wave. When an American orchestra tunes up, the oboe sounds the note that is called A-440 — A on the musical scale, and a frequency of 440 hertz. All the rest of the instruments use this as a basis for tuning. There is nothing sacred about this relationship between pitch and frequency; European orchestras do it differently. And even within an orchestra, the note called A on an oboe is called something else on a clarinet.

Definition of a musical scale begins with an interval called an *octave*. This is the interval between middle C and high C, for example. An increase in pitch of one octave represents a doubling of the frequency of the wave.

EXAMPLE 8.10 The lowest note on a piano is an A with a frequency of 27.5 hertz. What is the frequency of the A three octaves higher?

SOLUTION

Each octave up is double the frequency, so 27.5 hertz must be doubled three times, to get 220 hertz.

   Notes sound harmonious together when their frequencies are in simple whole-number ratios. The diatonic scale is built on such ratios. The group of three notes called a major triad, such as C-E-G or F-A-C, has its frequencies in the ratio 4:5:6. Figure 8.10 shows the frequencies of the notes of the diatonic scale in the key of C, with A taken as the standard orchestral value of 440 hertz. Other major triads can be formed by introducing sharps and flats into the scale. These frequencies are those played by a violonist; on a piano, they are altered slightly to make it possible to play in any key.

   A flute playing C sounds very different from a violin playing the same note. The difference is in the *tone quality* of the sound. The reason for the difference is that, when a violin string is bowed, it produces a great many frequencies simultaneously. The lowest, called the *fundamental,* is the one that specifies the pitch. The other frequencies are called the *overtones* of the note. Every instrument produces its characteristic mix of overtones, and that is what gives each its special character. A flute produces nearly a pure sine wave with few overtones, especially in its upper register, while a violin's voice has many overtones.

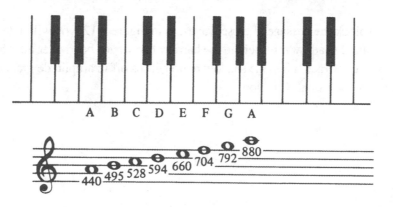

FIGURE 8.10

   The difference between a violin and a drum lies in the kinds of overtones. The overtones of a violin are simple whole-number multiples of its fundamental. When a cello plays a low A at 110 hertz, the string is also producing waves at 220 hertz, 330 hertz, 440 hertz, 550 hertz, and so on. We get a definite sense of pitch, represented by 110 hertz, the fundamental. But when a drum is struck, its overtones do not form such a simple series. If a drum produces its fundamental tone at 50 hertz, the overtone series looks like this: 79.7 hertz, 106.8 hertz, 114.8 hertz, 145.9 hertz, 179.9 hertz, and so on. In this sort of series, no definite sense of pitch is produced. This is noise.

CORE CONCEPT

Loudness of a sound corresponds to amplitude of the wave; pitch to frequency; tone quality to the mix of overtones.

TRY THIS
—6—

With the temperature at 22 degrees Celsius, a violin plays middle C at 264 hertz. What is the wavelength of the second overtone of the sound wave produced?

Diffraction

Waves are not hard-surfaced objects traveling in definite paths, like bullets or baseballs. They have a strong tendency to bend around corners. This phenomenon is called *diffraction.*

To study diffraction, we can make waves visible. Our model is the ripple produced on the surface of water by a point that is made to oscillate by means of a motor. This produces a series of circular waves, as in Figure 8.11. Alternate crests and troughs follow each other radially outward from the oscillator, forming traveling circular wave fronts.

If you try to block the passage of these waves, as in Figure 8.12, you will note that you cannot form a sharp shadow zone behind the barrier. The waves bend over into the shadow region. This wave diffracted into the shadow region drops off in amplitude gradually as you go farther into the shadow region.

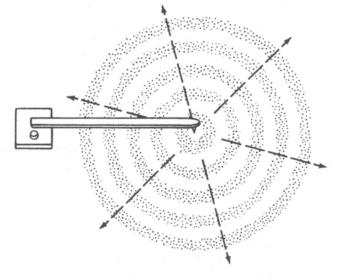

FIGURE 8.11

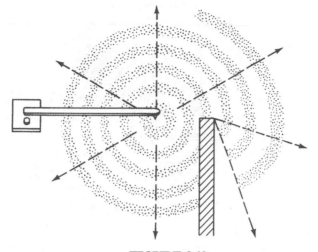

FIGURE 8.12

The amount of diffraction depends strongly on the wavelength. In Figure 8.13, the same source and barrier are used, but the frequency of the source has been speeded up so that the wavelength of the wave is smaller. In this case, the amplitude of the wave in the shadow zone drops off much more quickly. Longer waves diffract much more than short ones.

In Figure 8.14, a small barrier has been used—it is not much larger than a wavelength. Here the shadow region is very small; the waves diffract around the barrier and reunite on the other side. The net effect is just as though no barrier were there at all. To block a wave, you need a barrier substantially larger than a wavelength. This explains why a breakwater, designed to prevent the ocean's waves from washing sand down the beach, has to be so long.

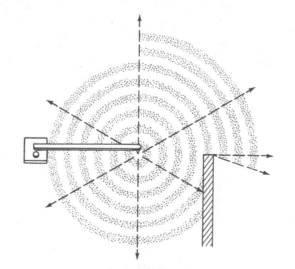

FIGURE 8.13

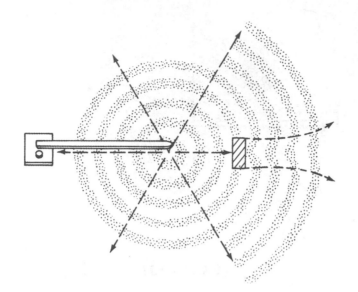

FIGURE 8.14

Figure 8.15 shows what happens when the wave comes to an opening in a barrier. The wave diffracts around the edges of the slit, forming arcs centered on the edges. If the slit is narrower, as in Figure 8.16, these diffracted arcs dominate the pattern, and the waves emerge on circular fronts. A narrow slit forms the same kind of wave as an oscillating-point source.

Since diffraction affects longer waves most, shorter waves tend to travel in straight lines, while longer ones bend around all corners. That is why a hi-fi speaker has only one large speaker ("woofer") for the low frequencies, but several small ones ("tweeters") pointing in different directions for the high frequencies.

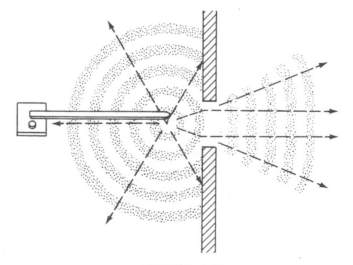

FIGURE 8.15

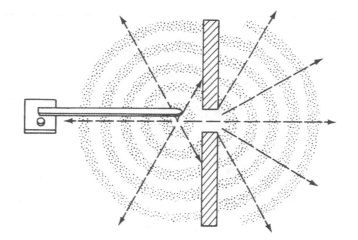

FIGURE 8.16

CORE CONCEPT

Waves diffract or bend around corners, and the effect is greater when the wavelength is longer.

TRY THIS
—7—

If you stand to one side of an open door listening to the music of an orchestra coming through the doorway, in what way will the sound you hear be distorted?

Shifting Frequencies

As you drive along in your car, you pass by a factory as its whistle blows. At the instant you pass the factory you hear a sudden drop in the pitch of the whistle. This is known as the *Doppler effect*.

Figure 8.17 shows why this happens. The concentric circles around the whistle represent zones of compression, spreading outward from the whistle. If you sit still, they will pass by you at a frequency equal to the frequency of the whistle. But if you are moving toward the whistle, your ear will intercept more of those compressions every second than are being produced by the whistle. You will hear a higher frequency than the whistle is producing. When you pass the factory, you will be moving away from the whistle, and the effect is just the reverse. You will hear a lower frequency.

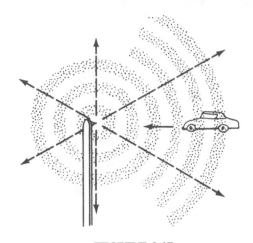

FIGURE 8.17

There is a similar Doppler frequency shift when it is the source of the sound that is moving. Again, you hear a frequency higher than that of the source as it approaches you, and lower as it recedes. But the reason is different, and the frequency shift is not exactly the same. Figure 8.18 shows the compression zones surrounding a source moving to the right. Each zone is a sphere, as usual, centered on the point at which it was produced. However, since the source is moving to the right, each compression originates farther to the right than the one before it. The result is that the wavelength of the sound is shorter in the direction in which the sound source is traveling, and longer behind it. Therefore the frequency shifts upward in front and downward behind.

If the source is moving faster than sound, something quite different happens, as can be seen in Figure 8.19. All the compression zones overlap each other, and the sound energy is concentrated mainly in one big compression, the cone-shaped envelope of all the compression zones. If an airplane flies overhead at a speed faster than sound, that enormous cone of compression trails along behind it, intersecting the ground at some distance behind the plane. Where the cone touches the ground, there is a "sonic boom"; windows break, children start to cry, and alarmed citizens call police stations to find out what caused the explosion.

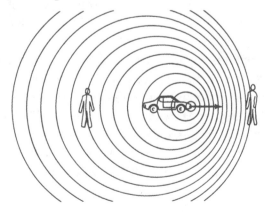

FIGURE 8.18

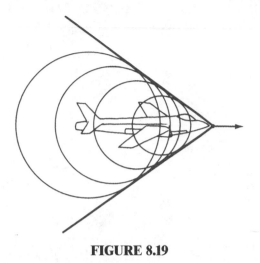

FIGURE 8.19

CORE CONCEPT

If an observer and a source of sound are in relative motion, the observed frequency is different from the emitted frequency.

TRY THIS
—8—

Why is the Doppler shift different when the source is moving than when the observer is moving?

Interference

One distinctive feature of waves is that it is possible for two waves, arriving simultaneously at the same place, to add up to nothing. This can occur with two sound waves if they arrive at some point exactly 180 degrees out of phase with each other. Then one of them is producing a compression while the other is producing a rarefaction, and the net result is that nothing at all happens. This process is called *destructive interference*.

To show this experimentally, get a twin pair of speakers and set them a short distance apart. This has to be done in an open field so there will be no reflection from walls. Connect both speakers to the same signal generator, so that they are producing a steady, single-frequency sound in phase with each other, as shown in Figure 8.20. A microphone moved around some distance from the speakers will be able to find points of destructive interference where the waves from the two speakers arrive out of phase with each other. The microphone will pick up no sound there.

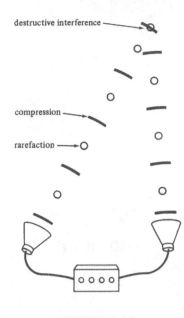

FIGURE 8.20

One way to locate such a point is to do a little geometry. If a point is just a half-wavelength farther from one speaker than it is from the other, the wave from the farther speaker will arrive a half-cycle later than the one from the near speaker. Then the two waves will be a half-cycle out of phase with each other and will interfere destructively.

EXAMPLE 8.11
Two speakers, producing identical sound waves in phase in an open field, feed into a microphone. If the microphone can detect no sound when it is 4.7 meters from one speaker and 5.2 meters from the other, what is the wavelength of the sound?

SOLUTION

If interference is destructive, the simplest solution is that one speaker is a half-wavelength farther from the microphone than the other speaker. Therefore, 5.2 m – 4.7 m = 0.5 m is a half-wavelength, and the wavelength of the sound is 1.0 m.

What happens to the energy of the sound waves? Surely it cannot just disappear! It turns up at a point of *constructive interference*. At any point equidistant from the two speakers, the two waves will arrive in phase and will reinforce each other. The sound is exceptionally loud there.

An easy way to hear the effect of interference is to get a couple of tuning forks, violin strings, or any other sources of a musical sound tuned very slightly apart from each other. If both are sounded together, the two waves arriving at your ear will go alternately in and out of phase with each other. This can be seen in Figure 8.21, which shows the pressure variation at your ear due to the simultaneous arrival of two waves of slightly different frequencies. You will hear *beats*—waawaawaawaa—as the waves alternate in constructive and destructive interference. The beat frequency will be the difference between the two sound-wave frequencies. Listening for beats is an extremely sensitive way to match frequencies, and musicians use it often in tuning instruments.

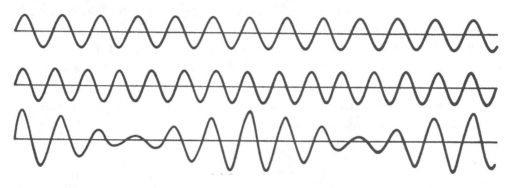

FIGURE 8.21

EXAMPLE 8.12 A piano tuner adjusts a string to sound C-264 hertz. In trying to match a second string to it, he sounds both together and hears three beats per second. What is the frequency of the other string?

SOLUTION

The difference between the two frequencies must be 3 hertz, so the other one is either 261 hertz or 267 hertz.

CORE CONCEPT

Waves interfere constructively at points where they arrive in phase and destructively at points where they arrive out of phase.

TRY THIS
—9—

Two speakers produce sound waves 2.0 meters long in phase. If a point of constructive interference is 12.5 meters from one of the speakers, how far is it from the one that is farther away?

Standing Waves

Remember the rope we used to demonstrate waves? Let's see what happens when the wave gets to the end of the rope.

First, assume that the end of the rope is tied to some fixed support. The moving rope cannot make the support shake, and the energy of the wave has nowhere to go except back into the rope. The wave is reflected from the fixed end, passing back along the rope in the opposite direction. Now we have two identical waves in the rope, one going to the right and the other to the left.

To see what happens in this case, study Figure 8.22 carefully. The uppermost wave is moving to the right, the second one is traveling to the left, and the bottom wave is the sum of the other two, added point by point. Successive pictures, from left to right, show what the rope looks like at intervals of one-eighth of a cycle. There are five points on the rope, marked by black dots, that are not moving at all. These are places where the two waves interfere with each other destructively. These points are called *nodes,* and they are spaced a half-wavelength apart down the whole length of the rope. The fixed end, unable to move, must be a node.

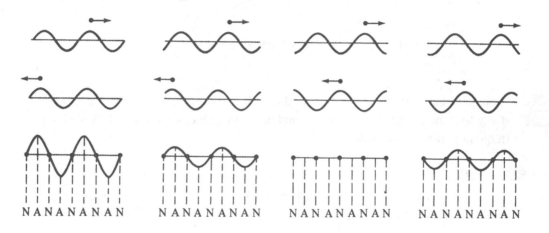

FIGURE 8.22

Halfway between the nodes are points of maximum vibration where the two waves interfere constructively. These points are *antinodes.* They are a half-wavelength apart. Alternate antinodes are in opposite phase, so that the distance between two antinodes in phase is one wavelength.

EXAMPLE 8.13   If a wave with a wavelength of 1.2 meters traveling in a rope is reflected from a fixed end, how far from the end are the first two antinodes?

SOLUTION

The end is fixed, so it must be a node. The first antinode is a quarter-wavelength in from the end, or 0.3 m, and the next one is a half-wavelength farther, at 0.9 m from the end.

This pattern of nodes and antinodes is called a *standing wave*. Within a given loop of a standing wave—from one node to the next—all points are in phase, but the amplitude varies from zero at the nodes to a maximum at the antinode.

CORE CONCEPT

A standing wave, consisting of nodes and antinodes alternating each quarter-wavelength, is formed when two equal traveling waves pass through each other in opposite directions.

TRY THIS
—10—

A rope is fastened at one end, and the other end is shaken with a frequency of 5.0 hertz. If the speed of the wave in the rope is 15 meters per second, how far from the attached end are (a) the first node; and (b) the first antinode?

Making Music

When a guitar player plucks a string, he confidently expects a note of a certain frequency to emerge. If he is any good, it will. The frequency of the sound that the guitar produces must be the same as the vibration frequency of the string, since that is what produces the sound wave. And when a string is plucked, it will vibrate with a number of frequencies, which depend on the length of the string and the speed of the wave in it.

This is what happens when a string is plucked: Waves of many frequencies start traveling in the string in both directions. When these waves reach the ends of the string, they are reflected back. Most of them die out almost at once. The only ones that remain are those that form stable standing waves in the string. The wavelengths of the standing waves are limited by the *boundary conditions* of the string: A standing wave can form only if its wavelength is such that it forms a node at each end of the string. This limitation is imposed because the ends of the string are clamped, so that the string cannot vibrate there.

The possible wavelengths are illustrated in Figure 8.23. The longest possible standing wave, shown in the first sketch, is one with a node at each end and an antinode in the middle of the string. Since the nodes of a standing wave are a half-wavelength apart, the wavelength of this standing wave is twice the length of the string. The frequency of this wave—and therefore of the sound wave that it produces—can be found by applying Equation 8-4; it is the speed of the wave divided by the wavelength.

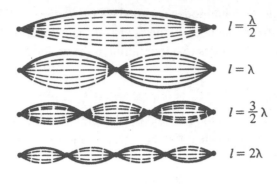

$l = \frac{\lambda}{2}$

$l = \lambda$

$l = \frac{3}{2}\lambda$

$l = 2\lambda$

FIGURE 8.23

EXAMPLE 8.14

What is the fundamental frequency sounded by a violin string 45 centimeters long if the speed of the wave in the string is 280 meters per second?

SOLUTION

At the fundamental frequency, the only nodes in the string are at the ends. Since these are a half-wavelength apart, the wavelength of the wave is twice the length of the string, or 0.90 meters. Then, from Equation 8.4,

$$f = \frac{v}{\lambda} = \frac{280 \text{ m/s}}{0.90 \text{ m}} = 311 \text{ Hz}$$

Other frequencies are possible because there are other standing waves that can fit into the string. The second sketch in Figure 8.23 shows another standing wave that meets the boundary conditions. It also has nodes at both ends, but it also has a node in the middle of the string. Its wavelength is thus the same as the length of the string. With half the wavelength of the first standing wave, it has twice the frequency. The other sketches show the basic rule that determines the wavelengths of the standing waves in a string: The length of the string must be some whole number of half-wavelengths, as calculated in Example 8.15.

EXAMPLE 8.15

What are the wavelengths of the four longest waves that can stand in a string 60 centimeters long?

SOLUTION

For the longest wave, the string vibrates in a single loop, with a node at each end. In this oscillation, the string is $\frac{1}{2}$ wavelength long, so

$$60 \text{ cm} = \frac{1}{2}\lambda; \quad \lambda = 120 \text{ cm}$$

Next, there is a node at each end and one in the middle, making the string 1 wavelength long:

$$60 \text{ cm} = \lambda$$

Vibrating in three loops, it is $\frac{3}{2}$ wavelength long:

$$60 \text{ cm} = \frac{3}{2}\lambda ; \qquad \lambda = 40 \text{ cm}$$

And in four loops:

$$60 \text{ cm} = 2\lambda; \qquad \lambda = 30 \text{ cm}$$

EXAMPLE 8.16   What are the four lowest frequencies of a 60-centimeter string if the speed of the wave in it is 240 meters per second?

SOLUTION

For each of the possible wavelengths, from Equation 8.4,

$$f = \frac{v}{\lambda}$$

Using the wavelengths calculated in Example 8.15, we have

$$f_1 = \frac{240 \text{ m/s}}{1.20 \text{ m}} = 200 \text{ Hz}$$

$$f_2 = \frac{240 \text{ m/s}}{0.60 \text{ m}} = 400 \text{ Hz}$$

$$f_3 = \frac{240 \text{ m/s}}{0.40 \text{ m}} = 600 \text{ Hz}$$

$$f_4 = \frac{240 \text{ m/s}}{0.30 \text{ m}} = 800 \text{Hz}$$

Thus, the frequencies that can be produced by a string depend on the length of the string and on the speed of a wave in it. This wave velocity increases as the tension in the string increases, as anyone who has ever tuned a violin or a guitar knows. Also, the speed of the wave depends on the thickness and density of the string. The low notes of a piano or a cello are produced by thicker, heavier strings. When a string is plucked or stroked, it produces, simultaneously, all those frequencies that are made by waves that can stand in the string. Example 8.16 shows that these frequencies form an overtone series in which all the frequencies are in simple, whole-number ratios. That is why the sound produced has a musical quality and a definite sense of pitch, which is determined by the lowest vibration frequency of the string.

CORE CONCEPT

The natural vibration frequencies of a string are specified by the boundary condition that there must be a node at each end.

TRY THIS
—11—

What are the first three frequencies produced by a plucked string 30 centimeters long if the speed of the wave in the string is 280 meters per second?

Pipes

An air column has its own natural vibration frequencies, and these serve as the basis for all wind instruments. The air in the pipe is set into vibration at one end, by some means or other. In a clarinet or an oboe, it is done by making a reed vibrate. The musician who plays a trumpet or a French horn does it by vibrating his lips against the mouthpiece. In a flute or an organ pipe, the vibration is produced by turbulent air passing across a narrow slit.

The boundary conditions in an organ pipe are not like those of a string. At the ends of the pipe, the trapped air is open to the outside air. Since the air is not constrained at the ends, rapid vibration is possible there, and the ends are antinodes, not nodes.

It is difficult to illustrate the standing wave pattern inside an organ pipe. It is a sound wave, so it is longitudinal, and its speed can be determined from the temperature alone, using Equation 8.5. Wavelength is determined by the boundary condition that there must be an antinode at each end. As you can see by the symbolic representation in Figure 8.24, this produces the same rule as for a string: The length of the pipe must be a whole number of half-wavelengths.

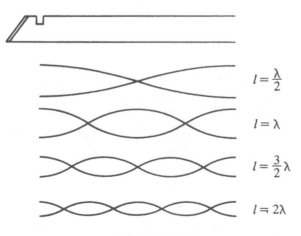

$$l = \frac{\lambda}{2}$$

$$l = \lambda$$

$$l = \frac{3}{2}\lambda$$

$$l = 2\lambda$$

FIGURE 8.24

EXAMPLE 8.17 With the temperature at 15 degrees Celsius, what are the three lowest frequencies sounded by an open organ pipe 40 centimeters long?

SOLUTION

The speed of the wave in the pipe is

$$v = 331\frac{m}{s} + \left(0.6\frac{m}{s}\right)(15) = 340 \text{ m/s}$$

For the fundamental:

$$40 \text{ cm} = \frac{1}{2}\lambda; \qquad \lambda = 0.80 \text{ m}; \qquad f_1 = \frac{340 \text{ m/s}}{0.80 \text{ m}} = 425 \text{ Hz}$$

First overtone:

$$40 \text{ cm} = \lambda; \qquad f_2 = \frac{340 \text{ m/s}}{0.40 \text{ cm}} = 850 \text{ Hz}$$

Second overtone:

$$40 \text{ cm} = \frac{3}{2}\lambda; \qquad \lambda = 0.80 \text{ m}; \qquad f_3 = \frac{340 \text{ m/s}}{0.2667 \text{ m}} = 1\,275 \text{ Hz}$$

The frequency of a wind instrument is changed by altering the length of the pipe. In a trumpet or a French horn, the buttons open valves that bring additional pipe lengths into the air column. The French horn player also controls the pitch by putting his hand into the bell of the instrument. In a flute or a bassoon, the buttons open holes in the side of the pipe, thereby establishing new antinodal points.

Some organ pipes are closed at the end, and the picture in these is quite different. The closed end is a node; the end that is energized is an antinode. Thus, the fundamental frequency, produced by the longest wave that can stand in the pipe, is fixed by the need for the wave to go from a node to an antinode—a quarter-wavelength. Therefore, at the fundamental frequency, the pipe is a quarter-wavelength long. The first overtone occurs when the pipe is $\frac{3}{4}\lambda$; the next at $\frac{5}{4}\lambda$; etc. (See Figure 8.25).

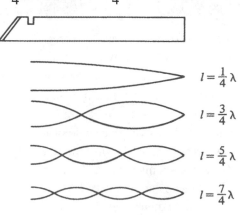

FIGURE 8.25

EXAMPLE 8.18 What are the fundamental and first overtone frequencies of a closed organ pipe 35 centimeters long when the temperature is 18 degrees Celsius?

SOLUTION

The speed of the sound wave in the pipe is 331 m/s + (0.6 m/s)(18) = 342 m/s. At the fundamental, the pipe is $\frac{1}{4}$ wavelength long, so

$$\frac{\lambda}{4} = 0.35 \text{ m}; \qquad \lambda = 1.40 \text{ m}$$

Then the frequency is

$$f = \frac{v}{\lambda} = \frac{342 \text{ m/s}}{1.40 \text{ cm}} = 244 \text{ Hz}$$

At the first overtone, the pipe is $\frac{3}{4}$ wavelength long, so the wavelength is $\frac{1}{3}$ as much as the fundamental. This makes the frequency three times as much, or 732 hertz.

CORE CONCEPT

Natural vibration frequencies of a pipe are fixed by the speed of sound in air and the boundary conditions of a node at a closed end and an antinode at an open end.

TRY THIS
—12—

At a temperature of 22 degrees Celsius, what are the first three frequencies of an open organ pipe 22 centimeters long?

How Loud?

The *loudness* of a sound depends on the power delivered to the ear by the sound wave. This quantity is called the *intensity* of the wave, measured in watts per square meter. A normal youthful ear can hear a sound that is delivering an intensity of only 10^{-12} watts per square meter. This is known as the *threshold of hearing*.

Loudness is not a physical quantity; it is a measure of the way the human ear perceives the intensity of the sound. To establish a scale of loudness, we must take into account the fact that it does not vary linearly with the intensity of the wave. The relationship is logarithmic. If a sound seems twice as loud, the intensity of the wave is 10 times as great; 3 times as loud means an intensity 100 times as much. The *threshold of pain* for the ear is about 1 watt per square meter, a trillion (10^{12}) times more than the threshold of hearing.

The conventional scale of loudness requires a logarithmic scale. It starts with the intensity of the sound wave at the threshold of hearing. This intensity, 10^{-12} W/m², is given the value, on the loudness scale, of 1 *bel* (named after the inventor of the telephone). On this scale, the threshold of pain is 12 bels. The usual unit of loudness is the *decibel*, a tenth of a bel.

Equation 8.6b represents the relationship between loudness and the intensity of the sound. I is the intensity of the sound wave arriving at the ear; I_0 is the intensity at the threshold of hearing, 10^{-12} W/m²; β (beta) is the loudness. The 10 is included to make the answer come out in decibels.

$$\beta = 10 \log\left(\frac{I}{I_0}\right)$$

(Equation 8.6a)

EXAMPLE 8.19

What is the decibel level of a sound whose intensity is 5×10^{-2} watt per square meter?

SOLUTION

From Equation 8.6a,

$$\beta = 10 \log\left(\frac{5 \times 10^{-10} \text{ W/m}^2}{1 \times 10^{-12} \text{ W/m}^2}\right) = 10 \log 500 = 27 \text{ dB}$$

A source of sound in open air is surrounded by spherical shells of sound waves, spreading out in all directions. The power emitted by the source is incorporated in the sound wave. The intensity of the wave at any distance from the source is the emitted power divided by the total area over which that power is spread. At any distance from the source, all the power is spread out over the spherical shell; at a distance r from the source, the area of that shell is $4\pi r^2$. Therefore, if P is the total power in the wave (the power emitted by the source),

$$I = \frac{P}{4\pi r^2}$$

(Equation 8.6b)

EXAMPLE 8.20

A jackhammer produces so much noise that the intensity is at the threshold of pain 1 meter from the source. What is the intensity at a distance of 10 meters?

SOLUTION

At the threshold of pain, the intensity is a trillion (10^{12}) times the threshold of hearing, or 1 watt per square meter. Since the intensity is inversely proportional to the square of the distance from the source,

$$\frac{I_2}{I_1} = \left(\frac{r_1}{r_2}\right)^2$$

$$I_2 = I_1\left(\frac{r_1}{r_2}\right)^2 = \left(1 \text{ W/m}^2\right)\left(\frac{1\text{m}}{10 \text{ m}}\right)^2 = 10^{-2} \text{ W/m}^2$$

which is still plenty loud.

CORE CONCEPT

Intensity of a sound wave in open air is inversely proportional to the square of the distance from the source; it is measured in decibels on a logarithmic scale.

TRY THIS
—13—

What is the decibel level 20 meters from a foghorn emitting sound waves with 3×10^{-5} watt of power?

Resonance

Every elastic object has its own special set of natural vibration frequencies. When you tap on a glass, knock on the door, or drop a screwdriver, the sound you hear is fixed in pitch by the natural frequencies of the object.

The natural vibration frequencies depend on two things: the speed of the wave in the object and the wavelengths that can stand stably in it. The speed is determined by the kind of material—its elasticity, its density, its thickness, and so forth. The wavelengths are determined by the boundary conditions. An edge that is open and unconstrained must be an antinode; any point that is clamped must be a node. The boundary conditions of a drumhead, for example, are that there must be nodes all along its edge. The tuning fork of Figure 8.26 has a node at the handle and antinodes at the ends.

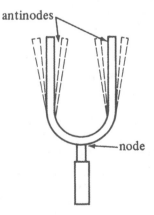

FIGURE 8.26

Every object has a special sensitivity to stimulation at its natural vibration frequencies. If you depress the pedal on a piano and strike any note, the vibration of the string will set into vibration all the other strings that have that natural frequency. A fine wine glass can be made to vibrate so energetically that it will crack, just by exposing it to sound at one of its own natural frequencies. This phenomenon is called *resonance*.

In a violin, there are many resonances. The upper plate has some, and there is a whole set of short-wavelength resonances between the f-holes. There are other resonances in the lower plate, in the trapped air mass, and even in the bridge. It is these resonances that make the difference between a Stradivarius and a cheap fiddle.

CORE CONCEPT

In every object, overtones are determined by the ratio of wave velocity to the wavelengths of standing waves, which are fixed by boundary conditions.

TRY THIS
—14—

What are the two lowest overtones of a tuning fork if each tine is 18 centimeters long and the speed of the wave in it is 420 meters per second? (These answers will be approximate; they take no account of the thickness of the metal.)

REVIEW EXERCISES FOR CHAPTER 8

Fill-In's. For each of the following, fill in the missing word or phrase:

1. The length of time it takes a point on a medium carrying a wave to go through one full cycle is the _____ of the wave.

2. The reciprocal of a wave's period is its _____.

3. The hertz is a measure of_____.

4. Two points on a wave that are a half-wavelength apart are_____ degrees out of phase with each other.

5. The maximum displacement of any point on a wave is called the _____ of the wave at that point.

6. In a real wave, the process by which a wave loses energy as it travels is called _____.

7. The distance between two successive points in phase is the _____ of the wave.

8. The product of frequency and wavelength gives the _____ of a wave.

9. In a longitudinal wave, the motion of the particles is _____ to the wave velocity.

10. The nature of the medium in which it travels largely determines the _____ of a wave.

11. A sound wave consists of alternate compressions and _____ spreading through space.

12. The speed of a sound wave in air depends only on _____.

13. The pitch of a musical sound depends on the _____ of the wave.

14. The frequency of a note two octaves below A-440 is _____ hertz.

15. Overtones determine the _____ of a musical sound.

16. A drum produces no definite sense of pitch because its _____ do not form a series with integral ratios.

17. Diffraction around corners is greater when the _____ is larger.

18. Because of diffraction, a narrow slit acts like a(n) _____ of waves.

19. If you are moving toward a source of sound, the _____ you perceive will be higher than that emitted by the source.

20. The wavelength of a sound wave varies from place to place if the _____ is moving.

21. Destructive interference occurs when two waves arrive at a point _____ with each other.

22. Beat frequency is the _____ of the frequencies of two sounds.

23. A point of maximum destructive interference is called a(n) _____.

24. In a standing wave in a rope, the distance from one antinode to the next is a(n) _____.

25. Stable standing waves can form in a string only at frequencies at which there is a(n) _____ at each end of the string.

26. At the fundamental frequency, the length of a vibrating string is _____ wavelength.

27. The closed end of an organ pipe must be a(n) _____ of the standing wave in the pipe.

28. To make an object resonate, it must be energized at one of its own _____.

29. Motion is simple harmonic when _____ is proportional to displacement.

30. When an object is in simple harmonic motion, it is in equilibrium when its velocity is _____ .

31. Increasing the intensity of a sound by a factor of 100 raises the loudness by _____ decibels.

32. The loudness of the sound from a siren is 50 decibels at a distance of 10 meters. The loudness at 100 meters will be _____ decibels.

Multiple Choice.

1. One hertz is equivalent to
   (1) one cycle per second
   (2) one second
   (3) one meter per second
   (4) one second per meter
   (5) one meter per second squared.

2. An object is in simple harmonic motion if its
   (1) acceleration is proportional to its velocity
   (2) velocity is proportional to displacement
   (3) period is proportional to acceleration
   (4) acceleration is proportional to displacement
   (5) velocity is proportional to period.

3. A standard pendulum could be the basis for an instrument used to measure
   (1) velocity
   (2) gravitational field
   (3) mass
   (4) tension
   (5) period.

4. If two points on a linear wave are in phase, the distance between them must be
   (1) exactly one wavelength
   (2) an odd number of half-wavelengths
   (3) some whole number of wavelengths
   (4) some number of half-wavelengths
   (5) one half-wavelength.

5. A sound wave in air is described as a longitudinal wave because the air molecules move
    (1) through a single dimension
    (2) through a very short distance
    (3) in simple harmonic motion
    (4) perpendicular to the direction of travel of the wave
    (5) parallel to the direction of travel of the wave.

6. If you are at an open-air concert and someone's head gets between you and the orchestra, you can still hear the orchestra because
    (1) sound waves pass easily though a head
    (2) a head is not very large compared with the wavelength of the sound
    (3) the sound is reflected from the head
    (4) the wavelength of the sound is much smaller than the head
    (5) the head resonates to the sound wave.

7. If a car sounding its horn is moving away from you, the sound you receive, compared with what it would be if the car were not in motion, has a
    (1) shorter wavelength and higher frequency
    (2) shorter wavelength and lower frequency
    (3) lower frequency but no difference in wavelength
    (4) longer wavelength but no difference in frequency
    (5) longer wavelength and lower frequency.

8. A sound with a wavelength of 2 meters is directed perpendicularly into a wall, so the reflected wave interferes with the incoming wave. The distance from the wall to the first node in the resulting standing wave is
    (1) 25 centimeters   (4) 1.5 meters
    (2) 50 centimeters   (5) 2 meters.
    (3) 1 meter

9. In tuning a piano, the piano tuner tests one of the strings against a 440-hertz tuning fork and hears 2 beats per second. The frequency of the string is
    (1) 404 hertz   (4) 442 hertz
    (2) 436 hertz   (5) either 438 hertz or 442 hertz.
    (3) 438 hertz

10. When a violin string is energized, which overtone is two octaves above the fundamental?
    (1) first   (4) fourth
    (2) secondx   (5) fifth.
    (3) third

Problems

1. A wave in a rope travels at 12 meters per second, and the wavelength of the wave is measured at 1.2 meters. Find (a) the frequency of the wave; and (b) its period.

2. A wave in the ocean is 30 meters long. A point on the wave front passes two boats that are 12 meters apart. What is the phase difference between the oscillations of the boats?

3. In a longitudinal wave in a spring, the distance from a compression to the nearest rarefaction is 35 centimeters. If the frequency of the wave is 4.0 hertz, how fast is the wave traveling?

4. How fast does a sound wave travel in air at a temperature of –6 degrees Celsius?

5. What is the wavelength of a sound wave produced by a violin string vibrating at 640 hertz if the temperature is 26 degrees Celsius ?

6. What is the frequency of the A two octaves below A-440?

7. With the temperature at 10 degrees Celsius, two loudspeakers are producing a 680-hertz tone in phase with each other. If you locate a node 15.8 meters from the nearer speaker, how far are you from the other one?

8. You are tuning a 12-string guitar and adjust one string to sound 220 hertz; when you strike this along with its matching string, you hear beats at a frequency of 4 hertz. You find that, if you tighten the matching string a little, the beats disappear. What was the frequency of the matching string?

9. A rope that transmits a wave at 8.0 meters per second is shaken at one end at a frequency of 6.0 hertz. If the other end is fastened down, how far from that end are the first two antinodes?

10. Determine the fundamental and the first two overtones of a violin string 35 centimeters long if the speed of the wave in it is 180 meters per second.

11. Determine the fundamental and the first two overtones of a closed organ pipe 35 centimeters long at a temperature of 20 degrees Celsius.

12. Determine the fundamental and the first overtone of a tuning fork whose tines are 18 centimeters long if the speed of the wave in the tuning fork is 60 meters per second.

13. To tune a piano on the equal-tempered scale, a piano tuner wants to adjust the high E to 1 318.5 hertz and does so by listening for beats that this string makes with the second overtone of A-440. When the adjustment is right, what beat frequency will the piano tuner hear?

14. If your loudness meter reads 15 decibels when you are 30 meters from a whistle, what is the output power of the whistle?

15. As you approach a factory whistle, your ears begin to hurt when you are 2 meters from it. How far would you have to be from it before you can no longer hear it? (Assume you have normal hearing and young ears.)

16. What power would a warning siren need to produce so that it will be audible at a distance of a kilometer?

17. A pendulum that marks seconds on the earth oscillates with a frequency of 2.4 hertz on another planet. What is the value of the gravitational field on that planet?

It's Electric

WHAT YOU WILL LEARN

The properties of static electric charges, and how they produce fields, potentials and energy.

SECTIONS IN THIS CHAPTER

- Another Kind of Force
- Can You Create Charge?
- Charged Conductors
- Charge and Force
- Electric Fields
- The Field Rules

- Electrostatic Induction
- Electricity Has Energy
- Potential
- The Charge on an Electron
- Charged Spheres

Another Kind of Force

When you take off a nylon shirt, why does it tend to cling to your body? How is it that you can pick up tiny bits of paper with a comb after you have used it on your hair? What produces that shock you get when you touch a doorknob after walking across a wool rug?

These are manifestations of a kind of force that is in many ways similar to the force of gravity. For one, it seems to act at a distance, through empty space. You can show this with an apparatus like that of Figure 9.1, which consists of a Ping-Pong ball coated with graphite and suspended from a very light thread. If you comb your hair and then bring the comb near the ball, you will find that you can attract the ball to the comb, from a distance.

But the force is not gravity. First of all, this force is far stronger than gravity. The gravitational attraction between a comb and a Ping-Pong ball is far too small to be detected; you need something the size of a 2-kilogram lead ball before you can find a measureable gravitational force. And, second, there seems to be something rather temporary about this new kind of force, for it would not be found if you failed to comb your hair before you tried the experiment. And if you wait a while, it will probably disappear. The comb must be given an *electric charge* before it will attract the ball.

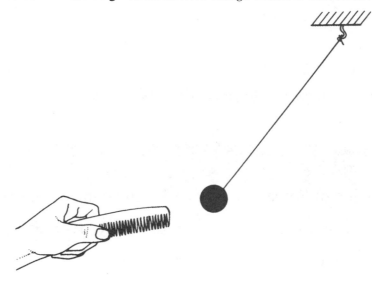

FIGURE 9.1

Now stroke the ball with the charged comb and see what happens. The ball jumps away from the comb; there is force of repulsion between the ball and the comb, something that never happens when the force is gravitational. If you hang two Ping-Pong balls side by side and stroke them both with the comb, they will repel each other, as shown in Figure 9.2.

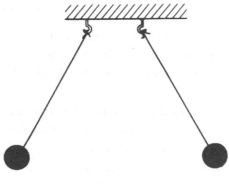

FIGURE 9.2

Now suppose you charge the second ball by a different method: Take a dry glass rod and stroke it with the kind of thin sheet plastic that is used to wrap food. Then stroke the second Ping-Pong ball with the glass rod. The two balls, one charged from the rubber comb and the other from the glass rod, will now attract each other.

We seem to have two different kinds of charge. If two objects are both charged from a rubber rod, or both from a glass rod, they repel each other. If you charge one from glass and the other from rubber, they attract.

CORE CONCEPT

Similar charges repel; dissimilar charges attract.

TRY THIS
—1—

Suppose you stroke a vinyl tile with fur and find that it repels a Ping-Pong ball that had been charged from a rubber rod. What could you conclude about the charge on the vinyl?

Can You Create Charge?

Charge one ball from rubber and the other from glass, and then touch the balls to each other. Sometimes, both charges will disappear completely. The two charges eliminate each other.

We can invoke a well-known mathematical rule to express this situation. The rule is this: If two quantities add up to zero, one is positive and the other is negative. It does not matter which of our two kinds of charge we call positive and which negative, as long as we are consistent. The usual convention, adopted long ago and never changed, is that the charge on a rubber rod is negative and that on a glass rod positive.

When you stroke a rubber rod with fur, the rubber becomes negative. And the fur becomes positive! You started with two neutral objects—total charge zero. After the stroking, you have a positive object and a negative one. The total charge is still zero. Similarly, when the glass rod becomes positive, the plastic film you stroked it with becomes negative. You cannot create a charge of one kind without producing the opposite kind at the same time. *In any process, the total amount of charge does not change.* This statement constitutes the *law of conservation of electric charge*. It is as fundamental to the theory of physics as the other conservation laws: mass–energy, momentum, and angular momentum.

Everything contains electrically charged particles, both positive and negative. These particles make up the atoms that compose the material universe. The size of these atoms is of the order of 10^{-10} meters; it would take several million of them, lined up, to reach across the dot that ends this sentence. Every atom is electrically neutral because it contains equal quantities of positive and negative charge.

The positive charge of an atom, as well as almost all its mass, is concentrated at the center of the atom in a *nucleus*. The nucleus is ten thousand times smaller than the whole atom; it is a marble sitting in the middle of a football field. The nucleus is a tightly bound cluster of two kinds of particles: *protons* and *neutrons*. The concentration of mass in these particles is far beyond imagining; the density of a nucleus is so great that a teaspoonful of nuclear matter would have a mass of several billion kilograms. There is one important difference between neutrons and protons: every proton carries a definite positive electric charge, whereas neutrons have none. The number of protons in the nucleus defines the chemical element to which the atom belongs. A hydrogen nucleus has 1 proton, an oxygen nucleus has 8, and a uranium nucleus has 92.

While most of the atom is empty space, the region surrounding the nucleus contains a definite number of negatively charged particles called *electrons*. An electron has little mass, but each electron has a negative charge equal in magnitude to the positive charge on the proton. Thus, a neutral atom has just as many electrons in the space surrounding the nucleus as there are protons inside the nucleus. The electrons are held near the nucleus by the electric force of attraction between opposite charges, just as a negative Ping-Pong ball is attracted to a positive one.

In different kinds of atoms the outermost electrons of an atom may be bound strongly or weakly. In chemical reactions a weakly bound electron may transfer to a different atom. An atom that gains an electron becomes a negatively charged *ion;* the atom that loses one becomes a positive ion. Many substances consist entirely of ions. A crystal of salt, for example, is made of equal numbers of positive sodium ions and negative chlorine ions.

Sometimes simple contact may cause a few electrons to move from one substance into another. When a rubber rod is stroked with fur, some electrons move from the fur into the rubber. The rubber rod may remove a billion electrons from the fur, leaving the fur positive. This sounds like a lot of electrons, but it can be compared to a drop of water in a mile-long lake. The total number of electrons in the rubber rod is around 10^{24}.

In metals, many electrons are loosely attached to the nucleus and move freely within the substance. If you prepare two oppositely charged Ping-Pong balls and then connect them through a metal rod, electrons from the negative ball will run into the rod; at the other end, electrons will run out of the rod into the positive ball. Both balls will be neutralized, since the charge can flow easily through the metal from one end to the other. A substance that allows charge to pass through it is called a *conductor.* (The graphite on the Ping-Pong balls is a conductor.)

An *insulator* is a material, such as rubber, plastic, or wood, that will not allow charge to flow through it. A charge placed at one end of a rubber rod stays there, even though all the accumulated electrons are repelling each other. Wires are wrapped in rubber, and tool handles in plastic, to keep the electric charge in the metal where it belongs.

In solids, it is only electrons that can move from place to place. But salt water, or the gas inside your fluorescent lamp, is an excellent conductor because it contains both negatively and positively charged ions that can move freely under the influence of an electric force.

CORE CONCEPT

An atom consists of a positively charged nucleus and negatively charged electrons. The electrons are mobile and account for chemical reactions and static electric charges.

TRY THIS
—2—

Explain why you cannot charge one end of a steel rod and leave the other end uncharged.

Charged Conductors

When a charge is placed on a conductor, the excess electrons can move freely through it. They repel each other, so they spread out until every electron is in equilibrium under the repulsive forces of all the others. Unless the repulsion is very strong, however, they cannot leave the surface because they are attracted by the stationary positive charges in the metal.

At equilibrium, *all* the excess charge is on the surface of the conductor. This is the arrangement when the electrons have repelled each other until they are as far apart as possible. There is never a charge inside a charged conductor. No matter how strongly charged the surface is, any charge added to the inside will immediately move to the outside.

This is how the Van de Graaff generator (Figure 9.3) is able to build up such a large charge. The moving belt picks up a small charge as it passes over a plastic pulley. The charge is carried up to metal fingers inside the spherical dome at the top. These fingers pick it up and deliver it to the inside of the dome. It immediately moves to the surface, so that more charge can be added to the inside. Large Van de Graaff generators may have domes up to 4 meters in diameter and can build up enormous charges on their surface.

The dome of a Van de Graaff generator is always spherical and must be smoothly polished. If these conditions are not met, the charge will not be distributed uniformly over the surface of the dome. On a conductor, charge will concentrate wherever the curvature of the surface is greatest. A sharp point concentrates charge enormously. This is how a lightning rod operates. Since it is pointed, all the excess charge accumulates at the point, thus leaks away, preventing a lightning strike to the building. Figure 9.4 shows how charge is distributed over the surface of an irregular conductor.

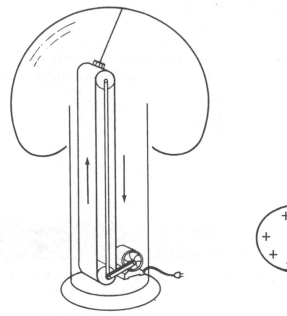

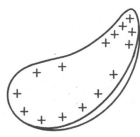

FIGURE 9.3   **FIGURE 9.4**

CORE CONCEPT

On a charged conductor, all charge is on the surface, and it concentrates at points of highest curvature.

TRY THIS
—3—

Explain why one safe place from which to operate the controls of a large Van de Graaff generator is inside the dome.

Charge and Force

Can charge be measured? Standards of measurement must be established, and they can best be defined in terms of the forces that the charges exert on each other.

Example 9.1 shows how a simple vector analysis enables us to measure the electric force exerted by the comb on the Ping-Pong ball of Figure 9.5. The tension in the string is exactly balanced by the (vertical) weight of the ball and the (horizontal) electric repulsion of the comb.

EXAMPLE 9.1   The 2.5-gram Ping-Pong ball of Figure 9.5 hangs at the end of a thread 40 centimeters long. When repelled by the comb, it is deflected 7.0 cm from its original position. How great is the electric force?

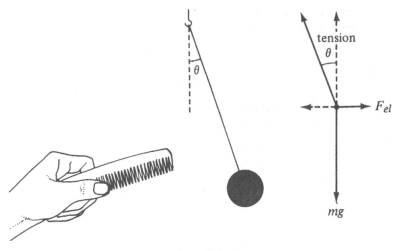

FIGURE 9.5

SOLUTION

As the vector diagram shows, the ratio of the electric force to the weight is the tangent of the angle that the string makes with the vertical. The sine of this angle is $7.0/40 = 0.1750$; its tangent is 0.1777. Therefore,

$$\frac{F_{el}}{mg} = 0.1777$$

$$F_{el} = (0.0025 \text{ kg})\left(9.8\frac{\text{m}}{\text{s}^2}\right)(0.1777)$$

$$F_{el} = 4.4 \times 10^{-3} \text{ N}$$

What we need is a way to determine how the force between two charged objects depends on the amount of charge. Fortunately, we can find this experimentally because there is a way to divide a charge into known fractions: (1) Coat two Ping-Pong balls with graphite to make them conductive, so that they will share equally any charge they may have; (2) charge them and touch them together so that they have the same charge; (3) measure the forces on them by the angle of deflection when they are suspended; (4) touch one ball with the finger—this allows its entire charge to pass off into your body; (5) touch the two balls together again, so that each now has half of its original charge; (6) measure the force again. Repeat many times, with as much variation as possible.

These experiments show that the force between the charges is proportional to the product of the charges and inversely proportional to the square of the distance between them. This is called *Coulomb's law:*

$$F_{el} = \frac{kq_1q_2}{r^2}$$
                                                                    (Equation 9.1)

in which the q's are the charges and r is the distance between them. The constant k is a universal constant whose value depends on how the units are defined. In the SI, the unit of charge is the *coulomb* (C) and $k = 9.00 \times 10^9$ newton-meter squared per coulomb squared.

The coulomb is an extremely large unit of charge, and the nanocoulomb (nC = 10^{-9} C) is more convenient in experiments of the kind described above. Examples 9.2 and 9.3 give rather typical values for such an experiment. When you create a spark by touching a doorknob after walking on a wool rug, you are discharging yourself of a charge that is probably no more than 50 nanocoulombs. A good-sized bolt of lightning discharges about 5 coulombs.

EXAMPLE 9.2 What is the force of attraction between two Ping-Pong balls whose centers are 10.0 centimeters apart if the charges on them are +12 nanocoulombs and –15 nanocoulombs, respectively?

SOLUTION

Applying Equation 9.1,

$$F_{el} = \frac{kq_1q_2}{r^2}$$

$$F_{el} = \frac{\left(9.00 \times 10^9 \ \frac{N \cdot m^2}{C^2}\right)\left(12 \times 10^{-9}C\right)\left(-15 \times 10^{-9}C\right)}{(0.100 \ m)^2}$$

$$F_{el} = -1.6 \times 10^{-4} \ N$$

The negative sign indicates a force of attraction.

EXAMPLE 9.3 Two small, equally conducting spheres are charged, touched together, and then separated until their centers are 12 centimeters apart. If they now repel each other with a force of 3.0×10^{-5} newtons, how much charge do they have?

SOLUTION

Since they are equal, they will share whatever charge they have equally. Call the charge on each q. Then Equation 9.1 becomes

$$F = \frac{kq^2}{r^2}$$

From which

$$q = r\sqrt{\frac{F}{k}} = 0.12 \ m \ \sqrt{\frac{3.0 \times 10^{-5}N}{9.0 \times 10^9 \ N \cdot m^2 / C^2}}$$

$$q = 6.9 \times 10^{-9} \ C, \quad \text{or} \quad 6.9 \ nC$$

The form of Coulomb's law (Equation 9.1) is exactly the same as the form of Newton's law of gravitation (Equation 4.9). It just substitutes charge for mass and uses a different universal constant. But there are important differences. First, mass is always positive, so the gravitational force always has the same sign. Since charge can be either positive or negative, the force can be positive (repulsion) or negative (attraction). Also, note that the constant k is larger than G by a factor of 10^{20}! The electrical force is much too strong ever to be mistaken for gravity.

CORE CONCEPT

Electric force is proportional to the product of charges and inversely proportional to the square of the distance between them:

$$F = \frac{kq_1q_2}{r^2}$$

TRY THIS
—4—

Two graphite-coated Ping-Pong balls are charged, touched together, and then separated by 15 centimeters. If each exerts a repulsive force of 6.0×10^{-5} newtons on the other, how much charge is there on each ball?

Electric Fields

Both Newton's law of universal gravitation and Coulomb's law of electric force describe the interaction between points—point masses or point charges. This means that they will work well as long as the distance between the objects is much larger than the objects themselves. However, in many cases it is difficult or impossible to apply Coulomb's law. Two small, charged objects exert substantial forces on each other. Each excess (or deficient) electron on one object exerts forces on all the charges on the other object. To find the total force that one object exerts on the other, we would have to calculate each of these forces separately, using Coulomb's law, and then add them all up. Impossible!

There is a way around this difficulty, a technique for calculating electric forces that works very well in certain cases, even for small objects close together. We have to start by separating the electric interaction into two parts: (1) a charged object affects the space around itself, creating an *electric field* in that space; and (2) the electric field exerts a force on any other charge placed in it. In this approach, we do not deal with attraction or repulsion of charges. It is the field that exerts the force.

An electric field is a property of space. It can be detected by placing a positive test charge at any point in space. If a force is detected acting on the charge, then there is an electric field at that location. By placing our test charge at many points, we can represent, using vectors, the properties of an electric field in any region, as in Figure 9.6. Each vector represents the field at the point where it is drawn, that is, the force that would act on a unit positive charge placed there.

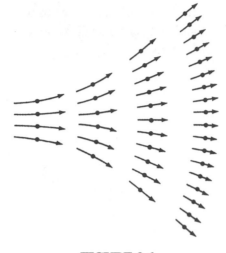

FIGURE 9.6

Figure 9.7 shows a simpler way to represent the same field. Like the vectors, the *field lines* represent the direction and magnitude of the field at any point. The direction is simply the direction in which the field line points at any position. The magnitude is the line density; the lines are closest together where the field is strongest. Thus, any positive charge placed in the field will be pushed in the direction of the field line that passes through that location. A negative charge will be pushed in the opposite direction. At any given point in space, the force the field exerts on a charge is proportional to the charge.

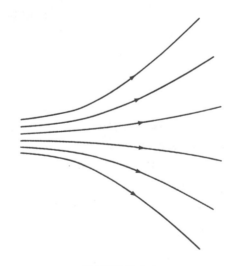

FIGURE 9.7

An electric field can be strong or weak. The *electric field strength ($\mathscr{E}$) is* defined as the electric force per unit of charge in the field:

$$\mathscr{E} = \frac{F_{el}}{q}$$

(Equation 9.2)

and the unit of electric field strength is the newton per coulomb. The field represented in Figure 9.7 has a definite strength at each point, greater where the field lines are closer together. If, for example, a charge of 30 nanocoulombs placed at some point experiences a force of 6×10^{-4} newton, the field strength at that point will be

$$\mathscr{E} = \frac{6 \times 10^{-4}\ \text{N}}{30 \times 10^{-9}\ \text{C}} = 2 \times 10^4\ \text{N/C}$$

EXAMPLE 9.4

What is the charge on a 0.20-gram pith ball if a downward field of 4.5×10^{-5} newtons per coulomb supports it in space?

SOLUTION

The force exerted by the field must be just enough to support the weight of the ball, which is

$$w = mg = (2.0 \times 10^{-4}\ \text{kg})\,(9.8\ \text{m/s}^2) = 1.96 \times 10^{-3}\ \text{N}$$

From Equation 9.5, the charge on the ball must be

$$q = \frac{F_{el}}{\mathscr{E}} = \frac{1.96 \times 10^{-3}\ \text{N}}{4.5 \times 10^5\ \text{N/C}} = 4.4 \times 10^{-9}\ \text{C}$$

Since the downward field exerts an upward force, the charge must be negative; it is –4.4 nanocoulombs.

CORE CONCEPT

An electric field may be mapped by means of field lines, indicating the direction and magnitude of the force exerted by the field on a positive test charge at any point.

TRY THIS
—5—

A charge of +5 nanocoulombs placed at a certain location in an electric field experiences a force of 2×10^{-5} newtons pushing it to the west. What are the direction and magnitude of the force that the field at the position will exert on a charge of –15 nanocoulombs?

The Field Rules

Any charged object is surrounded by a region of space in which a small test charge will experience a force. The charged object creates a field in the space around itself.

We can visualize the properties of these fields by using field lines. They must obey certain rules:

1. Field lines point away from positive charges and toward negative charges, since positive objects will repel our positive test charge. In fact, every field line starts on a positive charge and ends somewhere in a negative charge.

2. The more charge on an object, the more field lines attached to it. This can be made quantitative by drawing a number of field lines proportional to the charge.

    If the object is a conductor, there are further restrictions because the charges on a conductor will distribute themselves until all are in equilibrium. In this condition the following additional rules apply:

3. There is no field inside the object.

4. All field lines are perpendicular to the surface of the object. If any line were not, it would have a component parallel to the surface, and the charge in the surface would be set in motion.

5. The field lines are closest together at the surface where there are points and edges, since this is where the charge is most concentrated.

Using these rules, we can plot some fields. Figure 9.8 shows the field in the neighborhood of an irregular object with a positive charge, of two oppositely charged spheres, and of two negatively charged spheres. All are conductors, so they must obey all five rules.

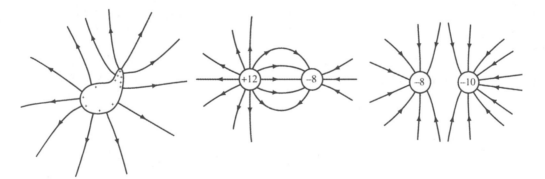

FIGURE 9.8

One field that is of special interest is the field between two oppositely charged flat plates set close together and parallel to each other. Since they are flat, the charge on them is evenly distributed (except near the edges), and so are the field lines. The lines are perpendicular to both plates—they run straight and parallel from one plate to the other, as shown in Figure 9.9. This is a *uniform field.* A charge placed anywhere between the plates will experience the same force, as long as it is not near the edge.

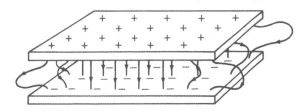

FIGURE 9.9

CORE CONCEPT

Charged objects are surrounded by electric fields, which obey certain definite rules.

TRY THIS
—6—

Using the known rules that control the formation of electric field lines, explain why the lines around a negatively charged sphere are radially inward and uniformly spaced.

Electrostatic Induction

A small metal needle, pivoted in the center like a compass needle, can be used to indicate the direction of an electric field. It works because an electric field will separate the charges in it.

How this works is illustrated in Figure 9.10. The field pushes positive charge in one direction and negative charge in the other. The positive charges cannot move, but the negative charges (electrons) move up the field, making one end negative and leaving the other end with an excess of positive charge. This effect, the separation of charge by the action of an electric field, is called *electrostatic induction.* Once the charges have been separated, the field exerts forces on the charges in opposite directions. As long as the forces are not lined up, they will result in a torque that tends to pull the needle into line with the field. The needle will oscillate around this position and eventually come to rest there.

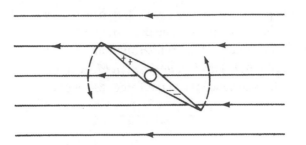

FIGURE 9.10

Electrostatic induction accounts for the commonly observed attraction of a neutral object to any charged object. The little ball in Figure 9.11 is in a *nonuniform* electric field. The field induces a charge separation in the ball. As shown, the positive end is in a weaker field than the negative end, so the force pushing the object up the field is stronger than the force on the positive end pushing it the other way. The object will move toward the stronger field. That is why you can pick up bits of paper with your comb.

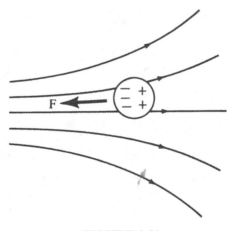

FIGURE 9.11

If you place two conductors in a field, in contact with each other, they will acquire opposite charges by induction. You can separate them and move them out of the field, and you will have a ball with a positive charge and another with a negative charge, as in Figure 9.12.

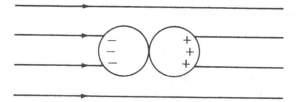

FIGURE 9.12

If the field is strong enough, electrostatic induction will produce a spark. The field separates the positive nucleus of air atoms from their negative electrons and sends them shooting through the air in opposite directions. After this happens once, the particles accelerate in the field until they hit another atom—causing it to split into positive and negative parts which repeat the process. In a very short time, air atoms in the trillions of trillions are giving up their electrons, and their tremendous speed agitates the surrounding air atoms to incandescence. Spark!

CORE CONCEPT

An electric field will separate the positive and negative charges of a neutral object.

TRY THIS
—7—

Explain why a neutral object placed between a pair of parallel charged plates may rotate but will not experience any net force.

Electricity Has Energy

To separate negative and positive charges from each other, work must be done against the force of attraction. Therefore separated charges are in a high-energy state. When the charges are brought together again, energy must be released. It may be in the form of a spark. When you plug in a lamp and turn it on, you are using this electric potential energy, converting it into heat and light.

A charge in an electric field has energy, just as a mass in a gravitational field has gravitational potential energy. Consider a small test charge in the electric field around an irregular charged object, as in Figure 9.13. The field is pushing it in the direction of the field lines. If some external force pushes it the other way, against the direction of the field, work is being done on it. Therefore, the positive charge has its greatest electric potential energy at the upper end of the field. Since the field pushes a negative charge the other way, its potential energy is greatest at the low end of the field. If released, positive charges fall down the field, and negative charges fall up it, losing potential energy as they go.

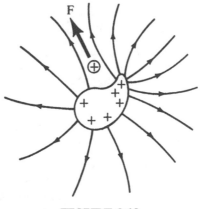

FIGURE 9.13

It takes no work to move a charge perpendicular to the field lines. This is like carrying a mass horizontally in a gravitational field; the force is perpendicular to the displacement. Equation 5.1 tells us that, when this is done, there is no change in the energy of the object, since the cosine of the angle between force and displacement is zero.

This gives us another method of plotting the geometry of an electric field. Figure 9.14 represents the same field as Figure 9.13, but a series of *equipotential* lines have been drawn on it. Each of the lines is drawn so that it is perpendicular to the field lines at all points. When a charge moves along an equipotential, its displacement is always perpendicular to the field lines, and thus perpendicular to the force on it. Therefore, it takes no work to move a charge along an equipotential. An equipotential is a line along which a charge always has the same electric potential energy.

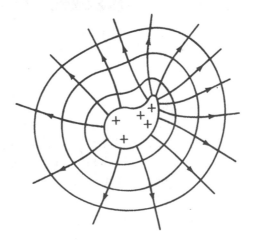

FIGURE 9.14

Note that the surface of any charged conductor is always equipotential, since the field lines must be perpendicular to the surface. Between the parallel plates of Figure 9.9, where the field is uniform, the equipotentials form a series of planes parallel to the plates and equally spaced between them. In a uniform field, this is the only shape that equipotentials can have so as to be everywhere perpendicular to the field lines.

EXAMPLE 9.5

A pair of charged plates like those of Figure 9.9 are 3.0 centimeters apart, and the upper plate is negative. The field in between the plates exerts a force of 2.0×10^{-5} newtons on a charged object between the plates, pushing it downward. (a) How much work would it take to move the object from the lower plate to the upper plate? (b) How much work would it take to move the object 6 centimeters horizontally? (c) Is the charge positive or negative?

SOLUTION

(a) From Equation 5.1, $W = F \,\Delta s \cos \theta$; here the force is in the direction of motion, so $\cos \theta = 1$. Therefore,

$$W = (2.0 \times 10^{-5} \text{ N})(0.030 \text{ m}) = 6.0 \times 10^{-7} \text{ J}$$

(b) $\theta = 90°$ and $\cos 90° = 0$, so no work is done.

(c) The upper (negative) plate is repelling the charge, so the charge must also be negative.

CORE CONCEPT

Equipotentials are surfaces of equal potential energy for any given charge and are always perpendicular to the field lines.

TRY THIS
—8—

If the field of Figure 9.9 exerts a force of 6.5×10^{-5} newtons on a charge, and the plates are 5.0 centimeters apart, what is the change in the electric potential energy of the charge as it moves from the left end of the lower plate to the right end of the upper plate?

Potential

The gravitational potential energy of an object depends on its mass and its location in a gravitational field. Similarly, its electric potential energy depends on its charge and its location in an electric field.

The crucial question about the location of an object in an electric field is this: On what equipotential is it located? It takes a definite amount of energy to move each unit of charge from one place to another. We define the *potential difference* between two points in the field as the work that would have to be done to move a unit charge from one of the points to the other:

$$\Delta V = \frac{\Delta E_{el}}{q}$$ **(Equation 9.3)**

where ΔV is the potential difference, ΔE_{el} is the difference in potential energy, and q is the charge. The SI unit of potential difference is the joule per coulomb, which is called a *volt* (V).

EXAMPLE 9.6

How much work does it take to move a charge of +30 nanocoulombs through a potential difference of 6.0 volts?

SOLUTION

From Equation 9.3:

$$\Delta E = q\Delta V = \left(30 \times 10^{-9}C\right)\left(6.0\frac{J}{C}\right) = 1.8 \times 10^{-7} \text{ J}$$

EXAMPLE 9.7

A charge of –4.5 nanocoulombs is moved in an electric field from a position where the potential is +2.0 volts to another where the potential is –15 volts. (a) Is the charge gaining or losing energy? (b) Must work be done on it, or does it release energy? (c) What is its change in energy?

SOLUTION

(a) A negative charge moving to a lower potential is gaining energy.
(b) Since it is gaining energy, work must be done on it.
(c) $\Delta E = q \, \Delta V = (-4.5 \times 10^{-9} \text{ C})(-17 \text{ V}) = +7.7 \times 10^{-8} \text{ J}.$

EXAMPLE 9.8

How much is the energy change of a charge of –20 nanocoulombs that moves in an electric field from an equipotential of +3 volts to an equipotential of +12 volts?

SOLUTION

From Equation 9.3:

$$\Delta E_{el} = q \, \Delta V$$

In going from a potential of +3 volts to one of +12 volts, the charge is going through a potential difference $\Delta V = +9$ volts. Therefore its energy change is

$$\Delta E_{el} = \left(-20 \times 10^{-9}C\right)\left(+9\frac{J}{C}\right)$$

$$\Delta E_{el} = -1.8 \times 10^{-7} \text{ J}$$

In moving to a higher potential, a *negative* charge *loses* energy.

The volt is a convenient unit for a number of reasons. For one thing, it can be measured directly by means of an instrument called a *voltmeter*. If you connect the terminals of a voltmeter to the terminals of an ordinary zinc–carbon dry cell battery, the voltmeter will read 1.5 volts. This means that the chemical activity inside the battery separates charge in such a way that it adds 1.5 joules of energy to each coulomb of charge it separates. The size of the battery does not matter; a big one will be able to provide more energy and do it more quickly, but the amount of energy per unit charge, the potential difference, is the same for a C cell as for an AA cell. Batteries that give a larger potential difference are made of combinations of cells, unless they use a different combination of chemicals to produce the electric energy.

EXAMPLE 9.9

How much energy must a battery add to a charge of +35 coulombs in a circuit to move it from a point at a potential of 60 volts to another point where the potential is 83 volts?

SOLUTION

The potential difference is 83 V – 60 V = 23 V. Then the increase in energy of the charge is $\Delta E = q\,\Delta V = (35\ \text{C})(23\ \text{V}) = 805\ \text{J}$.

EXAMPLE 9.10

To toast a couple of slices of bread, a toaster has to use 30 000 joules of energy, drawn from a 110-volt wall outlet. How much charge flows through the toaster?

SOLUTION

From Equation 9.3:

$$q = \frac{\Delta E_{el}}{\Delta V}$$

$$q = \frac{30\ 000\ \text{J}}{110\ \text{J/C}} = 270\ \text{C}$$

CORE CONCEPT

Potential difference, in volts, is the difference in electric energy per unit charge:

$$\Delta V = \frac{\Delta E_{el}}{q}$$

TRY THIS
—9—

If it takes 850 joules of energy in the starter motor, supplied by a 12-volt battery, to start your engine, how much charge flows through the starter motor?

The Charge on an Electron

With a voltmeter and a battery supplying an adjustable potential difference, there is a method of measuring extremely small charges. This experiment, when first performed, had a profound effect on the whole theory of physics.

Very tiny charges can be measured by determining the force exerted on them by a known electric field. The strength of the uniform field shown in Figure 9.9 can be determined simply and directly. Imagine that there is a charge q somewhere in the space between the plates. Then, according to Equation 9.5, the force acting on the charge is

$$F_{el} = \mathscr{E}q$$

Let Δs represent the distance between the plates. Then, if the charge is moved from one plate to the other, the change in its electric energy (the work done on it) is the electric force times the distance:

$$\Delta E_{el} = \mathscr{E}q\Delta s$$

Therefore, the electric field between the plates is

$$\mathscr{E} = \frac{\Delta E_{el}}{q\Delta s}$$

Equation 9.3 tells us that $\Delta E_{el}/q$ is the potential difference between the plates, so

$$\mathscr{E} = \frac{\Delta V}{\Delta s} \qquad \textbf{(Equation 9.4)}$$

Now we can use a ruler to determine Δs and a voltmeter to measure ΔV, and the electric field is found. Note how the labels work out. A volt is a joule per coulomb and a joule is a newton-meter. Therefore

$$\frac{V}{m} = \frac{J}{C \cdot m} = \frac{N \cdot m}{C \cdot m} = \frac{N}{C}$$

so a newton per coulomb is the same as a volt per meter. It is common practice, in fact, to express electric field strengths in volts per meter.

EXAMPLE 9.11

How much is the electric force on a pith ball with a charge of 22 nanocoulombs if the ball is between a pair of parallel plates 2.0 centimeters apart and the plates are connected to opposite terminals of a 60-volt battery?

SOLUTION

From Equation 9.4, the electric field strength between the plates is

$$\mathscr{E} = \frac{\Delta V}{\Delta s} = \frac{60 \text{ V}}{0.020 \text{ m}} = 3.0 \times 10^3 \text{ V/m}$$

Then from Equation 9.2

$$F_{el} = \mathscr{E}q = (3.0 \times 10^3 \text{ N/C})(22 \times 10^{-9} \text{ C}) = 6.6 \times 10^{-5} \text{ N}$$

This principle can be used to measure force if the charge is known, or charge if the force can be determined. An apparatus that can be used to measure extremely tiny charges is illustrated in Figure 9.15. Two flat plates, a few centimeters apart, are set in a horizontal position and attached to opposite terminals of an adjustable electric energy source. A voltmeter measures the potential difference between them, and a switch allows this field to be turned on and off. An atomizer produces tiny droplets of oil, which fall through a tiny hole in the upper plate into the uniform electric field between the plates. As the droplets fall, they are watched through a calibrated microscope. With a stopwatch, the experimenter can determine how long it takes a droplet to fall through a distance measured by the microscope.

FIGURE 9.15

To do the experiment, turn off the electric field and squeeze the atomizer bulb. When you see a falling droplet, time its passage between the marks in the microscope to determine the speed with which it is falling. Then turn on the electric field and adjust the potential difference until the droplet stands still. Read the voltmeter, and you will have enough information to calculate the charge on the droplet.

Here's how: First of all, when the electric field was turned off, the droplet was falling at its terminal velocity. Knowing the velocity, it is possible to calculate the viscous drag produced by its movement through the air. At terminal velocity, this is equal to the weight of the droplet. So, with the field turned off, its velocity tells us how much the droplet weighs.

Now turn the field on and adjust the potential difference between the plates until the droplet stands still. Now the electric force exerted on the droplet by the field is equal to the weight of the droplet, so we know how much the electric force is. We read the voltmeter and determine how much potential difference between the plates it takes to produce this much electric force.

EXAMPLE 9.12 What is the electric charge on an oil droplet whose weight is 2.9×10^{-15} newton if it is held stationary by the electric field between two horizontal plates separated by 5.0 centimeters when the potential difference between the plates is 90 volts?

SOLUTION

The electric force must be equal to the weight. The force times the distance between the plates must be equal to the work that would have to be done to move the charge from one plate to another, which is (from Equation 9.3) equal to $q\,\Delta V$. Therefore

$$(2.9 \times 10^{-15}\ \text{N})(0.050\ \text{m}) = q\left(90\,\frac{\text{J}}{\text{C}}\right)$$

from which $q = 1.6 \times 10^{-18}$ C.

The most remarkable outcome of this experiment was the discovery that every value of charge measured by this method turned out to be a definite integral multiple of the same small charge: 1.60×10^{-19} coulomb. This implied that charges can be added or subtracted from the droplet only in quanta, or steps that are all the same size. This electric charge quantum is the charge on an electron.

Since this experiment was first performed, over 100 subatomic particles have been found, and every one of them is either neutral or has exactly one quantum of charge, positive or negative. Electric charge is not infinitely subdivisible, and all charges everywhere consist of some whole number of quanta.

CORE CONCEPT

Electric charges exist only in whole-number multiples of the electron charge, 1.60×10^{-19} coulomb.

TRY THIS
—10—

In an oil drop experiment, a droplet is found to have a charge of $+8.0 \times 10^{-19}$ coulomb. How many electrons has it lost?

Charged Spheres

We started this chapter by studying the behavior of charged conductive spheres, the graphite-coated Ping-Pong balls. Let's take a closer look.

Consider first the field around a point charge q. At each point in that field is a unique value of electric potential. To find a formula for that electric potential, imagine that we place a unit positive test charge at a great distance from the point. The distance is so great that our test charge is unaffected by the charge q, so it is convenient to designate the potential at infinity as zero. What happens if we bring the test charge closer to q? If q is positive, work must be done; the amount of work done increases as the test charge gets closer. At any position, the work done so far is the electric potential energy of the test charge. If you know a little calculus, you can show that the work done is

$$\Delta E_{el} = \frac{kq_1q_2}{r}$$

(Note the similarity with Equation 5.4)

Dividing by the test charge, according to Equation 9.3, gives the potential at a distance r from a charge q:

$$V = k\frac{q}{r} \qquad \text{(Equation 9.5)}$$

Figure 9.16a shows the electric field around a point charge; the field lines emerge radially from the charge and are uniformly spaced. An equipotential at any distance from the charge is a spherical surface. Figure 9.16b shows the field around a charged conducting sphere. The field lines are radially outward (always perpendicular to the surface). Because the charges are distributed uniformly, the field lines are uniformly spaced. The two fields are identical. The potential on a sphere of radius r that has a charge q is identical to the potential at a distance r from a point charge q at the center of the sphere.

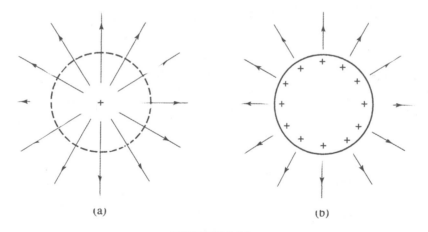

(a)   (b)

FIGURE 9.16

EXAMPLE 9.13   What is the potential at the surface of a spherical Van de Graaff generator terminal with a radius of 12 centimeters if the total charge on the terminal is 60 nanocoulombs?

SOLUTION
From Equation 9.5,

$$V = k\frac{q}{r} = \left(9.0 \times 10^9 \text{ N·m}^2 / \text{C}^2\right)\left(\frac{60 \times 10^{-9} \text{ C}}{0.12 \text{ m}}\right) = 4\,500 \text{ N · m/C} = 4\,500 \text{ V}$$

Suppose we take a hollow steel ball with a radius of 5 centimeters and touch it to that Van de Graaff terminal. How much charge will it acquire? Two constraints will enable us to find out how the charge is distributed between the two spheres: (1) the total charge must still be 60 nanocoulombs; and (2) when the two spheres are in contact, the combination must be equipotential, so $V_1 = V_2$.

EXAMPLE 9.14   If a conducting sphere 5 centimeters in radius is touched to the Van de Graaff terminal of Example 9.13, how much charge will each sphere have when they are separated?

SOLUTION

They must have the same potential, so

$$V = k\frac{q_1}{r_1} = k\frac{q_2}{r_2}$$

$$\frac{q_1}{12 \text{ cm}} = \frac{q_2}{5 \text{ cm}}$$

The spheres share the charge in proportion to their radii. The total charge cannot change, so

$$q_1 + q_2 = 60 \text{ nC}$$

Solving these two equations simultaneously gives $q_1 = 42$ nC and $q_2 = 18$ nC.

To find the potential at any distance from a spherical conductor, just use Equation 9.5, using r as the distance from the center. In fact, even if the charged object is not spherical, the equation can be used provided that r is large, as can be seen in the field around the object in Figure 9.17. If you get far enough from anything, it looks like a point.

CORE CONCEPT

Conductors in contact must have the same potential; potential around a spherical conductor is equal to kq/r.

TRY THIS
—11—

What is the surface charge density (charge per unit area) on a metal sphere 30 centimeters in diameter if the potential at the surface is 2 500 volts?

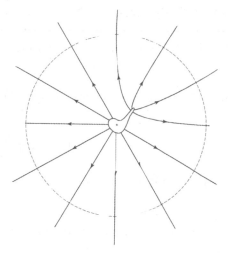

FIGURE 9.17

REVIEW EXERCISES FOR CHAPTER 9

Fill-In's. For each of the following, fill in the missing word or phrase.

1. Two similar electric charges _____ each other.

2. Comparing electric and gravitational forces, it will be found that the _____ is stronger.

3. The charge on a rubber rod is designated as_____.

4. When you comb your hair, the comb becomes negative because _____ have been transferred to it from your hair.

5. Metals are said to be good _____ because electric charge moves freely through them.

6. When a conductor is charged, all the charge is on its _____.

7. To confine electric charge and prevent it from escaping, it can be surrounded by a(n) _____.

8. On a charged conductor, the charge is most concentrated where the _____ is greatest.

9. The SI unit of electric charge is the_____ .

10. The electric force between two charges is directly proportional to their _____.

11. Doubling the distance between two charges changes the force between them by a factor of _____.

12. Electric field lines are _____ where the field is strongest.

13. An electric field will exert a force on a charge in the direction opposite the field if the charge is _____.

14. In a graphical representation of the electric field around a charged object, the number of field lines is proportional to the _____ on the object.

15. Electric field lines start on _____ charges.

16. On a charged conductor, the field lines are _____ to the surface.

17. The field between a pair of oppositely charged plates, located close together, is _____.

18. Electrostatic induction is the process by which a(n) _____ separates positive and negative charges from each other.

19. A charge moving through an electric field will have no change in its electric potential energy if it moves along a(n) _____.

20. Equipotentials are always perpendicular to _____.

21. Electric potential difference is measured in_____.

22. A 6-volt battery adds 6 joules of energy to each _____.

23. All electric charges are integral multiples of the charge on a(n) _____.

24. Charge passes through a wire in the form of _____.

25. The shape of the equipotential around a point charge is _____.

26. The spherical terminal of a Van de Graaff machine has a diameter of 20 centimeters and is charged to 20 000 volts. The potential 40 centimeters from the terminal is _____ volts.

Multiple Choice.

1. An important difference between the electrical and the gravitational force is that the gravitational force is always
   (1) downward
   (2) stronger
   (3) attractive
   (4) positive
   (5) removable.

2. Stroking a rubber balloon with a woolen mitten will put a charge on the balloon. If the mitten is then brought near the balloon,
   (1) the balloon will expand
   (2) no change will be noticed
   (3) the balloon will rise
   (4) the balloon will move away from the mitten
   (5) the balloon will move toward the mitten.

3. A glass rod stroked with silk becomes positively charged because
   (1) the silk removes electrons from the rod
   (2) the rod removes electrons from the silk
   (3) the friction creates a positive charge
   (4) the silk gains protons
   (5) the silk loses protons.

4. If an electric charge is placed on the sole of a flat iron,
   (1) potential will be greatest near the point
   (2) charge will be most concentrated near the point
   (3) charge and potential will both be greatest near the point
   (4) charge will concentrate near the point, but the potential will be smallest there
   (5) charge will be uniformly distributed, but the potential will be greatest near the point.

5. In charging the terminal of a Van de Graaff generator, the charge is always added to the interior of the terminal because
   (1) charge concentrates at the interior
   (2) there is never any charge on the interior surface
   (3) charge naturally flows more easily to an interior surface
   (4) there is no charge on the outer surface
   (5) charge distributes over both surfaces from the interior.

6. If an uncharged object is placed between a pair of charged parallel plates, the object will
   (1) move toward the positive plate
   (2) move toward the negative plate
   (3) move toward whichever plate has the stronger charge
   (4) not move toward either plate
   (5) move toward whichever plate has the weaker charge.

7. The energy of a charged particle in an electric field does not change if the particle moves
    (1) perpendicular to the field
    (2) in the direction of the field
    (3) opposite to the field
    (4) toward the positive charge
    (5) toward the negative charge.

8. Two charged pith balls are placed in contact, one above the other, in a uniform electric field directed upward. If they are separated,
    (1) neither ball will be charged
    (2) the upper ball will be negative and the lower ball will be positive
    (3) the lower ball will be positive and the lower ball will be neutral
    (4) the upper ball will be positive and the lower ball will be negative
    (5) both balls will be positive.

9. A volt is equal to a
    (1) newton-meter per coulomb
    (2) coulomb per newton-meter
    (3) newton per coulomb
    (4) coulomb per joule
    (5) coulomb per newton.

10. If you know the negative charge on an object, it is possible to calculate its
    (1) electric potential
    (2) electric potential energy
    (3) number of excess electrons
    (4) total number of electrons
    (5) number of excess protons.

Problems.

1. Two identical conductive spheres are charged at +32 nanocoulombs and –12 nanocoulombs, respectively. If they are touched together and then separated, how much charge will each have?

2. Two conductive spheres are separated by 15 centimeters, measured center to center. Sphere A is charged to +25 nanocoulombs and sphere B to +15 nanocoulombs. (a) How much force does A exert on B? (b) How much force does B exert on A?

3. At a certain location in space, the electric field exerts a force of 3.0×10^{-3} newton on a Ping-Pong ball bearing a charge of 12 nanocoulombs. What is the charge on an object in that position if the electric force on it is 2.5×10^{-5} newton?

4. A 1-centimeter plastic sphere and a 5-centimeter plastic sphere, both neutral, are placed in an electric field and then separated from each other. If the small sphere acquires a charge of –12 nanocoulombs, what is the charge on the large one?

5. How much work must be done to move a charge of +220 nanocoulombs from a place where the electric potential is +30 volts to another position where the potential is +5 volts?

6. How much energy is delivered to a motor if 6.0 coulombs passes through it from a 24-volt battery?

7. How much energy is there in the spark produced when a Van de Graaff generator terminal charged to 240 nanocoulombs discharges to a terminal at a potential 110 000 volts lower?

8. Two large, flat metal plates are set parallel to each other, separated by 12 centimeters, and connected to opposite terminals of a 240-volt power supply. A plastic sphere carrying a charge of +2.0 nanocoulombs is placed at the positive plate and released. Find (a) the energy lost by the sphere in falling to the other plate; and (b) the force exerted on the sphere by the electric field between the plates.

9. How many excess electrons are there on a Ping-Pong ball that has a charge on it of –8.0 nanocoulombs?

10. How much energy is lost by an electron when it is part of the discharge spark of a Van de Graaff generator that had been charged to a potential of 120 000 volts above the ground potential?

11. A wire 2.5 meters long is connected to opposite terminals of a 12-volt battery. What is the electric field strength in the wire?

12. What field strength is needed to support the weight of a 3.5-gram Ping-Pong ball if the ball has an electric charge of 8.0 nanocoulombs?

13. A pair of parallel plates are set in a vertical position and connected to opposite terminals of a 12-volt battery in a vacuum. A 0.25-gram pith ball with a charge of +22 nanocoulombs is released at the positive plate. How fast is it going when it reaches the negative plate? (Hint: The electric energy it loses becomes kinetic energy.)

14. How much electric force is exerted on a charge of 15 nanocoulombs if it is between parallel plates 4.0 centimeters apart that are connected to the opposite terminals of a 120-volt source?

15. How many excess electrons are there on a pith ball bearing a charge of –0.5 nanocoulombs?

16. What potential difference will be needed between a pair of parallel plates 10 centimeters apart if they are to exert a force of 4×10^{-17} newtons on an electron?

17. To charge a 12-volt battery, the charger must move 4.5×10^5 coulombs of charge from the positive plate to the negative plate. Find (a) the amount of energy stored; and (b) the number of electrons transferred.

18. Through what potential difference would an electron have to accelerate to reach a speed of 1.5×10^8 meters per second? (The mass of an electron is 9.1×10^{-31} kilograms.)

19. A Van de Graaff generator terminal with a radius of 20 centimeters is charged to +1 200 volts. What is its surface charge density?

20. How much work must be done in bringing a pith ball with a charge of +6.0 nanocoulombs up to the surface of the Van de Graaff generator of Problem 19?

21. If the pith ball of Problem 19 is 2.0 centimeters in diameter, how much charge will it have after it touches the Van de Graaff generator terminal?

22. How much is the electric potential at a point 2.0 centimeters from a proton?

23. A pith ball with a charge of +6.0 nanocoulombs is 12 centimeters from another with a charge of −3.5 nanocoulombs. What is the electric potential halfway between them?

The Power in the Wire

WHAT YOU WILL LEARN

The nature of electric current and how it delivers energy; electric circuits in use.

SECTIONS IN THIS CHAPTER

- Making Electricity Useful
- Producing Electric Current
- How Much Current?
- How Much Resistance?
- Delivering the Energy
- Superconductivity
- Big Batteries, Little Batteries
- Circuits
- Circuits with Branches
- Potential in the Branches
- The Series Circuit
- The Parallel Circuit
- Series-Parallel Combinations

Making Electricity Useful

During the past century, electricity has changed our lives. Learning how to use it has endowed us with a crucial faculty that we have never had before: the ability to transport large amounts of energy over long distances quickly and efficiently.

Electric charge has always existed. We make it useful by doing work on it to separate the negative charges from the positive charges, thus using the charge to store energy. When the charges come back together and neutralize each other, the energy must be released. The trick we have learned is how to use this released energy in a variety of ways, often at a great distance from the point at which the original separation of charge took place.

Although we can separate charges by stroking a rubber rod with fur, this is not a practical way to store large amounts of energy in the separated charges. What we need is a means of separating large amounts of charge and of doing it *continuously.* We need a gadget that will keep adding energy to electric charge while that energy is being used somewhere else.

There are two kinds of devices in common use that can do this: the battery and the generator. In a battery, a chemical process removes electrons from one terminal and deposits them on another, thus adding energy to the charge. A generator does the same thing by taking advantage of magnetic phenomena; you will see how this works in Chapter 11.

Any source of electric energy must get the energy from somewhere. The battery gets energy from the chemical reaction. The generator gets energy from the work done in turning it. Any such source can be rated in terms of the electric energy it produces for each unit of charge separated. This quantity—the energy per unit charge converted from some other form to electrical—is called the emf (pronounced ee-em-eff) of the source. The letters emf stand for "electromotive force," but this is a very poor name, left over from the days when electricity was not well understood. The emf of a source is not a force; it is a potential difference, a value of energy per unit of charge. Like all potential differences, the value of the emf of a battery or a generator is defined by Equation 9.3.

The common zinc–carbon dry cell, or the newer alkaline cell, produces an emf of 1.5 volts regardless of its size (Figure 10.1). A large cell produces more energy by separating more charges, but the energy added to each coulomb separated depends only on the kind of chemical reaction that does the job. If your radio uses a 9-volt battery, it is made of six dry cells, each adding 1.5 joules of electric energy to each coulomb that passes through it.

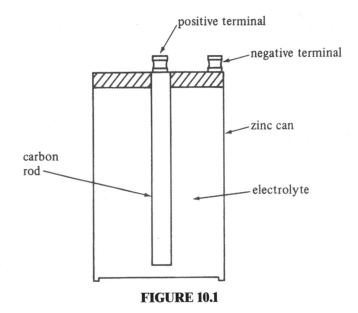

FIGURE 10.1

CORE CONCEPT

The emf of a source, measured in volts, is the energy per unit charge converted from some other form to electrical:

$$\text{emf} = \frac{\Delta E_{el}}{q}$$

TRY THIS
—1—

How many dry cells are there inside the 22.5-volt battery used to run a TV camera?

Producing Electric Current

A battery, with a concentration of electrons at one terminal and an excess positive charge at the other, is a storehouse of electric energy. If a metal wire is connected from one terminal to the other, electrons will run off the negative terminal into the wire, and out of the wire into the positive terminal. All the electrons in the wire will shift position, drifting toward the higher potential. As they travel, they lose electric energy, which is converted into random motion of molecules and electrons inside the wire. The wire gets hot.

With an instrument called an *ammeter* we can measure the rate at which the charge is flowing out of the battery, through the wire, and back into the battery. Figure 10.2 shows how this meter can be connected to measure the rate of flow of charge, and the diagram, using standard symbols, is the way the circuit is represented schematically. The ammeter indicates the rate at which the charge is flowing through it in coulombs per second, or *amperes* (A). This quantity is called *current* and is represented by the symbol I:

$$I = \frac{\Delta q}{\Delta t} \qquad \text{(Equation 10.1)}$$

See Example 10.1 for the use of this equation.

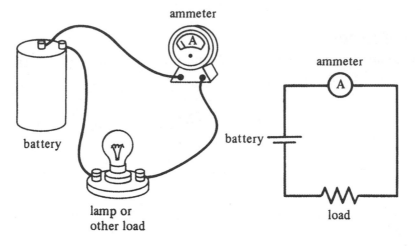

FIGURE 10.2

EXAMPLE 10.1   If a light bulb uses 0.50 amperes of current, how much charge flows through it in 5 minutes?

SOLUTION

Five minutes is 300 seconds, so from Equation 10.1,

$$\Delta q = I\,\Delta t = (0.50\ \text{A})(300\ \text{s}) = 150\ \text{C}$$

Compared with static charges, the current, even that in a flashlight bulb, is enormous. Charge on a Van de Graaff generator is measured in billionths of a coulomb, and we see a spectacular display of fireworks when it discharges. Yet the ammeter in Figure 10.2, registering the current when a small dry cell is used to operate a flashlight bulb, might well be a couple of amperes. But all this charge does not accumulate anywhere. The current flows out of the bulb as fast as it flows in—and the same can be said of the wires, the ammeter, and the battery. The ammeter tells us the rate at which the charge is flowing through every part of the circuit.

A current may consist of either kind of charge in motion. In a wire, it is the negative electrons that are moving; in a cyclotron it might be positive protons; in a plating solution or a battery, both kinds of charge move, in opposite directions. We define the direction of a current as the direction of flow of *positive* charge. Electrons flowing northward produce many of the same effects as positive charges flowing southward; in both cases the ammeter tells us that the direction of the current is southward. In the circuit of Figure 10.2, the current flows out of the positive terminal of the battery, through the ammeter, through the bulb, and then back into the negative terminal of the battery and through the battery to the positive terminal.

CORE CONCEPT

Current is the rate of flow of charge, measured in amperes (coulombs per second):

$$I = \frac{\Delta q}{\Delta t}$$

TRY THIS
—2—

An automobile battery can supply 50 amperes of current for 6 minutes before running down. How much charge does it separate?

How Much Current?

Charge flowing out of the positive terminal of a battery is at a high energy level; when it enters the negative terminal, its energy is low. The function of the battery is to boost the energy level of the charge.

Somewhere in the circuit, the energy of the flowing charge is converted into other forms. When a bulb is lit, the electric energy becomes heat and light. If you want to measure this conversion of energy, you will need a *voltmeter*.

Figure 10.3 shows a voltmeter in place for measuring the energy lost by the current as it passes through the bulb. A voltmeter, unlike an ammeter, is designed in such a way that very little current goes through it. Generally, we will be able to treat the voltmeter as though it is not there at all, ignoring any current it uses. The terminals of a voltmeter are connected to two different points in the circuit, and it registers the difference of potential between those two points.

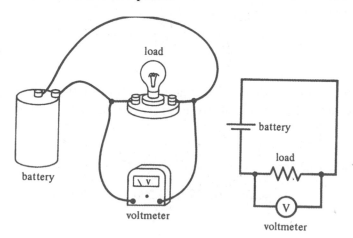

FIGURE 10.3

When you connect a wire to a battery, the amount of current in the wire depends on two things: the nature of the wire and the potential difference provided by the battery. A copper wire has a lot more current in it than an iron wire of the same dimensions; a linen thread has no current at all. There are good conductors and bad conductors.

Also, for a given wire, the current will be greater if the potential difference between its ends is greater. A wire, or any other electrical device, has a property called *resistance,* which tells how much current it will have for any given value of potential difference between its ends:

$$R = \frac{\Delta V}{I}$$ **(Equation 10.2)**

where I is current in a given conductor, R is resistance, and ΔV is potential difference between the ends. See Examples 10.2 through 10.5. Note that the unit of resistance, the volt per ampere, is called an *ohm* and is represented by Ω, the Greek letter omega.

 EXAMPLE 10.2 When a wire is connected to a 9-volt battery, the current is 0.020 ampere. What is the resistance of the wire?

SOLUTION

From Equation 10.2,

$$R = \frac{\Delta V}{I} = \frac{9 \text{ V}}{0.020 \text{ A}} = 450 \ \Omega$$

 EXAMPLE 10.3 How much is the current in a light bulb whose resistance is 350 ohms when the bulb is connected to a 110-volt outlet?

SOLUTION

From Equation 10.2,

$$I = \frac{\Delta V}{R} = \frac{110 \text{ V}}{350 \ \Omega} = 0.31 \text{ A}$$

EXAMPLE 10.4 In an electric circuit, a current of 0.025 ampere is flowing through a 2 200-ohm resistance. What is the potential difference between the ends of the object?

SOLUTION

From Equation 10.2,

$$\Delta V = IR = (0.025 \text{ A})(2 \ 200 \ \Omega) = 55 \text{ V}$$

EXAMPLE 10.5

A relay with a resistance of 12 ohms is in an electric circuit with one side at a potential of 85 volts and the other at 71 volts. How much current is in the relay?

SOLUTION

The potential difference across the relay is 85 V − 71 V = 14 V. From Equation 10.2,

$$I = \frac{\Delta V}{R} = \frac{14 \text{ V}}{12 \text{ } \Omega} = 1.2 \text{ A}$$

For a metal conductor, such as a wire, the resistance is constant at any given temperature. For other kinds of devices, such as transistors, solutions, plasmas, and vacuum tubes, the relationship between current and potential is much more complex, and resistance is not a constant.

A *resistor* is a conductor that is made especially to have a particular resistance. These devices are used in electronic circuits to control currents and potentials. Each resistor is calibrated according to its resistance; many are marked in kilohms (= 10^3 ohms) or megohms (= 10^6 ohms). Some are made of a clay–graphite mixture much like the "lead" in a pencil; others consist of a wire wrapped around a ceramic core. A variable resistor has a resistance that can be altered by turning a shaft. Many of the controls you turn when you use a radio or a television set are variable resistors.

EXAMPLE 10.6

In a circuit, a variable resistor is to be connected across a potential difference of 60 volts to allow the current to be adjusted in the range from 40 milliamperes to 100 milliamperes (mA = milliamperes, or 10^{-3} amperes). Over what range must the resistance vary?

SOLUTION

1 mA = 10^{-3} A; since $R = \Delta V/I$, the required range is

$$\frac{60 \text{ V}}{40 \times 10^{-3} \text{ A}} \quad \text{to} \quad \frac{60 \text{ V}}{100 \times 10^{-3} \text{ A}}$$

$$1\,500 \text{ } \Omega \quad \text{to} \quad 600 \text{ } \Omega$$

CORE CONCEPT

Resistance, measured in ohms, is the ratio of potential difference to current, and it is constant for metals and some other materials at constant temperature:

$$R = \frac{\Delta V}{I}$$

TRY THIS
—3—

A current of 0.15 amperes flows through a 20-ohm resistor away from a point in a circuit where the potential is 45 volts. What is the potential at the other end of the resistor?

How Much Resistance?

Metals are good conductors of electricity because they contain free electrons; that is, electrons that are not firmly attached to atoms of the metal. A potential difference between the ends of a wire establishes an electric field in the wire, exerting a force on the free electrons. The free electrons, moving up the field (they are negative, remember) is the current in the wire.

The current in the wire is proportional to the strength of the field; from Equation 9.4, it follows that

$$I \propto \frac{\Delta V}{l}$$

where l is the length of the wire. Since resistance is defined as $\Delta V / I$ (Equation 10.3), it follows that resistance is proportional to the length of the wire.

The amount of current also depends on the number of free electrons available to carry it; the thicker the wire, the more free electrons are available. Resistance is inversely proportional to the cross-sectional area of the wire.

The number of available free electrons also depends on the kind of metal in the wire. This, and other properties of the metal, enter into the property called the *resistivity* of the metal. The resistance of a wire depends on the physical dimensions of the wire and the resistivity ρ (rho) of the metal:

$$R = \frac{\rho l}{A} \qquad \text{(Equation 10.3)}$$

EXAMPLE 10.7 The resistivity of copper at room temperature is 1.72×10^{-6} ohm-centimeter. What is the resistance of a copper wire 35 meters long and 0.025 centimeter in cross section?

SOLUTION

From Equation 10.3,

$$R = \frac{\rho l}{A} = \frac{\left(1.72 \times 10^{-6}\,\Omega \cdot \text{cm}\right)\left(3\,500\ \text{cm}\right)}{0.025\ \text{cm}^2}$$

$$R = 0.24\ \Omega$$

**EXAMPLE
10.8**

We would like to make a 600-ohm resistor out of Nichrome wire with a cross-sectional area of 3.0×10^{-3} square centimeter. If the resistivity of Nichrome is 110×10^{-6} ohm-centimeters, how long should the wire be?

SOLUTION

From Equation 10.3,

$$l = \frac{RA}{\rho} = \frac{(600\ \Omega)(3.0 \times 10^{-3}\ \text{cm}^2)}{110 \times 10^{-6}\ \Omega\cdot\text{cm}} = 1.6 \times 10^4\ \text{cm, or 16 m}$$

For most metals, resistivity (and thus also resistance) increases with temperature. Typically, resistivity increases by a few parts per thousand for each degree above 20 degrees Celsius.

CORE CONCEPT

Resistance varies directly with the length of a wire and inversely with its cross-setional area; it increases with temperature.

TRY THIS
—4—

To test the resistivity of a metal, a wire with a cross-sectional area of 0.025 cm² is cut to a length of 10 m. A potential difference of 30 volts across the ends of the wire results in the flow of 16 amperes of current. What is the resistivity of the wire?

Delivering the Energy

A battery or a generator converts other forms of energy into electrical energy by separating positive and negative charges. When opposite termials of the battery or generator are connected through a complete conducting path, charge flows down the field, giving up the electrical energy. This energy must be converted into some other form of energy as it drops to the lower potential.

Potential difference between two points is the energy difference per unit charge; if the voltmeter in the circuit reads 1.5 volts, it means that every coulomb that passes through the bulb loses 1.5 joules of energy. Therefore, the total energy converted in the bulb, from Equation 9.3, is

$$\Delta E_{el} = q\ \Delta V$$

**EXAMPLE
10.9**

If a light bulb operates on 110 volts, how much energy does it convert from electrical to other forms when 150 coulombs of charge passes through it?

SOLUTION

From Equation 9.3,

$$\Delta E = q\,\Delta V = (150\text{ C})(110\text{ V}) = 16\,500\text{ J}$$

The ammeter and a clock can tell us the amount of charge that passes through the bulb, since (Equation 10.1) states

$$q = I\,\Delta t$$

Consequently, the total amount of energy converted to heat and light in the bulb is

$$\Delta E_{el} = I\,\Delta t\,\Delta V$$

If you examine a flashlight bulb carefully, you might find that it bears a label reading, perhaps, "1.5 V, 4 W," which is read as "1.5 volts, 4 watts." We previously met the watt as a unit of power, equal to one joule per second, and we used power as a measure of the rate of doing work. Now we must use the word in its more general sense— to mean the rate of any kind of energy conversion. The label on the bulb indicates that, if the bulb is connected to a 1.5-volt battery, it will convert 4 joules of electric energy into other forms every second.

Since power is the rate of energy conversion, it is equal to $\Delta E_{el}/\Delta t$. Therefore, the equation above becomes

$$P = I\,\Delta V \qquad\qquad \textbf{(Equation 10.4a)}$$

 EXAMPLE 10.10   An electric iron is rated at 900 watts on a 120-volt circuit. How much current does the iron draw from the outlet?

SOLUTION

From Equation 10.4a,

$$I = \frac{P}{\Delta V} = \frac{900\text{ W}}{120\text{ V}} = 7.5\text{ A}$$

Another expression, sometimes more useful, for the power converted by a resistance is found by substituting Equation 10.3 into Equation 10.4a:

$$P = I^2R \qquad\qquad \textbf{(Equation 10.4b)}$$

This equation is more than a matter of convenience. There are circumstances, particularly in AC circuits, in which Equation 10.4a gives the wrong answer. Equation 10.4b always works.

EXAMPLE 10.11   What is the most resistance a 950-watt electric iron might have if it blows a fuse on a 10-ampere circuit?

SOLUTION

Let R be the resistance that draws 10 amperes; any greater resistance will reduce the current below 10 amperes, and the fuse will not blow. Then, from Equation 10.4b,

$$R = \frac{P}{I^2} = \frac{950 \text{ W}}{(10 \text{ A})^2} = 9.5\Omega$$

EXAMPLE 10.12 A wall outlet, operating on 120 volts, delivers 4 amperes to an immersion heater for 10 minutes. If the heater is used to heat 2 000 grams of water, how much does the temperature of the water rise?

SOLUTION

The energy delivered is

$$\Delta E = I \, \Delta t \, \Delta V = (4 \text{ A})(600 \text{ s})(120 \text{ V}) = 288\ 000 \text{ J}$$

According to Equation 7.2, this must be equal to $cm \, \Delta T$:

$$288\ 000 \text{ J} = \left(\frac{4.19 \text{ J}}{\text{g}\cdot\text{K}}\right)(2\ 000 \text{ g})(\Delta T)$$

Therefore, $\Delta T = 34$ K.

CORE CONCEPT

Electric power, the rate of conversion of electric energy into other forms, is equal to the product of current and potential difference:

$$P = I \, \Delta V = I^2 R$$

TRY THIS
—5—

How much current flows through a 600-watt toaster in use on a 120-volt circuit?

Superconductivity

Notice that, in stating the resistivity of a conductor, the value at room temperature is usually used. This could give a wildly wrong answer if the wire is at some other temperature. The resistivity of metals increases substantially as the temperature goes up.

In 1911, the momentous discovery was made that, at extremely low temperatures, resistance may disappear altogether. Lead and mercury are not very good conductors, but when their temperature is about 4 kelvins, very near absolute zero, they become *superconductive*. In this state, a current that is introduced into a ring of the wire will continue to flow round and round forever. Since the discovery was made, certain alloys, such as niobium–tin, have been found to be superconductive at temperatures as high as 28 kelvins. Superconductivity has been in use in research laboratories for many years now. It is also used in certain industrial and medical equipment, especially because it makes possible extraordinarily strong electromagnets.

This changed dramatically in 1987, when it was discovered that certain ceramics, which are combined oxides of several metals, become superconductive at a much higher temperature—95 kelvins. This is still extremely cold, but it is above the temperature of liquid nitrogen. Liquid nitrogen is so cheap that it is feasible to simply vent the gas to the outside air and replace the liquid as it boils off. Cheap superconductivity became a definite prospect. Furthermore, this discovery opened the possibility that superconductivity could be achieved at higher temperatures, perhaps even at room temperature.

The superconducting ceramics are fabricated in the form of little pellets. Superconductivity is demonstrated by a peculiar property that is unique to this state: a pellet of superconducting material will float in air above a magnet. No way has yet been found to make use of these new high-temperature superconductors. They are extremely brittle, and have not been successfully fabricated into the form of a wire. Recent tests suggest that, even if such a wire could be made, it would not be able to carry very large currents. The old dream of wires carrying enormous currents across the countryside from generating station to consumer with no loss remains, so far, no more than a dream.

Meanwhile, both the theoretical and the experimental physicists are hard at work on superconductivity. There is as yet no theory that can suggest what new kinds of materials should be tested for high-temperature superconductivity. Cut-and-try methods have produced superconductivity at temperatures up to 220 kelvins. There is some hope that it will be possible to make a superconducting wire that will carry large currents at room temperature, and thus conserve the enormous amounts of energy now lost in heating up transmission lines. If this is ever achieved, it will probably be because the theoretical physicists have learned enough to tell the experimentalists what materials to try.

CORE CONCEPT

At very low temperatures, certain materials lose all electrical resistance and become superconductive. Recent results have raised the temperature at which superconductivity is achieved.

TRY THIS
—6—

What effect would room-temperature superconductivity have on the amount of oil imported?

Big Batteries, Little Batteries

The shutter of my camera operates on a 1.5-volt dry cell about the size of a dime. In the laboratory, I often have occasion to use a 1.5-volt dry cell that stands 15 centimeters high and weighs over a pound. Why can't I use my camera battery in the laboratory?

The emf of a battery is the energy per unit charge converted from chemical to electrical. It is 1.5 volts in both batteries. However, not all of this energy gets out of the battery. Some of it is converted into heat inside the battery itself.

If you connect a good voltmeter across the terminals of a dry cell, the voltmeter will indicate 1.50 volts. There is negligible current, since a good voltmeter has extremely high resistance. Even if the cell is dead, the voltmeter will indicate very nearly 1.50 volts. The voltmeter tells you how much energy would be lost by each coulomb of charge traveling from one terminal to the other. It says nothing about whether any charge actually moves.

If you connect an ammeter between the terminals, the story will be altogether different. Ammeters have extremely low resistance, and an ammeter is a short circuit, just as a short, thick copper wire would be. If the cell is one of those big #6 cells used in the laboratory, the ammeter will indicate about 30 amps if the cell is fresh. If the cell is a tiny dime-sized one, the current might be only a thousandth of that value. If either battery is old and used up, you might not even be able to detect this short-circuit current.

With only the voltmeter attached, there is no current, and the meter registers the emf of the cell. With only the ammeter attached, the only resistance in the circuit is the *internal resistance* of the battery. The current is found, then, from Equation 10.3, using the emf of the cell instead of any measurable potential difference. A small cell has more internal resistance than a large one, and a dead cell has a great deal more internal resistance than a live one. When the cell is short-circuited, by an ammeter or a short, thick wire, all the energy converted by the chemical process goes into heating up the interior of the cell.

EXAMPLE 10.13 A voltmeter connected across a small battery reads 9 volts, and an ammeter reads 16 amps. What is the internal resistance of the battery?

SOLUTION

The emf of the battery is 9 volts, and the internal resistance is the only resistance in the circuit when the ammeter is used. Therefore

$$R = \frac{\text{emf}}{I} = \frac{9 \text{ V}}{16 \text{ A}} = 0.56\Omega$$

The current supplied by a battery depends on both the internal and external resistance of the circuit. This current can be found from the equation

$$I = \frac{\text{emf}}{R_{\text{int}} + R_{\text{ext}}}$$ **(Equation 10.5a)**

If you measure the potential difference across a battery supplying a lot of current, the voltmeter will read *less than* the emf of the battery. This *terminal potential difference* is the energy per unit charge supplied to the external resistance; it does not include the energy used up inside the battery itself, which is IR_{int}. The equation above can be written as

$$IR_{\text{int}} + IR_{\text{ext}} = \text{emf}$$

and, since $IR_{\text{ext}} = \Delta V_{\text{term}}$, this becomes

$$\Delta V_{\text{term}} = \text{emf} - IR_{\text{int}}$$ **(Equation 10.5b)**

See Example 10.14.

EXAMPLE 10.14

The battery of Example 10.13 is connected to a resistance of 1.80 ohms. What will a voltmeter read if it is connected across the battery or the resistor?

SOLUTION

The current must be equal to the emf divided by the total resistance, according to Equation 10.7a:

$$I = \frac{\text{emf}}{R_{\text{int}} + R_{\text{ext}}} = \frac{9 \text{ V}}{0.56 + 1.80}$$

$$I = 3.8\text{A}$$

Then, according to Equation 10.5b,

$$\Delta V = \text{emf} - IR_{\text{int}} = 9 \text{ V} - (3.8 \text{ A})(0.56 \text{ }\Omega)$$
$$\Delta V = 6.9 \text{ V}$$

CORE CONCEPT

The terminal potential difference of a battery is its emf minus IR_{int}, the energy lost per unit charge by the current inside the battery:

$$\Delta V_{\text{term}} = \text{emf} - IR_{\text{int}}$$

TRY THIS
—7—

How much is the current in a 3.0-volt battery with an internal resistance of 0.20 ohms if the battery terminals are connected to an external resistance of 1.0 ohms?

Circuits

Current will flow only in a complete circuit consisting of a source of emf and a complete conducting path from its high-potential terminal to its low-potential terminal. The complete circuit shown in the circuit diagram of Figure 10.4 uses a battery with negligible internal resistance as its source of emf. Starting at the high-potential end (the longer line of the battery symbol), the current passes through a resistance (such as a lamp), a switch, and an ammeter and then back to the battery. The voltmeter is connected between the ends of the resistance wire and registers the potential difference (the *"IR* drop") between the ends.

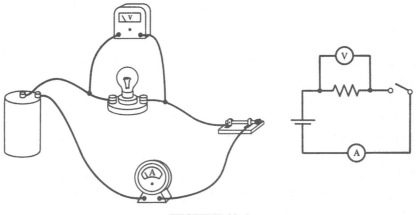

FIGURE 10.4

When the switch is open, electrons flow out of the battery, through the ammeter to the open switch. They can go no further. In a tiny fraction of a second, the potential on that side of the switch reaches the same level as the potential at the negative terminal of the battery, and there is no further flow of electrons. Similarly, the potential on the other side of the switch reaches the level of the potential at the positive terminal of the battery, and there is no current.

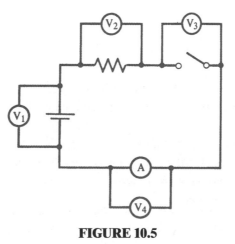

FIGURE 10.5

Now suppose we have four voltmeters in the circuit, as shown in Figure 10.5. V_1 reads the potential difference across the battery; this is nearly the battery emf. V_2 reads the potential difference across the resistance, which is IR; with $I = 0$, there is no potential drop across the resistor. V_3 reads the potential drop across the switch; this value will be the same as V_1, since each side of the switch is at the same potential as the battery terminal to which it is connected. V_4 registers the potential drop across the ammeter; this value will be zero, since there is no current and no resistance in the ammeter.

Now what happens if the switch is closed? The ammeter will indicate the current in all parts of the circuit, which must be equal to the emf of the battery divided by the total resistance of the whole circuit. The potential difference across the battery is its emf; across any other circuit element, the potential difference is IR. Since neither the switch nor the ammeter has any substantial resistance, V_3 and V_4 both read zero. V_1 and V_2 read alike.

EXAMPLE 10.15

In the circuit of Figure 10.5, $V_1 = 9.6$ volts. With the switch closed, what will the other voltmeters read?

SOLUTION

Each voltmeter must read IR, where R is the resistance across which the voltmeter is connected. Since both a switch and an ammeter have negligible resistance, V_3 and V_4 are both zero. Since there is no resistance between the battery and the resistor, the ends of the resistor must be at the same potentials as the battery terminals; so V_2 reads 9.6 volts.

EXAMPLE 10.16

A battery with an emf of 12.0 volts and little internal resistance is connected in a circuit like that of Figure 10.5. If the resistor is rated at 8.0 ohms, find (a) the current; and (b) the readings on the voltmeters when the switch is closed.

SOLUTION

(a) The total resistance of the circuit is 8.0 ohms. Therefore the current is, according to Equation 10.5a,

$$I = \frac{12.0 \text{ V}}{8.0 \text{ }\Omega} = 1.50 \text{ A}$$

(b) in the resistor, the potential difference is $\Delta V = IR = (1.50 \text{ A})(8.0 \text{ }\Omega) = 12 \text{ V}$. The switch and the ammeter have zero resistance, so V_3 and V_4 are both 0.

Each coulomb of charge makes a complete circuit. Each is given, in the battery, an amount of electric energy equal to the emf of the battery. This energy is converted to heat in all the resistances of the circuit: some in the internal resistance of the battery, the rest in the external resistances of the circuit.

CORE CONCEPT

Current flows only in a complete circuit; the energy it gains in the emf of the battery is lost in the resistances of the circuit.

TRY THIS
—8—

In a circuit like that of Figure 10.5, V_1 reads 4.50 volts when the switch is open. Find the resistance if the ammeter reads 0.12 amperes when the switch is closed.

Circuits with Branches

In Figure 10.5, the current goes from the battery through one light bulb and then returns to the battery. Often, the current has more than one possible route on its way out of the battery and back into it again.

Consider, for example, Figure 10.6, consisting of a network of seven resistors, each with an ammeter for measuring the current in it. If you know the readings on three of the resistors, is it possible to figure out what the other currents are?

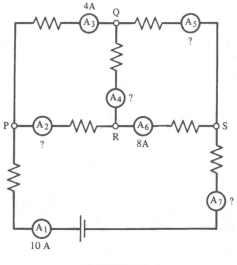

FIGURE 10.6

It is, if you remember a simple rule: Charge cannot accumulate anywhere, and its total quantity is conserved. This means that any current that flows into a point in the circuit must also flow out of it. This rule is called *Kirchhoff's law for currents*.

Ammeter A_1 shows that the current flowing into point P is 10 amperes, so 10 amperes must be leaving point P. And A_3 tells us that 4 amperes flows in the upper branch, to point Q. Therefore, the other 6 amperes must be going through A_2 to point R.

The current A_6, flowing out of point R, is 8 amperes, but only 6 amperes is coming to that point from P. Therefore, there must be 2 amperes coming down through A_4 from point Q to point R. That leaves 2 amperes going through A_5 and joining the 8 amperes from A_6 at point S. This adds up to 10 amperes that flows through A_7 and back into the battery—the same, of course, as the current leaving the other terminal of the battery.

CORE CONCEPT

The sum of the currents entering any circuit point is equal to the sum of the currents leaving it.

TRY THIS
—9—

In Figure 10.7, the ammeter A_1 reads 20 amperes, A_2 is 12 amperes, and A_3 is 9 amperes. Find the current in each of the seven resistors.

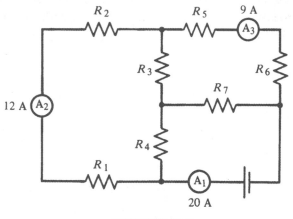

FIGURE 10.7

Potentials in the Branches

Figure 10.8 is another circuit that branches. This time voltmeters have been placed so as to measure the potential difference between the ends of each of the resistors. If we are given the readings on three of the voltmeters, can we figure out the other three?

Yes, by applying one simple rule called *Kirchhoff's law for potentials*. Every coulomb of charge that leaves the high-potential end of the battery (the left end as shown) has to lose 60 joules of energy before it goes back into the other end of the battery. We know this because the voltmeter V_1 shows that the potential is 60 volts higher at the left side of the battery than at the right side; 60 volts = 60 joules per coulomb. Then whatever path the charge takes through the circuit, all the potential differences in that path have to add up to 60 volts.

Consider, for example, the complete path through the three resistors connected across voltmeters V_2, V_4, and V_5. They form a complete path, so their potential differences have to add up to 60 volts. Since V_2 reads 22 volts and V_5 says 18 volts, the potential difference reading on V_4 must be 20 volts.

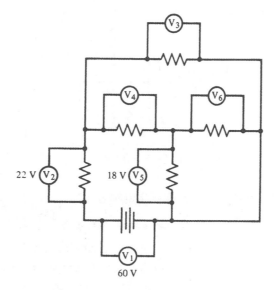

FIGURE 10.8

Now look at another path: the readings on V_2 and V_3 must add to 60 volts, so V_3 reads 28 volts. And going through V_2, V_4, and V_6 tells us that V_6 reads 18 volts.

CORE CONCEPT

Potential differences along any path through a complete circuit must add up to the terminal potential difference of the battery.

TRY THIS
—10—

In the circuit of Figure 10.9, the voltmeters give the following readings:
$\Delta V_B = 50$ volts; $\Delta V_1 = 20$ volts; $\Delta V_2 = 5$ volts; $\Delta V_4 = 4$ volts; $\Delta V_5 = 18$ volts. Find the potential differences across R_3, R_6, and R_7.

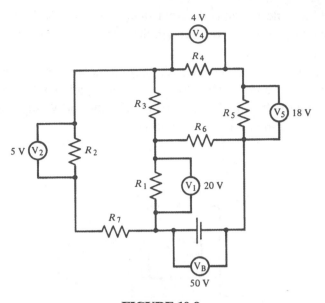

FIGURE 10.9

The Series Circuit

To determine the current in a circuit, you have to divide the emf of the source by the total resistance of the circuit. If the circuit contains a number of devices, the first task is to determine the total resistance of the combination of devices.

In a series circuit, like that of Figure 10.10, this is easy. There is no branching point in the circuit, so the same current must pass through every part of the circuit. An ammeter placed anywhere in the circuit will register this current. As the current passes through each conductor in turn, some of its energy is converted to heat.

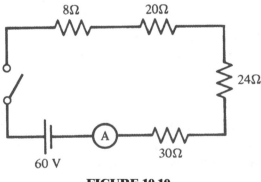

FIGURE 10.10

Since there is only one path through the circuit, the total potential drop in the entire circuit, the terminal potential difference of the battery or generator, must equal the sum of all the potential drops of the separate conductors:

$$\Delta V_{total} = \Delta V_1 + \Delta V_2 + \Delta V_3 + \ldots \qquad \textbf{(Equation 10.6a)}$$

so that

$$IR_S = IR_1 + IR_2 + IR_3 + \ldots$$

where R_S is the total resistance of a number of conductors connected in series. Since all the I's are the same,

$$R_S = R_1 + R_2 + R_3 + \ldots \qquad \textbf{(Equation 10.6b)}$$

EXAMPLE 10.17   In the circuit of Figure 10.10, find (a) the combined resistance; (b) the current; and (c) the potential difference across the 20-ohm resistor.

SOLUTION

(a) The combined resistance in a series circuit is the sum of the separate resistors: 8 Ω + 20 Ω + 24 Ω + 30 Ω = 82 Ω.

(b) The current in the whole circuit is $\Delta V/R$ = (60 V)/(82 Ω) = 0.73 A.

(c) The potential difference across the 20‑Ω resistor is IR = (0.73 A)(20 Ω) = 15 V.

EXAMPLE 10.18   Three resistors are connected in series to a 24-volt battery, and an ammeter in the circuit reads 0.50 amperes. The first resistor is rated at 22 ohms, and the second at 8 ohms. Find (a) the total resistance; (b) the resistance of the third resistor; and (c) the potential difference across the third resistor.

SOLUTION

(a) The total resistance of the circuit is

$$R = \frac{\Delta V}{I} = \frac{24 \text{ V}}{0.50 \text{ A}} = 48 \text{ Ω}$$

(b) In a series circuit, resistances add, so 22 Ω + 8 Ω + R_3 = 48 Ω, and R_3 = 18 Ω.

(c) For any circuit element, $\Delta V = IR$, so V_3 = (0.50 A)(18 Ω) = 9 V.

Which conductor dissipates the largest amount of energy? The rate at which energy is converted in a conductor is equal to I^2R (Equation 10.4b). While this equation is universally true, it is most conveniently applied in a series circuit in which all the I's are the same. The equation tells us that, in a series circuit, power is delivered in the greatest quantity to the largest resistance. This is why poor connections in a circuit, where the resistance is high, tend to overheat.

CORE CONCEPT

In series circuits, total resistance is the sum of all resistances:

$$R_S = R_1 + R_2 + R_3 + \cdots$$

TRY THIS
—11—

Three resistors, of 20 ohms, 40 ohms, and 60 ohms, are connected in series to a 24-volt battery of negligible internal resistance. Find (a) the current; (b) the potential difference across the 20-ohm resistor; and (c) the power dissipation in the 60-ohm resistor.

The Parallel Circuit

What is the resistance of a parallel circuit such as that in Figure 10.11? The chief characteristic of this circuit is that it branches; the current flows out of the battery, and some of it goes into each of the conductors.

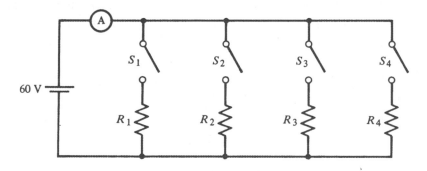

FIGURE 10.11

Unlike a series circuit, in a parallel circuit it is possible to turn on some of the circuit devices and not the others. With all four switches open, there is no current at all, and the emf of the battery will register on a voltmeter connected across any of the switches. If the switch S_1 is closed, there will be a complete circuit through R_1. The current in R_1, and in the ammeter, will be the battery potential difference ΔV divided by R_1.

Now what happens when both S_1 and S_2 are closed? There is now an additional conducting path in the circuit. Since there is a complete path from the battery through R_2 and back to the battery, the entire battery emf will be across R_2. The current in R_2 must then be $\Delta V/R_2$. *Both* currents flow through the ammeter, which must now read the sum of the two currents. Closing S_3 and S_4 provides additional conducting paths, and their currents $\Delta V/R_3$ and $\Delta V/R_4$ add to the total current. Adding additional current paths can increase the current indefinitely, limited only by the properties of the battery.

In a parallel circuit, every circuit element has the same potential difference as the source. Every time a new path is added, the total current goes up; each new current adds to that already coming out of the source. This means that every added circuit element *reduces* the total resistance of the circuit.

There is a rule for combining resistances in parallel. Since the current in each branch is $\Delta V/R$, and the total current is the sum of the branch currents,

$$\frac{\Delta V}{R_P} = \frac{\Delta V}{R_1} + \frac{\Delta V}{R_2} + \frac{\Delta V}{R_3} + \dots$$

where R_P is the combined resistance of conductors in parallel. Since all the ΔV's are the same, this becomes

$$\frac{1}{R_P} = \frac{1}{R_1} + \frac{1}{R_2} + \frac{1}{R_3} + \dots \qquad \text{(Equation 10.7)}$$

 EXAMPLE 10.19 In Figure 10.11, the same four resistors as in Figure 10.10 are connected in parallel. Find (a) the combined resistance; (b) the current in the battery; (c) the current in the 8-ohm resistor; and (d) the power dissipated in the 24-ohm resistor.

SOLUTION

(a) From Equation 10.7,

$$\frac{1}{R_P} = \frac{1}{8\ \Omega} + \frac{1}{20\ \Omega} + \frac{1}{24\ \Omega} + \frac{1}{30\ \Omega}$$
$$\frac{1}{R_P} = \frac{15 + 6 + 5 + 4}{120\ \Omega}$$

Therefore, $R_P = 4.0\ \Omega$.

(b) $I = \dfrac{\Delta V}{R} = \dfrac{60\ \text{V}}{4.0\ \Omega} = 15\ \text{A}$

(c) $I = \dfrac{\Delta V}{R} = \dfrac{60\ \text{V}}{8\ \Omega} = 7.5\ \text{A}$

(d) $I = \dfrac{\Delta V}{R} = \dfrac{60\ \text{V}}{24\ \Omega} = 2.5\ \text{A}$

$$P = I\,\Delta V = (2.5\ \text{A})(60\ \text{V}) = 150\ \text{W}$$

EXAMPLE 10.20

Three resistors are connected in parallel to a 24-volt battery, and the battery current is 3.0 amperes. The first resistor is rated at 20 ohms and the second at 40 ohms. Find (a) the total resistance; (b) the resistance of the third resistor; and (c) the current in the third resistor.

SOLUTION

(a) The total resistance is $R = \Delta V/I = 24\ \text{V}/3.0\ \text{A} = 8.0\ \Omega$.

(b) From Equation 10.7,

$$\frac{1}{8.0\ \Omega} = \frac{1}{20\ \Omega} + \frac{1}{40\ \Omega} + \frac{1}{R_3}$$

which becomes

$$\frac{1}{R_3} = \frac{1}{8.0\ \Omega} - \frac{1}{20\ \Omega} - \frac{1}{40\ \Omega}$$

$$\frac{1}{R_3} = \frac{5 - 2 - 1}{40\ \Omega}$$

$$R_3 = \frac{40\ \Omega}{2} = 20\ \Omega$$

(c) $I = \dfrac{\Delta V}{R} = \dfrac{24\ \text{V}}{20\ \Omega} = 1.2\ \text{A}$

In a household circuit, of course, all the lamps and appliances are connected in parallel. This way, each receives its rated potential difference of 115 volts and you can turn them on and off separately.

In a parallel circuit, the lion's share of the energy is delivered to the *smaller* resistances because they have the larger currents. Remember that the power is $I\,\Delta V$, and that all the ΔV's are the same.

CORE CONCEPT

In a parallel circuit, the inverse of the combined resistance equals the sum of the inverses of the separate resistances:

$$\frac{1}{R_\text{P}} = \frac{1}{R_1} + \frac{1}{R_2} + \frac{1}{R_3} + \dots$$

TRY THIS
—12—

Three resistors, of 20 ohms, 40 ohms, and 60 ohms, are connected in parallel to a 24-volt battery of negligible internal resistance. Find (a) the combined resistance; (b) the current; (c) the power dissipated in the 40-ohm resistance.

Series-Parallel Combinations

The combined resistance of a complicated circuit, like that of Figure 10.12a, can often be computed by treating it as a set of series and parallel circuits. You can repeatedly make combinations according to Equation 10.6a and Equation 10.7 until you arrive at a single value for the resistance of the whole circuit.

In this case, start by combining R_3 and R_4. Since they are in series with each other, their combined resistance is simply the sum of the separate resistances: $R_{3,4} = 60$ ohms. Replacing R_3 and R_4 by their 60-ohm equivalent makes the circuit look like that of Figure 10.12b.

Now we can calculate the combined resistance of the parallel combination of R_2, $R_{3,4}$, and R_5:

$$\frac{1}{R_{2,3,4,5}} = \frac{1}{20\ \Omega} + \frac{1}{60\ \Omega} = \frac{1}{60\ \Omega}$$

which gives $R_{2,3,4,5} = 12$ ohms. Now the equivalent circuit is Figure 10.12c.

$R_{2,3,4,5}$ is in series with R_1 and R_6, so these three can simply be added together to give $R_{total} = 30$ ohms. The current from the battery is therefore 12 volts/30 ohms = 0.40 ampere.

To find the currents in each of the conductors, we first have to find the potentials. The entire current of 0.40 ampere flows through both R_1 and R_6.

The *IR* drop across R_1 is 0.4 ampere × 10 ohms = 4 volts; through R_6 it is 3.2 volts. This leaves 4.8 volts across the parallel combination, which will be found across R_2, $R_{3,4}$, and R_5.

It is now easy to find the current in each of these resistances.

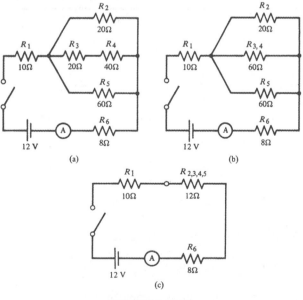

FIGURE 10.12

CORE CONCEPT

Net resistance can be found by making appropriate combinations of series and parallel circuits.

TRY THIS
—13—

Find the net resistance of the circuit of Figure 10.13.

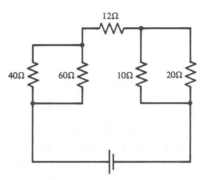

FIGURE 10.13

REVIEW EXERCISES FOR CHAPTER 10

Fill-In's. For each of the following, fill in the missing word or phrase.

1. Electric energy is produced by the _____ of charges.

2. The electric energy produced by a source per unit of charge separated is called the _____ of the source.

3. The unit of emf is the _____.

4. The rate of flow of electric charge is called _____.

5. A coulomb per second is called a(n) _____.

6. Current is measured with an instrument called a(n) _____.

7. Energy converted per unit charge is measured with an instrument called a(n) _____.

8. The rate at which energy is converted from one form to another is measured in _____.

9. The product of the current in a circuit element and the potential difference across it expresses the _____ in that element.

10. The current in any circuit element depends on the potential difference across the element and on its _____.

11. A volt per ampere is called a(n) _____.

12. At very low temperatures, certain materials lose all electrical resistance and become _____.

13. The purpose of a switch is to control the _____ in a circuit.

14. If there is no current, a voltmeter connected across a resistor will register _____.

15. At any point in a circuit, the _____ coming in must equal that which leaves.

16. Between any two points in a circuit, the sum of all _____ is the same through any pathway.

17. Combined resistance is the sum of separate resistances provided that the various conductors are connected in _____.

18. In a series circuit, the largest amount of power is delivered to the _____ resistance.

19. In a parallel circuit, each circuit element has the same _____.

20. Adding new circuit branches in parallel increases the total _____ and _____.

21. In a parallel circuit, the largest amount of power is delivered to the _____ resistance.

22. Copper is a preferred material for making wire because of its low _____.

23. High resistivity is a desirable feature of a wire being used to produce _____.

24. Ten meters of copper wire 0.10 millimeters in diameter will have the same resistance as _____ meters of copper wire 0.40 millimeters in diameter.

25. As an electric lightbulb heats up to operarting temperature, the current in it _____.

26. The terminal voltage of a battery may be lower than its emf because of the battery's _____.

27. A 12-volt battery connected to a 10-ohm load produces 1.0 amperes. The internal resistance of the battery is _____ ohms.

Multiple Choice.

1. The volt is a measure of the ratio of electric
   (1) force to charge
   (2) energy to charge
   (3) energy to force
   (4) force to field
   (5) energy to field.

2. A potential difference of 30 volts keeps 6 amperes flowing through a heating coil for 2 minutes. How much charge goes through the coil?
   (1) 6 coulombs
   (2) 13 coulombs
   (3) 300 coulombs
   (4) 720 coulombs
   (5) 3 600 coulombs.

3. Electric current and electric field are in opposite directions in a
   (1) voltmeter   (2) resistor   (3) solution   (4) battery   (5) switch.

4. A battery can make current flow because it
   (1) adds energy to existing charges
   (2) makes electrons
   (3) creates electric charge
   (4) produces heat
   (5) adds charge to electrons.

5. In an electric circuit, an electron will have its greatest potential energy where
   (1) the potential is zero
   (2) the current is zero
   (3) the potential is lowest
   (4) the current is greatest
   (5) the potential is highest.

6. In analyzing an electric circuit, the current can be considered negligible in a(n)
   (1) resistor   (2) ammeter   (3) battery   (4) switch   (5) voltmeter.

7. A 120-ohm resistor carries 0.5 amperes and has one end at a potential of +35 volts. The potential at the the other end of the resistor may be
   (1) +60 volts
   (2) +25 volts
   (3) −25 volts
   (4) −35 volts
   (5) −60 volts.

8. To produce 300 joules of heat every second on a 120-volt line, a heating element should have a resistance of
   (1) 2.5 ohms   (2) 6.3 ohms   (3) 48 ohms   (4) 190 ohms   (5) 36 000 ohms.

9. Kirchhoff's law for currents holds because
    (1) electric charge is conserved   (4) electric energy is conserved
    (2) current is invariant   (5) total energy is conserved.
    (3) momentum is conserved

10. Adding an additional resistor in parallel to an existing circuit has the effect of
    (1) increasing the net current
    (2) decreasing the net current
    (3) reducing the power consumption
    (4) increasing the overall potential difference
    (5) increasing the total resistance.

Problems.

1. What is the emf of twelve 2.2-volt storage cells connected in series?

2. The battery in Problem 1 delivers 4.0 amperes for 5.0 minutes. Find (a) the amount of energy the battery adds to each coulomb of charge; (b) the total amount of charge that passes out of the battery; (c) the total amount of energy supplied; (d) the power output.

3. In the circuit of Figure 10.14, which meter is correctly placed to measure (a) the current in the battery; (b) the potential difference across R_4; (c) the potential difference across R_3; (d) the current in R_2?

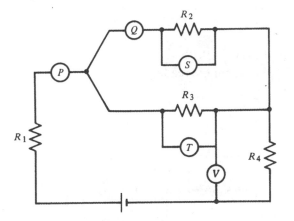

FIGURE 10.14

4. If an immersion heater delivers 11 amperes on a 115-volt circuit, how long will it take to heat 1 cup of water (250 grams) from 20 degrees Celsius to the boiling point?

5. What is the resistance, when hot, of a light bulb labeled "60 W, 115 V"?

6. A kilowatt-hour (kWh) is the energy of 1 000 watts delivered for 1 hour. If your television set draws 4.0 amperes on a 120-volt line, and electric energy costs you 8¢/kWh, how much does it cost to run the set for 8 hours?

7. What is the current in a 1 200-ohm resistor if its ends are at potentials of 180 volts and 30 volts, respectively?

8. What is the potential drop in a 40-ohm resistor if the current in it is 90 milliamperes?

9. If your electric iron uses 900 watts on a 115-volt line, how much current does it draw?

10. In the circuit of Figure 10.15, $A_5 = 8$ amperes, $A_2 = 3$ amperes, and $A_4 = 2$ amperes. What are the readings on the other three ammeters?

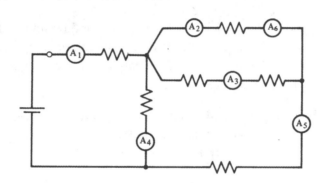

FIGURE 10.15

11. In the circuit of Figure 10.16, $V_2 = 6$ volts, $V_3 = 8$ volts, $V_5 = 3$ volts, and $V_7 = 20$ volts. What are the readings on the other three voltmeters?

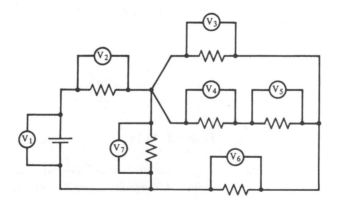

FIGURE 10.16

12. A 12-ohm resistor and an 18-ohm resistor are connected in series to a 24-volt battery. Find (a) the combined resistance; (b) the current; (c) the power dissipation in each of the resistors.

13. A variable resistor is connected in series with a heater whose resistance is 20 ohms, to be operated from a 160-volt power source. The heater is to operate at 850 watts. (a) How much current is needed? (b) What is the required resistance setting of the variable resistor?

14. A 600-watt toaster, a 150-watt lamp, and a 40-watt radio are all operating in parallel on a 120-volt line. Find (a) the current in each device; and (b) the total current.

15. Determine the net resistance of 100 ohms, 250 ohms, and 400 ohms connected in parallel.

16. Find the combined resistance of the circuit shown in Figure 10.17.

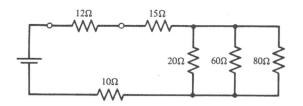

FIGURE 10.17

17. How many 50-watt, 120-volt lightbulbs can be connected in parallel to a circuit that can safely carry up to 15 amperes?

18. A 15-watt fluorescent lamp produces the same amount of light as a 60-watt incandescent lamp. If it burns for an average of 8 hours a day and electricity costs 9¢ per kilowatt-hour, how much money does the fluorescent lamp save in an average month?

19. A 120-volt steam vaporizer converts 500 cubic centimeters of water per hour into steam. If heat loss is negligible, what is the resistance of the heater?

20. At what rate is heat produced in a copper wire 150 meters long and 0.05 square centimeters in cross section if it is connected to the terminals of a 12-volt battery (ρ for copper $= 1.72 \times 10^{-8}$ ohm-meters)?

21. An ammeter indicates a current of 60 milliamperes. How many electrons flow through it every second?

22. A uniform Nichrome wire is stretched across the length of a meter bar and its ends are connected to a 12-volt battery. What will a voltmeter read if its terminals are connected to the wire at the 10 centimeter and 70 centimeter points?

23. A small motor is rated to produce 0.5 kilowatt of mechanical power when it is connected to a 120-volt line. If it draws 6.5 amperes from the line, what fraction of the electric energy input is wasted?

24. Three 1.5-volt dry cells are connected in series to a flashlight bulb. The bulb operates at 6.0 watts and the current in it is 1.5 amperes. What is the internal resistance of each dry cell?

25. A 24-volt storage battery is partially discharged, so its internal resistance is 2.0 ohms. How much current will it deliver to a motor rated at 60 watts, 24 volts?

Magnetism: Interacting Currents

WHAT YOU WILL LEARN

The nature of magnetism; its relationships to electric currents; magnetic fields, and permanent magnets.

The Mysterious Compass

Albert Einstein, surely one of the greatest physicists of all time, said that his fascination with the natural world began when he was a child; his father gave him a magnetic compass. He marveled at that tiny needle, which always seemed to know the orientation of the earth's coordinate system.

We call that little pointer a needle because at first it was in fact a needle. In the days when sailing ships were first setting out on the ocean to find new lands across the seas, every ship's captain carried a highly prized possession: a piece of lodestone. This is nothing but a rock, but a most unusual one. Chinese navigators had discovered that a piece of this rock, suspended from a thread, oscillated back and forth until it came to rest in a north-south orientation. The captain knew that if he took an ordinary steel needle and stroked it on his lodestone, it acquired that mysterious and highly useful property.

With the discovery that artificial magnets could be made of steel, experimentation with this strange phenomenon was sure to follow. In due time, several empirical laws emerged from all this study:

- Every magnet, if suspended and allowed to swing freely, orients itself approximately in a north-south direction. One end of the magnet is designated the **N-**, or north-pointing, pole; the other end is the **S-**, or south-pointing, pole.
- If an N- and an S-pole are placed near each other, they will attract each other and tend to move together. In contrast, two N- or two S-poles will repel each other.
- Only three common elements can be magnetized at room temperature: iron, cobalt, and nickel.
- These materials themselves become magnets if they are placed close to a magnet, a process called magnetic induction. In Figure 11.1a, the iron nail, in contact with a magnet, becomes a magnet capable of picking up a lot of tacks.
- These elements lose their magnetism as soon as they are removed from the influence of the magnet that magnetized them in the first place. Figure 11.1b shows what happens when the nail is removed from the magnet.

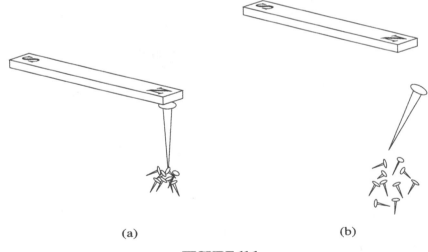

(a)   (b)

FIGURE 11.1

- A few materials, alloys and oxides of the magnetic elements, will retain their magnetism when removed from the source of induction. Alloys that can become permanent magnets include steel and alnico; oxides are various ceramics and lodestone (Fe_3O_4).

CORE CONCEPT
A permanent magnet has north- and south-pointing poles; like poles repel and opposite poles attract; certain materials can be magnetized by contact with a magnet.

What kind of pole is the north magnetic pole of the earth?

Space Magnetism

Every navigator and every Boy Scout knows that there is something special about the space around the earth. A compass placed anywhere will somehow line itself up in a particular direction. One end (the **N,** or north-pointing end) winds up, after some oscillation, settling down and pointing north, more or less. The compass has detected a special property of the space around the earth: a *magnetic field.* We can define the direction of the magnetic field at any location as the direction pointed out by the **N**-end of a compass needle.

The **N**-end of a compass does not always point to geographic north. Place the compass near a current, and you can make it point in any direction at all. Figure 11.2 shows how compasses point in the neighborhood of a bar magnet. The earth is by no means the only source of magnetic fields.

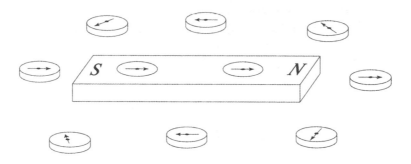

FIGURE 11.2

Those little compasses tell us the direction of the magnetic field at each point in space. Outside the bar, the **N**-pole of the compass points away from the **N**-pole of the bar and toward its **S**-pole. The two compasses inside the bar, however, tell the opposite story. The internal compass lines itself up with the polarity of the bar, its **N**-pole pointing to the **N**-pole of the bar.

The magnetic field has a unique direction and strength at every point in space. Its direction is that of the **N**-pole of the compass points. A compass can also give some idea of the strength of the field. In a strong field, the compass oscillates rapidly while coming to rest; in a weak field, it oscillates slowly.

Just as with electric fields, field lines can be conveniently used to represent the properties of the magnetic field. Figure 11.3 shows the magnetic field in the neighborhood of a bar magnet. The field is strongest where the field lines are closest together; in this case, it is at the poles of the magnet. There is an important difference between electric and magnetic field lines. Electric field lines terminate on charges, but magnetic field lines do not terminate anywhere. They always form continuous loops.

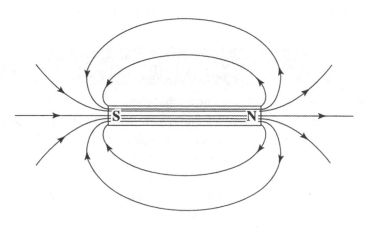

FIGURE 11.3

Figure 11.4 illustrates a fundamental property of a magnetic field. The field exerts a force on an electric current. The copper wire placed in the strong magnetic field between the poles of a horseshoe magnet will be completely unaffected by the field, until the switch is closed. As soon as current flows in the wire, an upward force acts on it, and the wire jumps up. If the current is reversed, the force is downward.

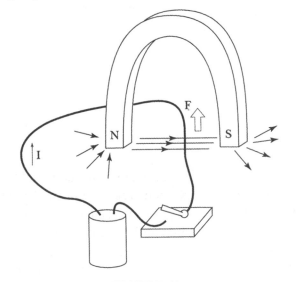

FIGURE 11.4

The mathematical definition of a magnetic field is based on the force the field exerts on a current in it. To measure the strength of a magnetic field, place a known current, in a wire of known length into the field and measure the force on the wire. That force is a measure of the strength of the field. The force depends on the product of current, length of wire, and strength of the magnetic field that produces the force.

However, there is a further complication; the force depends on the orientation of the current in the field. There is no force if the current is parallel to the field, and the force is maximum when the current is perpendicular to the field. The magnetic force is proportional to the *component of the field perpendicular to the current*. This can be summarized with the following equation, which is actually a definition of the magnetic field strength:

$$F_{mag} = IlB\sin\theta \qquad \textbf{(Equation 11.1)}$$

where I is the current, l is the length of the wire in the field, B is the magnetic field strength, and θ is the angle between the current and the field.

The SI unit of magnetic field, which you can discover from Equation 11.1, is the newton per ampere-meter. It is called a *tesla,* abbreviated T. The tesla is a very large unit. The MRI machine in your local hospital, used to explore the interior of the human body, uses a field of about 2 teslas. You have to be very careful around this machine; if you carry a screwdriver into the room when the MRI is running, the screwdriver will be snatched from your hand; it will fly through space and crash into a magnetic pole. The strongest magnetic fields ever created are about 40 teslas. The commonly used unit of magnetic field strength is the millitesla (mT), a thousandth of a tesla. The magnetic field of the earth is about 0.06 milliteslas.

EXAMPLE 11.1

What is the magnetic field that exerts a force of 2.4×10^{-4} newtons on a current of 12 amperes in a wire 30 centimeters long set perpendicular to the field?

SOLUTION

Since the current is perpendicular to the field, Equation 11.1 can be written as

$$B = \frac{F}{Il}$$

Therefore

$$B = \frac{2.4 \times 10^{-4} \text{ N}}{(12 \text{ A})(0.30 \text{ m})} = 6.7 \times 10^{-5} \text{ N/A} \cdot \text{m}$$

Since 1 mT is 10^{-3} N/A·m, the answer can be written as 0.067 mT.

EXAMPLE 11.2

A wire carrying 1.5 amperes has a length of 20 centimeters in a magnetic field of 40 milliteslas. If the wire is perpendicular to the field, how much force does the field exert on the wire?

SOLUTION

With the wire perpendicular to the field,

$$F = IlB = (1.5 \text{ A})(0.20 \text{ m})(0.040 \text{ T}) = 0.012 \text{ N}$$

An electric field exerts a force on a charge; a magnetic field exerts a force on a current. But there is a great difference: whereas the force exerted by an electric field (on a positive charge) is in the direction of the field, the force that a magnetic field exerts on a current *is perpendicular* to the field. The force is also perpendicular to the current. Figure 11.5 shows how you can find the direction of the magnetic force. Place your right hand out flat, with the fingers pointing in the direction of the field. Be sure to use your right hand even if you are left-handed. Point your thumb in the direction of the current. In the figure, the current is shown perpendicular to the field, but it does not have to be; the thumb can take various angles with respect to the fingers. Now push with the palm, upward in this case. That push is the magnetic force acting on the current. The smaller the angle between thumb and fingers, the weaker is the push. When the current is parallel to the field, there is no force at all.

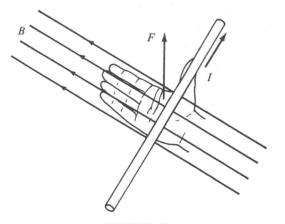

FIGURE 11.5

CORE CONCEPT

The magnetic force on a current, maximum when the current is perpendicular to the field, is perpendicular to both the field and the current and equal to the product of the field, current, and length of the wire:

$$\mathbf{F}_{mag} = IlB\sin\theta$$

TRY THIS
—2—

What are the direction and magnitude of the force that an upward-directed magnetic field of 2.0 milliteslas exerts on a north-flowing current of 10 amperes in a wire 15 centimeters long?

Loop Currents

Consider a rectangular loop of wire carrying a current through a magnetic field, as in Figure 11.6. If you apply the right-hand rule to each of the four sides of the loop, you will find that the force is always directly away from the center. The loop may stretch, but it will neither translate nor rotate.

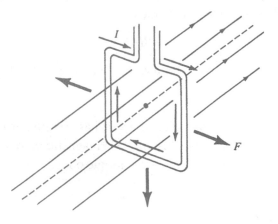

FIGURE 11.6

Now let's see what happens if the loop is oriented with its plane parallel to the field, as in Figure 11.7. The top and bottom wires are now, respectively, antiparallel and parallel to the field, so there is no force on them at all. The vertical wires have equal forces in opposite directions, as before, but the forces are no longer in line. The result is a torque.

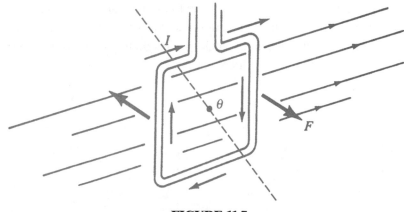

FIGURE 11.7

In Figure 11.7, the axis of the loop, the line drawn through its center, is perpendicular to the field. In this position, the torque is a maximum. When the axis of the loop is parallel to the field, as in Figure 11.6, there is no torque. The torque depends on the sine of the angle that the axis makes with the field.

The force that produces the torque is the magnetic force acting on the vertical wires. The torque due to each force is the force times the radius of rotation, which is half the length of the horizontal wire. Thus the total torque is the force on each vertical wire multiplied by the length of the horizontal wire. Therefore, the torque is $Bl_v Il_h$, where the l's are the lengths of the horizontal and vertical sides of the loop. Since the product of the sides is the area A, of the loop, the torque is BIA.

A general equation that includes the number of turns of wire and the angle that the axis of the loop makes with the field is thus

$$\tau = nBIA \sin \theta \qquad \text{(Equation 11.2)}$$

While this equation was derived for a rectangular loop, it holds for a loop of any shape.

EXAMPLE 11.3

A circular coil with a radius of 6.0 centimeters has 60 turns of wire; it carries a current of 24 milliamperes. It is in a 40-millitesla magnetic field with its axis at an angle of 30° to the field. How much is the torque on it?

SOLUTION

From Equation 11.2

$$\tau = (60)(40 \times 10^{-3}\,\text{T})(24 \times 10^{-3}\,\text{A})\pi(0.06\,\text{m})^2(\sin 30°)$$
$$\tau = 3.3 \times 10^{-4}\,\text{N·m}$$

A *solenoid* is a coil consisting of many loops in series. Figure 11.8 shows a solenoid at rest in its equilibrium position in a magnetic field. As in Figure 11.6, the plane of each loop is perpendicular to the field; the axis of the solenoid points in the direction of the field. If a solenoid, free to rotate, is placed in a field, it will oscillate and come to rest aligned with the field, just like the loop of Figure 11.6. The solenoid behaves like a compass. Can it be that a compass works as it does because it contains loop currents? In any event, we can, by analogy, designate the end of the solenoid that points in the direction of the field as its **N**-end.

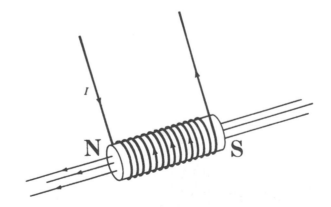

FIGURE 11.8

CORE CONCEPT

The torque on a loop in a magnetic field depends on the field strength, the area of the loop, the current, and the orientation of the loop.

TRY THIS
—3—

The loop of Figure 11.7 is 10 centimeters high and 20 centimeters wide, and it carries a current of 5 amperes in a field of 30 milliteslas. Find the torque on the loop.

How Do You Make a Field?

The earliest hint that there is some sort of relationship between electricity and magnetism was the observation that a current in a wire could deflect a nearby compass needle. It is electric currents that create magnetic fields.

The compasses in Figure 11.9 are exploring the field around the current in a wire. Each compass points in a tangential direction; that is, perpendicular to a line drawn to the wire. A field line joining all these compass headings forms a circle surrounding the wire. The entire field is a series of concentric circles, spaced further and further apart at greater distance from the wire. Figure 11.10 shows how your convenient right hand can tell you the direction of the field around the wire. Point your thumb in the direction of the current; your fingers point in the direction of the field.

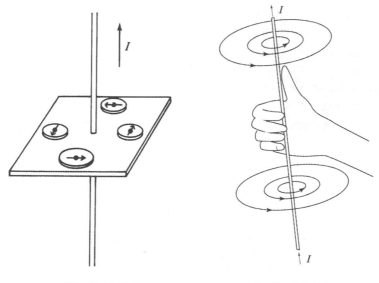

FIGURE 11.9   **FIGURE 11.10**

The current in a wire is commonly used to create strong magnetic fields.

Look at the loop of wire carrying a current as shown in Figure 11.11a. What is the direction of the field inside the loop?

Take out your right hand and apply the rule to the current on the left-hand side of the loop. You will find that the field inside the loop is directed into the page, away from you. If you use this rule, in fact, for the current in any part of the loop, you will get the same answer. Within the loop, all fields point away from you. The fields due to the current in all parts of the loop add up to produce a strong field inside the loop. If you want a still stronger field, you can use a longer wire and coil it into a lot of loops instead of just one, as in Figure 11.11b. In such a solenoid, the field is quite uniform, except near the ends.

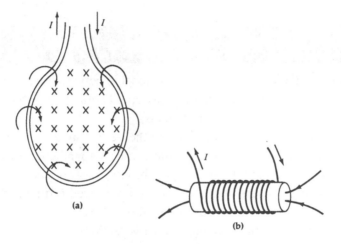

FIGURE 11.11

To find the direction of the field inside a solenoid, wrap your right hand around the solenoid with the fingers pointing in the direction of the current. Then your thumb points in the direction of the field lines inside the solenoid (see Figure 11.12). The field lines run through the center, parallel to the axis, as shown. If such a solenoid is suspended in an external field and is free to turn, it will come to equilibrium with its axis parallel to the external field. The end at which the field lines emerge will point in the direction of the external field, so that end is the **N**-pole of the solenoid. Your thumb points to the **N**-pole.

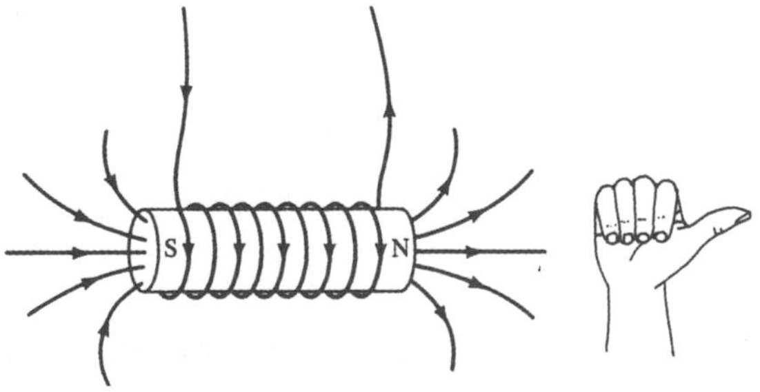

FIGURE 11.12

The strength of the uniform field near the center of a solenoid depends on the current and on how closely the wires are crowded together. This crowding can be expressed as the number of turns of wire per unit length of solenoid. The wire may be wrapped in more than one layer to increase this quantity.

Any electric current makes a field, and that field can exert a force on another current. In Figure 11.13, a current labeled I' is placed near and parallel to the current I. Using the method shown in Figure 11.10, the field lines are drawn around I. Use the right-hand rule of Figure 11.5 to find out what the field does to I'. The field pushes I' toward I'. *Parallel currents attract each other.* The same kind of analysis will show that antiparallel currents repel, and currents at right angles ignore each other.

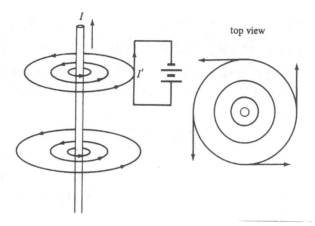

FIGURE 11.13

A couple of solenoids will help to make sense out of the well-known rule that two opposite magnetic poles attract each other. The two solenoids in Figure 11.14 have their windings and currents in the same direction. This produces magnetic fields with the polarities as shown. Now look at the directions of the currents in the two loops facing each other. The currents are parallel. A basic rule for the magnetic force is that parallel currents attract. If the current in the right-hand solenoid went in the opposite direction, the currents in the facing coils would be antiparallel and would repel. In other words, similar poles repel.

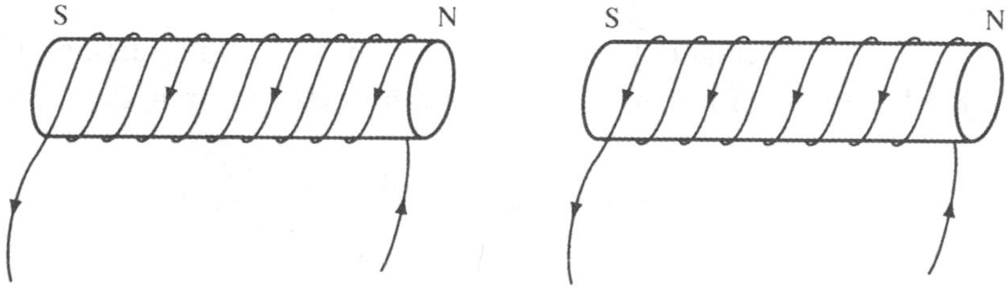

FIGURE 11.14

CORE CONCEPT
Electric currents create magnetic fields, which can be made strong by forming the wire into coils.

TRY THIS
—4—

Suggest three ways in which the field produced by a solenoid can be made stronger.

How Strong is the Field?

The magnetic field strength around the current in a long, straight wire is directly proportional to the current and inversely to the distance from the wire:

$$B = 2k'\frac{I}{r}$$ **(Equation 11.3)**

where k' is the *magnetic constant of free space*, exactly equal to 10^{-7} newton per ampere squared. The 2 is not included in the constant because this formulation turns out to be convenient in other contexts.

EXAMPLE 11.4   A magnetometer placed 1 centimeter from a wire indicates that the magnetic field at that position is 0.30 milliteslas. How much is the current in the wire?

SOLUTION
The field strength is 3.0×10^{-4} N/A·m. From Equation 11.3,

$$I = \frac{Br}{2k'} = \frac{\left(3.0 \times 10^{-4}\ \text{N/A·m}\right)(0.010\ \text{m})}{2 \times 10^{-7}\ \text{N/A}^2} = 15\ \text{A}$$

Now suppose a second wire, carrying current I_2, is placed a distance r from a wire carrying current I_1. The two wires will be attracted to each other. How much is the force on each wire?

The wire I_2 is perpendicular to the field B created by I_1. From Equation 11.1, the force on I_2 is BI_2l. Solving for B and setting the value equal to strength of the field produced by I_1 gives

$$B = \frac{F_{\text{mag}}}{I_2l} = \frac{2k'I_1}{r}$$

so

$$\frac{F_{\text{mag}}}{l} = \frac{2k'I_1I_2}{r}$$ **(Equation 11.4)**

This equation expresses the value of the force per unit length of wire between parallel currents. If the currents are antiparallel, the force is a repulsion; if the currents are perpendicular to each other, there is no force.

 EXAMPLE 11.5 The two wires of an electric cord are separated only by 0.10 centimeter of cotton insulation. If they carry 12 amperes in opposite directions, how much is the force on each centimeter of wire?

SOLUTION

From Equation 11.4,

$$\frac{F}{l} = \frac{2k'I_1I_2}{r} = \frac{(2 \times 10^{-7} \text{ N/A}^2)(12 \text{ A})^2}{1.0 \times 10^{-3} \text{ m}} = 2.8 \times 10^{-2} \text{ N/m}$$

which is 2.8×10^{-4} N/cm.

When a wire is formed into a loop, as in Figure 11.11, the field inside the loop is uniform. Many such loops, wound in series and placed alongside each other, constitute a solenoid. The field inside a solenoid can be quite strong since it consists of the combination of the separate fields of the coils. The field is also uniform, except that it weakens near the ends of the solenoid. The field inside a solenoid depends only on two factors: the current and how tightly the solenoid is wound. The more coils that are squeezed into each length of solenoid, the greater the field inside it. The field strength inside is given by

$$B = 4\pi k'I \frac{n}{l} \qquad \text{(Equation 11.5)}$$

where n/l is the number of turns of wire per unit length of solenoid. This may be increased by winding the coils closely and by adding additional layers of wire.

EXAMPLE 11.6 A solenoid is wound with wire in a single layer and connected to a battery to produce a field inside it that has a magnitude of 4.0 milliteslas. If the solenoid is rewound in two layers, making it half as long, how strong is the field inside it?

SOLUTION

The solenoid now has twice as many turns per unit length with the same current, so the field is twice as great: 8.0 milliteslas.

EXAMPLE 11.7 Two layers of wire, with 40 turns each, are wound onto a coil form 15 centimeters long, and a current of 0.30 ampere is passed through the wire. What is the magnetic field inside the solenoid?

SOLUTION

$$B = 4\pi k'I\frac{n}{l}$$

$$B = 4\pi(10^{-7} \text{ N/A}^2)(0.30 \text{ A})\left(\frac{80}{0.15 \text{ m}}\right)$$

$$B = 2.0 \times 10^{-4} \text{ N/A·m} = 0.20 \text{ mT}$$

The strongest magnetic fields in use, up to 20 teslas, are used in giant particle accelerators, such as the Tevatron in Illinois. The fields are made by solenoids, each carrying thousands of amperes. Such currents are possible only if resistance of the wire can be completely eliminated. The solenoids use superconducting niobium-tin or niobium-titanium alloy wire, kept at superconducting temperatures by baths of liquid helium at 4 kelvins. The creation of such supermagnets has, so far, been the most important application of the phenomenon of superconductivity.

CORE CONCEPT

Strong magnetic fields are produced by circulating currents:

$$B = 4\pi k'I\frac{n}{l}$$

TRY THIS
—5—

A solenoid is tightly wound in four layers of wire with a diameter of 0.20 millimeter. How much current is needed to produce a field of 1.5 milliteslas?

Permanent Magnets

A rod placed inside a solenoid might produce a drastic change in the magnetic field produced by the solenoid. Some materials weaken the field slightly; others make it a little stronger. And there are certain substances that can strengthen the field of a solenoid by a factor of hundreds, or even thousands. These are the *ferromagnetic* materials.

At room temperature, only three common elements are ferromagnetic: iron, nickel, and cobalt. These elements are special because of a peculiar property of their atoms, due to the arrangement of the electrons in them.

Every electron is like a little solenoid, or a compass. It possesses a negative charge and it is spinning. A crude model of such an electron is shown in Figure 11.15; without too much oversimplification of the strange and complex properties of this object, we can think of it as a tiny ball of negative charge, spinning on an axis. Using the left hand (because the charge is negative), we can apply the solenoid rule of the last section to determine that the electron has an **N**-pole and an **S**-pole. Placed in an external magnetic field, each electron will tend to come to equilibrium with its axis lined up with the field.

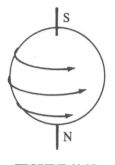

FIGURE 11.15

Electrons in an atom are under another constraint: they generally occur in pairs. The members of the pair are spinning in opposite directions, so that their magnetic fields nullify each other. Ferromagnetic elements are unique in having three or four electrons, in every atom, spinning in the same direction. This turns the whole atom into a tiny magnet, with its **N**-pole and **S**-pole.

In an ordinary piece of iron, the tiny atomic magnets are oriented at random, so that there is no net magnetic effect. When the iron is placed in a magnetic field, many of the atoms rotate and line up with the field. In a strong field, it is possible to line up all the atoms in the iron. Whatever atoms are aligned with the external field add their own magnetic fields to the external field and thus make it much stronger. A really strong electromagnet of this sort can be used to move junked automobiles from one place to another.

Certain materials, such as a lodestone or the steel needle Columbus used, retain their magnetism after the inducting field has been removed. These are called permanent magnets. Steel, an alloy of iron and carbon with other metals sometimes added, has been used for permanent magnets for centuries. The field of a steel bar magnet is formed by the loop currents of its revolving electrons; its magnetic field is just like that produced by the loop currents in a solenoid, as shown in Figure 11.16.

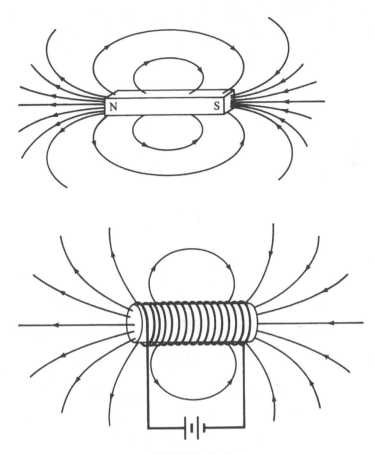

FIGURE 11.16

Better permanent magnet materials than steel have been made. Alnico is an alloy about half iron, with aluminum, nickel, and cobalt. It has long been the material of choice for permanent magnets. Since 1983, alnico has been widely replaced by an iron-boron-neodymium alloy that produces fields 50 times stronger than those of the best steel. Motors, loudspeakers, and telephones have been made smaller and longer lasting by these new rare earth magnets. In applications where a material is needed that does not conduct electricity or where cost is a factor, permanent magnets made of ferrites—oxides of iron and other metals—are in wide use. For strong temporary magnets, which can be turned on and off quickly, iron-nickel alloys are used as solenoid cores.

Just as in a solenoid, the magnetic field lines of a permanent magnet are continuous through the length of the magnet. The poles are points at which concentrations of field lines enter or leave. If a bar magnet is cut, as in Figure 11.17, the field lines emerge at the cut ends, and two new poles are created.

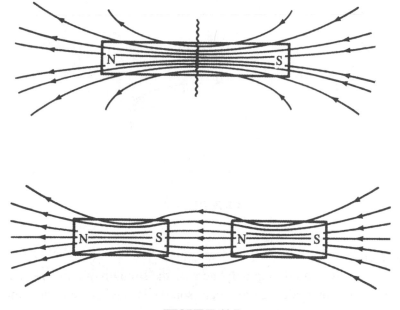

FIGURE 11.7

A bar magnet behaves like a solenoid because, in both, the magnetic properties are produced by circulating currents. This is the reason why a bar magnet has an **N**-pole at one end and an **S**-pole at the other. It also explains why two **N**-poles or two **S**-poles repel each other, whereas an **N**- and an **S**-pole attract. As in a solenoid, attraction results from parallel circulating currents. If the circulating currents are antiparallel, the magnets repel.

If an iron nail is brought near a permanent magnet, it will be attracted because the nail is in the magnetic field of the permanent magnet. This reorients the atoms of the nail, and it becomes a magnet. With its circulating currents oriented parallel to those of the permanent magnet, its pole faces the opposite pole of the permanent magnet.

If a magnet is to be used for lifting, the bar shape is not the best. The magnetic field is strongest at the poles, and is made much stronger by bringing the poles near each other. One way of doing this is to bend the magnet into the shape of a horseshoe. Another way, used in powerful electromagnets, is to make the poles concentric, as in Figure 11.18. One pole is the circular outer rim of the magnet, and the other is the cylinder at the center of the circle. The solenoid that energizes the magnet is in the space between the poles.

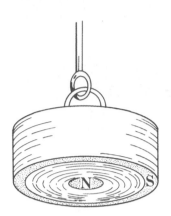

FIGURE 11.18

CORE CONCEPT

Ferromagnetic materials greatly increase the strength of a magnetic field because their individual atoms act as solenoids that can be rotated into alignment with the applied field.

TRY THIS
—6—

Draw the field lines in and near the horseshoe magnet of Figure 11.19.

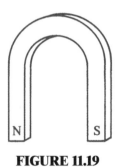

FIGURE 11.19

The Earth Is a Magnet

It is easy enough to plot the magnetic field of the earth. To explain it is quite another matter.

It is a most useful fact that, if a permanent magnet is suspended free to rotate, one end of it points north. This is only approximately true. The compass points to "magnetic north," which is a point now in the Canadian arctic, and moving northwestward. A compass at that position would point straight down; magnetic field lines enter the earth there. Note that the magnetic north pole is the **S**-pole of the earth; it is the point where field lines enter, not leave!

As Figure 11.20 shows, the field lines are not greatly concentrated at the magnetic north pole. The pattern looks like the pattern of field lines produced by a magnet 1 000 miles long, embedded in the earth and tilted at an angle of 15° from its axis. The field is horizontal near the equator only. Elsewhere it dips; that is, it makes an angle with the horizontal. The angle of dip reaches about 75° in the northern United States.

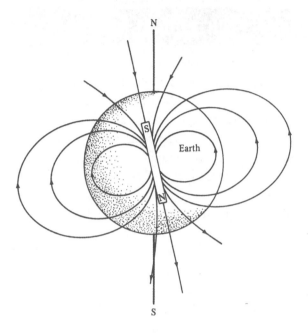

FIGURE 11.20

What produces the earth's field? Theoretical physicists are beginning to get a fairly good picture of where it comes from. The core of the earth is a solid mass of iron and nickel, but it is surrounded by hot liquid metal more than a thousand miles deep. Because the earth is hottest at the center, there are convection currents in this liquid metal. The spin of the earth distorts these currents so that they flow around the earth. A complicated feedback mechanism is set up in which small currents in the liquid metal produce magnetic fields, which amplify the currents, which adds to the fields, and so on. This theory has even begun to make some sense out of the fact that every few hundred thousand years the polarity of the earth's field reverses, taking about 1 000 years to change the direction of the field.

CORE CONCEPT

The earth's magnetic field has a poorly concentrated S-pole near the geographic north pole.

At a place where the magnitude of the earth's magnetic field is 5.3×10^{-5} tesla and the dip is 76°, how much is the horizontal component of the field?

Currents in Space

The picture on the screen of an older picture-tube television set is painted by a beam that consists of a thin stream of electrons, created by an *electron gun* at the base of the picture tube. When the beam strikes the fluorescent coating on the tube face, it produces a tiny spot of light. The coating is a mosaic of spots, each fluorescing in one of the primary colors, red, green, or blue. The beam moves across the screen to create a line of light, then moves down a little, and makes another line below the first. It covers the screen with 550 lines, 30 times every second. The beam creates a picture by turning on and off.

Figure 11.21 illustrates the principle of the electron gun. The filament is heated by a current. When it is red hot, it emits electrons from its surface, a process called *thermionic emission*. The anode is a metal cone, kept at a high positive potential, which attracts the electrons. They shoot through the hole in the anode in a thin beam. The electron beam then passes through a metal screen, the grid. Some of the picture information of the video signal is fed onto the grid. When the grid goes negative, it repels the electrons and shuts the beam off.

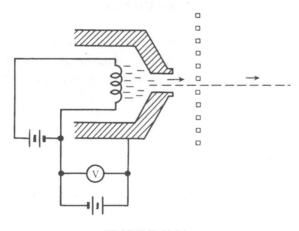

FIGURE 11.21

The electron beam paints a picture on your screen 30 times every second by moving across the screen. Where it hits the screen, the screen glows; the color depends on which of the three kinds of fluorescent spots on the screen the beam strikes. (Three kinds of spots are enough; see Chapter 14.) The grid turns the beam on and off as needed to make the picture and also as the beam returns from the end of one line to the start of the next. After making 550 lines, the beam starts all over at the top.

The beam is aimed by magnetic fields. A pair of *Helmholtz coils*, such as shown in Figure 11.22, consists of two coils of wire, with the distance apart equal to the radius of the coils. Parallel currents in the coils create a uniform magnetic field between them. The coils shown create a horizonal field. Any electron that flows through that field, unless it goes parallel to the field, will be deflected in a direction perpendicular to the field.

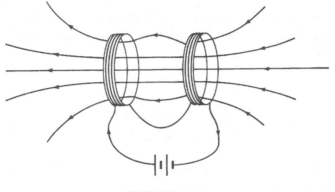

FIGURE 11.22

The neck of a television tube is surrounded by a yoke containing two sets of Helmholtz coils. Coils above and below the neck produce a vertical field that deflects the electrons horizontally. In these horizontal defletion coils, the current reverses 550 times a second to send the beam across the tube face. The other pair of coils sits alongside the neck of the tube. Its horizontal field deflects the electron beam vertically, moving it down a little after each line is made and sending it to the top of the tube after the picture is completed, all in a thirtieth of a second.

Equation 11.1 gives us the force on a current in a magnetic field; we can use it to derive the force on a single particle. Let us confine our attention to particles moving perpendicular to the field; then $\sin \theta = 1$ and the force on the current is IlB (current $\times$ length $\times$ magnetic field strength). But the current, according to Equation 10.1, is $\Delta q / \Delta t$, the charge passing through per unit time. Then the force becomes

$$F_{mag} = Bl \frac{\Delta q}{\Delta t}$$

But the l here is the distance the charge travels; divide this by the time it takes, and we get its velocity; $v = l/\Delta t$. Therefore, the magnetic force on a charged particle moving perpendicular to the field is

$$F_{mag} = Bqv \qquad \qquad \text{(Equation 11.6)}$$

EXAMPLE 11.8

A pith ball with a charge of +30 nanocoulombs is traveling at 12 meters per second through a magnetic field of 50 milliteslas. If the ball is moving perpendicularly to the field, how much is the force on it?

SOLUTION

When the velocity is perpendicular to the field,

$$F = qvB = (30 \times 10^{-9} \text{ C})(12 \text{ m/s})(0.050 \text{ T}) = 1.8 \times 10^{-8} \text{ N}$$

EXAMPLE 11.9

How strong a field is needed to exert a force of 5×10^{-11} newton on a proton traveling perpendicularly through the field at 2.0×10^8 meters per second? (See Appendix 2 for constants.)

SOLUTION

$$B = \frac{F}{qv} = \frac{5 \times 10^{-11} \text{ N}}{(1.60 \times 10^{-19} \text{ C})(2.0 \times 10^8 \text{ m/s})} = 1.6 \text{ T}$$

A moving charge is an electric current, and the direction of the force on the charge follows the same rule that applies to the force on a current: the force is always perpendicular to the field and also perpendicular to the velocity of the particle. To find the direction of the force, use the usual right-hand rule for currents, with the thumb pointing the way the particles are traveling. One important proviso: If the particles are negative, use your left hand.

EXAMPLE 11.10

What is the direction of the force that a vertical magnetic field, directed upward, will exert on an electron traveling eastward in it?

SOLUTION

Point your *left* fingers upward, since you are dealing with a negative charge. Rotate your hand until the thumb points east. Your palm will point northward, and that is the direction of the force.

EXAMPLE 11.11

In the picture tube in an old television set, a beam of electrons comes toward you from the rear of the tube and strikes the face of the tube, where it makes a bright spot. What must be the direction of the magnetic field in the tube to make the spot move from right to left across the screen?

SOLUTION

Use your left hand because the particles are negative. With your thumb pointing toward you and your palm facing to the right, the fingers will point upward, in the direction of the field.

CORE CONCEPT

Charged particles moving through a magnetic field experience a force perpendicular to the field and perpendicular to their velocity:

$$\mathbf{F}_{mag} = (q\,\mathbf{v}) \times \mathbf{B}$$

TRY THIS
—8—

What magnetic field would be needed to exert a force of 2.0×10^{-12} newtons on an electron moving perpendicular to the field at 2.5×10^{8} meters per second? (Remember the charge on an electron? See Appendix 2.)

Measuring Mass

A giant step forward in understanding electricity and charged particles was made when the mass of the electron was measured. This mass is extremely small, and it could be measured only because of the way a moving charge behaves in a magnetic field.

The magnetic force on a charged particle is always perpendicular to its velocity. If, for example, an electron is fired horizontally into a vertical electric field, the path of the electron will be diverted, either left or right depending on whether the field is pointed up or down. The diversion leaves the path still perpendicular to the field. The continued, steady force, always kept perpendicular to the velocity, drives the electron into a circular path. The magnetic force is supplying the electron's centripetal acceleration. The centripetal force is thus Bqv (Equation 11.6). Equation 4.4 tells us how any centripetal force is related to the radius of the circle. Thus

$$\frac{mv^2}{r} = Bqv$$

This equation tells us that the momentum of the electron is

$$mv = Bqr$$

The charge is known; we can send the electron beam into a known magnetic field and measure the radius of the circle in which it travels. The momentum of the electron is measurable.

The device that measured the mass of the electron depended on a simple fact of algebra. If we know the kinetic energy of a particle ($\frac{1}{2}mv^2$) and its momentum (mv), we can solve two simultaneous equations to find both its velocity and its mass. Figure 11.23 shows how this was done. In a vacuum, an electron gun fired a beam of electrons into a magnetic field, created by a pair of Helmholtz coils. If the dimensions of the coils and the current are known, the field between the coils can be calculated. The charge on the electrons is known, and the radius of the circle in which they travel can be measured. This gives Bqr, the momentum of the electrons.

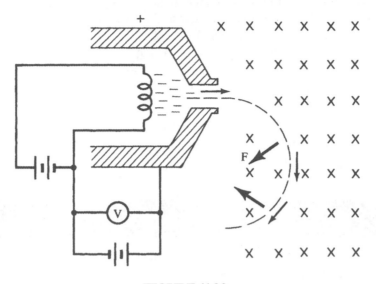

FIGURE 11.23

To find the kinetic energy of the electrons, the voltmeter measures the potential difference between the filament and the anode. Then the electric energy lost by each electron (from Equation 9.3) is $q\,\Delta V$. Since the electrons are traveling in a vacuum, all this energy is converted into kinetic energy. Example 11.12 shows how the mass of the electron is calculated from its momentum and its kinetic energy.

Note that the magnetic field cannot change the energy of the electrons, because the force that the field exerts is perpendicular to the velocity of the electrons. Just as you do no work in carrying something horizontally in a gravitational field, the magnetic force does no work when it exerts a force perpendicular to the velocity. The magnetic force is always exactly centripetal, changing the direction of the velocity, but not its magnitude.

EXAMPLE 11.12

An electron is accelerated through a potential difference of 150 volts and projected into a magnetic field of 25 milliteslas, where it goes into a circular path of radius 7.1 centimeters. What is its (a) kinetic energy; (b) momentum; (c) mass?

SOLUTION

(a) The kinetic energy of the electron is the electric energy it loses in falling through the field:

$$q\,\Delta V = (1.60 \times 10^{-19}\,\text{C})(150\,\text{V}) = 2.4 \times 10^{-17}\,\text{J}$$

(b) Its momentum, as above, is

$$Bqr = (0.025 \text{ Ts})(1.60 \times 10^{-19} \text{ C})(0.071 \text{ m}) = 2.84 \times 10^{-22} \text{ kg·m/s}$$

(c) Since its momentum $p = mv$ and its kinetic energy $E_{kin} = \frac{1}{2}mv^2$, it follows that

$$p^2 = m^2v^2 \qquad \text{(square the momentum equation)}$$

$$\frac{p^2}{2m} = \frac{mv^2}{2} = E_{kin} \qquad \text{(divide both sides by } 2m\text{)}$$

$$m = \frac{p^2}{2E_{kin}} \qquad \text{(solve for } m\text{)}$$

$$m = \frac{\left(2.84 \times 10^{-22} \text{ kg·m/s}\right)^2}{2\left(2.4 \times 10^{-17} \text{ J}\right)} = 1.7 \times 10^{-27} \text{ kg}$$

This method of determining the mass of charged particles, first used for the electron, is now the basis of a device called the *mass spectrograph,* which is used to analyze even complex mixtures of atoms. The material to be analyzed is heated to vaporize and ionize it, and then is accelerated through an electric field and sent into a magnetic field. The particles are allowed to traverse a half-circular path, whose radius is a measure of their mass. The particles are then picked up by a probe that measures the current at each radius. The current at any given radius is a measure of the number of particles having a given mass.

CORE CONCEPT

The circular path in which charged particles travel in a magnetic field can be used to determine their masses.

TRY THIS
—9—

A beam of protons, accelerated through a potential difference of 75 volts, enters a magnetic field of 3.7 milliteslas and travels in a circle of radius 0.30 meter. Find (a) the kinetic energy of the protons; (b) their momentum; (c) their mass. (Remember that the charge on a proton is the same in magnitude as the charge on an electron.)

Magnetism Makes Electricity

Consider the metal rod shown in Figure 11.24, placed in a magnetic field directed into the page. Let's push the rod to the right and see what happens.

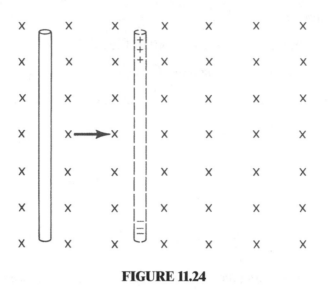

FIGURE 11.24

In the rod there are positive particles (nuclei) and negative particles (electrons). Both are moving to the right, since both are in the rod. If we apply the right-hand rule for positive particles moving through a magnetic field, we will find that the field exerts a force on them, pushing them upward. Use your left hand to find that the force on the electrons is downward. The nuclei cannot move, but the electrons can, since a metal contains free electrons. The result is a separation of charge. As long as the rod keeps moving through the field at right angles to the field lines, there will be a separation of charge in the rod. The bottom becomes negative, the top positive.

As we have seen, separation of charge is a means of storing electric potential energy. A battery does this by chemical action, and an electric field by electrostatic induction. When charge is separated by moving an object through a magnetic field, we call the process *electromagnetic induction.*

The energy stored per unit of charge is the *induced emf* in the rod. We can measure the emf by means of the setup shown in Figure 11.25. The rod moves over a pair of rails connected to a voltmeter. The voltmeter will read the induced emf, which depends on the strength of the field, the length of the rod, and the speed with which it is moving. If the field, the rod, and its velocity are all perpendicular to each other, the induced emf is

$$\text{emf} = Blv \qquad \text{(Equation 11.7)}$$

See Example 11.13.

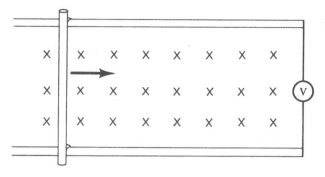

FIGURE 11.25

EXAMPLE
11.13

What emf is induced between the ends of a rod 25 centimeters long that is moving perpendicularly to a magnetic field of 350 milliteslas at 5.0 meters per second?

SOLUTION

$$\text{emf} = Blv = (0.35 \text{ T})(0.25 \text{ m})(5.0 \text{ m/s}) = 0.44 \text{ V}$$

Now suppose we remove the voltmeter and connect a wire between the rails, or a sensitive ammeter. Then the separated charges can reunite by flowing through the rails and the ammeter from one end of the rod to the other. We now have an induced *current*. As always, we get a current only when the source of emf is part of a complete circuit. The amount of current obeys Ohm's law, Equation 10.2.

You have just learned the principle of the electric generator, which supplies all the electricity that runs the country (except for the minute amount we get from batteries).

EXAMPLE
11.14

In a setup like that of Figure 11.26, the rod and rails have a resistance of 0.15 ohms and the voltmeter is replaced by a resistor of 0.50 ohms. The rails are 20 centimeters apart, and the rod is pushed along them at 8.0 meters per second. How strong a field would be needed to produce 0.10 milliampere of current?

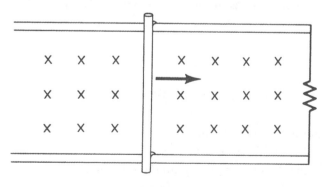

FIGURE 11.26

SOLUTION

From Equation 10.2,

$$\Delta V = IR = (1.0 \times 10^{-4}\,\text{A})(0.15\,\Omega + 0.50\,\Omega) = 6.5 \times 10^{-5}\,\text{V}$$

Then, from Equation 11.7,

$$B = \frac{emf}{lv} = \frac{6.5 \times 10^{-5}\,\text{V}}{(0.20\,\text{m})(8.0\,\text{m/s})} = 4.1 \times 10^{-5}\,\text{T}$$

Moving that rod, in Example 11.14, produced some electric energy and heated up the apparatus. The rate at which this energy was produced is the electric power output:

$$P_{out'} = I^2 R = (1.0 \times 10^{-4}\,\text{A})^2(0.65\,\Omega) = 6.5 \times 10^{-9}\,\text{W}$$

Where did this power come from? There is only one possibility: the work done in pushing the rod.

How much power was put into the rod? When the circuit is complete, there is current in the rod—0.10 millliampere. The magnetic field must be exerting a force on this current, equal to (Equation 11.1)

$$F = IlB = (1.0 \times 10^{-4}\,\text{A})(0.20\,\text{m})(4.1 \times 10^{-5}\,\text{T}) = 8.2 \times 10^{-10}\,\text{N}$$

Whoever is pushing that rod must work against this force, and the work done per unit time, the power input, is

$$P_{in} = \frac{F\Delta s}{\Delta t} = Fv = (8.2 \times 10^{-10}\,\text{N})(8.0\,\text{m/s}) = 6.5 \times 10^{-9}\,\text{W}$$

which is just equal to the power output. Hooray for the conservation of energy!

CORE CONCEPT

When a conductor moves through a magnetic field, an emf is induced across it:

$$emf = Blv$$

TRY THIS
—10—

The 15-centimeter rod of Figure 11.26 is moved to the left at 2.5 meters per second in a field of 50 milliteslas. The complete circuit has a resistance of 10 ohms. Find the direction and magnitude of the induced current.

The Law of Induction

Aside from moving a rod, there are many ways to induce an emf by using a magnetic field. The most general rule that tells us how is called *Faraday's law of induction.*

Take another look at Figure 11.25. We can think of the arrangement as a loop consisting of the moving rod, the rails, and the wires running to the voltmeter. Magnetic field lines thread through the loop. If the field is strong, the lines are close together. The total number of field lines depends on the strength of the field and on the area of the loop. The product of these two quantities, which is represented by the number of field lines passing through the loop, is called the *magnetic flux* in the loop. If the plane of the loop is perpendicular to the field lines, the magnetic flux is given by the equation

$$\phi = BA \qquad \text{(Equation 11.8)}$$

where ϕ (the Greek letter phi) stands for the flux, measured in *webers* (Wb).

Moving the rod across the field changes the amount of flux passing through the loop. It is only while the rod is moving—while the amount of flux passing through the loop is changing—that an emf is induced and the voltmeter gives a reading. Faraday's law of induction tells us that *the emf induced in a loop is equal to the rate of change of the magnetic flux in the loop.*

EXAMPLE 11.15

A rectangular loop of wire consists of 50 turns 10 centimeters wide and 20 centimeters long. It is in a field of 80 milliteslas, with its plane perpendicular to the field. If the loop is yanked out of the field, taking 0.050 second to leave the field, how much emf is induced in it?

SOLUTION

The magnetic flux in the loop is

$$\phi = BA = (0.080 \text{ T})(0.10 \text{ m})(0.20 \text{ m})$$
$$\phi = 0.0016 \text{ weber}$$

This flux reduces to zero in 0.050 s, so the rate of change of flux is (0.0016 weber)/(0.050 s) = 0.032 V in each coil. With 50 coils in series, the total emf is (50)(0.032 V) = 1.6 V.

There are many ways of changing the amount of magnetic flux in a loop. For example, take a solenoid and connect its ends to a galvanometer, which is an instrument that can measure extremely small currents (Figure 11.27). Now insert a bar magnet into the solenoid. A current will register on the galvanometer while the magnet is moving. The current flows one way as the magnet moves in, and the other way as it comes out. When the magnet is at rest, there is no induced emf, and thus no current in the circuit.

In Figure 11.28, an electromagnet has been inserted into the solenoid. As long as the current flows steadily in the electromagnet and nothing is moved, there will be no reading on the galvanometer. With this setup, there are several ways to change the flux inside the solenoid. Any one of them will produce a reading on the galvanometer *while the flux is changing*.

- Pull the electromagnet out, which has the same effect as pulling the bar magnet out.
- Push the electromagnet in. The galvanometer jumps in the other direction.
- Open the switch. The flux collapses, and there is a momentary spurt of current in the galvanometer.

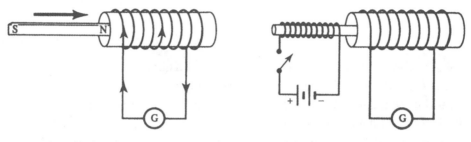

FIGURE 11.27   **FIGURE 11.28**

- Close the switch. While the flux is building, the galvanometer responds.
- Pull the iron core out of the electromagnet. This reduces the flux.
- Push the iron core in.

In all these cases, you get just a momentary surge of current.

Yet another method is possible, the one used in electric generators. Figure 11.29 shows a coil of wire in a magnetic field. In position (a), there is a lot of flux passing through the loop. At (b), there is none; all the field lines go right by the loop without going through it. Rotating the loop thus changes the amount of flux in it and induces an emf. If the loop is rotated continuously, the field in it is constantly changing; every half-cycle, the field reverses the direction in which it passes through the loop. This causes the current in the loop to reverse, and it changes according to the sine-wave pattern of Figure 11.30.

This is alternating current, or AC.

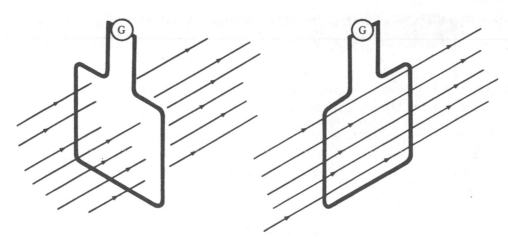

FIGURE 11.29

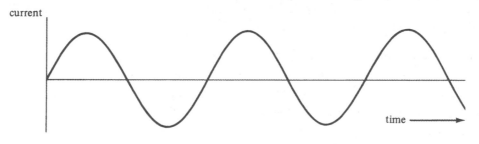

FIGURE 11.30

CORE CONCEPT

Emf is induced electromagnetically whenever the amount of magnetic flux passing through a loop of wire is changing:

$$\phi = BA$$

TRY THIS
—11—

A coil consisting of 50 turns of wire has an area of 150 square centimeters and is rotating in a field of 50 milliteslas. If it takes the coil 0.05 second to go from a position parallel to the field to a position perpendicular to it, what is the average emf induced in the coil?

Lenz's Law

Suppose you push a permanent magnet into a coil whose ends are connected to an ammeter, as in Figure 11.27. This will induce a current in the coil. What is the direction of the current? Here's how to figure it out.

- You have produced some electric energy; you must have done work in pushing the magnet in.
- If you did work in moving the magnet, you must have been pushing against a force holding it back.
- That force must be a magnetic force since it acts on a magnetic pole. The source of this force is the magnetic field produced by the current induced in the coil.
- Since the field of the coil opposes the motion of the magnet, it is repelling the magnet. The left end of the coil must be an **N**-pole.
- Using the right-hand rule for electromagnets, you find that in order to produce an **N**-pole on the left end of the coil, the current must flow in the direction shown. The galvanometer needle deflects to the left.

This scenario is one example of a general rule known as Lenz's law. It is a corollary of the law of conservation of energy. Lenz's law says that whenever a current is induced, the direction of the current must be such as to oppose the change that produced the induction. If you pull the magnet out of the coil, the current in the coil must produce a field that tries to pull the magnet back in. If a coil has a current induced by rotating in a magnetic field, the field must exert a torque opposing the rotation.

EXAMPLE 11.16 When the switch in Figure 11.28 is closed, what is the direction of the current through the galvanometer?

SOLUTION

As the current in the electromagnet builds up, it is producing an **N**-pole on the right. The field produced by the induced current must be in the opposite direction; its **N**-pole is on the left end. To produce this field, the current must flow in such a direction that it passes through the galvanometer to the left.

CORE CONCEPT

When a current is induced, its direction is such as to generate a field that opposes the change that induced it.

TRY THIS
—12—

How could you change the electromagnet in Figure 11.28 so as to make the galvanometer current go to the right when the switch is closed?

REVIEW EXERCISES FOR CHAPTER 11

Fill-In's. For each of the following, fill in the missing word or phrase.

1. Parallel currents _____ each other.

2. The force between currents is called the _____ force.

3. The direction of a magnetic field is the direction pointed to by the end of a compass needle that points _____ on the earth.

4. The N-pole of a compass points to the _____ pole of a permanent magnet.

5. The force that a magnetic field exerts on a current is always perpendicular to the _____ and to the _____ .

6. In a magnetic field pointing away from you, an electron traveling to the right will experience a force in the _____ direction.

7. To increase the kinetic energy of a charged particle, it is necessary to use a(n) _____ field.

8. When a charged particle travels perpendicularly to a magnetic field, freely in a vacuum, its path is a(n) _____ .

9. The radius of the circle in which charged particles travel in magnetic fields is used to measure their _____ .

10. Magnetic fields are produced by _____ .

11. If a magnetic field is 10 milliteslas at a point 1 centimeter from the current that produces it, the field 5 centimeters from the current is _____ milliteslas.

12. You are looking into a solenoid, at its S-pole, along its axis. From your viewpoint, the direction of the current in the solenoid is _____ .

13. Crowding the wires of a solenoid more closely together will _____ the strength of the field inside it.

14. _____ materials can drastically increase the strength of a magnetic field.

15. The special magnetic properties of magnetic materials result from the existence in their atoms of spinning _____.

16. Permanent magnets must be made of _____ of ferromagnetic elements.

17. The magnetic field of the earth points straight down at the _____ pole.

18. Separation of charge can be produced when a conductor is _____ in a magnetic field.

19. The amount of emf induced in a rod moving across a magnetic field depends on the _____ and the _____ of the rod.

20. An emf is induced in a loop of wire only while the _____ within the loop is changing.

21. The electric energy produced by moving a rod, part of a circuit, through a magnetic field is equal to the _____ in moving the rod.

22. A mass spectrometer measures the mass of particles by determining the _____ of the path they travel in a magnetic field.

23. A permanent magnet behaves like a solenoid because both contain currents in the form of _____.

24. If the current in a horizontal wire is flowing northward, the direction of the magnetic field above the wire is _____.

25. _____ field lines always form closed loops.

26. The most common ferromagnetic element is _____.

27. Magnetic field lines emerge from the _____ pole of a solenoid or a permanent magnet.

28. A wire loop carrying a current is in a magnetic field. The field will stretch the loop only if the axis of the loop is _____ to the field.

29. A wire is coiled into 50 turns to make a round loop 2 centimeters in diameter. It is connected to a battery and placed with its axis perpendicular to a magnetic field. A second loop of 50 turns 4 centimeters in diameter is connected to the same battery and placed in the same field. The torque on the second loop is _____ the torque on the first one.

30. You are looking down the axis of a solenoid, and the current from your position is clockwise. The end of the solenoid facing you is a _____ pole.

Multiple Choice.

1. Which of the following processes will *not* produce new magnetic poles?
    (1) cutting a bar magnet in half
    (2) turning on a current in a solenoid
    (3) running a current through a straight wire
    (4) placing an iron rod in contact with a magnet
    (5) inserting an iron rod into a solenoid carrying a current.

2. Magnetic field lines start
    (1) on N-poles   (4) on moving charges
    (2) on S-poles   (5) nowhere.
    (3) on current-carrying wires

3. Magnetic field lines form circles in the space
    (1) near a permanent magnet
    (2) around a current-carrying wire
    (3) inside a solenoid
    (4) inside a current-carrying loop
    (5) between the poles of a horseshoe magnet.

4. The magnetic field 1 centimeter from a current-carrying wire is 12 milliteslas. The field 3 centimeters from the wire will be
    (1) 6 milliteslas   (4) 1.3 milliteslas
    (2) 4 milliteslas   (5) 0.75 milliteslas.
    (3) 3 milliteslas

5. A tesla is equivalent to a
    (1) newton per coulomb   (4) newton per ampere-second
    (2) newton per ampere-meter   (5) newton per ampere squared.
    (3) ampere per newton

6. A vertical wire carries a current upward. The magnetic field north of the wire will be directed
    (1) upward   (2) eastward   (3) westward   (4) northward   (5) southward.

7. Nickel is classified as a ferromagnetic material because
   (1) it greatly strengthens an existing magnetic field
   (2) it can be made into a permanent magnet
   (3) it generates a magnetic field when a current is passed through it
   (4) it points northward in the earth's field
   (5) it contains iron.

8. A compass will point to the earth's true north pole if it is
   (1) anywhere on the earth
   (2) at the latitude of the magnetic north pole
   (3) at the earth's south magnetic pole
   (4) directly south of the magnetic north pole
   (5) in the northern hemisphere.

9. In a television tube, the function of the anode is to
   (1) accelerate electrons   (4) produce the electron beam
   (2) direct the electron beam   (5) produce a magnetic field.
   (3) control the color rendition

10. A magnetic field *cannot*
    (1) exert a force on a current
    (2) alter the direction of an electron beam
    (3) control the direction of a beam of protons
    (4) change the momentum of an electron
    (5) change the kinetic energy of an electron.

Problems.

1. What is the strength of a magnetic field that exerts a force of 2.5×10^{-4} newton on each meter of wire carrying 5.0 amperes of current perpendicular to the field?

2. A #26 wire has a linear density of 1.50 grams per meter and can carry no more than 2.0 amperes of current. How strong a magnetic field would you need to support the entire weight of a length of this wire carrying maximum current?

3. Find (a) the direction and (b) the magnitude of the force exerted on a proton traveling perpendicular to a magnetic field of 1.5 teslas, if the proton is going 1.8×10^7 meters per second. The proton is going north in an upward-directed field.

4. A charged particle going 2.6×10^7 meters per second enters a magnetic field of 0.75 teslas and goes into a circular path of radius 2.4 meters. If the particle has one elementary unit of charge (1.60×10^{-19} coulomb), what is its mass?

5. How fast is a proton traveling if a magnetic field of 0.25 tesla can make it travel in a circular path of radius 40 centimeters? (See Appendix 2 for constants.)

6. In Chicago, the earth's magnetic field is 0.055 millitesla, directly north, and dips down at 72°. Find the direction and magnitude of the force exerted on a wire 1.0 meter long if it carries a current of 8.0 amperes directly eastward.

7. What emf is produced by moving a 20-centimeter rod perpendicularly to a 1.2-tesla magnetic field at 3.5 meters per second?

8. A circular loop of wire 20 centimeters in diameter is set perpendicular to a magnetic field of 350 milliteslas. How much magnetic flux passes through the loop?

9. The loop of Problem 8 consists of 150 turns of wire, and its ends are connected to a potential-measuring device. If the magnetic field is turned off and drops to zero in 0.10 second, how much emf is induced in the loop?

10. A vertical wire carries a current of 20 amperes straight down. At a point 15 centimeters north of the wire, what is (a) the magnitude, and (b) the direction of the magnetic field produced by the current?

11. What current would have to flow in a wire to produce a field of 0.050 millitesla at a distance of 2 centimeters from the wire?

12. What is the field inside a solenoid wound tightly in two layers of copper wire 0.10 millimeter in diameter if the current is 0.50 ampere?

13. A solenoid is made with 200 turns of wire wound on a core 15 centimeters long. The resistance of the wire is 30 ohms, and it is connected to a 12-volt battery. (a) What is the field inside the solenoid? (b) If the same wire is wound in two layers 7.5 centimeters long and connected to the same battery, what is the field inside the solenoid?

14. A solenoid wound with two layers of wire is connected to a 12-volt battery and develops a field of 15 milliteslas. In an effort to increase the strength of the field, the amount of wire is doubled and wound on the same core in four layers. If the solenoid is now connected to the same battery, what will be the field inside?

15. A wire is 0.71 millimeters in diameter and has 0.053 ohm of resistance per meter. A solenoid is made of this wire, tightly wound in four layers on a core 15 centimeters long and 30 centimeters in diameter. What potential difference must be supplied to produced a magnetic field in it of 8.5 milliteslas?

16. A proton (mass 1.67×10^{-27} kilogram) is accelerated through a potential difference of 1 500 V and projected into a magnetic field of 0.80 tesla, perpendicular to the field. What is the radius of the circle in which it travels? (Hint: solve algebraically first; equate kinetic energy to energy lost in the electric field.)

17. A mixed stream of charged particles is sent through a tube in which there is a vertical electric field of 22 000 newtons per coulomb and a horizontal magnetic field of 35 milliteslas. What is the speed of the particles that pass straight through the crossed fields?

18. A circular coil 5.0 centimeter in diameter consists of 60 turns of fine wire. It is suspended between the poles of a powerful horseshoe magnet, where the magnetic field is 350 milliteslas. Its axis makes an angle of 20° with the field. If a current is sent through the coil, it experiences a torque of 3.2×10^{-3} newton-meters. How much is the current?

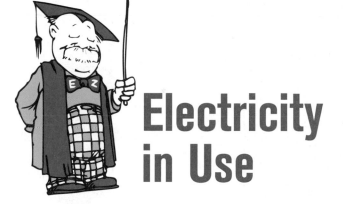

Electricity in Use

WHAT YOU WILL LEARN

How electricity is measured and used; alternating currents and power transmission.

SECTIONS IN THIS CHAPTER

- Measuring Electricity
- Electrical to Mechanical
- Manufacturing Electricity
- What Is AC?
- The Great Transformation

- Power Lines
- The Wires in Your Home
- Shock Protection
- Appliances

Measuring Electricity

Dozens of kinds of instruments, measuring many kinds of properties from temperature to velocity, are read by watching a needle move across the face of a dial. In the vast majority of these, the working part of the instrument is a *Weston galvanometer,* a highly sensitive device that accurately measures extremely small electric currents in the microampere range. It operates on the principle that the torque exerted on a solenoid by a magnetic field is proportional to the current in the solenoid (Equation 11.2).

The operation of a Weston galvanometer is illustrated in Figure 12.1. The solenoid is a coil of many strands of fine wire, wrapped around a soft iron core between the poles of a permanent magnet. This puts it in a strong magnetic field. The coil is free to rotate, and if there is current in it, the field will apply a torque to it. The solenoid rotates against the restraining torque of a fine coil spring. The greater the magnetic torque, the farther the solenoid rotates before being brought to rest by the spring. A needle attached to the coil reads against a scale.

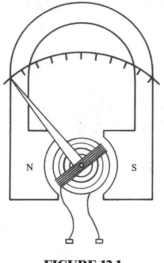

FIGURE 12.1

The Weston galvanometer is far too sensitive to measure currents in the ampere range. It is converted into an *ammeter* by attaching a *shunt* to its terminals, as shown in Figure 12.2. The shunt is a thick piece of material that is a good conductor, which has an extremely low resistance. When current is led into the ammeter, most of it goes through the shunt. The resistance of the shunt may be as little as one-millionth of that of the galvanometer movement. Then, for every ampere going into the ammeter, a microampere operates the galvanometer. Since the galvanometer takes a constant fraction of the current, the scale of the meter may be calibrated in amperes. Many ammeters have more than one scale, for measuring currents in different ranges. The range switch connects in a different shunt for each range.

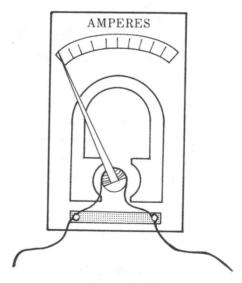

FIGURE 12.2

EXAMPLE
12.1

A Weston galvanometer has a resistance of 120 ohms; an 0.05-ohm shunt converts it into an ammeter. If the ammeter reads 1.2 amperes, how much current is in the galvanometer?

SOLUTION

The ammeter tells the truth only if its resistance is far larger than the resistance of the shunt, so practically all of the current goes through the shunt. The potential difference across the shut is $IR = (1.2 \text{ A})(0.05 \ \Omega) = 0.06$ V. Then the current in the galvanometer is $I = \Delta V/R = 0.06 \text{ V}/120 \ \Omega = 0.5$ mA. This is so small compared with the current being measured that it will not have any significant effect on the circuit.

The Weston galvanometer can also be converted into a voltmeter by connecting a high resistance in series with the movement, as in Figure 12.3. To measure the potential difference across the bulb, wires are run from the voltmeter terminals to the two sides of the bulb. Then the current through the voltmeter is proportional to the potential difference between the two sides of the bulb. The current is very small because it must flow through the resistor in series with the movement. Different ranges can be provided by switching in different series resistors. In using a voltmeter, it is important to keep in mind how it works. If, for example, you were to connect the voltmeter across a megohm resistor, the voltmeter might carry as much current as the circuit, or more. The meter would then distort the circuit parameters, and the meter would tell you lies.

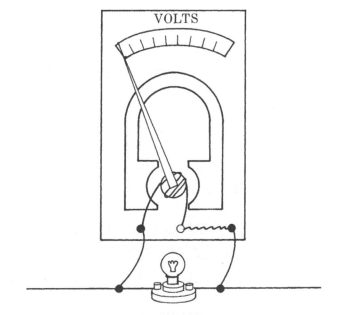

FIGURE 12.3

**EXAMPLE
12.2**   The needle of a Weston galvanometer goes to full scale when the current in its coil is 50 microamperes. What series resistor is needed to convert the galvanometer into a voltmeter with a 20-volt scale?

SOLUTION

At full scale, the current in the series combination of galvanometer coil and resistor must be

$$R = \Delta V/I = (20 \text{ V})/(50 \times 10^{-6} \text{ A}) = 4 \times 10^{5} \ \Omega$$

Nearly all of this resistance will be in the external resistor.

An ohmmeter contains a small battery, which produces a current in the resistance to be measured. The current operates a Weston galvanometer. The greater the resistance, the less is the current. This is the reason why, on a multimeter, the ohms scale reads from right to left.

The temperature gauge on your car's dashboard is also a Weston galvanometer, connected to a *thermocouple* buried in the engine. A thermocouple is a pair of wires made of two different metals, twisted together. When hot, it produces a small current, which is fed into the dashboard galvanometer to tell you when your engine is overheated.

The Weston galvanometer has been the working part of many kinds of instruments for the better part of a century. It can measure any parameter that can be made proportional to a tiny current. In recent years, new and better instruments have been designed, using electronic circuitry and digital readout. It is likely, however, that the Weston movement will continue in use for many years to come.

CORE CONCEPT

The Weston galvanometer, which forms the working part of many instruments, measures tiny currents according to the torque exerted on a solenoid in a magnetic field.

TRY THIS
—1—

What would a device have to do to convert a Weston galvanometer into an instrument for measuring velocity?

Electrical to Mechanical

We employ many devices for making electricity do work for us. One of the simplest, which is nothing but a solenoid with a loose-fitting iron core, sounds the chimes that, in many households, announce the arrival of visitors. The solenoid is in a vertical position, under the chime, and the core is partly inside the solenoid and partly below it. When the switch is closed, the magnetic field of the solenoid pulls the core upward, throwing it against the chime. Similar devices lock the doors of automobiles, trigger the shutter of a camera from a remote location, and so on.

The more common type of doorbell, a buzzer, is a little more complex. As shown in Figure 12.4, it consists of an electromagnet that attracts a spring steel armature when the switch is closed. However, the current also flows through the armature to a contact point. When the armature is attracted to the magnet, the circuit is broken at that point. The hammer strikes the gong, and the spring pulls the armature back. This completes the circuit, so the cycle starts again.

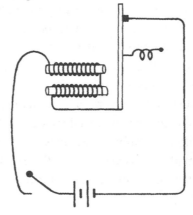

FIGURE 12.4

A great deal of the electrical energy produced is converted to mechanical energy, almost entirely by electric motors. The highly oversimplified diagram in Figure 12.5 shows how a motor works. It uses the same principle as a Weston galvanometer: a magnetic field will exert a torque on a loop current. However, if the current is just fed into the loop, it will not make the loop rotate continuously. The loop will turn into its equilibrium position, with its plane perpendicular to the magnetic field. When it over-shoots this position, the torque reverses and brings it back. The loop will oscillate around the equilibrium position until it comes to rest there, like a compass needle in a magnetic field.

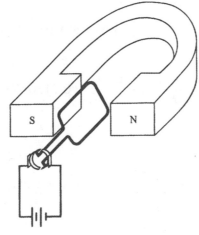

FIGURE 12.5

The loop can be made to turn continuously by feeding the current to it through a *split-ring commutator*. This is a ring divided in half. The ends of the loop are connected to the two halves of the commutator, which turns with the loop. Current is fed into the loop through the commutator, by a pair of fixed *brushes* that ride on the commutator as it turns. When the loop reaches its equilibrium position, each brush passes the split and makes contact with the other half of the commutator, thereby feeding the current through the loop in the opposite direction. As a result, the current does not reverse when the loop passes equilibrium, and the loop turns continuously.

In a real motor, the rotating armature is not a single coil, but a large number of coils. The commutator is split into many segments; each coil is fed by a pair of segments, with the two segments on opposite sides of the commutator. The torque on a coil is greatest when the plane of the coil is parallel to the field. The coils and commutator are arranged so that each coil is fed its current as it approaches this position.

The magnetic field that pushes the current in the armature is supplied by a *field magnet*. Until recently, field magnets were usually electromagnets, and this is still the practice in some motors. With the coming of the new inexpensive, strong, permanent rare earth magnets, many small motors are made with permanent field magnets. This greatly improves the efficiency of the motor since it is no longer necessary to supply current to operate the field magnets.

EXAMPLE 12.3

A dumbwaiter with a mass of 35 kilograms is lifted by an electric motor, going a vertical distance of 14 meters in 20 seconds. If the motor uses 2.8 amperes on a 120-volt line, what is the efficiency of the system?

SOLUTION

The power output of the system is the increase in potential energy per unit time:

$$P_{out} = \frac{(35 \text{ kg})(9.8 \text{ m/s}^2)(14 \text{ m})}{20 \text{ s}} = 240 \text{ W}$$

The power input is electric power:

$$P_{in} = I\,\Delta V = (2.8 \text{ A})(120 \text{ V}) = 336 \text{ W}$$

So the efficiency is

$$eff = \frac{P_{out}}{P_{in}} = \frac{240 \text{ W}}{336 \text{ W}} = 0.71, \text{ or } 71\%$$

CORE CONCEPT

An electric motor makes use of the torque that a magnetic field exerts on a loop current to convert electrical energy to mechanical energy.

At what rate is heat produced by a motor that operates at 78 percent efficiency, using 2.0 amperes on a 240-volt circuit?

Manufacturing Electricity

Few scientific discoveries have ever made an impact on society equal to that of Michael Faraday's observation that, when a wire is moved through a magnetic field, the electric charges in it can be separated (Chapter 11). It was this principle that made possible the production of electric energy in large quantities.

The principle is simple. Take a loop of wire and place it in a magnetic field, as in Figure 12.6. As the loop is turned, the amount of magnetic flux passing through it is constantly changing. The flux in the loop follows a sine-wave pattern, as in Figure 12.7, reaching its maximum when the plane of the loop is perpendicular to the field. The rate of change of the flux in the loop (Figure 12.7a) is the slope of this graph and is plotted in Figure 12.7b. According to Faraday's law, this rate of change is equal to the emf induced in the loop. That sine-wave emf produces an alternating potential difference at the terminals of the loop.

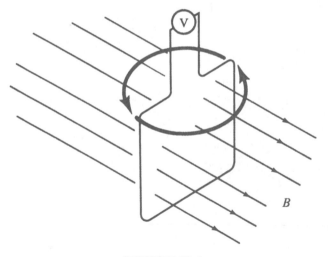

FIGURE 12.6

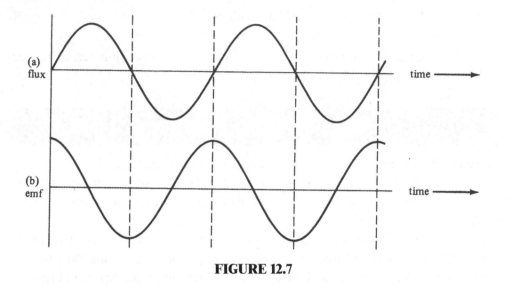

FIGURE 12.7

Attaching the ends of the loop to a split-ring commutator, as in Figure 12.8, brings this potential difference into an external circuit in the form shown in Figure 12.9. Every half-cycle, the external circuit is connected to a different end of the loop. The result is that the output potential difference is always in the same direction, although it fluctuates wildly.

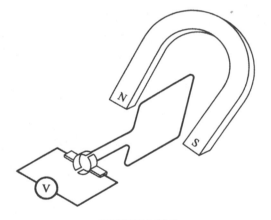

FIGURE 12.8

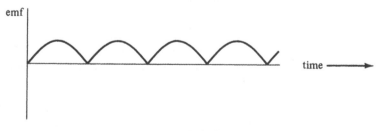

FIGURE 12.9

In a real direct-current (DC) generator, there are many coils of wire in the rotating armature, connected to many segments of a split-ring commutator. Only that coil which, at any moment, has its plane parallel to the magnetic field feeds its output to the brushes. The result is an emf that fluctuates a great deal less, as in Figure 12.10.

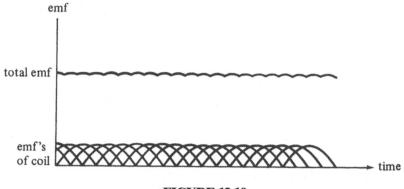

FIGURE 12.10

Early in the electric power game, it was noticed that the alternating emf in the armature had some distinct advantages and that the split-ring commutator, which produced a DC output, was unnecessary. Today, all power-plant generators take the emf directly out of the armature. It is alternating current (AC) that is distributed to the consumer. Large generators are turned by steam turbines powered by burning coal, oil, or gas, or by steam produced in a nuclear reactor or by windmills. The energy released in chemical or nuclear reactions is turned into heat and then into electrical energy to operate lights and machinery.

A generator may be turned quite easily if its output is not connected into a complete circuit. In this case, a high potential difference may exist at the generator terminals but no current will flow; the generator is producing no electric energy. As soon as a load is added—that is, a circuit is completed—current flows through the armature. Lenz's law now tells us that the armature current must oppose the force that turns the armature. The energy that the generator produces cannot be more than the work done in turning the armature. The output will always be less than the work done since some of the work is busy overcoming the friction in the system and producing heat due to the resistance of the armature wire.

CORE CONCEPT

Electric energy is produced by generators which operate on the principle that a conductor moving through a magnetic field has its charges separated.

TRY THIS
—3—

Explain why a rotating loop produces AC.

What Is AC?

If connected into a circuit, the alternating emf of a generator will produce an alternating current. As the potential reverses 120 times a second, the electrons throughout the circuit oscillate back and forth. Energy is transferred in a wavelike pattern, just as the oscillations of air molecules transfer energy in the form of a sound wave.

How shall we measure the size of an alternating current? No single number can express it, since it is constantly changing, varying sinusoidally with time as shown in Figure 12.11a. The average current is not a useful concept; since current flows as much in one direction as in the other, the average is always zero. We need some measure of current that will tell us how to calculate the energy used in a light bulb when it is fed with AC.

We know that the power consumed in an electric circuit element, the rate at which it uses energy, is given by Equation 10.4b.

$$P = I^2 R$$

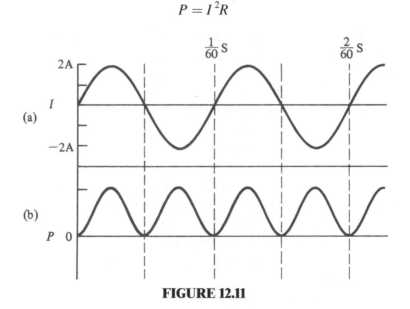

FIGURE 12.11

The graph of Figure 12.11b shows how I^2R varies with time when the maximum current is 2 amperes (as in Figure 12.11a) and the resistance is 5 ohms. Whether I is positive or negative, P is always positive. The current delivers power to the bulb no matter which way it is flowing. The power goes from zero to 20 watts and back to zero 120 times a second. In special circumstances, it is possible to show that a light bulb grows dim and bright in just this rhythm.

The average power delivered to the bulb is a useful concept; it is, in fact, just half of the maximum power. To relate this power to the current, we can invent a quantity called the *effective value* of the current, which is defined by this equation:

$$P_{av} = I_{eff}^2 R \qquad \text{(Equation 12.1a)}$$

Now we can calculate power in ways familiar from our work with DC circuits.

EXAMPLE 12.4

How much is the effective AC current if it is delivering 120 watts to a 30-ohm resistance?

SOLUTION

From Equation 12.1a,

$$I_{eff} = \sqrt{\frac{P_{av}}{R}} = \sqrt{\frac{120 \text{ W}}{30 \text{ }\Omega}} = 2.0 \text{ A}$$

From Figure 12.11, it is clear that

$$P_{max} = I_{max}{}^2 R$$

and

$$P_{av} = \frac{P_{max}}{2}$$

so that

$$P_{av} = \frac{I_{max}{}^2 R}{2}$$

and combining this with Equation 12.1a gives the relationship between effective and maximum AC current:

$$I_{eff} = \frac{I_{max}}{\sqrt{2}}$$

(Equation 12.1b)

EXAMPLE 12.5

An AC ammeter, reading effective values, records a current of 2.5 amperes. How much is the maximum current?

SOLUTION

From Equation 12.1b,

$$I_{max} = I_{eff}\sqrt{2} = (2.5 \text{ A})\left(\sqrt{2}\right) = 3.5 \text{ A}$$

EXAMPLE 12.6

What is the maximum power being delivered to a light bulb operating on 120 volts AC and drawing 0.3 ampere AC?

SOLUTION

For a pure resistance, like a light bulb, Equation 10-5a applies, using effective values to get average power. Thus $P_{av} = I\,\Delta V = (120 \text{ V})(0.3 \text{ A}) = 36 \text{ W}$. The maximum power is twice this, 72 W.

A similar line of argument gives the relationship between effective and maximum AC potential difference:

$$\Delta V_{\text{eff}} = \frac{\Delta V_{\text{max}}}{\sqrt{2}}$$

(Equation 12.1c)

When a household circuit is rated at "120 V," this means that the effective value of the potential difference is 120 volts. This calls for some caution. For example, if you were to use wire whose insulation is adequate protection at 150 volts DC, you could not use it on 120 volts AC, since there the potential rises to $120\sqrt{2} = 170$ volts.

And a further warning: It is not always correct to calculate power in an AC circuit as the product $I_{\text{eff}} \Delta V_{\text{eff}}$. There are even cases in which a circuit element can have substantial AC current and potential difference without using any average power at all! Stick to $I_{\text{eff}}^{2}R$; it always works.

A word on notation. In everyday usage, whenever AC values of potential or current are given, it is understood that effective values are being used. AC voltmeters and ammeters are calibrated in effective values. In theoretical discussions, however, it is important to distinguish between effective values and instantaneous values. To make this distinction clear, it is customary to specify that effective values are meant by writing the symbol for current or potential with an overbar. Thus, $\overline{I}$ means effective current and $\Delta\overline{V}$ means effective potential difference. If the overbar is not used, instantaneous values are indicated.

CORE CONCEPT

Effective AC current, used in calculating power, is maximum current divided by $\sqrt{2}$:

$$P_{\text{av}} = \overline{I}^{2}R; \quad \overline{I} = \frac{I_{\text{max}}}{\sqrt{2}}; \quad \Delta\overline{V} = \frac{\Delta V_{\text{max}}}{\sqrt{2}}$$

TRY THIS
—4—

3.0 amperes AC (effective value) is flowing in a 20-ohm heater. Find (a) the maximum current; and (b) the power delivered.

The Great Transformation

AC has replaced DC chiefly for one reason: the transformer. This is a device that can use electric power at one potential to produce an almost equal amount of power at another potential. If you go to Europe, you can take along a transformer that will change 220 volts AC to the 120 volts AC you need to run your hair dryer. An old-fashioned television set contains transformers that can change the 120 volts AC you put into it into dozens of different potentials, all the way up to the 30 000 volts you need to run the picture tube. Your electric utility company ships energy to you at 7 000 volts and changes it down to the 120 volts you use at home by means of a transformer near your home, either on a utility pole or underground. And transformers work only on AC.

The principle of the transformer is illustrated in Figure 12.12. A coil is wrapped around an iron core, and AC is fed into it. This *primary coil* sets up a magnetic field inside the iron, oscillating at 60 cycles or whatever the AC frequency happens to be. A second coil is also wrapped around the iron core and connected to an energy-using load of some sort. The second coil (called the *secondary*) surrounds a changing magnetic field; according to Faraday's law of induction, there must be a potential difference induced in the coil. If it becomes part of a complete circuit, there will be a current in the secondary.

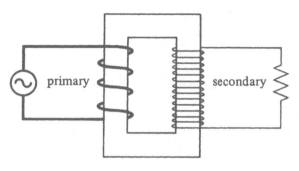

FIGURE 12.12

In the transformer, every turn of wire, whether in the input coil or the secondary, encloses the same magnetic flux. Therefore, the potential emf induced in each turn of wire is the same in both coils. To make the emf in the secondary larger than the potential difference across the primary, all we have to do is to use more turns of wire in the secondary than in the primary. The two potential differences, in fact, will be in the same ratio as the number of turns of wire in the two coils:

$$\frac{\Delta V_s}{\Delta V_P} = \frac{n_s}{n_P} \qquad \textbf{(Equation 12.2a)}$$

A transformer can increase (step up) or decrease (step down) potentials strictly according to the relative number of turns in the two coils. Some transformers have several secondaries, so that a number of different potentials can be obtained in the output. A variable transformer, like the one that feeds toy electric trains, has a movable tap that can contact a variable number of turns in the secondary. The possibilities are endless, once the principle has been established.

A transformer is a highly efficient device, and many of them can produce, in the secondary, well over 90 percent of the energy put into the primary. But we get no energy for nothing. However, if we neglect the small loss due to heating of the wires of the coils, the power output of a transformer will be equal to the power input. In a transformer, $\overline{I}\,\Delta\overline{V}$ is a valid expression for power, so we can write

$$\overline{I}_s\Delta\overline{V}_s = \overline{I}_p\Delta\overline{V}_p$$

Combining this with Equation 12.2a tells us that

$$\frac{\overline{I}_P}{\overline{I}_s} = \frac{n_s}{n_P} \qquad \textbf{(Equation 12.2b)}$$

Thus, if you want to step up potentials to higher values, you must expect to put correspondingly more current into the primary than you get out of the secondary.

EXAMPLE 12.7 A transformer has 500 turns in the primary and 3 000 turns in the secondary. In the primary, the potential difference is 120 volts AC and the current is 150 milliamperes. Find (a) the secondary potential difference; and (b) the secondary current.

SOLUTION

(a) The potentials are in the same ratio as the number of turns:

$$\frac{\Delta\overline{V}_s}{\Delta\overline{V}_P} = \frac{n_s}{n_P}$$

$$\frac{\Delta\overline{V}_s}{120\ \text{V}} = \frac{3\ 000}{500}$$

$$\Delta\overline{V}_s = 720\ \text{V}$$

(b) The currents are in an inverse ratio to the number of turns:

$$\frac{\overline{I}_P}{\overline{I}_s} = \frac{n_s}{n_P}$$

$$\frac{150\ \text{mA}}{\overline{I}_s} = \frac{3\ 000}{500}$$

$$\overline{I}_s = 25\ \text{mA}$$

If you have a radio or any other electrical device that is marked "AC only," it is because the current you send into it is being transformed to use the power at different potentials. If you plug this appliance into a DC source, feeding power into a transformer, the current will be enormous and something will burn out.

CORE CONCEPT

A transformer converts potential differences and currents in an inverse ratio:

$$\frac{n_s}{n_s} = \frac{\Delta \overline{V}_s}{\Delta \overline{V}_P} = \frac{\overline{I}_p}{\overline{I}_s}$$

TRY THIS
—5—

A radio works on 15 volts, obtained from a transformer that plugs into a 120-volt wall outlet. It uses 3.0 watts. Find (a) the turns ratio in the transformer; and (b) the current in the primary and secondary.

Power Lines

It was the transformer that made possible the power grid that carries electricity to the most remote parts of the country. Without the transformer, and the AC generator that feeds it, we would still need a powerhouse no more than $\frac{1}{2}$ mile from our homes, and we would still not get very much power.

To see why this is so, consider a modest home that needs, from time to time, 10 kilowatts of power to keep its inhabitants happy and comfortable. To keep the numbers simple, let's assume that we will supply this power at a potential difference of 100 volts. Then, from Equation 10.4a,

$$\overline{I} = \frac{P}{\Delta \overline{V}} = \frac{10\ 000\ \text{W}}{100\ \text{V}} = 100\ \text{A}$$

so we have to send 100 amperes to the house.

Let's send this current through some good-sized aluminum wires. If we assume it is about 1 centimeter thick and 1 mile long, each wire will have a resistance of about 0.5 ohm. The two transmission wires and the house now constitute a series circuit connected to the powerhouse, as illustrated in Figure 12.13. This current of 100 amperes has to flow through the generator and the wires, as well as through the house.

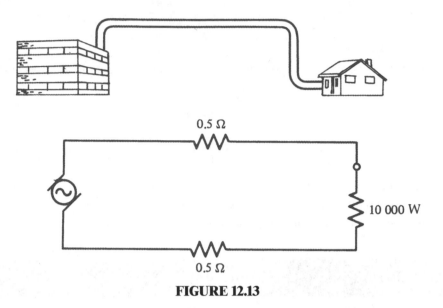

FIGURE 12.13

As the current flows through the aluminum wires, it loses power; the wires have resistance, which converts electric energy into heat. The power loss in the wires is

$$P = \bar{I}^2 R$$
$$P = (100 \text{ A})^2 (2 \times 0.5 \text{ } \Omega) = 10\,000 \text{ W}$$

We are using just as much energy to heat the transmission line as the house is using, and this is terribly wasteful.

EXAMPLE 12.8

How much power is wasted in supplying 30 000 watts to a small factory at a potential difference of 240 volts if the transmission lines have a resistance of 0.2 ohms in each direction?

SOLUTION

From Equation 10.4, the current is

$$\bar{I} = \frac{P}{\Delta \bar{V}} = \frac{30\,000 \text{ W}}{240 \text{ V}} = 125 \text{ A}$$

The waste occurs when this current passes through 0.4 ohms in the transmission lines; from Equation 12.1a,

$$P = \bar{I}^2 R = (125 \text{ A})^2 (0.4 \text{ } \Omega) = 6\,300 \text{ W}$$

There is a way out of this problem. Suppose we supply the 10 kilowatts to the house at 1 000 volts instead of 100 volts. Then the house will get the same power by drawing only 10 amperes of current from the generator. The power loss in the lines will now be (10 amperes)^2(1 ohm) = 100 watts. We have saved 99 percent of the wasted power.

The catch to this scheme is that no one wants a potential difference of 1 000 volts in his electrical outlets. It is too dangerous. The transformer must come to the rescue. Power companies use very high potentials—and thus small currents and small line losses—in sending electric power over long distances. Locally, transformers reduce the potentials and increase the current, for safer conveyance of the power for short distances, where the wire resistance is small.

CORE CONCEPT

Power losses in transmission lines are minimized by the use of high potentials and small currents.

TRY THIS
—6—

A school uses 140 kilowatts of power, supplied at 120 volts from a transformer nearby. The line from the transformer has a resistance of 0.002 ohms in each direction. How much power is lost in the lines?

The Wires in Your Home

Electricity can be dangerous. Precautions must be taken against two kinds of hazard: fire and shock. Proper installation of all electrical equipment is essential. Standards are set by the *National Electrical Code,* a fat book that tells, in the greatest detail, what is and is not acceptable. Local communities may give this code the force of law, and they often make their own codes even more restrictive.

Unfortunately, every wire has resistance and converts some of the energy it carries into heat. When current flows, the wire warms up and keeps getting hotter until it loses heat to its surroundings at the same rate at which it is being made. Three factors determine the rate at which the wire can give up heat: (1) thickness—a thick wire has more surface area than a thin one, so there is a larger region over which heat can escape; (2) insulation—thick insulation tends to inhibit the escape of heat; (3) temperature—the hotter the wire is, the greater the rate at which it can radiate away its heat.

For any given wire, the controlling factor is temperature. The temperature of the wire will rise until it is in a thermal steady state, with heat escaping as fast as it is produced. If there is too much current, the wire may not reach this condition until it is dangerously hot. Then, the wire may melt, or it may set fire to the house.

The current that any given wire can safely carry is limited by its thickness and its insulation. Since insulation is absolutely necessary, it comes down to this: a fat wire can carry more current safely than a skinny one. Not only does the fat wire have less resistance, but it also has a larger surface area to help dispose of heat. The #14 wire (diameter 1.628 millimeters) commonly used in household wiring, made of copper and insulated, carries 15 amperes safely. Some household circuits, particularly those feeding kitchens, are wired with #12 wire (diameter 2.053 millimeters), which can carry 20 amperes. If you have an electric oven, it will need its own circuit; it operates on 240 volts, may use as much as 30 amperes, and thus calls for #10 wire (diameter 2.588 millimeters).

To protect your home against any circuit carrying too much current, every circuit incorporates a fuse or a circuit breaker. This device is placed in series with the line, so all the current in the circuit goes through it. A fuse is simply a strip of metal which has a low melting point. When the current gets too high, the fuse melts and interrupts the circuit. You must then find the problem, correct it, and replace the fuse.

A circuit breaker serves the same function, but it does not have to be replaced. It is simply a switch that opens when the current gets too high, and it can be reset after the problem is corrected. One kind of circuit breaker is based on an electromagnet. Strong current strengthens the field until it can pull on an iron core strongly enough to open the switch.

A fuse or circuit breaker must be matched to the wire it is designed to protect: On a circuit wired with #14 wire, for example, you must use a 15-ampere fuse.

CORE CONCEPT
Electric circuits produce heat and must be protected against overload.

TRY THIS
—7—

Considering the values given above, can you guess what gauge wire is protected by a 40-ampere fuse?

Shock Protection

The shock from a 120-volt line is dangerous and can be fatal. There are two lines of defense against shock: (1) Keep the electricity in the wires where it belongs; and (2) arrange the circuits so that in case the electricity escapes it does no harm.

A person receives a dangerous shock when a substantial current passes through his or her body. The current causes spastic contraction of muscles. If it goes through the heart, it can kill. This situation occurs when a hand touches a high-potential wire while the feet are at ground potential. Bare feet, especially if wet, are particularly dangerous; rubber shoes insulate the body from ground and prevent the formation of a complete circuit from high potential to ground. It is only the current passing through the body that counts. The white (neutral) wire of your household wiring may be carrying a lot of current, but it is at nearly ground potential, and it is not dangerous to touch it. Careful, though! Not every house is wired correctly!

Good insulation keeps the potentials inside the wire where they belong. But insulation may crack with age and may also wear off if it happens to touch a moving part. So even if wiring is properly installed at first, it may go bad and let the electricity leak out to a metal part. Allowance must be made for this eventuality.

The key to protection against shock is grounding. This means that circuits must be planned in such a way that a failure of the insulation of the high potential will result in a direct, low-resistance path from the high potential to the earth.

Suppose, for example, that the insulation fails inside your electric mixer, so that its metal casing makes contact with the high-potential wire. Then you will be in danger of a severe shock if you touch the casing, for your body provides a path from high potential to ground. However, if the shell of the mixer is connected to ground through a wire as in Figure 12.14, this pathway will have a much lower resistance than your body, and that is the path the current will follow.

All electric codes call for careful grounding. Connections from one wire to another must be enclosed in a metal box. Every switch, every outlet, and every lamp connection must be surrounded by metal. All these metal boxes should be connected to each other by grounding wires in the cables that run through the walls of the house. Some codes even insist that all the cables be enclosed in metal sheathing. The entire electric circuitry is thus enclosed in metal, and the system grounded somewhere, either by connecting it to a water pipe or to a separate rod driven 10 feet into the earth.

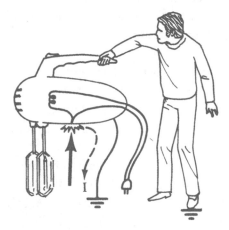

FIGURE 12.14

Modern installations call for three prongs in every electric outlet. The third, connected to the green wire of the electric cable, is the grounding terminal. The cord of your mixer contains a third wire, running from the metal shell of the mixer to the third terminal of the outlet.

CORE CONCEPT
Accidental shock is prevented by enclosing all circuitry in grounded metal.

TRY THIS
—8—

Explain why no grounding wire is needed in an electric drill made with a plastic case.

Appliances

Electrical appliances do many things for you, from brushing your teeth to waking you up in the morning. Whatever the gadget, however, the electrical part of it falls into one of four categories: (1) resistance heating, (2) operation of a motor, (3) fluorescent lighting, or (4) electronic control. The last of these, which includes your radio, television set, computer, and cell phone, is a whole world of its own, and we will not go into it.

Resistance heating devices include your toaster, electric iron, incandescent lamps, and electric heating system, if you have one. The principle of operation is simple: Just send a great deal of current through a wire, and it will convert the electric energy to heat at a rate equal to I^2R. The choice of metal for the heating element is crucial. You need something that will neither melt nor break at very high temperatures, and that has only moderate resistivity. For light bulbs, tungsten is used; for toasters and electric coffee pots, the choice is likely to be an alloy such as Nichrome.

If you stop to count the electric motors in your home, you will be surprised to discover how many there are. Many of them are of the universal type, which work on either AC or DC. They feed the current into the armature through brushes and a commutator, as described earlier. However, there are other kinds. An induction motor, which is probably the kind in your dishwasher, works on AC only and has neither brushes nor commutator. Its armature has no electrical connections at all. The current in the armature is induced by the alternating field produced in the field coils. It is like a transformer, with the secondary allowed to rotate when torque is applied to it by the field of the primary.

Electric clocks contain a device called a synchronous motor. This is a small permanent magnet placed inside an alternating field. The magnet spins in keeping up with the rapid changes in the field. On a 60-cycle AC, therefore, the permanent magnet spins at 60 revolutions per second. Since this frequency is accurately controlled by the utility company, the electric clock keeps excellent time. The newer clocks and watches, battery-operated, work on an entirely different principle. Their motor is controlled by an oscillating quartz crystal.

The fluorescent lightbulb is unique. It operates on the special electrical properties of an ionized gas, or plasma, which it contains. The bulb is largely evacuated, but it contains a small amount of mercury vapor. Hot filaments at the ends of the tube release electrons, which are accelerated toward the other end of the tube by an applied electric field. The electrons strike mercury atoms. The energy they add to these atoms increases the electron energy within the atom. These electrons within the mercury atom then give up their excess energy in the form of ultraviolet rays. The rays strike the coating on the inside of the tube, which is a special material that converts ultraviolet rays into visible light.

When Thomas Edison introduced the first practical incandescent lamp in 1879, he revolutionized people's lifestyles. The dim glow of a candle or a gas flame, which limited people's nighttime activity, became a thing of the past. Life became a 24-hour activity.

Unfortunately, the incandescent lamp is an extremely low-efficiency device. Efficiency of 7 percent is now considered the minimum acceptable level. In other words, for each 100 watt-hours of electric energy you buy, 93 turn into heat and only 7 into light. This is the main reason that people are replacing their incandescent bulbs with the new compact fluorescents, which screw into the same socket as the one used by the incandescents. There is no loss of light if you replace a 150-watt incandescent bulb with a 30-watt fluorescent. Furthermore, a fluorescent bulb remains cool, and lasts up to five times as long as an incandescent.

CORE CONCEPT
Electricity provides resistance heating, kinetic energy from motors, and other valuable services.

TRY THIS
—9—

Count the electric motors in your home.

REVIEW EXERCISES FOR CHAPTER 12

Fill-In's. For each of the following, fill in the missing word or phrase.

1. A Weston galvanometer is basically an instrument that measures small
    _____.

2. Adding a series resistor to a galvanometer converts it into a(n) _____.

3. The needle of a Weston galvanometer is moved by the _____ produced by
    a magnetic field on a wire loop.

4. An ammeter can be used with a wide range of scales by providing it with a
    wide selection of _____.

5. A(n) _____ is needed in a motor to reverse the current in the armature
    every half cycle.

6. The armature in a motor rotates within a(n) _____ field.

7. To produce DC, the output of a generator must be fed through a(n) _____.

8. In any generator, the current in the armature is of the _____ type.

9. The average value of an alternating current is always _____.

10. AC power is equal to I^2R if the value of I used in the equation is the _____ current.

11. If a transformer is used to step up potentials, the secondary current is _____ than the primary current.

12. In a transformer, energy is transferred from primary to secondary by means of a changing _____.

13. Power loss in transmission lines is minimized by the use of high _____.

14. Long-distance transmission of electric power is possible only because potentials can be changed by using a(n) _____.

15. A typical value for the potential of a long distance electric transmission line is _____ volts.

16. Wiring is protected from overload by a(n) _____ or a(n) _____.

17. As the temperature of a wire rises, the rate at which it loses heat _____.

18. Electric circuits are grounded to protect against accidental _____.

19. The grounding wire of an electric installation can be recognized because it is _____.

20. AC electric clocks operate on _____ motors.

Multiple Choice.

1. An important part of a Weston galvanometer movement is a(n)
    (1) permanent magnet   (4) commutator
    (2) armature   (5) series resistor.
    (3) electric field

2. The Weston galvanometer is essentially an instrument that measures
    (1) small potentials   (4) large potentials
    (2) small resistances   (5) large currents.
    (3) small currents

3. A battery is needed in order to convert a galvanometer into a(n)
   (1) voltmeter
   (2) ohmmeter
   (3) temperature gauge
   (4) speedometer
   (5) ammeter.

4. The current in the armature of a motor is reversed every half cycle due to the action of a(n)
   (1) armature
   (2) field coil
   (3) brush
   (4) external power source
   (5) commutator.

5. The current in a generator armature is AC because
   (1) the magnetic field reverses at intervals
   (2) the current in the field coils is AC
   (3) the rotation of the armature causes the field through it to reverse
   (4) the commutator feeds current into it in opposite directions every half cycle
   (5) the generator is designed to produce AC because it is more useful than DC.

6. An effective AC potential difference of 100 volts is placed across a 10-ohm resistor. The average current in the resistor is
   (1) 10 amperes
   (2) 14 amperes
   (3) 7 amperes
   (4) 5 amperes
   (5) zero.

7. Compared with its value in the primary of a step-up transformer, the value in the secondary
   (1) of potential difference is lower
   (2) of power is lower
   (3) of power is higher
   (4) of current is lower
   (5) of current is higher.

8. Very high potentials are used in long-distance transmission lines because
   (1) they deliver more current
   (2) they reduce power loss in the line
   (3) they are easier to produce at the powerhouse
   (4) they are less damaging to the environment
   (5) they can be produced with less power loss.

9. Wires with thicker insulation must be used whenever
   (1) more current must be carried
   (2) potentials are higher
   (3) more heat must be dissipated
   (4) more power must be delivered
   (5) line loss must be reduced.

10. In a properly wired household circuit, which combination of wires is least likely to produce a serious shock if both are touched at the same time?
   (1) black and white
   (2) black and green
   (3) green and white
   (4) none; all are equally dangerous
   (5) all three; any combination of two is reasonably safe.

Problems.

1. An alternating current in a 60-ohm resistor has a peak value of 0.50 amperes. Find (a) the peak power; (b) the effective potential difference.

2. On a 120-volt AC line, a 300-watt bulb is burning. Find (a) the effective current; (b) the peak power.

3. A transformer is to be used to step 120 volts AC down to 6 volts AC. If it has 2 400 turns of wire in the primary, how many must there be in the secondary?

4. A 60-watt bulb is designed to operate on 12 volts and is to be connected through a transformer into a 120-volt circuit. Find (a) the secondary current; (b) the primary current; (c) the maximum potential difference the insulation in the primary circuit must withstand.

5. A large motor is needed to provide energy at the rate of 6.5 kilowatts, and it is to be supplied with current through a transmission line with 0.20 ohms of resistance. How much power would be saved by designing the motor to operate at 1 500 volts instead of 120 volts?

6. Determine the amount of power wasted in providing a factory with 45 kilowatts of electric power at 560 volts through a transmission line with a resistance of 1.5 ohms.

7. What is the greatest amount of power that can be delivered through a #14 wire operated at 120 volts?

8. A large power transformer converts 325 000 volts to 7 500 volts, at the rate of 2.5 million watts. Find (a) the primary current; (b) the secondary current.

9. What potential difference must be withstood by the insulation around the wire that delivers AC current at 30 000 volts to the picture tube of your television set?

10. A galvanometer with a resistance of 70 ohms reads full scale when the current in its coils is 0.60 milliamperes. What is the resistance of a shunt needed to use the galvanometer as an ammeter that reads 10 amperes at full scale?

11. If the galvanometer of Problem 10 is to be used as a voltmeter, what resistance must be connected to it to produce a voltmeter with a 100-volt scale?

Electrons in Control

WHAT YOU WILL LEARN

About electronics, processes that use electricity to control information rather than power; the devices that provide this control.

SECTIONS IN THIS CHAPTER

- What Is Electronics?
- Resistors
- Capacitors
- Capacitance in AC
- Inductance

- Inductance in AC
- The *R-L-C* Circuit
- Conductors
- Semiconductors
- Transistors

What is Electronics?

Electricity delivers energy, but it can also be used for another purpose: to carry and process information. This includes many things, including many of the devices that have produced the unique flavor of life in the twentieth century: radios, telephones, televisions, computers, robot controls, advanced weapon systems, and so on. For the most part, modern electronic systems use very small currents and do not place any substantial load on our power supplies.

The first use of electricity for information processing came with the invention of the electric telegraph. The telephone came a half-century later. Modern electronics is the result of three separate revolutionary discoveries. The first of these was the discovery that the hot filaments of an electric lightbulb release electrons, which form a cloud in the space around the filament. Electronics began with the invention of a positive terminal, a "plate," inserted into the bulb, which attracted the electrons. If the air was removed from the bulb, electrons flowed from the hot filament to the plate. If a circuit is completed from plate to filament, "plate current" will flow in it. The current is strictly one-way; it cannot go in the reverse direction across the vacuum.

This first vacuum tube was called a diode because it had just two elements: plate and filament. It provided a useful means of "rectifying" the current, that is, changing AC to DC. The tubes became enormously more useful when another element was inserted between the plate and filament. This "grid" could be given a potential that controls the amount of current. A small varying potential on the grid could affect the plate current so as to produce a much amplified effect on the plate current.

Three kinds of circuit elements could be wired onto the vacuum tube to control the currents: resistors, capacitors, and inductors (including transformers). The vacuum tube with its associated elements could be wired into circuits with many different functions: rectifiers, amplifiers, switches, and oscillators that produced AC with frequencies in the kilohertz range. This made radio possible and then television, radar, and many other devices.

All these devices operated on small currents and used little electric energy, except for one part of the whole system. The filaments of the vacuum tubes had to be heated. The first programmable digital computer used 18 000 vacuum tubes in a system 8 meters high and 24 meters long. It had to have its own power lines brought in, and the large room in which it was housed needed elaborate air-conditioning.

The second electronics revolution changed all that. It came with the invention of the transistor. A transistor can perform nearly all the functions of the vacuum tube, and it does not have to be heated. Gone are the special power lines, the air-conditioning, the whole room; electronic devices came down to a size that made them practical everywhere.

The third revolution carried this process to an ultimate extreme. The main elements of electronic circuits—connecting wires, resistors, capacitors and transistors—are now microscopic in size and use minute amounts of energy. A computer chip the size of a dime holds an integrated circuit that contains 20 million transistors and all the associated resistors and capacitors. A modern laptop has thousands of times the capacity of that first programmable computer.

The integrated circuit is far beyond the scope of this book. We can get an understanding of how electronic circuits work by studying the functions of the four elements of the second generation: resistors, capacitors, inductors, and transistors.

CORE CONCEPT

Electronics is the science of controlling small currents so as to process information.

TRY THIS
—1—

What potential develops across a 16 000-ohm resistor in the plate circuit of a vacuum tube if the plate current is 2.5 milliamperes AC?

Resistors

A resistor is nothing but a device that has a standard, calibrated resistance. Some are made by winding a length of wire on a core, but most of them are just little cylinders of a carbon-clay mixture. They are calibrated according to resistance and are available with values from a few ohms up to 10 megohms (million ohms) or more. The resistance is encoded in colored bands on the shell.

Every resistor produces heat, equal to I^2R, when it is in use. If this heat is allowed to accumulate, the temperature of the resistor rises, and it could burn out. Resistors are rated, not only by their resistance, but also by their allowed wattage. A 1-watt resistor perhaps 4 centimeters long is much larger than a $\frac{1}{4}$-watt resistor. The reason is that the 1-watt resistor is able to get rid of four times as much heat. The heat has to pass out through the surface; a larger surface gives a bigger area through which heat can be dissipated.

Of all the circuit elements, resistors are unique in that they obey Ohm's law at any frequency. Whether the current is DC or a gigahertz AC (that's 10^9 hertz), $\Delta \overline{V} = \overline{I}R$ in a resistor.

CORE CONCEPT

A resistor obeys Ohm's Law at any frequency.

TRY THIS
—2—

In selecting a resistor, what wattage should be chosen for a 2 000-ohm resistor that has to carry 20 milliamperes of 1 500-kilohertz AC?

Capacitors

Storage of electric charge is a familiar phenomenon to anyone who has ever used a charged comb to pick up bits of paper. In many applications, charge can be stored in a device called a *capacitor* (often still referred to by its older name, *condenser*).

If you have ever used a camera with an electronic flash gun, you have used a stored charge. The flash, as a rule, cannot be fired any more frequently than once every 5 seconds; it takes quite a lot of charge to produce the flash. During the 5-second recycle time, the charge is being built up in a capacitor. Typically, a 6-volt battery feeds into circuits that boost the voltage to several hundred volts to store charge in the capacitor. When the picture is taken, a couple of hundredths of a coulomb shoot through the lamp, producing a flash that lasts about 0.2 millisecond.

A capacitor consists of two conducting plates separated by an insulator. When the plates are connected to opposite terminals of a battery or other electric source, as in

Figure 13.1a, charge flows into one plate and out of the other. As the charge accumulates, the potential difference between the plates rises until the potential difference between the plates is equal to that of the battery. The capacitor can then be disconnected from the battery, and the potential difference between the plates will remain, theoretically forever. If the plates of the charged capacitor are connected through a conducting path, a flow of current will discharge it (Figure 13.1b).

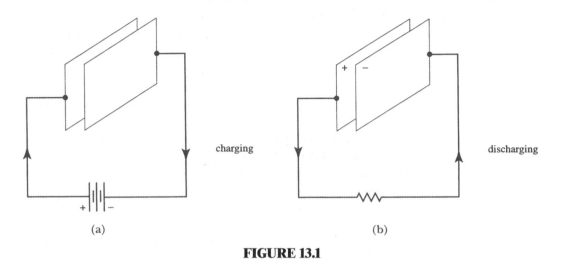

charging

discharging

(a)

(b)

FIGURE 13.1

There are many kinds of capacitors. One common kind consists of layers of tinfoil, separated by waxed paper and rolled up. Another type is a series of interlocking metal plates with nothing but air between them. This type is adjustable, for one set of plates can move in and out of the spaces between the other set. This type of adjustable capacitor is used to tune a radio to different stations. In an integrated circuit, a capacitor may be microscopic in size, and is nothing more than two thin metallic layers with an insulating layer between them.

The amount of charge that a capacitor stores is proportional to the potential difference between its plates. For each capacitor, the ratio depends on how the capacitor is made. If the plates are large, they will store more charge for a given potential difference. Also, the closer the plates are to each other, the more charge will be stored. In addition, the amount of charge that can be stored depends on what kind of insulator separates the plates. The physical properties of a capacitor determine its characteristic property, its *capacitance (C)*, which is defined as

$$C = \frac{q}{\Delta V}$$ **(Equation 13.1)**

The unit of capacitance is the *farad* (F), equal to a coulomb per volt. This is a truly enormous unit, and the commonly used units are the microfarad ($= 10^{-6}$ farad) and the picofarad ($= 10^{-12}$ farad).

EXAMPLE 13.1 A 20-picofarad capacitor is connected to the terminals of a 150-volt DC generator. How much charge is stored?

SOLUTION

From Equation 13.1,

$$q = C \Delta V = (20 \times 10^{-12} \text{ F})(150 \text{ V}) = 3.0 \times 10^{-9} \text{ C} = 3.0 \text{ nC}$$

When the charged plates of a capacitor are connected to an external circuit, the charge flows through the circuit until the capacitor reaches zero potential difference.

EXAMPLE 13.2 How much heat is generated if the plates of the charged capacitor in Example 13.1 are connected to opposite sides of a 50-ohm resistor?

SOLUTION

Equation 9.3 tells us that the energy stored in the capacitor, which is converted into heat in the resistor, is

$$\Delta E = q \Delta V = (3.0 \times 10^{-9} \text{ C})(150 \text{ V}) = 4.5 \times 10^{-7} \text{ J}$$

The size of the resistance does not matter; if it is a large resistance, the capacitor will take longer to discharge. This property of a resistance–capacitance combination is used in circuits as a timing device. The larger the resistance and capacitance in the combination, the longer it will take for the capacitor to charge up to some predetermined level.

The property of capacitance is a kind of inertia of potential. In a circuit, if a potential is suddenly applied, the potential of any capacitor will take time to build up. Similarly, if potentials drop, the capacitor will take time to discharge, thus preventing a rapid drop in potential.

Every capacitor is rated not only by its capacitance but also by its breakdown voltage. If the potential difference exceeds this value, the insulation between the plates will break, and the capacitor will spark over and discharge through its insulator. A thin insulator increases capacitance, but it also decreases the breakdown voltage.

A capacitor is an open circuit, so no current flows through it. However, as it charges, there is a brief flow of current into it and an equal flow out the other side. In an AC circuit, current flows alternately in and out of a capacitor, repeatedly charging and discharging it. An AC meter on either side of the capacitor will indicate a flow of alternating current; while nothing actually passes through it, an AC can seem to do so.

A capacitor uses no energy, even on AC. Whatever charge is stored during one-quarter of a cycle is returned as the capacitor discharges during the next quarter.

CORE CONCEPT

A capacitor is a device that stores an amount of electric charge proportional to the potential difference between its plates.

TRY THIS
—3—

A capacitor is rated at 30 picofarads with a breakdown at 80 volts. What is the largest charge it can store?

Capacitance in AC

Suppose a capacitor is connected through a switch to a battery. When the switch is closed, what happens?

Figure 13.2 shows what happens. As soon as the switch is closed, the current rushes in and the potential difference builds up rapidly. The current is feeding charge into the capacitor, so the rate at which the capacitor potential grows is proportional to the current. As the capacitor potential rises toward the battery potential, the current drops off; when the battery potential is reached, the current stops.

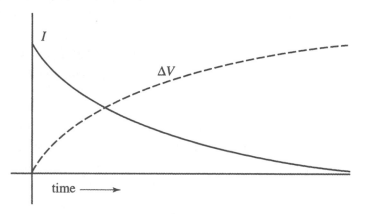

FIGURE 13.2

Now, what happens if the capacitor is connected to an AC generator? The dashed line of Figure 13.3 shows how an AC current changes with time. At all points, the slope of the potential difference graph (the rate of change of potential difference) is proportional to the current. At (a), the current is a maximum and the potential difference graph has its greatest slope. At (b), the current is zero and the potential difference graph is horizontal; its slope is zero. At (c), the slope of the potential difference curve is a negative maximum and so is the current.

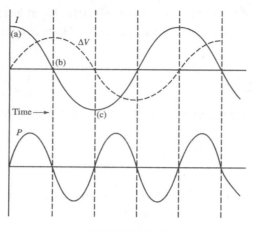

FIGURE 13.3

The crucial lesson of these graphs is that the current reaches its maximum value a quarter cycle before the potential difference does. *In a capacitor, the phase of the potential difference is always 90° behind the current.*

A second lesson is found in the power graph. The product $I\Delta V$ is zero whenever either I or ΔV is zero. When both are either negative or positive, the power is positive. When I or ΔV is negative and the other is positive, the power is negative; instead of taking energy from the circuit, the capacitor is returning it. During one quarter cycle, the electric field of the capacitor is storing energy; in the next quarter cycle, the energy is returned to the circuit. The net power is always zero; capacitors do not use power. Note that $\bar{I}^2 R$ is also zero; there is no resistance.

Suppose a capacitor of capacitance C is connected to an AC generator producing an effective potential difference of $\Delta \bar{V}_{eff}$. How much current flows? The current depends very strongly on the frequency of the AC. Remember, the current is proportional to the rate of change of the potential difference; higher frequencies are changing more rapidly, so there will be more current. The current also depends on the capacitance; larger capacitors store more charge and return it to the circuit. Of course, the effective current is also proportional to the effective potential difference:

$$\bar{I} = 2\pi f C \Delta \bar{V}$$

It is convenient to define a quantity for AC currents that is a ratio of potential difference to current. This is like Ohm's law, except that instead of resistance in the equation, there is a quantity called *capacitive reactance (X_C)*. The previous equation converts into this definition of capacitive reactance:

$$X_C = \frac{\Delta \bar{V}}{\bar{I}} = \frac{1}{2\pi f C}$$ **(Equation 13.2)**

Since reactance is the ratio of potential difference to current, its unit is the ohm.

| EXAMPLE 13.3 | An electronic oscillator produces 4.5 volts effective AC at a frequency of 1 850 kilohertz, feeding it into a 12-picofarad capacitor. How much current flows? |

SOLUTION

From Equation 13.2,

$$\bar{I} = 2\pi fc\Delta\bar{V} = 2\pi(1\ 850 \times 10^{3}\ \text{Hz})(12 \times 10^{-12}\ \text{F})(4.5\ \text{V}) = 0.63\ \text{mA}$$

CORE CONCEPT

In a capacitor, potential difference trails current by 90°, so no power is used; current is proportional to frequency.

TRY THIS
—4—

An audio amplifier produces a sound system with a mixture of frequencies at equal potentials over a broad sound spectrum. If the current flowing through a capacitor is 40 milliamperes at 500 hertz, how much is it at 2 000 hertz?

Inductance

Take a long piece of wire and connect it, through a switch, to a battery. When the switch is closed, the potential drops across the length of the wire; the drop is equal to *IR*, which is the potential difference of the battery. Charge flowing down the potential gradient gives up its energy to the resistance of the wire, generating heat.

Now take that same wire and wind it into a coil, wrapped around an iron core. Something different happens. As soon as the current starts to flow, a magnetic field grows. Faraday tells us that this changing magnetic field must induce a potential difference across the coil. This is called the *self-induced potential difference*. Lenz tells us that the direction of this potential difference must be such as to oppose the buildup of the current. If it is a 10-volt battery, closing the switch makes the current start to rise; it rises so rapidly at first that the self-induced potential difference is 10 volts. The energy lost by the flowing charge does not heat the wire; it is stored in a magnetic field.

This situation changes rapidly. The best way to visualize what happens next is to think of that wire in two parts: a resistance in series with a resistanceless coil. As soon as there is current, some of the energy of the flowing charge is lost in the resistance of the wire. The potential drops across the resistance by an amount *IR*, leaving less for the self-induced potential difference across the coil.

The inductance of a coil is a kind of inertia of current; it tends to prevent changes in the current. It slows the buildup of current when the switch is closed. When the switch is opened, the magnetic field collapses and induces current back into the coil. This induced current must counteract any *change* in the current; it will make the current continue to flow. The energy that was stored in the magnetic field appears as a spark at the opening switch, where the current continues to flow for a moment.

Self-induced potential difference is proportional to the rate of change of the current:

$$\Delta V_{\mathrm{L}} = -L \frac{\Delta I}{\Delta t}$$

(Equation 13.3)

where ΔV_{L} is the self-induced potential difference and L is the *inductance* of the coil. The inductance of the coil depends on the number of turns of wire per unit length of the coil and on the nature of the core. The SI unit of inductance is the *henry* (H) and the millihenry (mH) is commonly used.

EXAMPLE 13.4

A 15-millihenry coil with a resistance of 160 ohms is connected to a 24-volt battery. When the current is 80 milliamperes, at what rate is it increasing?

SOLUTION

The potential drop across the resistance is

$$\Delta V = IR = (80 \times 10^{-3}\,\mathrm{A})(160\,\Omega) = 13\,\mathrm{V}$$

so the self-induced potential difference is

$$24\,\mathrm{V} - 13\,\mathrm{V} = 9\,\mathrm{V}$$

Then from Equation 13.3,

$$\frac{\Delta I}{\Delta t} = \frac{V_L}{L} = \frac{9\,\mathrm{V}}{15 \times 10^{-3}\,\mathrm{H}} = 600\,\mathrm{A/s}$$

Any real coil is made of wire, so it has resistance. What happens when the current turns on in a real coil?

Think of a real coil as a series combination of inductance and resistance. At the instant the switch is closed there is no current, so there is no potential drop across the resistance. The entire potential difference is across the inductance; the self-induced potential difference is equal to the potential difference supplied by the battery. Suppose our coil has 20 millihenries of inductance and 80 ohms of resistance. If we are using a 10-volt battery, the rate at which the current starts to increase when the switch is closed can be found from Equation 13.3:

$$\frac{\Delta I}{\Delta t} = \frac{\Delta V}{L} = \frac{10\,\mathrm{V}}{20 \times 10^{-3}\,\mathrm{H}} = 500\,\mathrm{A/s}$$

The current cannot keep on increasing at the rate of 500 amperes per second. As soon as there is current, part of the 10-volt battery potential difference is used up by the resistance. The rate of increase slows down; in a short time, the current becomes steady, determined only by the resistance. It settles down to a steady rate of

$$I = \frac{\Delta V}{R} = \frac{10 \text{ V}}{80 \ \Omega} = 0.13 \text{ A}$$

Once the current stops changing, there is no self-induced potential difference across the inductance of the coil.

CORE CONCEPT

A changing current in a coil induces a potential difference in such a direction as to oppose the change in the current.

TRY THIS
—5—

What is the inductance of a coil if the current in it gets to 20 milliamperes in the first microsecond after it is connected to a 24-volt battery? (Assume the resistance is large enough so that it takes 20 microseconds to reach the $\Delta V/R$ current.)

Inductance in AC

Figure 13.5 shows what happens when an AC potential is applied to an inductance, where the potential difference is proportional to the rate of change of the current. At (a), the current is rising rapidly and the potential difference is at its maximum. At (b) the current is at its maximum, the slope of the graph is zero, and there is no potential difference.

The power graph shows that the average power delivered to an inductance is zero, just as in a capacitance. The energy is stored in a magnetic field for a quarter cycle and returned to the circuit as the field collapses in the next quarter cycle.

Figure 13.5 may look much like Figure 13.3, but there is an important difference. *In an inductance, the potential difference is always 90° ahead of the current. In a capacitor, you remember, the potential difference is behind the current.*

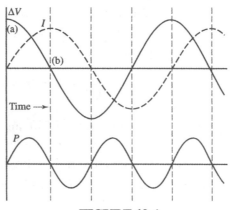

FIGURE 13.4

Just as in a capacitance, the impedance of an inductance depends on the frequency. However, the dependence is just the reverse. Inductance is a tendency to oppose changes in current; the faster the changes in current, the greater the impedance. The low frequencies go through an inductance most easily. The *inductive reactance* of a coil is given by the formula

$$X_L = 2\pi f L \qquad \text{(Equation 13.4)}$$

Note that for DC, when $f = 0$, there is no reactance, and the impedance of a coil will be just the resistance of the wire. In a capacitor, you remember, the reactance is infinite in DC, and there is no current.

Inductors differ from capacitors in another important respect. Except for superconducting magnets kept at 4 kelvins, every coil has resistance as well as inductance.

The relationship between current and potential difference in an AC circuit is complicated by the fact that they may be out of phase with each other. The current and potential difference in a resistance are in phase, but in inductance, they are 90° out of phase. If you think of a coil as a combination of inductance and resistance, the current in the inductance is 90° out of phase with that in the resistance. The ratio of potential difference to current is a combination of resistance and inductive reactance. It is called the *impedance* of the circuit, and is defined by this equation:

$$Z = \frac{\Delta V}{I} = \sqrt{R^2 + X^2} \qquad \text{(Equation 13.5)}$$

Where Z, in ohms, is the *impedance* of the combination.

EXAMPLE 13.5 How much is the current in the coil of an AC relay operating on 120 volts, 60 hertz AC if its resistance is 45 ohms and its inductance is 75 millihenries?

SOLUTION

First, find the reactance:

$$X_L = 2\pi f L = 2\pi(60 \text{ Hz})(75 \times 10^{-3} \text{ H}) = 28 \ \Omega$$

Then determine the net impedance:

$$Z = \sqrt{R^2 + X^2} = \sqrt{(45 \ \Omega)^2 + (28 \ \Omega)^2} = 53 \ \Omega$$

so the current is

$$\bar{I} = \frac{\Delta \bar{V}}{Z} = \frac{120 \text{ V}}{53 \ \Omega} = 2.3 \text{ A}$$

Every coil that carries current gets warm because there is always some part of the electric power that is delivered to the resistance of the wire. Since no net power is consumed by the inductance of the coil, the power delivered is $\bar{I}^2 R$ not $\bar{I}^2 Z$.

CORE CONCEPT

An inductor's impedance in an AC circuit increases with frequency, 90° out of phase with its resistance.

TRY THIS
—6—

At what rate is heat produced in a 3.0-henry coil with a resistance of 120 ohms if the coil is operating on 40 volts, 600 hertz AC?

The *R-L-C* Circuit

What is the impedance if a resistor and a capacitor are in series? Since the current and potential difference are in phase in the resistor and 90° apart in the capacitor, Equation 13.5 will give the correct impedance. The situation gets more complicated if a series circuit contains resistance, inductance and capacitance.

When both inductive and capacitive reactances are in the circuit, what value of **X** do we use in solving Equation 13.5? In the inductance, the potential difference is 90° *ahead* of the current; in a capacitor, the potential difference is 90° *behind* the current. In a series circuit, both have the same current, so the two reactances are just 180° apart. When one is positive, the other is negative, and vice versa. The net reactance is the difference between the two. Then Equation 13.5 becomes

$$Z = \sqrt{R^2 + (X_L - X_C)^2}$$ **(Equation 13.6)**

EXAMPLE
13.6

A 200-ohm resistor, a 42-microhenry coil, and a 250-picofarad capacitor are in series in a 1 200-kilohertz circuit. If the potential difference across the combination is 75 volts, find (a) the current; and (b) the rate of heat production.

SOLUTION

This is tedious, but there is no shortcut. First, find the capacitive reactance:

$$X_C = \frac{1}{2\pi fC} = \frac{1}{2\pi (1.20 \times 10^6 \text{ Hz})(250 \times 10^{-12} \text{ F})} = 530 \ \Omega$$

Then find the inductive reactance:

$$X_L = 2\pi fL = 2\pi(1.20 \times 10^6 \text{ Hz})(42 \times 10^{-6} \text{ H}) = 320 \ \Omega$$

Then the total impedance is

$$Z = \sqrt{R^2 + (X_L - X_C)^2} = \sqrt{(200 \ \Omega)^2 + (320 \ \Omega - 530 \ \Omega)^2} = 290 \ \Omega$$

So the current is

$$\overline{I} = \frac{\Delta \overline{V}}{Z} = \frac{75 \text{ V}}{290 \ \Omega} = 0.26 \text{ A}$$

The power consumed is

$$P = \overline{I}^2 R = (0.26 \text{ A})^2 (200 \ \Omega) = 13 \text{ W}$$

which is the rate of heat production, 13 joules per second.

The combination of inductance and capacitance can be used to separate a given frequency out of a mixture. It is this kind of combination that makes it possible for you to tune your radio to the frequency on which your favorite station broadcasts.

At low frequencies, capacitive reactance dominates. At high frequencies, the capacitive reactance drops and the inductive reactance dominates. Remember that they are 180° out of phase with each other; it is the *difference* between them that contributes to the impedance of the circuit. At some frequency, the two reactances are equal and will cancel each other out. Impedance is then a minimum, equal to the resistance of the circuit, and current is a maximum. That frequency is the *resonant frequency* of the circuit. The total impedance of the *R-L-C* series circuit at the resonant frequency is nothing but the resistance of the coil.

The condition for resonance is $X_L = X_C$:

$$2\pi f L = \frac{1}{2\pi f C} \text{ so that}$$

$$f = \frac{1}{2\pi\sqrt{LC}} \qquad \textbf{(Equation 13.7)}$$

EXAMPLE 13.7 What is the resonant frequency of a 35-millihenry coil and a 12-picofarad capacitor?

SOLUTION
From Equation 13.7

$$f = \frac{1}{2\pi\sqrt{LC}} = \frac{1}{2\pi\sqrt{(35 \times 10^{-3} \text{ H})(12 \times 10^{-12} \text{ F})}} = 246 \text{ kHz}$$

CORE CONCEPT
Inductance and capacitance in an AC circuit are in opposite phase, which must be considered in calculations.

TRY THIS
—7—

What capacitance is needed to tune a 5.0-millihenry coil to a frequency of 1 560 megahertz?

Conductors

All these circuit elements must be connected by pathways that carry electrons from one element to another. This is the function of connecting wires, usually of copper. Silver and gold conduct better but are fiscally undesirable.

What makes a material a conductor of electricity is the structure of its atoms. Every atom consists of a nucleus surrounded by electrons organized into concentric shells. The outermost shell determines the electrical properties of the atom. This shell contains anywhere from one to eight electrons, the number being unique to each element. Metals have one or two, sometimes three, electrons in the outer shell. In good conductors, one or more of these outer electrons is so loosely bound to its atom that it floats freely in the space between atoms instead of staying attached to its parent. A piece of solid copper is full of a sea of free electrons, moving at random and ricocheting off the atoms fixed in their crystal lattices.

In copper, there is an average of one free electron per atom. Suppose we have a 1-centimeter cube of copper. How many free electrons are there in it? The mass of that sample is about 9.0 grams, and the atomic mass of copper is 64 grams per mole. Since there are 6.0×10^{23} atoms to a mole (Avogadro's number), the number of atoms, and therefore also the number of free electrons in our sample, is

$$\left(\frac{9 \text{ grams}}{64 \text{ grams/mole}} \right) (6.0 \times 10^{23} \text{ atoms per mole}) = 8 \times 10^{22} \text{ atoms}$$

We can find out how fast all these electrons are moving by treating these free electrons as though they were the molecules of a gas. Equation 7.7 gives the average random kinetic energy of a molecule—or an electron in this case—as $3/2kT$, where T is the kelvin temperature and k is Boltzmann's constant. Then the kinetic energy of our electrons at room temperature is

$$3/2(1.38 \times 10^{-23} \text{ J/K})(300 \text{ K}) = 6.2 \times 10^{-21} \text{ J}$$

Since this must equal $\frac{1}{2}mv^2$ (Equation 5.3) and we know the mass of an electron, we can get the speed of the electrons:

$$v = \sqrt{\frac{2E}{m}} = \sqrt{\frac{2(6.2 \times 10^{-21}\text{ J})}{9.1 \times 10^{-31}\text{ kg}}} = 1.2 \times 10^5\text{ m/s}$$

which is very fast indeed.

Now suppose we draw this cubic centimeter of copper out into a wire 10 meters long and 1 square millimeter in cross section. We will connect this wire to the terminals of a 0.17-volt battery, just enough to make 1 ampere flow. The battery establishes an electric field in the wire, gently nudging the electrons up toward the positive terminal.

EXAMPLE 13.8
How much force does this field exert on each electron?

SOLUTION

The electric field in the wire is

$$\frac{0.17\text{ V}}{10\text{ m}} = 0.017\text{ V/m} = 0.017\text{ N/C}$$

Then the force on each electron (Equation 9.2) is

$$F_{\text{el}} = \mathscr{E}q = (0.017\text{ N/C})(1.6 \times 10^{-19}\text{ C}) = 2.7 \times 10^{-21}\text{ N}$$

These electrons are feeling this force pushing them up the field; they do not accelerate very far before bumping into an atom. As they bounce around, they gradually make their way to the positive terminal. Now let's find out how fast this drift velocity is.

The current is 1 ampere, so 1 coulomb of charge leaves the end of the wire every second. The electron with the furthest distance to travel goes 10 meters; by the time it gets to the end, all the free electrons in the wire have passed out of it. The question, then, is how long does it take all the electrons to pass out of the wire at the rate of 1 coulomb per second?

The total charge on all these electrons is the number of electrons times the charge per electron:

$$(8 \times 10^{22}\text{ electrons})(1.6 \times 10^{-19}\text{ C/electron}) = 1.3 \times 10^4\text{ C}$$

Passing out of the wire at the rate of 1 coulomb per second, the charge will all leave the wire in 1.3×10^4 seconds. The furthest electron travels 10 meters in this time, so it is drifting along at

$$(10\text{ m})/(1.3 \times 10^4\text{ s}) = 8 \times 10^{-4}\text{ m/s}$$

If you throw a switch to turn on a lamp a couple of meters away, it takes 40 minutes for an electron to get from the switch to the lamp. What turns on the lamp is a pulse, each electron pushing those ahead of it until the pulse arrives at the lamp. The pulse travels at a speed approaching the speed of light, so you can start reading immediately.

If you are feeding the lamp with AC, it is not a pulse that travels but a longitudinal wave. Each electron vibrates, passing its energy along to the next, just like molecules of air passing a sound wave. The wave travels at a speed close to the speed of light.

EXAMPLE 13.9

What is the wavelength of the AC current wave in a 60-hertz transmission line?

SOLUTION
From Equation 8.4,

$$\lambda = \frac{v}{f} = \frac{3.0 \times 10^8 \text{ m/s}}{60 \text{ Hz}} = 5 \times 10^6 \text{ m}$$

At 60 hertz, Washington, D.C. and San Francisco are in phase. On the other hand, at the enormously high frequencies of microwave radar, for example, differences in phase occur over short distances and must be taken into account.

CORE CONCEPT
Electric current in a wire is carried by the movement of free electrons.

TRY THIS
—8—

At what frequency is a microwave oven operating if the wavelength is 10 centimeters?

Semiconductors

Metals are good conductors of electricity. Their outermost electron shells contain, generally, only one or two electrons out of the eight possible. These electrons are loosely held and drift off the atom to join the sea of free electrons. Any electric field can supply enough energy to get an electron moving, until it caroms off an atom in the crystal lattice. As the temperature rises, the atoms vibrate more strongly, interfering with the motion of the free electrons. That is why the resistance of a metal increases with temperature.

Elements with six to eight electrons in the outer shell keep them firmly bound and so do all those chemical compounds that we call insulators, such as glass, rubber, or polyethylene. It takes a strong electric field to supply enough energy to remove one of these electrons. When the field is strong enough, an electron is pulled off with high energy. It smashes into another atom, releasing one of its electrons. The whole process cascades until a current flashes through the insulator. This is what happens when the insulator between the plates of a capacitor breaks down or when lighting strikes.

Then there are those elements with just four electrons in their outer shell, such as carbon, silicon, and germanium. In these *semiconductors,* the electrons are bound, but not very strongly. Ordinary thermal agitation of the atoms is enough to release a few electrons, about one electron per billion atoms in silicon at room temperature. These become free electrons and can be moved as easily as those in a metal. As the temperature rises, more electrons will join the free-electron sea. This makes the resistance of a semiconductor decrease as it heats up, just the opposite of a metal.

The ability of a silicon semiconductor to carry a current can be increased by "doping" the silicon. In this process, a small amount of arsenic, for example, is mixed with the silicon. Arsenic has five outer electrons; the crystal lattice of silicon uses only four electrons for each atom. The arsenic atoms join the lattice by releasing the extra electron into the free-electron sea. This can raise the number of free electrons from one in a billion to one in a million atoms. Because of its excess negative charge, this kind of semiconductor is called an *n*-type semiconductor.

Doping silicon with gallium has the opposite effect. Gallium has only three electrons in its outer orbit. When it joins the crystal lattice, it leaves a hole. This hole represents a missing electron, so it has an effective positive charge. An electron can move into the hole, leaving a positive hole somewhere else. This *p*-type semiconductor behaves as though there is a sea of positive holes, free to move when subjected to an electric field.

The device diagrammed in Figure 13.5 is a solid-state diode. It consists of a block of *p*-type silicon in contact with a block of *n*-type, with conducting metal plates on the outside. If these plates are connected to a battery as shown in (a), current flows. Electrons flow from the battery into the *n*-type side, and holes come into the *p*-type side. At the junction, the holes and electrons eliminate each other, and new holes and electrons keep coming in from the battery. Current flows through the diode.

Figure 13.5b shows what happens when the battery is connected in the opposite direction. Electrons are attracted to the positive plate and holes to the negative plate. The two kinds of charge do not meet at the junction, and no current flows through the diode. Current can flow through the diode only in the $p \rightarrow n$ direction.

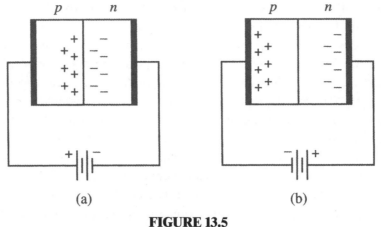

(a)   (b)

FIGURE 13.5

Until the invention of doped semiconductors, the common way to establish this one-way traffic through a circuit was by using a vacuum tube diode. The solid-state diode needs no heated filament and so uses no energy; it is small and can even be a microscopic dot on an integrated circuit chip.

A diode can be used to convert AC to DC. If an AC potential is applied to the diode, current will flow only while the potential is in the $p \rightarrow n$ direction. Figure 13.6 shows how two diodes can be used to make full use of both halves of the AC cycle. The diodes face each other in opposite directions so that one or the other is always passing current. The two currents are united to produce a current that always goes in one direction although it is fluctuating wildly. The fluctuations are smoothed out by means of an inductance-capacitance filter.

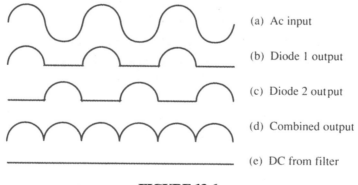

(a) Ac input

(b) Diode 1 output

(c) Diode 2 output

(d) Combined output

(e) DC from filter

FIGURE 13.6

Another type of diode makes use of the fact that there is a limit to the one-way flow property. If the voltage in the $n \rightarrow p$ direction gets too high, current will surge through the diode in the wrong direction. The potential is thus prevented from ever exceeding the limit. Diodes of this sort are used to control voltages.

CORE CONCEPT

Doped silicon semiconductors can be used to form diodes that can rectify AC.

TRY THIS
—9—

Why does current flow through a diode in the $n \rightarrow p$ direction if the potential difference in that direction gets very high?

Transistors

Adding a third layer to the solid-state diode produces a transistor. The n-p-n transistor diagrammed in Figure 13.7 consists of a sandwich of two n-type semiconductors separated by an extremely thin layer of p-type, called the base. Currents are organized through this transistor in such a way that a small AC potential introduced into the base produces a substantial AC current between the two pieces of "bread" that make up the sandwich. When that current flows through the load resistor, it develops an AC potential difference much larger than the one on the base.

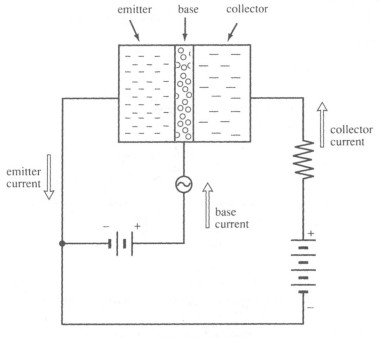

FIGURE 13.7

In the circuit as shown, a small potential keeps the base positive with respect to the emitter. The collector is much more strongly positive. Electrons will flow from the emitter into the base. Because the base is extremely thin and because of the high potential on the collector, most of these electrons flow on into the collector. The three currents shown (flowing in the directions opposite to the electron flow) must add up: the emitter current is the sum of the other two.

Now see what happens if an AC signal is introduced into the potential between emitter and base. When the emitter becomes more negative, more electrons flow from emitter to base. Most of them go from the base into the collector. The result is that the collector current increases greatly; the change in this current can be hundreds of times greater than the change in the base current. The transistor has amplified the signal.

An audiotape is a magnetic device. Its surface is magnetized in such a way that the magnetization is an analog of the sound wave that has been recorded on it. As it passes through the reading head, the magnetic fields of the tape induce a tiny current into a coil in the head. That current is an AC analog of the sound wave. The current creates the potentials that feed into the base of a transistor, which amplifies these potentials, probably several times. The final output is passed into the loudspeakers. Whether the original signal comes from a radio, a record player, a CD, or a telephone, the principle is the same: the weak AC sound analog signal is amplified by transistors. Nowadays, the transistors are probably nothing but dots on a chip.

CORE CONCEPT

Solid-state diodes and transistors serve as rectifiers, voltage controllers, and amplifiers in electronic circuits.

TRY THIS
—10—

How would the circuit of Figure 13.7 have to be different if a *p-n-p* transistor were used an amplifier?

REVIEW EXERCISES FOR CHAPTER 13

Fill-In's. For each of the following, fill in the missing word or phrase.

1. Electronics is the use of electricity to process _____.

2. An electronic vacuum tube must use energy to heat its _____.

3. A resistor is calibrated in terms of its resistance and its _____.

4. The only electronic device that obeys Ohm's law at all frequencies is the _____.

5. A layer of insulating material is an essential part of a(n) _____.

6. The farad is a measure of _____.

7. A capacitor stores energy in the form of a(n) _____.

8. In a capacitor, the thinner the insulator, the less the _____.

9. Winding a wire into a coil will change its _____.

10. The potential difference induced in an inductor exists only while the current in it is _____.

11. At a given potential difference, the DC steady current in an inductor depends on its _____.

12. Rapid changes of current are opposed by the property of a circuit called its _____.

13. Rapid changes of potential are opposed by the property of a circuit called its _____.

14. The number of free electrons determines the property of a circuit element called its _____.

15. Semiconductors characteristically have _____ electrons in their outermost orbits.

16. The average speed of electrons in a wire carrying a current is called their _____.

17. An alternating current in a wire is carried by a(n) _____ wave.

18. A(n) _____ is a solid-state device used as a rectifier.

19. For an AC signal to be amplified, it is introduced onto the _____ of a transistor.

20. Adding gallium to silicon produces a(n) _____ semiconductor.

21. As a filter, a capacitor passes the _____ frequencies.

22. If a 20-microfarad capacitor is connected to a 120-volt, 400-cycle AC generator, the power loss in the capacitor is _____.

23. As the frequency of the AC in a capacitor increases, the phase difference between potential and current _____.

24. A resistor and a capacitor are connected in series to an AC generator. As the frequency increases, the potential difference across the resistor _____.

25. In an AC circuit, the ratio between effective values of potential difference and current is called _____.

26. A variable inductance is connected to an AC generator. As the inductance increases, the current _____.

Multiple Choice.

1. The functions of a vacuum tube have largely been replaced by those of
   (1) capacitors (2) transistors (3) resistors (4) inductors (5) conductors.

2. A high-wattage resistor is made large because a large resistor has
   (1) less resistance   (4) bigger volume
   (2) more resistance   (5) more surface.
   (3) higher breakdown voltage

3. A farad is equivalent to one
   (1) volt per ampere   (4) coulomb per ampere
   (2) coulomb per volt   (5) ampere per coulomb.
   (3) volt per coulomb

4. No DC passes in either direction through a(n)
   (1) transistor   (4) diode
   (2) inductor   (5) capacitor.
   (3) semiconductor

5. Low frequencies pass more easily than high ones through a(n)
   (1) capacitor (2) diode (3) inductor (4) vacuum tube (5) resistor.

6. An inductor stores energy in a(n)
   (1) current (2) coil (3) insulator (4) magnetic field (5) electric field.

7. The henry is a measure of the relationship between
   (1) current and potential
   (2) inductance and resistance
   (3) potential and rate of change of current
   (4) inductance and capacitance
   (5) current and rate of change of potential.

8. Resistance rises with temperature in a conductor because
   (1) increased atomic vibrations interfere with the passage of electrons
   (2) fewer electrons are available at higher temperatures
   (3) expansion of the metal hinders the flow of current
   (4) higher temperatures tend to reduce potential differences
   (5) more positive holes appear at higher temperatures.

9. When a diode is carrying current
   (1) electrons and holes eliminate each other
   (2) more electrons flow in than out
   (3) more electrons are produced in the emitter
   (4) more electrons are produced in the collector
   (5) more holes are produced in the emitter.

10. The positive charge in a *p*-type semiconductor is a(n)
    (1) positive electron
    (2) silicon nucleus
    (3) gallium nucleus
    (4) electron from an arsenic atom
    (5) empty space in a crystal.

Problems.

1. A quarter watt resistor is 2 centimeters long and 4 millimeters in diameter. If a 2-watt resistor made of the same material is to be made 4 centimeters long, what must its diameter be?

2. A 40-microfarad capacitor has a breakdown voltage of 12 volts. If a steady current of 2.0 milliamperes flows into it, how long will it take for the capacitor to break down?

3. A typical lightning flash discharges a capacitance consisting of clouds and earth separated by air as an insulator. The current in the flash averages about 200 kiloamperes, and it lasts for 20 microseconds. Air breaks down electrically when the field is about a million volts per meter. If the flash is 5 kilometers long, what is the capacitance?

4. What is the largest potential difference that can be placed across a 200-ohm resistor rated at a quarter watt?

5. A 3.6-ampere current is flowing into a 2-microfarad capacitor. At what rate is the potential difference on the capacitor increasing?

6. A 40-millihenry coil is connected to a 12-volt battery, and the current increases to a steady 0.50 amperes. At what rate is the current increasing when it is 0.20 amperes?

7. When a 25-millihenry coil is connected to a 24-volt battery, the current is increasing at 400 amperes per second when it is 250 milliamperes. When a steady current is reached, at what rate is heat produced?

8. An audio signal passes through a capacitor. Compare the currents at 1 000 hertz and at 50 hertz if both frequencies have the same potential difference.

9. What is the capacitance of a capacitor if an AC voltmeter placed across it reads 120 volts and an AC ammeter in series with it reads 2.5 milliamperes at 1 000 hertz?

10. With the overall potential difference at 60 volts, how much resistance must be placed in series with a 4.0-picofarad capacitor to keep a 25-megahertz current at 30 milliamperes?

11. What potential difference is needed to keep 250 milliamperes of 1 200 megahertz AC flowing through a 20-microhenry coil if its resistance is 75 ohms?

12. What is the current if a 1 800 megahertz AC potential of 12 volts is applied to a 550-picofarad capacitor in series with a 5.0-microhenry coil that has a resistance of 40 ohms?

13. What capacitance is needed to tune an FM radio station broadcasting on 94 megahertz if the tuning coil has an inductance of 1.2 microhenries?

14. In a 400-hertz AC circuit, a 0.25-henry coil with 30 ohms resistance is in series with a 1.0-microfarad capacitor. To keep the current constant at 220 milliamperes with an applied potential difference of 100 volts, how much resistance should be added in series?

Energy in Space

WHAT YOU WILL LEARN

How electric and magnetic fields produce electromagnetic waves; properties of these waves, including visible light.

Field Interactions

In 1865, the Scottish physicist James Clerk Maxwell did for electromagnetism what Newton had done two centuries earlier for mechanics. He developed a set of equations from which everything known about electricity and magnetism could be derived. His conclusions not only resulted in a deeper understanding of the phenomena but paved the way for a technological development that changed our world. It also set the stage for the revolution that Albert Einstein made with the theory of relativity.

Maxwell's four equations use a form of mathematics beyond the level of this book, but they can be expressed, roughly, in ordinary English:

1. Electric field lines originate on positive charges and terminate on negative charges.

2. Magnetic field lines always form closed loops.

3. A changing magnetic field produces an electric field.

4. An electric current or a changing electric field produces a magnetic field.

The first two equations formalize in mathematical terms the laws of field lines that had been developed by Michael Faraday. The third equation is Faraday's law of induction in a different form. The part of the fourth equation dealing with current is Oersted's discovery of the effect of a current on a magnet, made quantitative. The only new thing in these equations, really only an assumption by Maxwell, was the concept that when an electric field is changing, a magnetic field results.

Consider the effects produced by an electric charge. The first law says that it is surrounded by an electric field. The fourth law says that if the charge is in motion, thus constituting a current, the electric field is changing and will produce a magnetic field as well.

The surprise comes in considering what happens if the electric charge is *accelerated*. Then, the magnetic field produced by its motion becomes a changing magnetic field. This must, in turn, produce a new electric field. If the charge is oscillating, in simple harmonic motion, for example, it is constantly producing new, changing electric fields, which produce changing magnetic fields, and so on. Each field reinforces the other; they chase each other out into space, away from the oscillating charge. The result is an electromagnetic wave, the necessary product of an accelerated charge.

Maxwell was able to calculate the speed with which this electromagnetic wave would travel in free space. His theory of the electromagnetic wave tells us that it must travel with a speed that depends on the electric and magnetic properties of empty space. These are the well-known constants of Coulomb's law of static electricity (Equation 9.1; $k = 9.00 \times 10^9$ newton-meters squared per coulomb squared) and the law of production of a magnetic field (Equation 11.3; $k' = 10^{-7}$ newton per ampere squared). The speed of an electromagnetic wave turns out to be

$$c = \sqrt{\frac{k}{k'}} = \sqrt{\frac{9.00 \times 10^9 \ \text{N} \cdot \text{m}^2/\text{C}^2}{10^{-7} \ \text{N/A}^2}}$$

$$c = 3.00 \times 10^8 \ \text{m/s}$$

EXAMPLE 14.1   The moon is 3.8×10^8 meters from the earth. How long does it take a radio signal to get to the moon and back to a receiver on earth?

SOLUTION

Going both ways, the signal travels 7.6×10^8 meters at speed c, so it takes

$$\Delta t = \frac{\Delta s}{v} = \frac{7.6 \times 10^8 \text{ m}}{3.0 \times 10^8 \text{ m/s}} = 2.5 \text{ s}$$

The nature of light, in those days, was one of the great mysteries of science. Light was known to be a wave; the speed and wavelengths of visible light had been measured with considerable accuracy. Still, no one had any idea what kind of wave it is. The electromagnetic wave postulated by Maxwell had a theoretical speed identical with the known speed of light. Mathematical deduction explained to us that we can finally understand the nature of the waves that illuminate our world and support life by using the energy of the sun.

It took only a few years for experimenters to produce other kinds of electromagnetic waves. Today, they bring us sound and images to entertain us; satellites in the sky use them to communicate with earth; we send them billions of miles into space to control the probes exploring our solar system.

CORE CONCEPT

Light is an electromagnetic wave traveling through empty space at a constant speed c.

TRY THIS
—1—

What is the frequency of a light wave whose wavelength is 5.5×10^{-7} meter?

Electromagnetic Waves

The simplest kind of electromagnetic wave has a sine-wave form, and it is produced by sending an alternating current into an antenna. As the charge in the antenna oscillates back and forth, it is accelerated, and therefore it generates an electromagnetic wave. For best results, the antenna should be a half-wavelength long.

Suppose, for example, that an FM station wants to send a signal into space at a frequency of 98 MHz (megahertz, or 10^6 hertz). This is 98 on the dial of your FM radio. The station generates an AC at that frequency and sends it down a transmission line consisting of a pair of parallel wires. The wires end in an antenna; in the simplest case, this is simply the same wires bent at right angles to the transmission line, as shown in Figure 14.1.

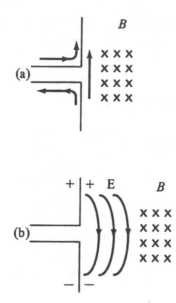

FIGURE 14.1

At a given moment, the current in the system might be as shown in Figure 14.la. Looked at in the antenna only, it appears as a current flowing from one end to the other. When it reaches the end, it can go no further. In a quarter cycle, charge piles up at the other end of the antenna. Remember that an AC in a line is a wave. Since there can be no current at the ends of the antenna, current nodes form there. If we make the antenna just a half-wavelength long, we can set up a stable standing wave of current in the antenna. We know that the wave travels in the wire at almost the speed of light, so we can easily figure out how long the antenna should be.

EXAMPLE 14.2   How long should the antenna be to broadcast at 98 megahertz?

SOLUTION

The wavelength, from Equation 8.4, is

$$\lambda = \frac{c}{f} = \frac{3.0 \times 10^8 \text{ m/s}}{98 \times 10^6 \text{ Hz}} = 3.06 \text{ m}$$

The most efficient antenna is half this length, or 1.53 meters.

It is the accelerated charges, moving back and forth in the antenna in a standing wave pattern, that produce the electromagnetic wave. In Figure 14.1a we have caught the current at a moment when it is going upward in the antenna. Using our familiar right-hand rule, we can see that, at that moment, there is a magnetic field around the antenna. In the space to the right of the antenna, the direction of the field is into the page, as shown by the x's, the tail ends of the field vectors.

Figure 14.1b shows the situation a quarter-cycle later. The current has dropped to zero. However, charge has piled up at the upper end of the antenna, leaving the lower end negative. This means that there must be an *electric* field around the antenna, its field lines running out at the top and into the antenna at the bottom. In the space to the right, the electric field lines point downward.

Both the electric and the magnetic fields are constantly changing, so each one reinforces the other. The final result is shown in Figure 14.1c. Magnetic fields and electric fields, both alternating in direction, chase each other through space, moving away from the antenna in all directions, at the speed of light. This is an electromagnetic wave.

To set the record straight, it should be noted that the fields alternate at first, 90° out of phase with each other. As they travel, they gradually come into phase. At all times, the electric and magnetic fields are perpendicular to each other.

Your receiving antenna, set parallel to the transmitting antenna, detects these alternating electric and magnetic fields and responds by setting up an alternating current, a standing wave of its own. This antenna is also a half-wavelength long, so that it resonates at its natural frequency. The AC signal is amplified electronically, and other electronic circuits extract the information from the signal and convert it to audio-frequency AC signals that operate the loudspeaker.

Electromagnetic waves of many different wavelengths are produced by a number of different processes. Here is a breakdown of the whole electromagnetic spectrum:

Radio waves, produced by electronic circuitry: wavelengths from 1 000 meters to 10^{-3} meter (1 millimeter). The longest waves are used in the AM broadcast band; then come, in sequence, short waves, the very high-frequency (VHF) waves used in FM and television, UHF waves, and microwaves.

Infrared rays, emitted by all warm objects, are produced by the random thermal motions of molecules. Wavelengths are from 10^{-3} meters down to about 7×10^{-7} meters.

Visible light occupies the tiny part of the spectrum between 7×10^{-7} meters (700 *nanometers,* or 700×10^{-9} meters) to about 400 nanometers. It is produced by changes in the energy levels of the outer electrons in atoms.

Ultraviolet rays are produced in much the same way, but their wavelength is too short to enter the eye; they cannot pass through the lens of the eye. Their wavelengths range from 400 nanometers to about 5 nanometers.

X-rays and *gamma rays* are produced by energy transitions in the innermost electrons of large atoms, or in nuclei. Wavelengths are from 5×10^{-9} meters down to about 10^{-13} meters, smaller than an atom.

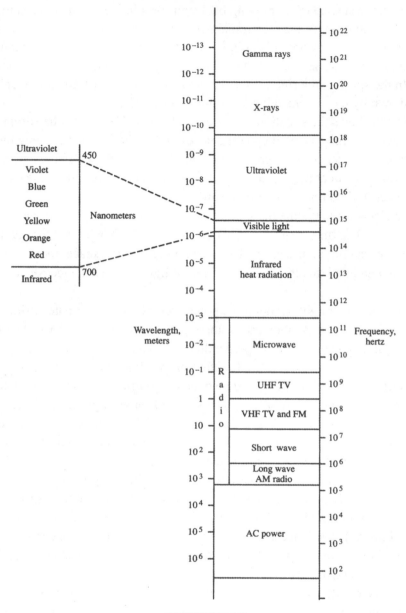

FIGURE 14.2

EXAMPLE
14.3

Compare the frequency of the longest radio waves with that of the shortest gamma rays.

SOLUTION

Longest radio waves:

$$f = \frac{c}{\lambda} = \frac{3.0 \times 10^8 \text{ m/s}}{1\,000 \text{ m}} = 3.0 \times 10^5 \text{ Hz, or 300 kHz}$$

Shortest gamma rays:

$$\frac{3.0 \times 10^8 \text{ m/s}}{10^{-13} \text{ m}} = 3 \times 10^{21} \text{ Hz}$$

which is a hundred million billion times as great.

CORE CONCEPT

Electromagnetic waves are produced by accelerated charges in a wide range of wavelengths.

TRY THIS
—2—

What is the best length for a microwave antenna for use on a frequency of 4 500 megahertz?

Straight Lines or Curves?

Basically, the difference between one electromagnetic wave and another lies in their frequencies. The frequency difference produces profound differences between one kind of wave and another.

For example, we learned in Chapter 8 that waves tend to bend around corners (diffract) and that the amount of bending increases with wavelength. The waves of AM radio, from 60 meters to 100 meters long, easily bend around buildings and other barriers and find their way to your receiving antenna. They can travel for a thousand miles or more, following the curvature of the earth. In contrast, the range of shortwave radio, in straightline transmission, is 100 miles or less. Short waves, however, bounce off the ionosphere, an ionized layer of air that lies above the stratosphere. That is why, especially at night, short waves may reach remote parts of the world.

FM and television signals, at wavelengths of a few meters, can be received only up to 20 or 25 miles from the transmitter and are likely to be blocked by buildings or mountains. And the propagation of microwaves at wavelengths of a few centimeters is essentially line of sight. Microwave relay stations, now a familiar sight in the country, are placed on hilltops in view of each other. Each antenna is a half-wave long, no more than a couple of centimeters. It is backed by a parabolic dish that reflects the transmitted waves into a narrow beam aimed at the next relay tower. This is how long-distance telephone signals travel across the country.

At the extremely short wavelengths of visible light, diffraction is usually negligible. That is why we find that light travels in straight lines, and that is why an object placed near a point source of light casts a sharp-edged shadow, as in Figure 14.3. Even with visible light, however, effects due to diffraction can be found. Pierce a piece of aluminum foil with a fine needle and pass light through the hole. If you examine the spot of light that falls on a screen after passing through such a hole, you will see that it is surrounded by a series of bright and dark rings. Light is spilling out beyond the spot that would form according to the straight-line propagation rule. The dark bands are produced by destructive interference where light from different parts of the hole arrives out of phase. See Figure 14.4.

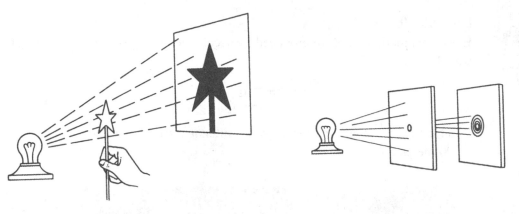

| **FIGURE 14.3** | **FIGURE 14.4** |

CORE CONCEPT
Electromagnetic waves diffract, and the effect is greater at longer wavelengths.

TRY THIS
—3—

Explain why a microscope is unable to show the existence of objects no bigger than the wavelength of the light.

Measuring Wavelength

The wavelengths of light, small as they are, can be measured with extremely high accuracy. The method takes advantage of the phenomena of diffraction and interference.

As we saw in Chapter 8, a small opening, not much bigger than a wavelength, acts as a point source of waves. It is more convenient to use a thin slit rather than a point. If light is passed through such a slit, as in Figure 14.5, and then through two slits parallel to each other and close together, the light emerging from the double slit and falling on a screen forms a series of alternate light and dark bars, as shown. The bright bars are regions of constructive interference, and the dark bars are regions of destructive interference. From the geometry of the system, the wavelength of the light can be found.

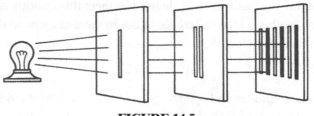

FIGURE 14.5

The geometry is diagramed in Figure 14.6. The two slits serve as point sources of light, oscillating in phase with each other. Any point that is the same distance from both slits will therefore receive crests from both slits in phase. Such a point is an *antinode*. All these points lie on the antinodal line marked $n = 0$ on the diagram; this line is perpendicular to the line joining the slits, and halfway between them. Bright light travels down this antinodal line, forming a bright bar on the screen.

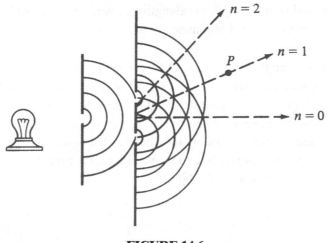

FIGURE 14.6

There are other antinodes. The point marked P is closer to the upper slit than to the lower slit. It is, in fact, *exactly one wavelength closer*. Thus, a crest leaving the lower slit takes longer to get to P than a crest leaving the upper one—exactly one period longer. The crest from the upper slit arrives at point P at the same moment as the crest that left the lower slit one period earlier. All points on the antinodal line marked $n = 1$ are precisely one wavelength farther from the upper slit than from the lower one. At all such points, the interference is constructive. Bright light travels down the line, lighting up another bar on the screen.

Halfway between the antinodal lines marked $n = 0$ and $n = 1$, a crest from the upper slit arrives at the same moment as a trough from the lower slit, and the interference is destructive. There is a nodal line between the anti-nodal lines $n = 0$ and $n = 1$, and it marks a dark bar on the screen.

There are yet other antinodes. All of them must meet this condition: The path difference to any antinode, that is, the difference between the distances to the two slits, must be a whole number of wavelengths:

$$s_2 - s_1 = n\lambda \qquad \text{(Equation 14.1a)}$$

where n is any whole number and λ is the wavelength. Each of the bright bars is the end of an antinodal line, and each can be numbered, positive and negative, from 0 (the central line) to some higher value.

EXAMPLE 14.4   In the interference pattern produced by a pair of slits acting on red light of wavelength 720 nanometers, how much farther is the fifth bright bar from one slit than from the other?

SOLUTION

At the fifth bright bar, a crest from one slit meets a crest that left the other slit 5 periods earlier and so traveled 5 wavelengths farther. The path difference is therefore 5×720 nanometers $= 3\,600$ nanometers.

In practice, the device that makes use of this principle is a *diffraction grating*. It consists of a set of scratch marks made on a sheet of glass—a great many of them, accurately drawn with a diamond, straight and parallel. There may be 700 or 800 lines to a millimeter in such a grating. When light is passed through such a grating, it forms a series of nodes and antinodes, exactly like those of a double slit.

With a little geometry, it is possible to arrive at a useful expression that will make it possible to find the wavelength of the light:

$$\sin \theta_n = \frac{n\lambda}{d} \qquad \text{(Equation 14.1b)}$$

where d is the distance between the slits, n is a whole number, λ is the wavelength, and θ is the angle that any antinodal line makes with the central $n = 0$ line, as shown in Figure 14.7. The equation works well either for a pair of slits or for the multiple slits of a diffraction grating.

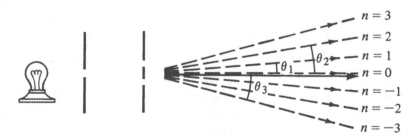

FIGURE 14.7

EXAMPLE 14.5 If light with a wavelength of 580 nanometers is passed through a pair of slits 0.004 millimeter apart, at what diffraction angle will the first two bright bars appear on either side of the central bar?

SOLUTION

For the first bar, $n = 1$, so

$$\sin \theta = \frac{\lambda}{d} = \frac{580 \times 10^{-9} \text{ m}}{0.004 \times 10^{-3} \text{ m}}$$

which gives $\theta = 8.3°$. For the next bar, $n = 2$ and

$$\sin \theta = \frac{2(580 \times 10^{-9} \text{ m})}{0.004 \times 10^{-3} \text{ m}}$$

and $\theta = 16.9°$.

EXAMPLE 14.6 What is the wavelength of monochromatic light that makes a diffraction angle of 26.5° with a grating ruled 640 lines per millimeter?

SOLUTION

This is presumably the first maximum, so $n = 1$ and, from Equation 14.1b,

$$\lambda = d \sin \theta$$

$$\lambda = \left(\frac{1}{640} \text{ mm}\right)(\sin 26.5°)$$

$$\lambda = 6.97 \times 10^{-4} \text{ mm, or 697 nm}$$

A *spectrometer* is an instrument for measuring the wavelengths in a beam of light. The configuration of a grating spectrometer is shown in Figure 14.8. The light to be analyzed passes through a narrow slit, and from there onto a diffraction grating. A screen is placed a distance L from the grating. The light that passes straight through the grating, along the axis of the instrument, forms a vertical line *(n = 0)*. Lines formed by diffraction appear to the sides of the axis. Each wavelength forms a separate $n = 1$ line; only the $n = 1$ lines on one side are used. The distance x from the axis to the $n = 1$ line depends on the wavelength of the light. If L is large, the ratio x/L is very nearly equal to the value of $\sin \theta$ in Equation 14.1b. Therefore, the wavelength of the light can be found from the expression

$$\lambda = d\frac{x}{L} \qquad \text{(Equation 14.1c)}$$

Since d and L are fixed dimensions of the spectrometer, the wavelength of any spectral line is directly proportional to its distance from the axis.

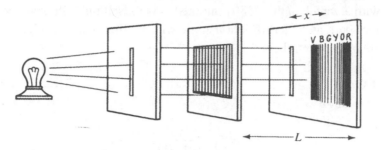

FIGURE 14.8

When white light is passed through a spectrometer, its spectrum is a complete rainbow of colors. One end of the spectrum is violet, and it shades through intermediate hues into blue, green, yellow, orange, and finally red. The human eye can distinguish about 100 subtle color differences in the spectrum of white light. The wavelengths range from about 450 nanometers at the violet end to 720 nanometers at the red end (1 nanometer = 10^{-9} meters).

EXAMPLE 14.7   If a grating is ruled 580 lines per millimeter and the screen is 65 centimeters from the grating, how far from the axis will the violet line of a spectrum appear?

SOLUTION

Since there are 580 lines per millimeter, the distance between lines is $(10^{-3}$ m$)/580$. Then, from Equation 14.4c,

$$x = \frac{\lambda L}{d} = \left(4.5 \times 10^{-7} \text{ m}\right)(0.65 \text{ m})\left(\frac{580}{10^{-3} \text{ m}}\right) = 1.70 \times 10^{-1} \text{ m}$$

CORE CONCEPT

The interference pattern produced by a diffraction grating provides a means of measuring wavelength.

TRY THIS
—4—

A spectroscope is made with a grating ruled 640 lines per millimeter, and the screen is 38 centimeters from the grating. What is the wavelength, in namometers, of a beam of yellow sodium vapor light if its spectrum is a single line 14.3 centimeters from the axis?

Coherence

There are many differences between the radio wave produced in an antenna and the light wave produced in a bulb, aside from wavelength. For one thing, the radio wave is *monochromatic;* that is, it consists of waves of a single frequency, while a lightbulb produces a mixture of frequencies.

Another difference is that the radio wave is *coherent.* Just what this means is illustrated in Figure 14.9. If you put a receiving antenna at any point near a transmitting antenna, you can detect a regular oscillation of the electric field. It has a definite value at any moment. In contrast, the electric fields of the light from a bulb are all mixed up; every oscillating atom produces its own wave, and these waves arrive at any point in every conceivable phase. Some cancel others, and some reinforce others.

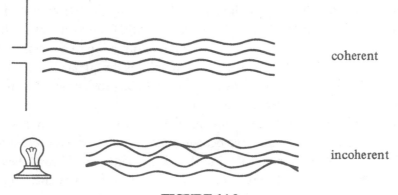

coherent

incoherent

FIGURE 14.9

Another point is that the radio wave is continuous, and the light wave is pulsed. Each beam of light consists of enormous numbers of short bursts of waves, lasting perhaps 10^{-7} second each. It is like the difference between a stream of water flowing smoothly from a tap, and a shower.

A *laser* is a device that produces a continuous, coherent, and monochromatic light beam. A beam of this sort has a number of extremely useful properties that are not available with ordinary light. An ordinary light beam, no matter how well focused, spreads out as it travels. A laser beam, on the other hand, can remain confined to a single thin line for distances of many miles. This quality has made laser beams useful in building bridges, leveling fields, aiming artillery, and measuring distances. A laser beam aimed at the moon spread so little that the spot it formed was visible from earth, and made it possible to measure the distance to the moon accurate to a few centimeters.

That thin laser beam can deliver substantial amounts of energy to a tiny spot. Industry uses it to cut and weld even good-sized pieces. The beam can be focused down to a spot that drills microscopic holes with extreme accuracy. A laser beam replaces the scalpel in some types of surgery, for it can cut human tissue and simultaneously seal off blood vessels. A focused beam cuts the grooves that store music on compact disks; in your home, another laser reads the digital message stored in the grooves. A biologist can send a laser beam through her microscope and use it to destroy a single chromosome. The world's most powerful laser occupies a whole building, and is used to deliver an enormous burst of energy to a deuterium pellet in order to produce nuclear fusion.

Because laser light is coherent, it can be used to make *holograms,* which are nothing more than stored light waves. A hologram is a piece of photographic film, but there is no picture on it. Instead, there are many squiggly lines — a record of the actual light reflected from an object. When laser light goes through this hologram, it forms a faithful copy of the light that came from the object. The result is a full three-dimensional image that can be viewed from various directions.

The laser beam, because it is coherent, can be broken up into extremely short pulses called *bits.* Information is stored in digital form by turning the bits on and off. For transmission by optical cable, a laser the size of a grain of salt switches the bits. Each fiber can carry 46 million bits per second, enough information to transmit 4 000 conversations simultaneously. A similar tiny laser amplifies the signal every few kilometers. A much more powerful laser converts music into bits for cutting the information into the master from which compact disks are made.

The laser was invented in 1960, following a theory first proposed by Albert Einstein. Since then, many kinds have been invented for all sorts of uses, and they have become a routine part of our life. In the supermarket or the library, they read bar codes. A laser printer makes 100 pages per minute of the highest quality copy. Visible laser beams entertain us at many kinds of shows. A credit card is made with a hologram, to make it difficult to forge. If you develop a detached retina, your ophthalmologist will weld it back into place with a laser beam. If you serve in the military, you will use a laser to aim weapons. Lasers have found their way into every kind of science and technology, from astronomy to zoology.

CORE CONCEPT

A laser, producing monochromatic, continuous, coherent light, has many practical uses.

TRY THIS
—5—

Determine the frequency of a red laser whose wavelength is 6.9×10^{-7} meters.

Polarization

There is yet another difference between the electromagnetic wave produced by a radio antenna and that of a lightbulb.

As Figure 14.10 shows, a vertical antenna produces a wave with a vertical electric field. This field varies both in time and in space, but it is always vertical. In contrast, the molecular vibrations that produce a light wave in a lamp are in every direction. At any given point in space, each member of this melange of waves oscillates in its own direction—always, to be sure, perpendicular to the direction of travel, but oriented at random in the plane perpendicular to that direction. A wave like that of the radio antenna, which has all its electric vibrations in the same direction, is said to be *polarized*.

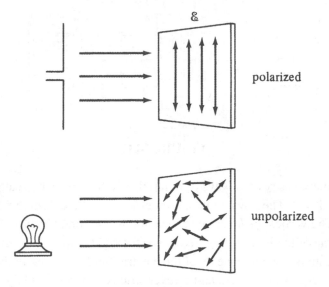

FIGURE 14.10

Light waves can be polarized. The simplest method is to use a polarizing filter; this filter looks like a sheet of gray plastic, but it is made with a thin layer of long, thin molecules, all lined up in the same direction. If light passes through such a filter, as in Figure 14.11, the electric vibrations parallel to the molecules of the filter will get through, but the electric vibrations perpendicular to the axis of the filter will be blocked. Thus the light that passes through the filter, about half of the total, will be polarized along the axis of the filter.

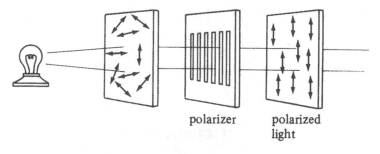

FIGURE 14.11

To the human eye (but not to that of a bee!) polarized light looks like any other light. You can tell whether or not light is polarized by passing it through a polarizing filter. Rotate the filter, as in Figure 14.12. If the light is polarized, it will be blocked when the axis of the filter is perpendicular to the electric field of the light. Rotating the filter through 90° makes the light entering your eye go from bright to dark, and another 90° restores the original brightness. Unpolarized light entering the filter looks equally bright no matter how the filter is oriented.

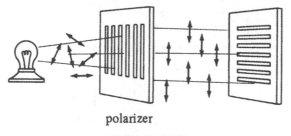

polarizer

FIGURE 14.12

Our environment has several ways of polarizing light. One is by scattering, as illustrated in Figure 14.13. The bowl contains some sort of colloidal mixture; milk diluted in water will do. The vertical beam coming in contains horizontal electric fields in random orientation. Some of the light is scattered out to the right; call it the x-direction. This beam cannot contain any components of the electric fields oriented in the x-direction. Electromagnetic waves are *always* and forever strictly transverse. Only electric fields perpendicular to the path of the beam, in the y-direction, can be in this beam. Similarly, the y-beam has only x-electric fields. The blue of the sky is produced by the scattering of sunlight. If you look at the sky in a direction at right angles to your line to the sun, you are looking at polarized light.

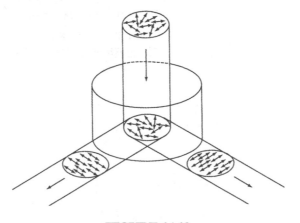

FIGURE 14.13

Reflection can also polarize light. In Figure 14.14, the beam coming from the right onto a horizontal surface is striking the surface at such an angle that the vertical electric fields in the reflected beam would be parallel to the path of the beam. This cannot happen. The emerging beam is polarized horizontally. That is why your vertically polarizing sunglasses can eliminate the glare from a road.

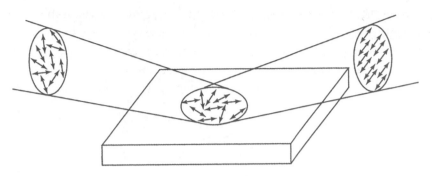

FIGURE 14.14

CORE CONCEPT
Light can be polarized, and the polarization can be detected by rotating a polarizing filter.

Explain how you would test the blue light of the sky to determine its direction of polarization.

Color

We tend to think of color as somehow being a property of an object: The book is red. But it is impossible to understand what that simple statement means unless we take into account the instrument that is making that decision: the human eye.

What is the meaning of the statement that the book is red? This: In ordinary sunlight, or any other source of light that provides a mixture of all visible wavelengths, the book absorbs certain wavelengths of light and reflects others. Those that are reflected enter the eye, where they stimulate certain cells in the retina, but not others. This particular mixture of wavelengths sets off a set of nerve impulses going to the brain, and the brain sees the color red.

The perception of "red" does not correspond to a unique set of wavelengths. Your repertory of visual perceptions is much more limited than the wavelength mixes that reach your eye. The human eye can distinguish several hundred colors and shades, but all these perceptions result from combinations of only three separate kinds of visual signals.

The nerve signals that tell the brain what color you are seeing originate in just three kinds of color-sensitive cells in the retina. These three kinds of cells are distinguished by their possession of one of three different visual pigments. Each of these pigments is sensitive in varying degree to different wavelengths of light. Figure 14.15 shows the sensitivities of these pigments. One, called "red," responds to light at the longest visible wavelength, about 760 nanometers, if the light is very bright. This pigment is most sensitive at about 600 nanometers. The "blue" pigment reaches, weakly, down to about 380 nanometers and is most sensitive at 430 nanometers. The "green" pigment is most sensitive in the middle of the spectrum at 520 nanometers and reaches nearly to both ends.

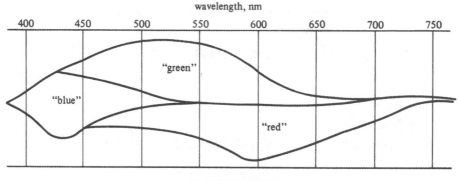

FIGURE 14.15

Under ordinary conditions, the color you see depends on the degree to which each of these pigments is stimulated. From 690 to 760 nanometers, things look red, since only the "red" pigment reacts in this range. The monochromatic light of a sodium vapor lamp, at 570 nanometers, acts on both the "red" and the "green" pigment almost equally; this combination produces the sensation we call yellow. Any combination of wavelengths that hits the red and green pigments equally looks yellow. Butter looks yellow because it absorbs light at the short-wavelength end, up to about 500 nanometers, and reflects the rest. When the reflected light hits the retina, it is thus received by the green and red pigments, but not the blue.

Wavelengths at three different parts of the spectrum can stimulate each of the three pigments selectively and thus can produce any sensation of color whatever. The screen of a color television set contains a mosaic of just three kinds of dots: red (long wave), blue (short wave), and green (medium wave). All the colors you see are produced in your eyes by proper combinations of these three wavelength ranges.

Understanding of the three pigments is only the beginning of a knowledge of color vision. Perception of color may often depend on the wavelengths reflected by nearby objects, as well as those reflected by the object itself; your eye reacts to contrast. Also, there is a mechanism, now partially understood, by which the retina can make allowance for differences in illumination. In the early morning, sunlight is quite red, and a color picture taken at that time will look red. But your retina has a mechanism to compensate for the imbalance in illumination, and objects will appear to you in their normal colors.

CORE CONCEPT

Perception of color depends on the relative stimulation of three kinds of color-sensitive cells in the retina.

TRY THIS
—7—

A magenta filter passes into the eye only the light that affects the "red" and "blue" pigments. What wavelengths are absorbed by the filter as white light passes through it?

Thin Films

A soap bubble in the sunlight and an oil slick on the road are sources of beautiful color displays. Such thin film colors result from the interference of two sets of light beams: the beam reflected from the top of the film and the beam that went through the film and reflected from the bottom. If these two reflected beams emerge in phase, the light is enhanced; if they emerge out of phase, interference will cancel the light. Just what happens depends on the wavelength of the light and the thickness of the film. Since these films are not uniformly thick, different wavelengths will be enhanced or canceled at different places. That is what produces all those colors.

Figure 14.16 shows what happens when a monochromatic light beam arrives perpendicularly at a thin film, such as an oil slick, that is just a half-wavelength thick. At the first surface, the beam impinges on a closed, resistant surface. The electric field cannot transfer energy to the oil because it it is not nearly strong enough to set molecules of the oil into motion. This is like what happens at the end of a closed organ pipe. The wave forms a node at that surface; beam 1 emerges from the surface 180° out of phase with the incoming beam.

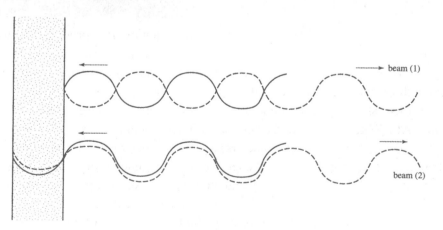

FIGURE 14.16

At the other surface, things are different. The wave arrives facing an open, completely unresisting boundary, like that of an open organ pipe. No energy is transferred because there is nothing to transfer it to, but there is no restriction on the oscillation of the electric field. The wave forms an antinode; beam 2 emerges in the same phase as the incoming beam.

The distance beam 2 travels is twice the thickness of the oil, which is just one wavelength. When the two reflected beams recombine, they might be 180° out of phase and thus cancel each other out. If a mixed beam of light shines on this film, the light with a wavelength in the oil just twice the thickness of the film will not be in the reflected beam. On the other hand, the light whose wavelength is a quarter the thickness of the film will travel just a half-wavelength further than the first beam. Because of the phase reversal at the first surface, the two beams leave the film in phase.

If the film is just one wavelength thick, the second beam travels two wavelengths, and interference at the surface is destructive. The same thing happens any time the thickness of the film is a whole number of half-wavelengths. Interference is constructive at a thickness of $\lambda/4$, $3\lambda/4$, or any other odd number of quarter wavelengths. The rainbow effect appears because the film is not uniformly thick.

CORE CONCEPT

Thin film colors are produced by constructive and destructive interference of the two reflected beams. They depend on the wavelength of the light and the thickness of the film.

TRY THIS
—8—

What color will be missing in the light reflected from a soap bubble 2.4×10^{-7} meters thick? The speed of light in water is 2.2×10^{8} meters per second.

Red Hot, White Hot

Every warm object emits infrared electromagnetic radiation. It is produced, as usual, by accelerated charges. Atoms are in constant oscillation, as we saw earlier. The energy of the oscillation is proportional to the temperature. It is these oscillations that constitute the accelerated charge producing the radiation.

Figure 14.17 shows the rate of radiation at all frequencies, for several different temperatures. At 500 kelvins, the radiation is all infrared, ranging from a frequency of about 3×10^{12} hertz ($\lambda = 10^5$ nanometers) to 2×10^{14} hertz ($\lambda = 1\ 500$ nanometers). At 1 000 kelvins, the total amount of energy radiated is much more—in fact, 16 times as much. Also, both the top frequency and the frequency of maximum radiation are higher. There is even a little at 4×10^{-4} hertz, in the visible range. An object begins to glow red. When the temperature is raised to 4 000 kelvins, a lot of energy is released over the entire visible range, and the object becomes white hot. There is even a little ultraviolet light. This is the temperature at which electric lightbulbs operate.

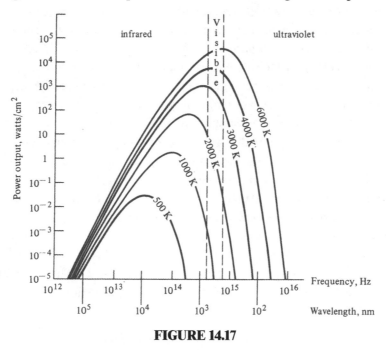

FIGURE 14.17

These curves apply to what is called a *blackbody*—an ideal object that absorbs all the light that falls on it, reflecting nothing. A blackbody is also a perfect radiator; any electromagnetic waves generated within it pass out through its surface freely. No real object is perfectly black, but some things come quite close. Analysis of the radiation of a blackbody is especially simple because it does not matter what the object is made of.

Each curve, at any temperature, has the same shape. Starting at the left end and moving to the right, we find that the amount of radiation is greater at the higher frequencies, until the curve reaches a peak at a certain frequency level. Above this point, the amount of radiation becomes smaller at higher frequencies, until there is none at all above a certain upper limit.

There is an old theoretical explanation for the fact that more energy is radiated at higher frequencies. It is assumed that the atomic oscillators set up standing electromagnetic waves within an object, something like the standing waves that an AC sets up in a radio antenna. It is these waves that produce the traveling waves that escape into the outside world. Now, within any given space, a lot more high-frequency (short-wavelength) waves can fit than low-frequency waves. In a guitar string 1 meter long, for example, there are 4 overtones with wavelengths between 10 and 15 centimeters, but 11 overtones with wavelengths between 5 and 10 centimeters. If the energy of the atomic oscillators is equally distributed among all possible wavelengths, there will be a lot more high frequencies than low ones, and the energy will concentrate at the upper end of the frequency spectrum.

This theory explains the left-hand side of the curve nicely, but it does not explain why the curve does not continue to rise toward the right forever. Clearly, there is something preventing energy from feeding into the higher frequencies.

The reason the higher frequencies do not get their fair share of the energy is this: *The energy of an oscillator cannot increase gradually; it goes up in steps.* An atomic oscillator cannot pick up just a little energy at a time; to jump up to the next step, it has to get enough energy in a package to move up a step. Furthermore, the steps are greater at higher frequencies. At very high frequencies, an oscillator is stuck at a low energy level unless it can get an enormous chunk of energy all at once. The higher the frequency, the more unlikely it is that the oscillator will get enough energy in one piece to jump up to the next level. It is easier at higher temperatures, when more energy is available, but at any temperature, there is an upper limit.

Mathematical analysis of measured blackbody radiation curves led to an equation for determining the size of the steps at any given frequency:

$$\Delta E = hf = hc/\lambda \qquad \text{(Equation 14.2)}$$

where ΔE *is the size of the quantum of energy at a frequency* f. The constant h is Planck's constant, which has a fundamental place in every equation of subatomic physics. Its value is

$$h = 6.62 \times 10^{-34} \text{ joule-second}$$

The joule is an inconveniently large unit of energy in dealing with electrons and photons. A much smaller unit has been devised that takes advantage of the fact that electrons, protons, and many other particles all have exactly one elementary unit of electric charge. The new energy unit is the *electron volt* (eV). It is defined as the energy change when one elementary unit of charge falls though a potential difference of one volt. Since $E_{elec} = q\Delta V$ (Equation 9.3), the electron volt is

$$1 \text{ eV} = (1.602 \times 10^{-19} \text{ J/V})(1 \text{ V}) = 1.602 \times 10^{-19} \text{ J}$$

Putting Planck's constant into this convenient unit,

$$h = 4.135 \times 10^{-15} \text{ eV·s}$$

 EXAMPLE 14.8   An atomic oscillator is vibrating with a frequency of 3.0 x 10^{14} hertz. What is the amount of energy that will boost its energy by one step?

SOLUTION

The quantum of energy at that frequency is

$$\Delta E = hf$$
$$\Delta E = (4.14 \times 10^{-15} \text{ eV·s})(3.0 \times 10^{14} \text{ Hz})$$
$$\Delta E = 1.24 \text{ eV}$$

EXAMPLE 14.9 What is the size of the energy step by which an atomic oscillator drops when it emits a package of red light whose wavelength is 710 nanometers?

SOLUTION
Since the frequency of the light is $\frac{c}{\lambda}$, Equation 14.2 can be written as

$$E = \frac{hc}{\lambda} = \frac{\left(4.14 \times 10^{-15} \text{ eV·s}\right)\left(3.00 \times 10^8 \text{ m/s}\right)}{710 \times 10^{-9} \text{ m}}$$

$$E = 1.75 \text{ eV}$$

In an electric light bulb, we put in electric energy that sets the molecules oscillating. They are going fast enough to generate electromagnetic waves in every part of the visible spectrum. The oscillators radiate out energy by dropping to a lower energy level, only to be soon pushed back up by the electric current. Every time an oscillator drops down one energy step, it gives off a tiny packet of light. If the size of the step is ΔE, the frequency of the light can be determined from Equation 14.2.

CORE CONCEPT
A radiating hot object gives off energy in packets, each with one quantum of energy, which is proportional to the frequency of the oscillation.

TRY THIS
—9—

How much energy is there in each packet of light emitted by a hot filament in the ultraviolet range with a wavelength of 240 nanometers?

Photoelectricity

Blackbody radiation studies showed that light is emitted in the form of tiny packages. Other studies have shown that, when light is absorbed, it must be taken into a material as tiny packages and not in a continuous stream. This model of light, as consisting of discrete packages of energy, explains so many observable phenomena that it must be taken as a useful way to look at light. The packages of light are called *photons*. Each has an amount of energy that depends only on the frequency of the light, according to the equation

$$E_{\text{photon}} = hf \qquad \text{(Equation 14.3)}$$

What is the energy of a violet photon if the wavelength of the light is 470 nanometers?

SOLUTION

The frequency of the light, from Equation 8.4, is

$$f = \frac{c}{\lambda}$$

$$f = \frac{3.0 \times 10^8 \text{ m/s}}{470 \times 10^{-9} \text{ m}} = 6.38 \times 10^{14} \text{ Hz}$$

Therefore, the energy of the photon is

$$hf = (4.14 \times 10^{-15} \text{ eV·s})(6.38 \times 10^{14} \text{ Hz}) = 2.64 \text{ eV}$$

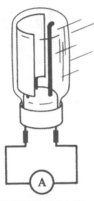

FIGURE 14.18

The nature of photons can be explored in the process of *photoelectric emission*. When the proper kind of light strikes the right kind of metal, electrons are given off. In a device called a *photoelectric cell,* or "electric eye," the emitted electrons come off a metal plate shaped like a half-cylinder, as shown in Figure 14.18. They are collected by a metal rod. If the rod and the plate are connected through an external circuit, current will flow in the circuit and can be measured by a sensitive ammeter. A photoelectric cell can be used to control other circuits, such as the one that opens a door when you interrupt a beam of light.

If we think of light as a wave reaching a surface and making electrons oscillate, we are led to expect a certain kind of behavior. The wave makes the electrons oscillate in resonance, like a tuning fork being stimulated by a sound wave. Electrons build up energy gradually until they have enough to get out of the metal. The brighter the light, the sooner the electrons escape and the more energy they have when they leave.

But that is not the way it works. If you shine visible light on a silver surface, no electrons leave, and the ammeter remains at zero no matter how bright the light is or how long you leave it shining. To get a photoelectric current, you have to use ultraviolet light. This works because each ultraviolet photon has enough energy to release one electron. If the kind of light you use produces photoelectric emission, making it brighter produces a larger photoelectric current because the metal surface releases more electrons per second. But to get any photoelectric emission at all, you must bombard the surface with photons that have enough energy to release the electrons.

Every metal has its characteristic electron binding energy. This is the amount of energy that has to be added to a surface electron to release it from the surface. Deeper electrons need more energy. The minimum energy needed to release an electron is called the *work function* of the metal. Metals such as sodium and potassium, which do not bind their electrons strongly, have smaller work functions. Orange light, at a wavelength of 530 nanometers, will produce a photoelectric current in potassium. Any shorter wavelength will do the same. But to get electron emission from lithium, you need a wavelength no longer than 430 nanometers, in the blue region. Most metals will not emit at all in the visible range, but need ultraviolet light.

 EXAMPLE 14.11 Lithium can emit electrons using light of any wavelength less than 430 nanometers. What is the work function of lithium?

SOLUTION

The work function is the energy of the least energetic photon that produces photo-emission:

$$E_{photon} = hf = \frac{hc}{\lambda}$$

$$E_{photon} = \frac{(4.14 \times 10^{-15} \text{ eV} \cdot \text{s})(3.00 \times 10^8 \text{ m/s})}{430 \times 10^{-9} \text{ m}} = 2.88 \text{ eV}$$

If the incident photon has more than enough energy to release electrons, the excess will become the kinetic energy of the electrons. This energy can be measured by adding a variable power supply and a voltmeter to the circuit, as shown in Figure 14.19. Note that the *negative* terminal of the battery is facing the collecting rod. To get back to the emitting plate, the electrons have to go through the battery in the wrong direction, thus losing energy as they travel. If the battery potential is turned up high enough, it will allow the battery to absorb completely the energy of the electrons, and the current will stop. This *cutoff potential* is the energy per unit charge of the most energetic electron as it leaves the plate, or, from Equation 9.3,

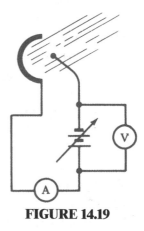

FIGURE 14.19

$$E_{kin} = q \, \Delta V_{cutoff}$$

This energy of the fastest electron is the energy of the photon that released the electron, less the work function of the metal:

$$q \, \Delta V = hf - W \qquad \text{(Equation 14.4)}$$

EXAMPLE 14.12 An ultraviolet photon of wavelength 140 nanometers strikes the surface of a metal with a work function of 1.26 electron volts. What is the maximum energy of the emitted electron?

SOLUTION

The energy of the photon is

$$hf = \frac{hc}{\lambda} = \frac{\left(4.14 \times 10^{-15}\ \text{eV}\right)\left(3.00 \times 10^8\ \text{m/s}\right)}{\left(140 \times 10^{-9}\ \text{m}\right)} = 8.87\ \text{eV}$$

The maximum energy is the photon energy minus the work function:

$$8.87\ \text{eV} - 1.26\ \text{eV} = 7.61\ \text{eV}$$

EXAMPLE 14.13 What is the cutoff potential for stopping the photoelectric current produced by the electrons of Example 14.12?

SOLUTION

Since the maximum electron energy is 7.61 eV, the current will be cut off with 7.61 volts.

EXAMPLE 14.14 How much reverse potential will cut off the photoelectric current from a lithium surface (work function 2.89 electron volts) illuminated by ultraviolet light at 2.2×10^{15} hertz?

SOLUTION

The energy of each photon is

$$hf = (4.14 \times 10^{-15}\ \text{eV·s})(2.2 \times 10^{15}/\text{s}) = 9.11\ \text{eV}$$

Subtract the work function to get the energy:

$$9.11\ \text{eV} - 2.89\ \text{eV} = 6.22\ \text{eV}$$

which will be stopped by a reverse potential of 6.22 volts.

Note that the cutoff potential does not depend on how bright the light is. This potential depends only on the photon energy, that is, on the frequency of the incident light. Both the work function of the metal and Planck's constant can be determined by measuring the cutoff potential at various frequencies.

CORE CONCEPT

A photoelectron is released by absorbing the energy of a single photon, which must have more energy than the work function of the metal:

$$E_{\text{photon}} = hf; \quad q\,\Delta V = hf - W$$

What is the work function of a metal if ultraviolet light at a frequency of 3.51×10^{-15} hertz produces a photoelectric current that can be cut off with a reverse potential of 9.2 volts?

The Energy of Photons

The energy of the ultraviolet photon in Example 14.14 is on the order of a few electron volts. This is no coincidence. In ordinary chemical reactions, energy changes per molecule are about that much. That is why electric batteries produce one or two volts per cell. The reason you cannot see infrared rays is that the energy of each infrared photon is too small to produce a chemical change in your retina. Neither can it produce photosynthesis or affect a photographic plate (with one minor exception).

You cannot see in the ultraviolet region because the energy of these photons—3 electron volts and more—is so great that it could damage your retina. The lens of your eye filters them out.

We have become more cautious, in recent years, about taking X-ray pictures, because we understand that the photon energies of these rays, which are thousands of electron volts, can be extremely damaging to living tissue. And the photon energies of the gamma rays emitted by radioactive materials reach into the millions of electron volts.

Depending on the nature of the experiment, we elect to use either a wave model or a particle model for light. The wave model is necessary when we look at interference effects, or diffraction. In studying the photoelectric effect, or blackbody radiation, we must resort to the language of photons. It is futile to ask which model is "correct." We cannot settle the question by examining light in transit, for light can be studied only by letting it strike something. Then, it may show interference effects, or it may produce photoelectric emission. It may do both. Whatever it does, it no longer exists once it lands. Examine it and it is gone.

The models we use are free creations of the human mind, allowing us to think about light in familiar terms derived from experience with large objects. Which model is best might well depend on the amount of energy in the photons. It is quite impossible to detect photon effects in radio waves, but the gamma rays emerging from radioactive nuclei are never treated as waves. We count the separate photons.

You tend to think of a barrel of water as a continuous material. It can be poured, and it is measured in liters. You probably know that it consists of separate molecules, but they are so small that they rarely enter into our consideration. When you fill the same barrel with cannonballs, you treat them differently. The separate packages are so large that you cannot avoid dealing with them individually. Radio waves are like the water; X-rays are like the cannonballs.

Visible light comes in between, like a barrel of sand. You can pour it and measure it in liters, or in cubic yards. Yet you will surely notice that it consists of separate grains if you get one in your eye.

CORE CONCEPT

Particle properties of electromagnetic waves become noticeable at about 1 electron volt per photon and become more important as the photon energy increases.

TRY THIS
—11—

What is the wavelength of a 22 000-electron-volt X-ray photon?

Photons are Particles

At the beginning of this chapter, we found that light consists of electromagnetic waves. At the end of it, light seems to consist of a flow of particles called photons. Are they really good, substantial particles, like electrons or protons?

When a fast-moving particle strikes a solid object it exerts a force. Figure 14.20 illustrates a classic experiment that measured the force exerted by photons. The horizontal bar, suspended at its center from a long, thin fiber, has a mirror at one end and a black plate at the other. Light shining on the black plate is absorbed; at the mirror it is reflected.

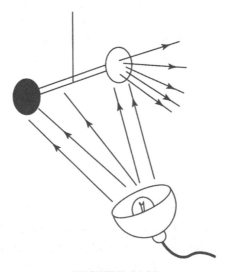

FIGURE 14.20

If a photon has momentum, it must exert a force when it strikes something. The change of its momentum, according to Equation 4 .3, must be equal to the impulse it exerts. If the photon has momentum p, this momentum drops to zero when the photon is absorbed into the black plate. At the mirror, the momentum changes from $+p$ to $-p$, a change of $2p$. The impulse exerted on the mirror is twice as much as on the black plate. Light shining on this system makes the bar swing; the force on the mirror is greater than on the black plate. The force that light exerts is very small, but it exists.

How much is the momentum of a photon? Momentum is defined as mass times velocity. A photon is usually described as a massless particle, but this calls for some explanation. A photon has no *rest* mass; it is never at rest. But it does have *relativistic* mass, since it has energy. Its mass is thus E/c^2 (Equation 5.7); multiplying this by its velocity c makes its momentum equal to E/c. Since the energy of a photon is hf (Equation 14.3), its momentum is hf/c. But $c = f\lambda$, so the expression for the momentum of a photon reduces, very simply, to

$$p = \frac{h}{\lambda}$$

(Equation 14.5)

EXAMPLE 14.15

How much impulse does a violet photon of wavelength 450 nanometers exert in being reflected from a mirror?

SOLUTION

From Equation 14.5, the momentum of the photon is

$$p = \frac{h}{\lambda} = \frac{6.6 \times 10^{-34}\,\text{J} \cdot \text{s}}{450 \times 10^{-9}\,\text{m}} = 1.5 \times 10^{-27}\,\text{N} \cdot \text{s}$$

Since the momentum changes from $+p$ to $-p$, the change of momentum is twice the original momentum, or 3.0×10^{-27} N·s. This is equal to the impulse exerted.

The final proof that photons are particles came with the demonstration that they could make elastic collisions with electrons. To show this, we need a photon that has approximately the same mass as an electron; nothing can be learned by bouncing a table tennis ball off a bowling ball.

EXAMPLE 14.16

What is the wavelength of a photon that has the same mass as an electron?

SOLUTION

The table in Appendix 2 tells us that the mass of an electron is 9.1×10^{-31} kg. This corresponds to energy of

$$E = mc^2 = (9.1 \times 10^{-31}\,\text{kg})(3.0 \times 10^8\,\text{m/s})^2 = 8.2 \times 10^{-14}\,\text{J}$$

Equation 14.2 says this must be equal to hf; and $f = c/\lambda$.

Therefore

$$\lambda = \frac{hc}{E} = \frac{\left(6.6 \times 10^{-34} \text{ J} \cdot \text{s}\right)\left(3.0 \times 10^{8} \text{ m/s}\right)}{8.2 \times 10^{-14} \text{ J}} = 2.4 \times 10^{-12} \text{ m}$$

This is a highly energetic X-ray photon.

The experiment was done by firing a beam of X-rays into a graphite block and measuring the wavelength of the X-rays that came out. Those that went straight through were unchanged. However, some of the emerging X-rays came out of the block at various angles, and they always had longer wavelengths. This means that their energies were smaller. The results could be explained in mathematical detail by assuming that some of the photons made elastic collisions with electrons, losing some of their energy in the collision. The laws of energy and momentum conservation predict that those photons deflected furthest would have lost the largest amounts of energy, and that was what happened.

Then what is light? It depends on how you look at it. As light travels through space, it has all the properties of a wave, as shown by the interference experiments. However, when it interacts with matter on an atomic scale, it behaves like any other particle. And that's the way it is.

CORE CONCEPT

Photons have energy and momentum, and behave in all ways like particles.

TRY THIS
—12—

How many ultraviolet photons of wavelength 250 nanometers would it take to deliver a microjoule of energy?

REVIEW EXERCISES FOR CHAPTER 14

Fill-In's. For each of the following, fill in the missing word or phrase.

1. An electromagnetic wave is produced by a(n) _____ electric charge.

2. One piece of evidence suggesting that light is an electromagnetic wave is that its _____ agrees with a calculated value.

3. The ideal length for a broadcasting antenna is a(n) _____.

4. In an electromagnetic wave, the magnetic field is _____ to the electric field.

5. The shortest electromagnetic waves are the _____.

6. Ultraviolet waves are _____ in wavelength than visible light.

7. The _____ wavelengths of electromagnetic waves travel in straighter lines.

8. The phenomenon of _____ can be noticed when a wave passes through a hole not much larger than a wavelength.

9. In the diffraction pattern of a double slit, the path difference at the $n = 3$ line is 3 _____.

10. A bright bar on a diffraction pattern is the part of a standing wave called a(n) _____.

11. A diffraction grating pattern separates light into colors because the colors differ in _____.

12. The spectral color that is diffracted to the largest angle is _____.

13. Coherent light is produced by a(n) _____.

14. _____ light has all its electric field vectors oriented in the same direction.

15. No light passes through two polarizing filters if their axes are _____ to each other.

16. All color perception is produced by combinations of stimuli affecting just _____ kinds of nerve cells.

17. Equal stimulation of the red and green pigment cells produces a sensation of _____ color.

18. The "blue" pigment responds mostly to the _____ wavelength end of the spectrum.

19. A warm object produces _____ radiation.

20. As a blackbody increases its temperature, both the _____ and the _____ of the radiation increase.

21. The higher frequencies of a blackbody radiator do not receive as much energy as the low frequencies because the energy is _____.

22. The size of energy quanta is proportional to _____.

23. In the photoelectric effect, electrons are emitted from a metal under the influence of _____.

24. Photoelectric emission works only if the light has a high enough _____.

25. The minimum energy needed to release an electron from a metal is the _____ of the metal.

26. The cutoff potential of a photoelectric current is a measure of the _____ of the electrons.

27. The photon model of light is most appropriate at _____ wavelengths.

28. If an electromagnetic wave is produced by a vertical antenna, the orientation of the magnetic field in the wave is _____.

Multiple Choice.

1. A proton falling through a vacuum generates
   (1) an electric field only
   (2) a magnetic field only
   (3) no fields
   (4) stationary electric and magnetic fields
   (5) an electromagnetic wave.

2. One property that all electromagnetic waves have is that they
   (1) are polarized
   (2) have very high frequencies
   (3) are transverse
   (4) are visible
   (5) can produce a sound.

3. A dipole antenna 50 centimeters long is designed for a frequency of
   (1) 150 megahertz
   (2) 200 megahertz
   (3) 300 megahertz
   (4) 600 megahertz
   (5) 1 200 megahertz.

4. The wave property of light is best demonstrated by showing that light waves
   (1) form interference patterns
   (2) carry energy
   (3) interact with electrons
   (4) can exert a force
   (5) all travel at the same speed.

5. The wavelength of red light is
   (1) longer than infrared and shorter than violet
   (2) longer than X-rays and shorter than green light
   (3) longer than infrared and shorter than microwaves
   (4) longer than green light and shorter than microwaves
   (5) longer than ultraviolet and shorter than X-rays.

6. When white light is passed through a diffraction grating, it is separated into colors because wavelength determines the
   (1) velocity of light in a grating
   (2) angle of constructive interference
   (3) degree of absorption
   (4) amount of reflection
   (5) rotation of the plane of polarization.

7. Light from an incandescent bulb is not monochromatic because the charges that produce it oscillate in random
   (1) pulses   (2) frequencies (3) directions   (4) phases   (5) amplitudes.

8. Strong evidence for the particle nature of light came from the elastic collision of light with
   (1) electrons   (2) protons   (3) nuclei   (4) each other (5) mirrors.

9. Making a red light brighter will change its
   (1) energy per photon   (4) wavelength
   (2) total energy   (5) frequency.
   (3) momentum per photon

10. Maxwell's analysis of electromagnetism solved the long-standing problem of
    (1) the nature of the electric charge
    (2) the origin of magnetic fields
    (3) the nature of light
    (4) the meaning of electromagnetic induction
    (5) the source of magnetic poles.

11. The colors produced by a thin film result from
    (1) interference   (4) diffraction
    (2) dispersion   (5) reflection.
    (3) transmission

12. The first application of Planck's constant came from the study of
    (1) thin films   (4) radiation from hot bodies
    (2) protons   (5) radiation from an antenna.
    (3) photons

Problems.

1. State whether each of the following will produce (A) an electric field only; (B) both electric and magnetic fields; (C) an electromagnetic wave; (D) none of these: (a) a proton falling in a vacuum; (b) an electron at rest; (c) a neutron traveling at constant speed in a straight line; (d) an electron traveling in a circle at constant speed; (e) a proton moving at constant speed in a straight line.

2. (a) What is the wavelength of a television station broadcasting on a frequency of 106 megahertz? (b) What is the best length for an antenna for receiving this station?

3. What color is light whose frequency is 7.1×10^{14} hertz?

4. (a) What is the difference in path length to two slits for the tenth antinode if the slits are acting on green light of wavelength 530 nanometers? (b) If the diffraction angle for this antinode is 12.0°, how far apart are the slits?

5. A diffraction grating ruled 580 lines per millimeter is illuminated with white light. What is the range of all diffraction angles for the visible spectrum, with wavelengths from 380 to 760 nanometers? (Only the first-order spectrum is used.)

6. Show that there is no red light in the third-order spectrum produced by a diffraction grating ruled 530 lines per millimeter.

7. Referring to Figure 14.15, determine what color will be perceived if the eye receives an equal mixture of monochromatic light at 650 nanometers and 540 nanometers.

8. What wavelengths of light are blocked by a red filter?

9. What is the wavelength of the light at which energy radiated by a blackbody is greatest at a temperature of 4 000 kelvins? (See Figure 14.17.)

10. What is the size of the quantum step of energy at a frequency of 10^{14} hertz?

11. What is the photon energy of an ultraviolet photon if the wavelength of the light is 140 nanometers?

12. What is the longest wavelength of light that will release an electron from the surface of a metal whose work function is 3.1×10^{-19} joule?

13. What is the work function of a metal if a retarding potential of 22.0 volts is just enough to cut off the photoelectric current when the surface is illuminated by ultraviolet light whose frequency is 6.8×10^{15} hertz?

14. What is the maximum energy of an electron released from a surface whose work function is 1.6 electron volts by a yellow photon whose wavelength is 580 nanometers?

15. What is the cutoff potential of a photoelectric current produced by shining a blue light at 470 nanometers onto a surface whose work function is 2.2 electron volts?

16. A laser produces an intense beam of green light. If the beam exerts a force of 3.8×10^{-9} newtons on a mirror, at what rate are photons striking the surface?

17. If the beam of Problem 16 strikes a black surface, how much heat is generated in 10 minutes?

18. A beam of green light (wavelength 530 nanometers) passes perpendicularly through a sheet of glass 2.0 millimeters thick. Its speed in the glass is 65 percent of its speed in a vacuum. How many wavelengths of the light fit within the glass?

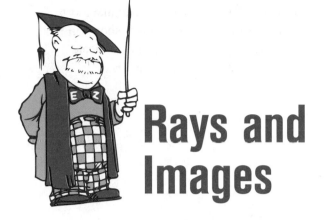

Rays and Images

WHAT YOU WILL LEARN

How light rays behave under different circumstances; how lenses and mirrors form images; the nature of color.

SECTIONS IN THIS CHAPTER	
• What's a Ray?	• Focusing the Rays
• Illumination	• Forming Images
• Rays That Bounce	• Virtual Images by Lenses
• Seeing Things	• Algebraic Solutions
• Bending Rays	• The Lens in Your Camera
• Trapped Light	• Curved Mirrors
• Colors by Prism	• Making It Big

What's a Ray?

Long before photons were dreamed of, even before there was an acceptable wave model for light, there were useful rules governing the behavior of light. These rules were based on a model that is still highly useful under many conditions. Light was visualized as consisting of rays, emerging from a lamp and traveling in straight lines unless deflected in some way.

Consider the lamp doing its job of emitting light in Figure 15.1a. The arrows are vectors representing the direction of travel and the strength of the light. Light travels out from the lamp in all directions, getting weaker as it goes. This is a vector field; like most such fields, it can be represented more simply by the use of field lines. This is shown in Figure 15.1b. These are field lines with the usual properties of such lines: they indicate the direction of the field at all points; they are closest together where the field is strongest.

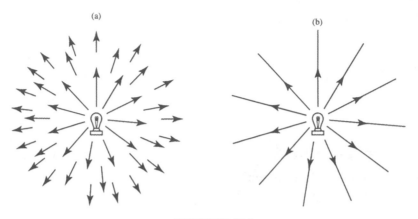

FIGURE 15.1

These field lines are what we call light rays. They are straight in empty space but can be bent by interaction with matter. At all points, the light rays are perpendicular to the wave fronts.

The ray model, however, is not always useful. When we consider the production or absorption of light, only the photon model provides meaningful explanations. And when light is diffracted through narrow slits or around small objects, its behavior can be accounted for only on the basis of a wave model. With these limitations, we can use a ray model effectively without concerning ourselves with the question of whether light is "really" a wave or a stream of particles.

CORE CONCEPT
A ray model is useful in dealing with light except where diffraction is important, or where light is absorbed or emitted.

TRY THIS
—1—

Using a ray model, explain the difference between the light coming from a lightbulb and the light of a laser.

Illumination

Some lamps are brighter than others. We can think of the difference this way, if we like: There are more rays emerging from a bright lamp than from a dim one.

The total amount of light emerging from a lamp is called the *luminous flux* that the lamp produces. Luminous flux is measured in *lumens,* a unit that is carefully defined in the SI in terms of the light radiated under carefully controlled conditions by platinum at its melting point. An ordinary 100-watt light bulb produces about 1 600 lumens of luminous flux.

The lumen is not strictly a physical unit, since it depends on the sensitivity of the human eye. The eye is most sensitive to light at the center of the spectrum, to green light. The lumen measures how bright the light appears to the eye. In terms of the actual electromagnetic wave, or photon energy of the light, if the light is red or violet, it takes many more watts of power to constitute a lumen of luminous flux than if the light is green. A 1-watt green electromagnetic wave has 680 lumens of luminous flux; for a red or violet wave to appear as bright, and thus have the same amount of luminous flux, it would need more than ten times as much actual power. The luminous flux of ultraviolet and infrared rays is always zero, regardless of how much energy they deliver.

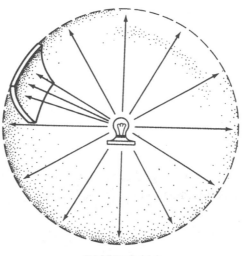

FIGURE 15.2

We are often concerned with the amount of light that falls on a surface. How much light must shine on a book, for example, to make it possible to read in comfort? How much light is needed to take a picture? This quantity is the *illumination* of the surface, and it is measured in lumens per square meter. A lumen per square meter is called a *lux*.

A simple application of the ray model enables us to determine the relationship between the amount of luminous flux produced by a lamp and the illumination falling on a surface any distance from the source. For simplicity, we will consider only a source that is considerably smaller than the distance from the source to the illuminated surface—a point source.

The relationship is illustrated in Figure 15.2. Think of the source as being at the center of a large sphere of radius r. Then all the luminous flux produced by the lamp spreads out radially, in straight light rays, to the inner surface of the sphere. The area of the sphere is $4\pi r^2$, so all this flux spreads out over that surface uniformly. The amount of flux per unit area, the illumination I, is therefore

$$I = \frac{F}{4\pi r^2}$$

(Equation 15.1)

This equation gives the illumination on any surface a distance r from a source emitting luminous flux F, provided that the surface is perpendicular to the rays reaching it.

An electric lightbulb is extremely inefficient; about 97 percent of the electric energy put into it becomes invisible infrared rays. A 100-watt lightbulb produces about 1 600 lumens, representing only about 3 watts of electromagnetic power.

EXAMPLE 15.1 What is the illumination on a book that is 2.5 meters from a lamp that produces 1 200 lumens of flux?

SOLUTION

From Equation 15.1,

$$I = \frac{F}{4\pi r^2}$$

$$I = \frac{1\ 200\ \text{lumens}}{4\pi(2.5\ \text{m})^2} = 15\ \text{lumens/m}^2 = 15\ \text{lux}$$

EXAMPLE 15.2 How much luminous flux must a lightbulb produce if it is to be placed 3.7 meters from a surface where the illumination needed is 20 lux?

SOLUTION

From Equation 15.1,

$$F = 4\pi r^2 I = 4\pi(3.7\ \text{m})^2(20\ \text{lux}) = 3\ 400\ \text{lumens}$$

CORE CONCEPT

Illumination, in lux, is inversely proportional to the square of the distance from the source:

$$I = \frac{F}{4\pi r^2}$$

A fluorescent bulb is much more efficient than an incandescent lamp. What is the illumination at a distance of 3.5 meters from a 40-watt fluorescent lamp that operates at 16 percent efficiency? (Figure that, on the average, 1 watt of electromagnetic energy spread out over the visible spectrum produces about 500 lumens of luminous flux.)

Rays That Bounce

The physical system that comes closest to the mathematical line known as a light ray is the fine beam produced by a laser. It can be used to study the behavior of light rays. The beam can be made visible in space by providing lots of dust in the space.

If the beam of a laser is shone onto a flat mirror, it can be seen to reflect off the surface, as in Figure 15.3. The incident beam (the one coming toward the surface) and the reflected beam will both lie in a plane perpendicular to the surface of the mirror.

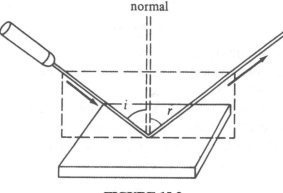

FIGURE 15.3

In discussing the geometry of rays, it is customary to measure all angles, where rays strike a surface, with respect to a line drawn perpendicular to the surface. This line is called the *normal*. As shown in Figure 15.3, the *angle of incidence* is the angle that the incident ray makes with the normal. The *angle of reflection* is the angle that the reflected ray makes with the normal. On a flat, mirrored surface, these two angles are equal. See Example 15.3.

EXAMPLE 15.3 A beam of light strikes a mirror at an angle of incidence of 40°. What is the angle between the incident and reflected beams?

SOLUTION
The normal bisects the required angle, and each half is 40°. The angle is thus 80°.

When light strikes an ordinary object, three things may happen to it:

1.   It may be transmitted through the object; this is what happens when light strikes a pane of glass. If the glass is frosted, the light will pass through, but will be scattered, or diffused, in all directions instead of passing straight through.

2.   It may be reflected off the surface; when the surface is mirrorlike and flat, the reflection obeys the rules discussed above and is said to be *specular*. Most surfaces are not of this type, and they reflect the light diffusely, in all directions, as illustrated in Figure 15.4.

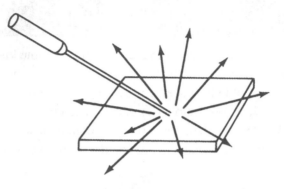

FIGURE 15.4

3.   The light may be *absorbed* by the material, its energy being converted to heat. An ideal perfectly black object is one that absorbs all the light that comes to it.

Typically, objects transmit, reflect, and absorb light selectively. A red filter is one that transmits long-wavelength rays (red) and absorbs the rest of the visible spectrum. A green book *reflects* the central part of the spectrum and absorbs both ends, so that the reflected light that reaches your eye appears green.

CORE CONCEPT

Light may be selectively transmitted, reflected, or absorbed; when reflection is specular, the angle of incidence equals the angle of reflection.

TRY THIS
—3—

If you are to look into a vertical mirror and see your feet, a ray must reflect from your feet, and then reflect specularly from the mirror to your eye. What is the smallest vertical mirror in which you can see your feet if you are 5 feet tall?

Seeing Things

When you look at a book, as in Figure 15.5, you can see it because light diffusely reflected from it enters your eye. *Each point* on the book scatters light in all directions, acting much like a point source of light. Some of the scattered light enters your eye. The rays coming into your eye from any single point on the book form a cone. These rays are divergent; that is, they spread apart from each other as they travel. When all these cones of divergent rays, from every point on the book, come into your eye, you see a book.

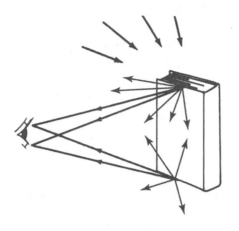

FIGURE 15.5

You can also see the book by looking at its reflection in a mirror. In this case, the cone of rays that passes from a point on the book to your eye is folded at the surface of the mirror, as shown in Figure 15.6. Your eye cannot detect the folding; it tells you that the rays are diverging from some point behind the mirror, as shown by the dashed lines in the figure. The point from which the light appears to be coming is called an *image point*. Every point on the object (the book) has a corresponding image point, so there is an *image* of the book behind the mirror. Since there is no light actually at the image, that image is said to be *virtual*.

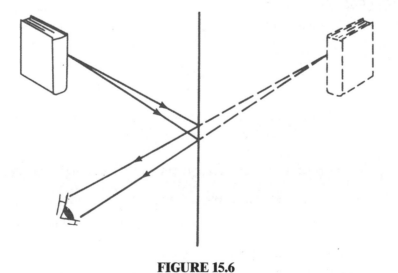

FIGURE 15.6

For each ray, the angle of incidence is equal to the angle of reflection. A little geometry will show that this fact has a consequence: The image point is just as far behind the mirror as its corresponding object point is in front of it, and it lies on the normal drawn from the object point. This means that the virtual image in a plane mirror is the same size as the object and is just as far behind the mirror as the object is in front of it. The image has these properties regardless of the position from which it is viewed.

You are taking a picture of yourself in a mirror while standing 12 feet from the mirror. At what distance should you focus your camera?

SOLUTION

Set it at 24 feet; your image is 12 feet behind the mirror, and thus 24 feet from you. If your eye cannot tell the reflected rays from rays coming directly from an object, neither can a camera.

Contrary to common opinion, the mirror image is not reversed right to left. Try this: Write something on a piece of transparent paper and hold it in front of you as if to read it. Do this while standing in front of a mirror and look in the mirror. You will find that the writing is still oriented in the usual way, and that you can read it in the mirror without difficulty. If the paper is not transparent, the only way you can read it in the mirror is to turn it around so that the writing faces the mirror. Then it appears reversed. But the mirror did not reverse it; you did!

The mirror image is reversed front to back, not right to left. If you look at yourself in a mirror, the image of your nose is closer to the mirror than the image of your ears. If you are facing north, your image is facing south. But if you raise your eastern hand, the eastern hand of your image waves back at you.

CORE CONCEPT

The image in a plane mirror is virtual, the same size as the object, as far behind the mirror as the object is in front of it, and reversed front to back.

TRY THIS
—4—

You are watching a friend walking toward a vertical plane mirror. What happens to the size of his image as he approaches the mirror?

Bending Rays

The experienced fish archer of Figure 15.7 knows that, if he aims his arrow at the place where the fish appears to be, he will miss it. The light from the fish, coming to his eye, bends as it emerges from the water. This phenomenon, called *refraction,* is due to the fact that the light travels faster in air than it does in water.

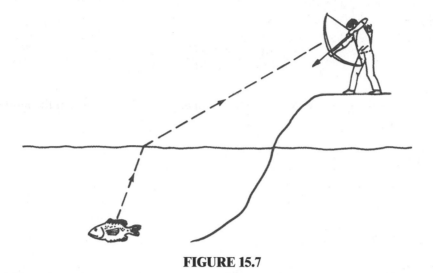

FIGURE 15.7

Exploring this phenomenon with a laser beam, as shown in Figure 15.8, yields *Snell's law,* a rule that has been known for four centuries. When the beam enters a transparent medium from the air, making an angle of incidence i on the way in, it bends toward the normal, so that the angle of refraction r is smaller than the angle of incidence. Note that, as always, the angles are measured from the normal. As the angle of incidence is changed, the ratio between the sines of the two angles remains constant:

$$\frac{\sin i}{\sin r} = n$$

where n is a characteristic of the medium known as its *index of refraction.*

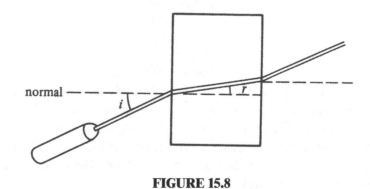

FIGURE 15.8

EXAMPLE 15.5 A light ray enters a piece of crown glass at an angle of 57° and is refracted to 31° inside the glass. What is the index of refraction of the glass?

SOLUTION

$$n = \frac{\sin i}{\sin r} = \frac{\sin 57°}{\sin 31°} = 1.63$$

EXAMPLE 15.6 If a ray enters the glass in Example 15.5 at an angle of 26°, what will the angle of refraction be?

SOLUTION

$$\sin r = \frac{\sin i}{n} = \frac{\sin 26°}{1.63}$$

from which $r = 16°$.

Geometric analysis of the passage of the light into a new medium, coming from a vacuum, shows that the index of refraction of the medium is an expression of the way the light slows down as it enters. It is easily shown that

$$n = \frac{c}{v}$$

where c is the speed of light in a vacuum ($= 2.998 \times 10^8$ meters per second) and v is the speed of light in the medium whose index of refraction is n. See Example 15.7.

EXAMPLE 15.7 What is the speed of light in water, which has an index of refraction of 1.33?

SOLUTION

$$v = \frac{c}{n} = \frac{2.998 \times 10^8 \text{ m/s}}{1.33}$$
$$v = 2.25 \times 10^8 \text{ m/s}$$

The most convenient form of Snell's law would be one that makes it possible to calculate the angles of incidence or refraction at the interface between *any* two media—glass and water, Lucite and glycerine, benzene and alcohol, whatever. This rule even makes it unnecessary to specify which angle is the angle of incidence and which the angle of refraction, since the rays are always reversible. If the two media that form the interface are designated A and B as in Figure 15.9, and their indices of refraction are n_A and n_B, we can just call the angles in the two media θ_A and θ_B, and Snell's law takes this form:

$$\frac{\sin \theta_A}{\sin \theta_B} = \frac{n_B}{n_A} \qquad \textbf{(Equation 15.2)}$$

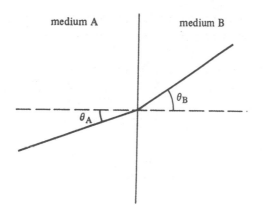

medium A   medium B

θ_B

θ_A

FIGURE 15.9

EXAMPLE 15.8 A ray of light passes from water ($n = 1.33$) into a sheet of flint glass ($n = 1.61$), making an angle of incidence of 65°. What is the angle of refraction?

SOLUTION

Apply Equation 15.2, letting the water be medium A and the glass be medium B,

$$\sin\theta_B = \sin\theta_A\left(\frac{n_A}{n_B}\right)$$

$$\sin\theta_B = \sin 65°\left(\frac{1.33}{1.61}\right)$$

$$\theta_B = 48°$$

EXAMPLE 15.9 A block of crown glass ($n = 1.63$) is immersed in an unknown liquid. A ray of light is measured to make an angle of incidence within the unknown liquid of 48° as it approaches the glass. The angle in the glass is 36°. What is the index of refraction of the liquid?

SOLUTION

Using subscript g for glass and subscript l for liquid, we have

$$\frac{\sin\theta_g}{\sin\theta_l} = \frac{n_l}{n_g}$$

$$\frac{\sin 36°}{\sin 48°} = \frac{n_l}{1.63}$$

which gives $n_l = 1.29$.

CORE CONCEPT

When a ray strikes an interface, the sines of the angles of incidence and refraction are inversely proportional to the indices of refractions of the two media:

$$\frac{\sin\theta_A}{\sin\theta_B} = \frac{n_B}{n_A}$$

TRY THIS
—5—

A diamond ($n = 2.42$) is in water ($n = 1.33$), and a ray of light shines on it, making an angle of incidence of 55°. What is the angle of refraction inside the diamond?

Trapped Light

Consider the light rays emerging from a lamp that is under water, as in Figure 15.10. Ray *A* strikes the upper surface of the water along the normal; it speeds up as it emerges, but it does not bend. Ray *B* strikes at an angle and bends away from the normal, so that the angle of refraction is larger than the angle of incidence. Ray *C* is more so. Ray *D* is refracted so much that it emerges from the water along the surface; the angle of refraction is 90°.

Now look at ray *E*, which makes an angle of incidence of 50°. Since light travels in air practically as fast as in a vacuum, $n_{air} = 1$. We calculate the angle of refraction in the usual way:

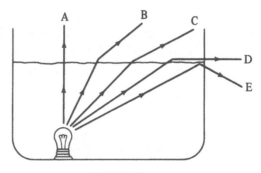

FIGURE 15.10

$$\frac{\sin 50°}{\sin r} = \frac{n_{air}}{n_{water}} = \frac{1}{1.33}$$

$$\sin r = (1.33)(0.766) = 1.019$$

Don't try to find r in your trig tables; there is no such angle. The largest possible angle of refraction is 90°.

Ray E undergoes *total internal reflection*. All rays that strike the surface at an angle larger than the *critical angle of incidence* are trapped inside the water. Reflection of this sort at an interior surface is more nearly complete and involves less loss of light than the best mirror can provide. That is why prisms like those in Figure 15.11 are used, rather than mirrors, in good binoculars, spotting scopes, and the finders of single-lens reflex cameras.

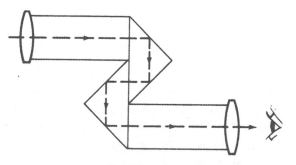

FIGURE 15.11

This kind of total reflection is always *internal;* that is, it occurs only when light is emerging into a medium of lower index of refraction. Only then is the ray bent away from the normal, and only then is there an angle of incidence that produces refraction at 90°. Total internal reflection, when it occurs, follows the usual rule of reflection: The angle of incidence equals the angle of reflection.

It is easy to find the critical angle of incidence; it is the angle at which the angle of refraction is 90°. For water, for example, using Equation 15.2,

$$\frac{\sin i_C}{\sin 90°} = \frac{n_{air}}{n_{water}} = \frac{1}{1.33}$$

$$i_c = 49°$$

Thus, any ray inside the water that strikes the surface at more than 49° will be reflected back.

EXAMPLE 15.10

What is the critical angle of incidence for crown glass ($n = 1.63$)?

SOLUTION

Apply Equation 15.2 to glass and air, with the angle in air taken as 90°:

$$\frac{\sin \theta_g}{\sin \theta_a} = \frac{n_a}{n_g}; \quad \frac{\sin i_C}{\sin 90°} = \frac{1}{1.63}$$

which gives $i_c = 38°$.

EXAMPLE 15.11   What index of refraction must a transparent material have in order for the total internal reflection to take place at all angles larger than 55°?

SOLUTION
Again,

$$\frac{\sin i_C}{\sin 90°} = \frac{1}{n}$$

Therefore

$$n = \frac{1}{\sin 55°} = 1.22$$

The phenomenon of total internal reflection has become the basis for a whole new technology, fiber optics. Fine glass fibers are used to carry beams of light. If the curvature of the fiber is not too great, total internal reflection prevents the light from escaping. Optical fiber cables are rapidly replacing copper wire for carrying telephone conversations. The light is produced by a tiny laser, and modulated into a digital code that carries the voice information. A typical cable is a couple of centimeters across and consists of 144 glass fibers embedded in a matrix with a low index of refraction. The cable is quite flexible. New kinds of ultraclear glass are being developed that will carry the signal for many kilometers before it gets so weak that it must be amplified.

CORE CONCEPT

Total internal reflection occurs when the angle of incidence on an internal surface exceeds the angle at which the angle of refraction is 90°.

TRY THIS
—6—

What is the critical angle of incidence of diamond ($n = 2.42$)? (Can you see why a diamond sparkles?)

Colors by Prism

Figure 15.12 shows a beam of white light entering a triangular prism at a high angle of incidence. It is refracted downward toward the normal on entering the glass. It leaves at a surface slanted in the opposite direction and is refracted away from the normal—again downward, because of the different orientation of the surface. The emerging ray will be separated into the colors of the spectrum. Red refracts least, violet most.

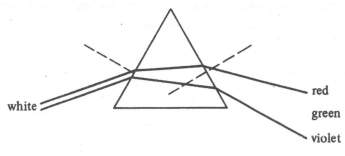

FIGURE 15.12

In passing through a glass prism, light is refracted because it travels more slowly through the glass. Since we observe that the high-frequency end of the spectrum is refracted more than the low-frequency end, it is obvious that violet light slows up more than red light in passing through the glass. Glass *disperses* the light. Any medium in which the higher frequencies travel more slowly than the lower frequencies is said to be a *dispersive* medium.

Before the development of high-quality diffraction gratings, prisms were used in spectroscopes to analyze the wavelength content of light. They are by no means as good as gratings, however. Also note that the spectrum produced by a prism is the reverse of that made by a grating; in a grating, it is the red light that is deflected at the greatest angle from its original path.

Dispersion of light does not have much practical value, aside from producing the beautiful sparkling of a diamond or a crystal chandelier. It is a problem in optics. Until lens makers learned how to overcome it, everything seen in a microscope was surrounded by a rainbow-hued halo.

CORE CONCEPT
Light is dispersed into colors by a prism because the different frequencies travel at different speeds in the glass.

TRY THIS
—7—

In silicate flint glass, the index of refraction is 1.61 for red light and 1.66 for violet light. Find the angles of refraction for red and violet light if a beam of white light approaches the surface of such a piece of glass at an angle of 65°.

Focusing the Rays

Suppose we set a couple of prisms base to base, as shown in Figure 15.13, and send a couple of rays of light into them. The upper ray bends downward (toward the normal) on entering the prism, and downward (away from the normal) on leaving. The lower ray bends upward twice, and the two rays converge at a point.

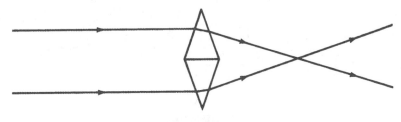

FIGURE 15.13

We can set up a system of this sort in which not just two but all the rays that may enter the system will converge at a point. A piece of glass or plastic that can do this is called a *converging lens*. Its surface (or both surfaces) is spherical—a small part of a large sphere. With such a *thin lens* entering light rays can be focused approximately at a point, as shown in Figure 15.14.

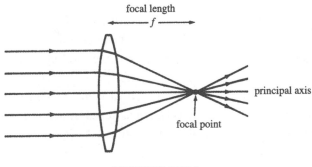

FIGURE 15.14

Lenses of this sort obey certain mathematical rules. First, we will define a line called the *principal axis* of the lens. This line is the axis of symmetry of the lens, drawn normal to its surface at the center. All rays of light that enter the lens parallel to this principal axis converge at a point called the *principal focus* of the lens. The distance from this focal point to the center of the lens is called the *focal length* of the lens. This distance enters into all lens equations, where it is represented as f. At the same distance on the other side of the lens is another focal point, where rays entering the lens in the other direction, parallel to the principal axis, converge.

Suppose a light source is placed at the principal focus of a converging lens, as shown in Figure 15.15a. Following the rule that in any optical system the light rays are reversible, we know that these rays will emerge from the lens parallel to the principal axis—and thus parallel to each other. Such a beam is said to be *collimated;* this is how a narrow beam is produced in some searchlights. No lens can do a really good job of collimating the light from an electric lamp. The only really first-rate collimated beams are produced by putting a lens in front of a laser.

The effect of the lens in Figure 15.15a is to collect a lot of divergent rays and reduce the divergence until the rays are made parallel. Now suppose the light source is placed closer to the lens than the principal focus. The rays reaching the lens are more strongly divergent, as shown in Figure 15.15b. The lens reduces the divergence, but not enough to bring the rays into the parallel condition. On emerging, the rays are still divergent, but less so than when they entered.

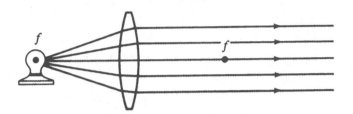

(a) Source at distance *f* beam collimated

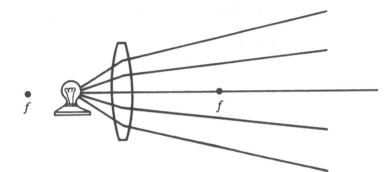

(b) Source at less than distance *f* beam divergent

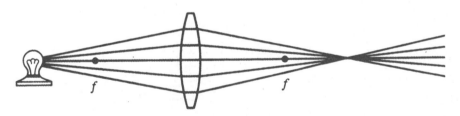

(c) Source beyond distance *f* convergent beam

FIGURE 15.15

In Figure 15.15c, the light source is farther from the lens than one focal length. Now the rays entering the lens are *less* divergent than they would be if they had come from a source at the principal focus. The lens will now converge the light *more* than enough to make it collimated, and the emerging rays will converge at some point beyond the principal focus on the other side.

When the light source is very far away, the entering rays are essentially parallel, and the light converges at the principal focus. Every star, even the closest star, is so far away that all the light coming from it forms a point image at the principal focus of the lens. Astronomers never have to refocus their instruments.

Figure 15.16 shows what happens when light enters a *diverging lens*. Any lens that is thinner at the middle than at the edges *increases* the divergence of the light that enters it. Light coming in parallel to the principal axis diverges, as shown in Figure 15.16a. In Figure 15.16b you can see that all the rays emerging from the lens *appear* to be diverging from some point behind the lens. The point from which all these rays *appear* to diverge is a virtual focus; if the rays entered the lens parallel to its principal axis, they will appear to diverge from the principal focus of the lens. The focal length of a diverging lens is, again, the distance from the principal focus to the center of the lens; it is always given as a negative number.

(a)

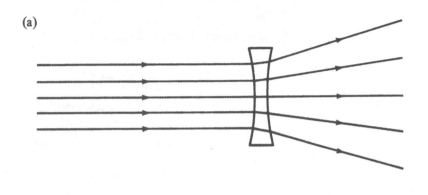

(b)

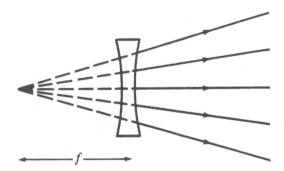

f

FIGURE 15.16

CORE CONCEPT

All rays entering a converging lens parallel to its principal axis converge at a distance of one focal length from the lens; for diverging lenses, the distance is that from which the parallel rays seem to diverge.

Explain the difference between lenses whose focal lengths are, respectively, +20 centimeters and –30 centimeters.

Forming Images

When you point your camera at someone's face, the lens focuses the light reflected from that face to form a real image of the face on an image sensor. The sensor is a chip, smaller than a dime, that is made of millions of cells that function much like photoelectric cells. Each cell is sensitive to one of the three primary colors, red, blue, or green. When the proper color of light falls on the cell, it responds by releasing a few electrons; it produces a pulse of current that is converted into a stored bit of information. When the picture is displayed, each of these bits produces a pixel, a tiny spot of color, on the display screen. These millions of pixels are a mosaic that constitutes the picture.

Figure 15.17 shows how this process of forming a real image works. Each point on the object acts like a point source of light, since it is diffusely reflecting light in all directions. Some of this light enters the lens and is brought to a focus at some point beyond the principal focus of the lens. Object point A emits light that focuses at image point A'; object point B forms an image point at B'; and so on.

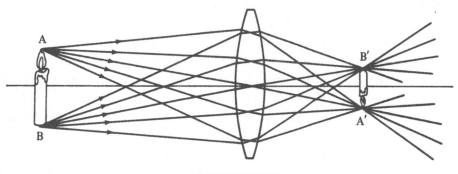

FIGURE 15.17

There are at least three different ways in which you can observe this real image. First, you could stand some distance behind it. After the rays converge, they cross each other and diverge, forming the kind of light rays your eye can respond to. You will see the image at the position from which the light is actually diverging. Since the rays cross, the image is upside down. A second way would be to place a screen at the image position. Then, after the converging rays strike the screen, they will be scattered off in all directions. Your eye can intercept these diverging, scattered rays to enable you to see the image—like a slide projected onto a screen. The third thing you might do is place an image sensor at the image position to make the bits that define the picture.

Exactly where is the image formed? Where do you have to put the screen or the film? To find out, we can locate the image of a single object point; the other image points will lie in the same plane. Further, we can make the problem simple by tracing exactly two rays from an object point to find out where they intersect to form an image point. All the other rays from that object point will converge at the same place.

Figure 15.18 is a *ray diagram* used to find the location of an image. The *object distance (D_o) is* the distance from the object to the lens; f is the focal length of the lens; and we are looking for D_i, the *image distance.* We will use as our object point the tip of the arrow and trace exactly two rays from that point.

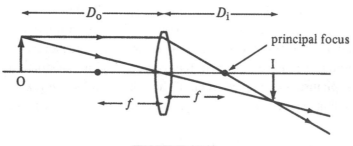

FIGURE 15.18

One ray goes from the object point to the lens parallel to the principal axis of the lens. We know that this ray must pass through the principal focus of the lens, on the other side, after passing through the lens. The second ray that is easily traced is the one that goes directly through the center of the lens. This will be deflected a little on the way in, but its refraction on the way out will exactly compensate and we can treat the ray as though it passed straight through the lens. We will make very little error by assuming that it passes straight through the center of the lens without deflection. In the ray diagram, these two rays are drawn to find out where they intersect. That intersection is the image point, and the rest of the image can be sketched in.

The image shown in Figure 15.18 is considerably smaller than the object, as it would be in a camera. It is a real, inverted image. If you try the same ray-plotting procedure with objects placed closer to the lens, you will find that the image gets farther away from the lens, and larger. In a projector, the object may be a 35-millimeter slide that projects onto a screen to form an image a few meters across. This is accomplished by placing the object (the slide) just a little beyond the principal focus of the lens. If the object is *at* the principal focus, the rays from any object point will emerge from the lens parallel to each other and no image will be formed.

EXAMPLE 15.12   Plot a ray diagram, to scale, for finding the location of the image of a lamp 20 centimeters from a +15.centimeter lens.

SOLUTION

Plot just two rays, one parallel to the principal axis and the other going straight through the center of the lens: See Figure 15.19.

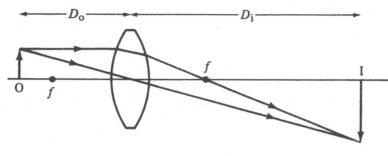

FIGURE 15.19

CORE CONCEPT

Converging lenses form real, inverted images of objects at any distance greater than the focal length of the lens.

TRY THIS
—9—

By careful construction to scale, find (a) the image distance and (b) the size of the image of a card 15 centimeters high if it is placed 40 centimeters from a lens whose focal length is +20 centimeters.

Virtual Images by Lenses

Everyone has used a magnifying glass at some time or another. You hold something close to it and look through. What you see is an enlarged image, considerably farther behind the lens than the object is. If you are using the lens to read a printed page, the image is behind the page. Obviously, it must be virtual, for there is no light coming to your eye from behind the page.

Figure 15.20 is a ray diagram showing how this enlarged, erect, virtual image is formed. With the object closer to the lens than one focal length, the rays coming out of the lens are still divergent—but less so than when they went in. To find the image, trace the rays back to their apparent point of origin. Follow the usual rules: One ray goes straight through the center of the lens; the other enters the lens parallel to the principal axis and is refracted so as to pass through the principal focus.

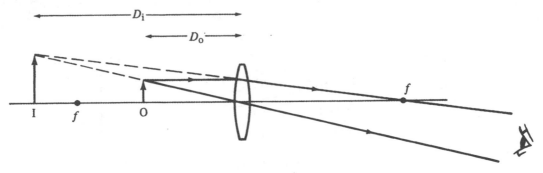

FIGURE 15.20

A diverging lens also forms images. The image formed by a diverging lens is always virtual, right side up, smaller than the object, and closer to the lens than the principal focus. The ray diagram is shown in Figure 15.21. One ray goes right through the center in usual form. The other, entering the lens parallel to the principal axis, is refracted *away* from the axis. Its path is from its location within the lens on a straight line away from the focal point on the *same side* as the object. The lens is a "demagnifying" glass; if you look through it, everything looks small.

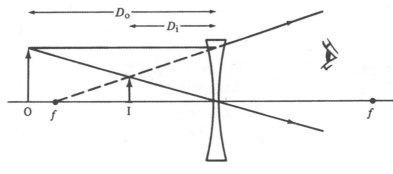

FIGURE 15.21

CORE CONCEPT
Lenses form virtual images when the light emerging from them is divergent.

TRY THIS
—10—

Construct a ray diagram for finding the image formed by a diverging lens of focal length –30 centimeters when the object is 20 centimeters from the lens.

Algebraic Solutions

From the geometry of a ray diagram, two equations can be derived that make it possible to calculate the size and position of any image formed by a lens of any kind:

$$\frac{1}{f} = \frac{1}{D_o} + \frac{1}{D_i}$$

(Equation 15.3a)

in which f is the focal length of a lens (always negative for a diverging lens, remember!), D_o is the object distance, and D_i is the image distance. With this equation, you can find out where the image is. You can also tell whether it is real or virtual, for D_i is negative for virtual images.

 EXAMPLE 15.13 How far from the lens is the image formed of a lamp 35 centimeters away from a lens of focal length + 20 centimeters?

SOLUTION

From Equation 15.3a,

$$\frac{1}{D_i} = \frac{1}{f} - \frac{1}{D_o}$$

Therefore

$$\frac{1}{D_i} = \frac{1}{20 \text{ cm}} - \frac{1}{35 \text{ cm}} = \frac{7 - 4}{140 \text{ cm}}$$

and $D_i = 47$ cm.

EXAMPLE 15.14 How far from the lens is the image formed by a + 20 centimeters lens if the object is 12 centimeters from it?

SOLUTION

Use Equation 15.3a:

$$\frac{1}{f} = \frac{1}{D_o} + \frac{1}{D_i}$$

$$\frac{1}{20 \text{ cm}} = \frac{1}{12 \text{ cm}} + \frac{1}{D_i}$$

$$\frac{1}{D_i} = \frac{1}{20 \text{ cm}} - \frac{1}{12 \text{ cm}} = \frac{6 - 10}{120 \text{ cm}}$$

$$D_i = -30 \text{ cm}$$

The negative sign indicates that the image is virtual.

EXAMPLE **15.15** What focal length lens is needed to form a virtual image 12 centimeters from the lens when the object is 35 centimeters from the lens?

SOLUTION

With a virtual image, D_i is negative, so

$$\frac{1}{f} = \frac{1}{35 \text{ cm}} + \frac{1}{-12 \text{ cm}} = \frac{12 - 35}{(12)(35) \text{ cm}}$$

and $f = -18$ cm. A diverging lens is needed.

To find the size of the image, follow the rule that object and image sizes are always in the same ratio as their respective distances from the lens. This is a good approximation for small distances, but the relationship breaks down when the object distance is large.

$$\frac{D_o}{D_i} = \frac{S_o}{S_i}$$

(Equation 15.3b)

 EXAMPLE **15.16** Find the size of the image of a book 18 centimeters high placed 30 centimeters from a diverging lens of focal length –20 centimeters.

SOLUTION

First we will have to find out where the image is, using Equation 15.3a:

$$\frac{1}{D_i} = \frac{1}{f} - \frac{1}{D_o}$$

$$\frac{1}{D_i} = \frac{1}{-20 \text{ cm}} - \frac{1}{30 \text{ cm}} = \frac{-3 - 2}{60 \text{ cm}}$$

$$D_i = -12 \text{ cm}$$

Now apply Equation 15.3b:

$$S_i = S_o\left(\frac{-D_i}{D_o}\right)$$

$$S_i = (18 \text{ cm})\left(\frac{-12 \text{ cm}}{30 \text{ cm}}\right) = -7.2 \text{ cm}$$

For virtual images, both size and distance are negative.

CORE CONCEPT

Image sizes and distances can be calculated algebraically.

$$\frac{1}{f} = \frac{1}{D_o} + \frac{1}{D_i}; \quad \frac{D_o}{D_i} = \frac{S_o}{S_i}$$

TRY THIS
—11—

A projector is to be placed 3.0 meters from a screen to project an image 50 centimeters wide from a 35-millimeter slide. (a) How far from the lens must the slide be placed? (b) What focal length lens is needed?

The Lens in Your Camera

A glance at Equation 15.3a tells you this: When the object distance is very large, the image distance is equal to the focal length. That is why when your camera is focused at infinity, the distance from the center of the lens to the image sensor is equal to the focal length of the lens. As the object distance decreases, the image distance increases. Therefore, when you focus your camera on something close, the lens moves farther from the image sensor.

You might have a "fixed-focus" camera, in which everything from 5 feet to infinity is "in focus." How can that be? The situation is as in Figure 15.22, where the image of the tree is in front of the image of the girl. What your camera—or any camera—does is to make a compromise. You place the sensor between the two images. Each image point, whether for the tree or the girl, is not really a point at all, but a little circle, called the "circle of confusion." If the circle of confusion is too big, the picture looks blurred, "out of focus." By compromising, you get both the tree and the girl in acceptable focus.

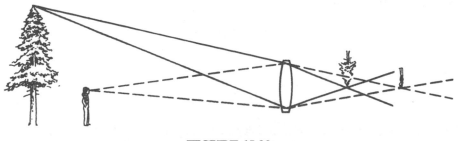

FIGURE 15.22

There is no such thing as a perfect lens; in even the best of cameras, each object point is represented by a circle of confusion on the sensor. A sharp picture is one in which the circle of confusion is small. When a photographer says that the "depth of field" is from 12 feet to 40 feet for a particular camera setting, he is saying that, for all objects in this range, the circle of confusion on the sensor is acceptably small. In a fixed-focus camera, the focal length of the lens is short, and this produces a large depth of field.

The ordinary lens of the usual 35-millimeter camera has a focal length of 45 or 50 millimeters. A "wide angle" lens is one with a shorter focal length, perhaps 28 mm. According to our lens equations, a short focal length means a small image distance. With a short focal length, simple, fixed-focus cameras can be made very thin. But these lenses form small images. This can be an advantage if you want to get a wide panorama onto the sensor, but it makes closeups impossible. Longer focal length lenses (say, 300 millimeters for a 35-millimeter camera) produce larger images for distant objects; these are "telephoto" lenses.

Every lens is rated according to its *focal ratio,* or just f-number. This is the ratio between the focal length of the lens and its diameter. It tells how effective the lens is in gathering light. A large diameter, of course, means that more light comes in; a long focal length lens spreads the light out over a bigger sensor area. Thus, the smaller the f-number, the less light the camera needs to take a picture. Fixed-focus cameras are usually f:11, while expensive 35-millimeter cameras may have f-numbers as low as f:1.2.

A low focal ratio brings with it problems. As you can see in Figure 15.23, the cone of light forming the image point is much more divergent when the lens is large. This means that a slight displacement of the sensor from the image position produces an unacceptably large circle of confusion. You have to focus very carefully with a fast lens, and the depth of field is very small. Every camera with a fast lens incorporates a *diaphragm,* a black screen with an adjustable hole in the center, to reduce the effective diameter of the lens. An f:1.2 lens can be "stopped down," perhaps as far as f:22, for use when there is ample light. Then only the center of the lens is used. This makes the cone of light smaller, and focusing is less critical.

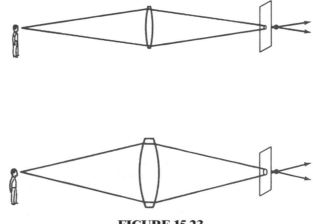

FIGURE 15.23

| EXAMPLE 15.17 | A lens is marked f = 45 mm, focal ratio f:3.5. (a) What is the diameter of the lens? (b) By what factor is the light coming into the lens cut down when the diaphragm is closed down to f:8? |

SOLUTION

(a) The focal ratio is the ratio between the focal length of the lens and its diameter; (45 mm)/diam = 3.5, so diam = 12.9 mm. (b) Closing down the diaphragm to f:8 reduces the effective diameter of the lens by a factor of 3.5/8 = 0.44 — it is a little less than half-size. However, the amount of light coming in depends on the area of the lens, not its diameter. Since the area is proportional to the square of the diameter, the ratio of the two areas is $0.44^2 = 0.19$. Only about one-fifth as much light comes in.

Every lens has the problem of *chromatic aberration,* the effect caused by the dispersion of light. The focal length for violet is always less than it is for red. This difficulty is overcome by use of a doublet lens, the only kind used in cameras today. A doublet lens consists of a converging lens cemented to a diverging lens. The function of the diverging lens is to defocus the violet light more than the red, so both come to the same focus. This works because the two lenses are made of different kinds of glass.

There are many other kinds of aberrations, and all of them get worse as the lens gets wider. Those nice, neat formulas are derived for *thin* lenses. A wide, f:1.2 lens is getting kind of fat, and the rays that pass through the edge of the lens focus closer in than those at the center. This is called *spherical aberration.* It, and many other kinds of aberration, are corrected by making fast lenses out of combinations of individual lenses made of different kinds of glass. A fast lens in an expensive camera might be made of as many as 14 separate pieces of glass, as shown in Figure 15.24.

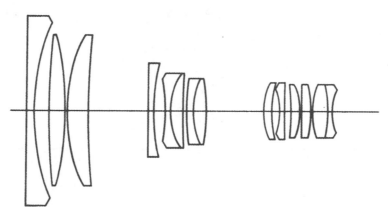

FIGURE 15.24

CORE CONCEPT

Lens equations are only approximations, and real lenses are made by combining simple lenses.

TRY THIS
—12—

What happens to the amount of light that enters a lens as the diaphragm is opened up from f:16 to f:8?

Curved Mirrors

The right-side mirror on a car or truck gives you a wide-angle view of the space behind you, but cars behind you are closer than they seem to be. The reason is that the mirror is curved. Its surface is spherical, bulging out toward you.

This *convex mirror* diverges the light it reflects, so it forms the same sort of images as a concave lens. Note the geometry of the spherical convex mirror in Figure 15.25. Point c is the center of the curvature of the mirror; cm and cn are radii. The radius of a circle is always perpendicular to the circumference, so cm and cn are normals. One ray parallel to the principal axis of the mirror is shown. It is reflected at an angle equal to the angle of incidence. The virtual ray, the extension behind the mirror of the reflected ray, passes through point f. Any ray incident on the mirror parallel to the principal axis would have a virtual ray passing through f, so f is the principal focus of the mirror. Since angles i and r are equal, the distance fn is just half of cn. The focal length of the mirror is just half of its radius of curvature.

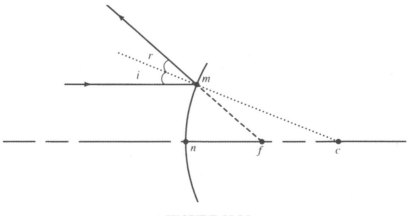

FIGURE 15.25

Figure 15.26 shows image formation by a convex mirror. The usual two rays are traced, one parallel to the principal axis and the other striking the mirror on the principal axis. The intersection of their virtual rays shows the size and location of the image. The image is virtual, erect, smaller than the object, and behind the mirror at a distance less than the focal length. It can be analyzed algebraically with the same equations used for a lens. A convex mirror diverges the light, so it behaves like a concave lens. In any equation, its focal length is negative. Also, the image forms behind the mirror, so it must be virtual. Distances behind the mirror are negative.

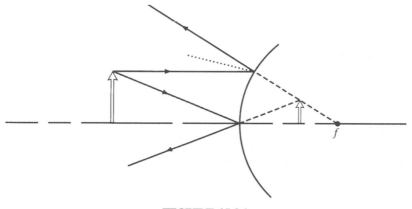

FIGURE 15.26

EXAMPLE 15.18 A car 2.2 meters high is 3.5 meters from a convex mirror with a radius of curvature of 90 centimeters. Find the size and position of the image.

SOLUTION

The focal length of the mirror is –45 centimeters, half the radius of curvature. It is negative because a convex mirror diverges light. From Equation 15.3a:

$$\frac{1}{D_i} = \frac{1}{f} - \frac{1}{D_o} = \frac{1}{-0.45 \text{ m}} - \frac{1}{3.5 \text{ m}}$$

$$D_i = -0.40 \text{ m}$$

Equation 15.3b gives the size of the image:

$$S_i = S_o\left(\frac{D_i}{D_o}\right) = (2.2 \text{ m})\left(\frac{-0.40 \text{ m}}{3.5 \text{ m}}\right) = -0.25 \text{ m}$$

The negative sign tells you that the image is virtual and on the opposite side of the mirror from the object. You can see it by looking into the mirror.

A concave spherical mirror converges light, so it forms the same kinds of images as a convex lens. When it is used as a shaving or a makeup mirror, the object (a face) is closer than the focal length, so the image is virtual, enlarged, erect, and behind the mirror. Concave mirrors are the optical objectives of astronomical telescopes. There, the object distance is essentially infinity and the image distance is the focal length of the mirror—half its radius of curvature. The image is real and is in front of the mirror. Many modern cameras can be fitted with "mirror lenses," which are actually concave mirrors. They are used to produce telescopic effects, and form real images on the film.

EXAMPLE 15.19

A concave mirror is to be used in a camera so that the image of an object 1.5 meters high will fill a frame of 35-millimeter film when the object is at a distance of 8.0 meters. What must the radius of curvature of the mirror be?

SOLUTION

First find the image distance, from Equation 15.3b:

$$D_i = D_o\left(\frac{S_i}{S_o}\right) = 8.0 \text{ m}\left(\frac{0.035 \text{ m}}{1.5 \text{ m}}\right) = 0.187 \text{ m}$$

Then, from Equation 15.3a:

$$\frac{1}{f} = \frac{1}{D_o} + \frac{1}{D_i} = \frac{1}{8.0 \text{ m}} + \frac{1}{0.187 \text{ m}}$$

so
$$f = 0.180 \text{ m}$$

and the radius of curvature must be twice this, or 36 centimeters.

CORE CONCEPT

Curved mirrors form the same kinds of images as lenses —concave mirrors like convex lenses, and vice versa. The focal length of a spherical mirror is half the radius of curvature.

TRY THIS
—13—

What is the radius of curvature of a convex mirror if it forms an image 45 centimeters behind the mirror of an object that is 3.0 meters in front of it?

Making It Big

A magnifying glass is a useful instrument for reading the small print. You ordinarily use one by placing it directly in front of your eye and looking through it at something you want enlarged, as a jeweler uses a loupe. Figure 15.20 shows how a converging lens forms an enlarged, virtual image. To understand how a magnifying glass works, it is necessary to combine this information with an analysis of the optical properties of the eye.

The lens of your eye is unique in an important respect. Its focal length adjusts to the needs of the situation. If you are looking at something a great distance away, the muscles that control the thickness of the lens are relaxed. If that something moves toward you, you can keep it in focus by changing the thickness of the lens. Starting when it is about 10 meters away, lens muscles act to thicken the lens, shortening its focal length. A normal eye can keep this process going until the object is about 25 centimeters away. This distance is called the *near point;* it is the limit on the closeness to which your eye can focus.

As that something moves toward you, its real image on your retina gets bigger and bigger. When you use a magnifying glass, you are looking at the virtual image formed by the lens. You get the greatest available magnification by placing that virtual image at the near point of your eye, 25 centimeters away. As shown in Figure 15.27b, this means that the image distance of the virtual image formed by the lens is −25 centimeters. The negative sign indicates that the image is virtual.

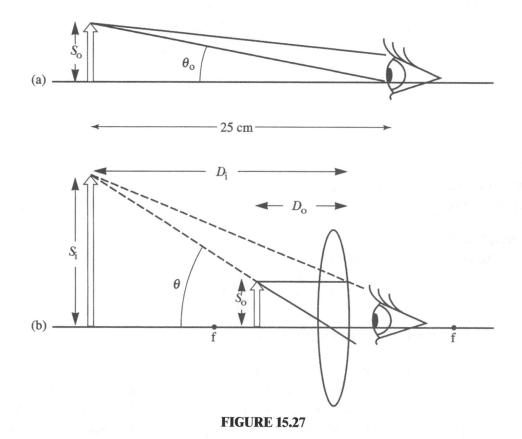

FIGURE 15.27

How much is the magnification? It is the ratio of the sizes of the image on your retina with and without the lens. In Figure 15.27a, the object is at the near point. Its principal ray (the one passing undeflected through the center of the lens of your eye) makes an angle θ_0 with the principal axis of the lens. In Figure 15.27b, a magnifying glass is used to form a virtual image at the near point. The principal ray makes a much larger angle, θ, with the principal axis. It forms a correspondingly larger real image inside the eye.

Now we can calculate the magnification produced by the lens. Comparing the two figures shows that the ratio of these angles is the same as the ratio S_i/S_o. Equation 15.3b tells us that this is equal to D_i/D_o, so this is the magnification M. From Equation 15.3a,

$$D_i D_o = f D_o + f D_i$$

$$\frac{D_i}{D_o} = \frac{D_i - f}{f} = M$$

For maximum magnification, the image is at the near point, −25 centimeters so $D_i = -25$ cm and

$$M_{max} = \frac{-25 \text{ cm}}{D_o} = \frac{-25 \text{ cm} - f}{f}$$

$$M_{max} = 1 + \frac{25 \text{ cm}}{f} \qquad \text{(Equation 15.4)}$$

EXAMPLE 15.20

A philatelist is using a magnifying glass with a focal length of 8.0 centimeters to examine a postage stamp 1.3 centimeters wide. What is the largest apparent size of the stamp he can get with the lens?

SOLUTION

The maximum magnification is

$$M_{max} = 1 + \frac{25 \text{ cm}}{8.0 \text{ cm}} = 4.1$$

So the apparent size of the stamp is 4.1×1.3 cm $= 5.3$ cm.

If the object in Figure 15.27b is moved further from the lens, the angle θ gets smaller, as does the image on the retina. There is a limit. When the object is at the principal focus, rays emerging from the lens are parallel; beyond that, they are convergent, and a real image is formed. Your eye cannot form an image of converging rays. If you place your magnifying glass farther from the lens than the focal length, you will see only a badly blurred picture. The object near the focal point produces the smallest magnification possible with the lens. At that point, the magnification is

$$M_{min} = \frac{25 \text{ cm}}{f}$$

CORE CONCEPT
A magnifying glass produces its greatest magnification when the image is at the near point and its smallest when the object is near the focal point.

REVIEW EXERCISES FOR CHAPTER 15

Fill-In's. For each of the following, fill in the missing word or phrase.

1. The ray model of light is useful when the effects due to _____ are too small to detect.

2. In studying the emission or absorption of light, the _____ model is often most useful.

3. The amount of light emitted by a lamp can be measured in _____.

4. The eye is most sensitive to _____ light.

5. Illumination of 1 lumen per square meter is called a(n) _____.

6. When light rays strike an interface, we always measure the angles with respect to the _____.

7. The angle of reflection is always equal to the _____.

8. A red object reflects the _____ wavelength end of the visible spectrum.

9. Reflection is said to be _____ if a collimated beam of light, on striking a surface, is collimated when it leaves.

10. An image is said to be _____ if there is no light actually coming from the image position.

11. The image formed in a plane mirror, compared with the object, is of _____ size.

12. On entering glass from air, a ray of light bends _____ the normal.

13. To determine the _____ of a plastic, you would measure the angle of incidence and the angle of refraction of a ray of light entering the plastic.

14. Light bends on entering glass from air because its _____ is less in the glass.

15. At a refracting interface, the ratio between the _____ of the two angles is a constant.

16. Total internal reflection occurs when the angle of incidence is more than the _____.

17. To get total internal reflection, the ray must approach an interface with a medium with a(n) _____ index of refraction.

18. In a dispersive medium, the speed of light is a function of _____.

19. When a spectrum is formed by a prism, it is the _____ light that is refracted most.

20. A(n) _____ lens is thicker in the middle than at the edges.

21. Rays that enter a converging lens parallel to the principal axis will emerge and pass through the _____ of the lens.

22. To make a spotlight, the light source must be placed at the _____ of the lens.

23. If an image can be observed by casting it onto a screen, we know that the image is _____

24. In order for a converging lens to form a real image, the object distance must be more than _____.

25. A ray diagram enables us to locate images by plotting exactly _____ rays emerging from a single object point.

26. A(n) _____ lens can form only virtual images.

27. A converging lens forms a virtual image when the object distance is less than the _____.

28. The ratio of object distance to image distance is the same as the ratio of object _____ to image _____.

29. In an algebraic solution for the location of an image, a negative value indicates a _____ image.

30. In a camera, the distance from the center of the lens to the image sensor is the focal length of the lens when the camera is focused at _____.

31. Chromatic aberration results from _____ of light by the lens.

32. The light-gathering power of a lens is indicated by its _____.

33. To form a large image of a distant object, a camera needs a lens with a(n) _____ focal length.

34. The image on the image sensor is blurred when the _____ is large.

35. Real images can be formed by a(n) _____ mirror.

36. If the radius of curvature of a mirror is 40 centimeters, its focal length is _____ centimeters.

37. A lens whose focal length adjusts automatically is found in the _____.

38. The closest distance at which a normal eye can focus on an object is called the _____.

Multiple Choice.

1. A new plastic is discovered in which all wavelengths of visible light travel at the same speed. A prism made of this plastic would be unable to produce
    (1) total internal reflection   (4) dispersion
    (2) an image   (5) refraction.
    (3) diffusion

2. When a beam of light strikes an interior surface at the critical angle, the angle of refraction is
    (1) the critical angle   (4) zero
    (2) more than the critical angle   (5) 90°.
    (3) less than the critical angle

3. A virtual image may be formed by
    (1) a convex mirror only   (4) either kind of curved mirror only
    (2) a plane mirror only   (5) any of the above.
    (3) a concave mirror only

4. An ordinary object can be treated geometrically as though it were a collection of point sources because the object
    (1) generates light   (4) focuses light
    (2) reflects diffusely   (5) disperses light.
    (3) reflects specularly

5. A light ray makes an angle of incidence of 20° with a mirror. The angle between the incident and reflected rays is
    (1) 20° on any mirror
    (2) 40° on any mirror
    (3) 70° on any mirror
    (4) 20° on a plane mirror only
    (5) 40° on a plane mirror only.

6. Light from a point source passes through a lens of focal length $+f$. The emerging light will be divergent if the distance from source to lens is
    (1) less than f
    (2) exactly f
    (3) more than f
    (4) anything at all
    (5) nothing, no distance will work.

7. As you walk toward a plane mirror, your image in the mirror becomes
    (1) closer to the mirror and smaller
    (2) closer to the mirror and larger
    (3) closer to the mirror with no change in size
    (4) further from the mirror with no change in size
    (5) further from the mirror and smaller.

8. The ray model of light cannot be used to explain
    (1) diffraction
    (2) magnification
    (3) reflection
    (4) refraction
    (5) image formation.

9. A physical quantity whose value depends on the properties of the human eye is
    (1) frequency
    (2) wavelength
    (3) illumination
    (4) focal length
    (5) index of refraction.

10. A beam of light can be collimated by placing a point source of light at the principal focus of a
    (1) concave mirror
    (2) concave lens
    (3) convex mirror
    (4) convex lens
    (5) more than one of the above.

Problems.

1. How much luminous flux must a lamp produce to provide 20 lux of illumination at a distance of 1.5 meters?

2. Approximately what is the luminous flux output of a 200-watt lamp that operates at 4 percent efficiency?

3. The image of a post in a plane mirror is 3 meters high and 12 meters behind the mirror. Find the size and position of the image if the object distance is halved.

4. What is the index of refraction of Lucite if light travels in it at 2.00×10^8 meters per second?

5. Find the index of refraction of an unknown liquid if a ray of light enters it at an angle of 45° and is refracted to 32°.

6. A layer of benzene ($n = 1.50$) is floated on top of a mass of water ($n = 1.33$). If a light ray enters the benzene from above, making an angle of incidence of 55°, what is the angle of refraction inside the water?

7. A 45° prism is to be used to produce total internal reflection at two surfaces, as in Figure 15.11. What is the minimum index of refraction needed for the glass?

8. Determine the critical angle of incidence of a plastic if a ray entering it at 62° is refracted to 47°.

9. The index of refraction of borate flint glass is 1.590 for violet light and 1.566 for red. If a violet ray is refracted to 65.0° on entering the glass, what is the angle of refraction for a red ray?

10. Determine the size and location of the image in a magnifying glass of focal length + 5.0 centimeters if it is being used to examine a beetle 1.2 centimeters long placed 2.0 centimeters in front of the lens.

11. What focal length lens would be needed to form a virtual image at a distance of 6.5 centimeters from the lens when the object is 9.0 centimeters from it?

12. Find the size and location of the image formed by a + 20-centimeter lens when a vase 40 centimeters tall is placed 50 centimeters from the lens.

13. A lamp and a screen are 3.0 meters apart, and an image of the lamp is to be formed on the screen by a +50-centimeter lens. What are the two possible distances from lamp to lens at which a real image will form on the screen?

14. A photographer, using a + 80-millimeter lens, wants to take a portrait of the uppermost 1.00 meter of a subject, to form an image on the sensor 6.5 centimeters high. How far from the lens must the subject be seated?

15. What is the diameter of a lens marked "f:3.5, focal length 80 mm"?

16. If a convex mirror has a radius of curvature of 40 centimeters, what size image will it form of a candlestick 35 centimeters high that is placed 1.5 meters in front of the mirror?

17. Where must a vase be placed with respect to a concave mirror whose radius of curvature is 78 centimeters in order to form a real image one-half the size of the vase?

18. A pencil 10.0 centimeters long is placed 30 centimeters from a concave mirror. A real image of the pencil 6.5 centimeters long is found between it and the mirror. The pencil is then moved to a position 5.0 centimeters from the mirror. Describe the position, size, and nature of the image formed.

19. The rearview mirror on the right side of a car produces an image of oncoming traffic at 15 centimeters in front of the mirror. What is its radius of curvature?

20. What focal length is needed to have a lens that can be used to magnify the print on a page by a factor of 5?

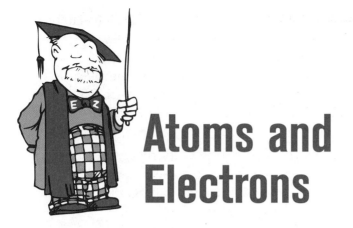

Atoms and Electrons

WHAT YOU WILL LEARN

The properties of electrons and their incorporation into atoms; the beginnings of quantum mechanics.

SECTIONS IN THIS CHAPTER

- The Discovery of the Electron
- Finding the Nucleus
- Physics in Trouble—Again
- The Big Breakthrough: The Hydrogen Atom
- Measuring the Energy Levels
- Matter Waves
- The Probability Wave
- Experiment and Reality
- The New Hydrogen Atom
- Other Elements
- Making X-rays

The Discovery of the Electron

Benjamin Franklin, and many physicists after him, strongly suspected that electrical effects were produced by the motions of tiny, separate units of electric charge. This was finally proved in 1897 by J. J. Thomson.

Thomson's experiment was based on the study of cathode rays. They had been observed in a glass tube, partially evacuated, with electrodes at each end. If the electrodes are connected to a source of high potential difference, a greenish glow appears around the cathode, the negative terminal. Thomson was able to show that the glow is produced by a flood of particles, all with identical charge and mass, flowing out of the cathode. He called these particles *electrons*.

A simplified form of Thomson's apparatus is shown in Figure 16.1. At the left end is an electron gun. The hot filament releases electrons. A pair of positive plates with holes in them make an electric field that forms the particles into a beam and accelerates them. The electron beam passes to the other end of the tube. If a tiny amount of air is left in the tube, the path of the beam will be seen as a glowing line. When the beam strikes a fluorescent screen, it produces a glowing spot. This device graduated into the TV picture tube.

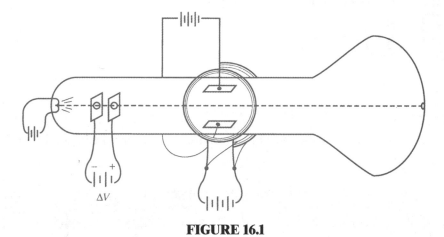

FIGURE 16.1

Thomson found that either an electric or a magnetic field could deflect the beam. As shown in Figure 16.1, a pair of plates placed on opposite sides of the tube could set up an electric field. The magnitude of the field, according to Equation 9.2, is the potential difference on the plates divided by the distance between them. Since the field exerts a force on the particles, they must be charged. According to Equation 9.2, the electric force on the particles is the field times the charge on the particles, $\mathscr{E}q$. The beam deflects toward the positive plate, so the particles must be negative.

If the beam consists of a stream of charged particles, it ought to be—and it is—deflected by a magnetic field also. The horizontal magnetic field set up by a pair of vertical Helmholtz coils outside the tube deflects the beam vertically. The magnitude of the field can be found from the dimensions of the coils and the current in them. Equation 11.6 tells us that the force is the product of the charge and velocity of the particles, and the magnetic field they are in, qvB.

Now see what happens if the current in the Helmholtz coils is adjusted so as to make the beam go straight through the two fields. The net force on the particles is zero, so the electric and magnetic forces must be equal in magnitude:

$$\mathscr{E}q = qvB$$

so that

$$v = \frac{\mathscr{E}}{B}$$

In other words, when the two fields allow the beam to pass straight through, the velocity of the particles can be found from the two field strengths. The charge does not matter, so long as there is one.

Now it is possible to find the ratio of mass to charge of the particles. Equation 9.3 tells us that the electric energy lost as the particles travel between the accelerating plates is the potential difference on the plates times the charge, or $q\,\Delta V$. This becomes the kinetic energy of the particles as they emerge from the gun, so

$$q\Delta V = \frac{1}{2}mv^2$$

Therefore, if the velocity of the particles is known, the ratio of charge to mass can be found:

$$\frac{q}{m} = \frac{v^2}{2\Delta V}$$

EXAMPLE 16.1 A stream of particles is accelerated through a potential difference of 500 volts, and then passes straight through a pair of crossed electric and magnetic fields. If the electric field is 2 600 newtons per coulomb and the magnetic field is 12 milliteslas, what is the charge-to-mass ratio of the particles?

SOLUTION

First find the velocity of the particles. Since the beam passes through without deflection, the two forces on the particles are equal, and

$$v = \frac{\mathcal{E}}{B} = \frac{2\ 600 \text{ N/C}}{12 \times 10^{-3} \text{ N/A·m}} = 2.17 \times 10^5 \text{ m/s}$$

Then the kinetic energy of the particles is equal to the electric energy they lost in the gun, so $q\,\Delta V = \frac{1}{2}mv^2$, and

$$\frac{q}{m} = \frac{v^2}{2\Delta V} = \frac{\left(2.17 \times 10^5 \text{ m/s}\right)^2}{2 \times 500 \text{ V}} = 4.7 \times 10^7 \text{ C/kg}$$

With this value, this could be a stream of alpha particles.

Thomson's experiment did not determine either the mass or the charge on the electron. It did show that the stream consists of separate particles, all having the same definite mass and charge. When Millikan used the principles of electrostatics to measure the charge on the electron (see "The Charge on an Electron" in Chapter 9), it was then a simple matter to find its mass.

CORE CONCEPT

The existence of the electron was proved in an experiment in which its charge-to-mass ratio was measured in a cathode-ray tube.

TRY THIS
—1—

What would happen to the electron beam if the accelerating potential were increased?

Finding the Nucleus

Everything contains electrons; they can be transferred from one object to another; and they flow freely through metals. Electron charge has been measured, as well as electron mass.

Since electrons are negative, and ordinary objects have no electric charge, everything must also contain positive charges equal in magnitude to the charge on all their electrons. How is that positive charge distributed within a material? Just where is it located?

The discovery of the atomic nucleus provided an answer to these questions. The crucial experiments were done by Ernest Rutherford. In 1911, he advanced a theory of the atom that was considerably better than any other up to that time. Like many good theories, however, it eventually gave way to a better one.

Rutherford worked out a method of studying the structure of matter by bombarding it with alpha particles. No one was really sure what these bits of matter ejected by radioactive materials are. However, their mass, charge, and energy were measurable, and this was enough to go on with.

In the original experiment, a stream of alpha particles was obtained by enclosing radioactive polonium inside a small lead box with a hole in it (Figure 16.2). Alpha particles streamed out of the hole and struck an extremely thin sheet of gold foil. On the other side of the foil was a screen coated with a chemical salt that produced a flash of light every time it was struck by an alpha particle. The whole apparatus was enclosed in high vacuum.

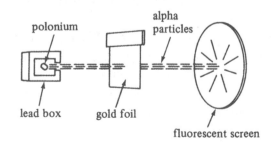

FIGURE 16.2

The first discovery was that the gold foil was mostly empty space; 99.99 percent of all the alpha particles went straight through the foil, 400 atoms thick, just as though there was nothing there. An occasional particle bounced back in the direction it had come from. Whatever it is that makes up gold, it is concentrated in many tiny lumps separated by distances about 1 000 times their own size—nuclei.

Most of the particles that did not go straight through the gold foil did not bounce straight back but were deflected to one side or another of their straight-line path. They struck the screen off to the side of the main, straight-line path. Mathematical analysis of this deflection, shown in Figure 16.3, made it possible to determine the nature of the force the nuclei exerted on the particles. It turned out to be an inverse-square force of repulsion and agreed perfectly with values calculated from Coulomb's law of electric force, Equation 9.1. It was evident that the positive charge in the gold foil was concentrated just where the mass is found, in the nucleus.

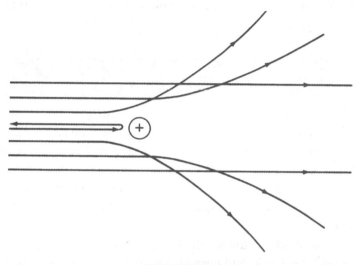

FIGURE 16.3

When this experiment was repeated with other kinds of foil, it was found that each chemical element had its own characteristic electric charge on the nucleus. Since each atom as a whole is electrically neutral, the charge on the nucleus must be equal in magnitude to the charge on some whole number of electrons. The nucleus contains an integral number of positive elementary charges, and that number, unique for each chemical element, is the *atomic number* of the element.

Out of these experiments came Rutherford's classic model of the atom. Nearly all the mass of an atom is concentrated in its tiny nucleus, about 10^{-14} meters in diameter. If 10 billion nuclei could be lined up together, they would stretch a distance equal to the thickness of a hair. A teaspoonful of nuclear substance would have a mass of a billion metric tons. Surrounding each nucleus, at vast distances, are electrons, revolving around the nucleus. The electrons are negative and are kept in orbit by their electrical attraction to the positive nucleus just as gravity keeps the planets in their orbits around the sun. Here is a comparable ratio: the nucleus inside the atom is like a pea in a football field. The number of electrons is equal to the number of positive electric charges on the nucleus, so the atom as a whole is neutral. Rutherford's model of the nucleus has held up, but his picture of electrons in orbit, like planets around the sun, had some fatal flaws.

CORE CONCEPT

An atom consists of a nucleus, containing most of its mass and a quantized positive charge, surrounded by negative electrons whose number is the atomic number of the element.

TRY THIS
—2—

How much is the positive charge, in coulombs, in the nucleus of an atom of tin, atomic number 50?

Physics in Trouble—Again

When the twentieth century dawned, physics was faced with a number of puzzling phenomena, which did not seem to have any explanations. The answers, when they came, involved a complete revision of the most fundamental concepts in physics, such as mass, length, energy, and time. Even the notion of causality and the meaning of physical laws had to undergo a violent revolution before the answers to many of these problems could be incorporated into the body of the science. Here are a few of the puzzles demanding solutions:

No known process could account for the fantastic rate at which the sun produces energy. It seemed that the sun would burn out in a few thousand years, yet geological evidence indicated that its age would be measured in billions of years, not thousands.

Rocks could be found in the earth that kept on producing tiny, energetic particles, even though no energy was being put into them.

The axis of the orbit of the planet Mercury was changing its position, in violation of all the known laws of mechanics.

As indicated previously, Rutherford's model of the atom had a fatal flaw. Electrons are charged particles. If they are going in circular paths, they are accelerated charges. As we saw in "Field Interactions" in Chapter 14, an accelerated charge must generate electromagnetic waves, sending its energy off into space. Therefore, Rutherford's atom would collapse in a tiny fraction of a second.

There was no explanation for the blackbody radiation curves shown in Figure 14.17. The existing theory explained the lower end of the graph perfectly, but it could not explain why the graph curved downward. The theory predicted that more and more energy would be concentrated as the frequencies go higher, without limit. This was called the "ultraviolet catastrophe."

There was no explanation for the fact that the spectrum of the light emitted by a gas, when excited by an electric current, consists of separate frequencies. The spectrum of a gas is a set of discrete lines, unlike the continuous spectrum of an electric lamp or the sun.

In short, physics was ready for a revolution. Two entirely fresh ideas were introduced: the theory of relativity and quantum theory. Both required physicists to reconsider their longest standing and most basic concepts. Many were reluctant to do so, but the force of experimental evidence was irresistible. Twentieth-century physics is a whole new ball game.

It is a remarkable fact that the most basic concepts of both new theories were published in 1905 in four papers by the same author; he was Albert Einstein, a little-known German patent examiner in Switzerland. One of the papers produced the theory known as the *special theory of relativity,* which revised Newton's long-standing laws of mechanics. It accounted perfectly for the strange behavior of Mercury. In a second paper, published later that year, Einstein analyzed this theory further and produced his famous equation $E = mc^2$. Eventually, this accounted for the energy of the sun and of the rocks that were producing energetic particles (this will be discussed further in the next chapter).

As we saw in "Red Hot, White Hot" in Chapter 14, the blackbody radiation curve was explained on the assumption that energy levels are quantized. The hot object can reduce its energy only in jumps, not in a continuous flow. The higher the frequency of the emitted light, the bigger is the energy jump that produces it. In one of the four papers published in 1905, Einstein concluded that light itself exists in lumps of energy, in photons. This was expressed in his second simple, fundamental equation, which gives the energy of a photon: $E = hf$. He then predicted the results of experiments on the photoelectric effect ("Photoelectricity" in Chapter 14), which were later verified. This led to an explanation for the line spectra of gases, and a better model of the atom.

The simplest line spectrum is produced by atomic hydrogen. If hydrogen is heated to a high temperature and subjected to an electric current, its molecules dissociate into separate atoms, and the gas glows. When the light is analyzed by a spectroscope, it is found to consist of a series of separate wavelengths, shown in Figure 16.4. The lines of the spectrum obey a simple equation, which eventually yielded its deep meaning when it was written in terms of the energy of the photons emitted by the hot hydrogen:

$$E = hf = 13.6 \text{ eV}\left(\frac{1}{n_1{}^2} - \frac{1}{n_2{}^2}\right) \qquad \textbf{(Equation 16.1)}$$

where the ns are whole numbers. If $n_1 = 2$, values of n_2 from 3 up give the energies of all the photons of visible light shown in Figure 16.4. Other values of n_1 give series of lines in the ultraviolet and infrared.

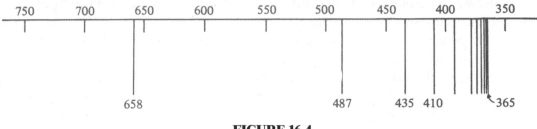

FIGURE 16.4

EXAMPLE 16.2   What is the wavelength of the line in the visible spectrum of atomic hydrogen specified by the value $n_2 = 5$?

SOLUTION

In the visible spectrum, $n_1 = 2$, so

$$E = 13.6 \text{ eV}\left(\frac{1}{2^2} - \frac{1}{5^2}\right) = 2.86 \text{ eV}$$

Since $E = hf$ and $c = f\lambda$,

$$\lambda = \frac{ch}{E} = \frac{(3.00 \times 10^8 \text{ m/s})(4.15 \times 10^{-15} \text{ eV·s})}{2.86 \text{ eV}} = 4.35 \times 10^{-7} \text{ m}$$

This is the third line in the visible part of the spectrum.

A regularity like this one cries out for an explanation. When it came, it answered a lot of questions and produced a great many new ones.

CORE CONCEPT

Unanswered questions, including the meaning of the line spectra of gases, called for a complete reexamination of the fundamental laws of physics.

TRY THIS
—3—

What is the highest frequency in the visible spectrum of atomic hydrogen?

The Big Breakthrough: The Hydrogen Atom

Serious problems call for drastic solutions. Niels Bohr found a solution to the problem of the line spectra of gases. To do it, he had to assume that electrons could travel in circular paths without radiating away their energy. This violated a basic law of physics, but it worked.

Bohr assumed that the spectral lines represent the energy emitted by electrons orbiting the nucleus. The hydrogen electron could receive energy, as from an electric field, and jump out to an orbit farther from the nucleus. It does not release any energy while in orbit but only as it falls back to a smaller orbit. It falls back in jumps. In each jump from a larger orbit to a smaller one, a photon of energy is released. Only certain orbits are possible, so the number of possible jumps is limited. The spectral lines are the photons released in all possible jumps. The wavelengths of these lines represent the energies of the photons that make them.

What defines the possible energy levels of the hydrogen electron? Bohr accounted for the lines of the hydrogen spectrum by making one assumption: the angular momentum of the single electron of the hydrogen atom is quantized. Angular momentum is the product of mass, speed, and the radius of the circle, mvr. Bohr assumed that, for the electron, it must obey the equation

$$mvr = \frac{nh}{2\pi} \qquad \textbf{(Equation 16.2)}$$

where h is Planck's constant and n is any whole number.

With this expression and two well-known equations from classical physics, Bohr produced an expression for the energy of the electron. He began by using an equation for the force on the electron. The force that holds the electron in orbit is the Coulomb attraction (Equation 9.1). The charges are the charge on the electron and the charge on the nucleus. Since they are equal, we can write q^2 instead of $q_1 q_2$ The force is centripetal, so it must be (Equation 4.4)

$$\frac{mv^2}{r} = \frac{kq^2}{r^2}$$

What are the possible energies of these different electron orbits? Multiplying through by $r/2$ gives an expression for kinetic energy:

$$E_{\text{kin}} = \frac{mv^2}{2} = \frac{kq^2}{2r}$$

The equation for electric potential energy has exactly the same form as the gravitational potential energy equation (Equation 5.4). Just change the constant and replace mass by charge, and the equation for electric potential energy is

$$E_{\text{pot}} = -\frac{kq^2}{r}$$

Add the potential and the kinetic energy to get the total energy:

$$E_{\text{tot}} = \frac{kq^2}{2r} - \frac{kq^2}{r} = -\frac{kq^2}{2r}$$

Look again at the centripetal force equation; multiply it by mr^3 to get

$$m^2v^2r^2 = kq^2rm$$

and, since mvr is the quantized angular momentum,

$$kq^2rm = \frac{n^2h^2}{4\pi^2}$$

Then use this to eliminate r from the energy equation:

$$E = \frac{2\pi^2k^2q^4m}{n^2h^2}$$

All these quantities are known:

> k = electric constant of free space = 9.0×10^9 newton-meters squared per coulomb squared
> q = electronic charge = 1.60×10^{-19} coulombs
> m = mass of electron = 9.1×10^{-31} kilograms
> h = Planck's constant (in SI) = 6.6×10^{-34} joule-seconds

Inserting these values yields

$$E = -\frac{2.18 \times 10^{-18} \text{ J}}{n^2} = -\frac{13.6 \text{ eV}}{n^2}$$

The energy is negative because the particle is bound. When $n = 1$, called the ground state, an additional 13.6 electron volts of energy will release the electron; at infinite separation the energy rises to zero. It had been known for some time that the ionization energy of hydrogen is 13.6 electron volts.

The possible excited states of the electron are represented by values of n, the *principal quantum number*. Clearly the change in energy as the electron jumps from one state to another is given by Equation 16.1. When the jump is made, a photon is emitted, carrying away the excess energy. Figure 16.5 shows the first few energy levels of the hydrogen atom, from $n = 1$ to $n = 6$.

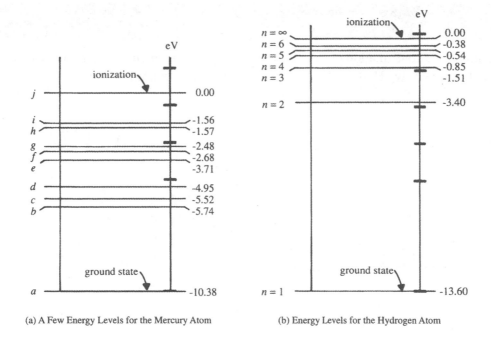

(a) A Few Energy Levels for the Mercury Atom   (b) Energy Levels for the Hydrogen Atom

FIGURE 16.5

This second model of the atom—not by any means the last—was the second great triumph for the new quantum theory. Other atoms have much more complex line spectra, because they have more than one electron, and they interact with each other; some of the levels in mercury are shown in Figure 16.5. Bohr's great discovery remains: the energy of the emitted photons is the difference between the quantized energy levels in the atom.

EXAMPLE 16.3   How much is the energy of a hydrogen electron in the third excited state?

SOLUTION

The energy is negative; it is the amount of energy that would be needed to raise the energy to zero, when $n = \infty$. Then the energy is simply $-13.6 \text{ eV}/3^2 = -1.5 \text{ eV}$.

CORE CONCEPT

For the hydrogen atom, a simple expression, calculated from simple assumptions, gives the quantum states of the electron.

TRY THIS
—4—

What is the wavelength of the photon emitted when a mercury electron drops from level f to level c?

Measuring the Energy Levels

A remarkable confirmation of Bohr's theory came in experiments done by James Franck and Gustave Hertz in 1914. Their apparatus is diagrammed in Figure 16.6.

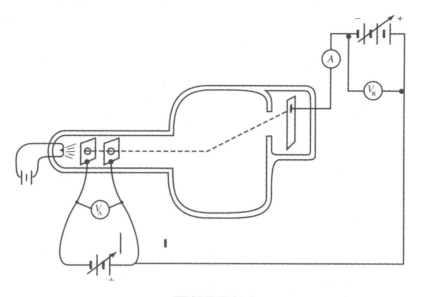

FIGURE 16.6

The electron gun on the left can accelerate electrons to give them kinetic energy. The voltmeter labeled V_A measures the accelerating potential. If the voltmeter reads 3 volts, for example, the electrons leave the gun with 3 electron volts of kinetic energy. They are fired into an evacuated chamber containing a small amount of mercury vapor.

On the other side of the chamber, but not in line with the gun, is an opening through which the electrons can pass. Some of the electrons that collide with mercury atoms will find their way through the hole to strike a plate bearing a negative potential. There is an ammeter (A) in the collecting plate circuit. The energy of the electrons arriving at the plate is measured by increasing the retarding potential (V_R) until the current is cut off. For example, if it takes a retarding potential of 3 V to stop the flow of current, the electrons are arriving at the plate with 3 electron volts of energy.

Electrons leaving the gun at 3 electron volts arrive at the collecting plate with 3 electron volts of energy. They have been deflected by colliding with mercury atoms, but the collisions were elastic, and the electrons have not lost any energy. An analogy is a Ping-Pong ball striking a bowling ball and bouncing off in a different direction.

Now what happens if the accelerating potential is gradually increased? When the accelerating potential exceeds 4.64 volts, something dramatically different happens. The energy of the electrons at the collecting plate suddenly drops, and it drops by just 4.64 electron volts. If the accelerating potential is 5.0 volts, a retarding potential of 0.36 volts will cut off the collecting plate current. And something else happens: the gas begins to glow. It emits photons with just 4.64 electron volts of energy.

What has happened is clear. It takes 4.64 electron volts to raise electrons of the mercury atoms from the ground state to level *b* of Figure 16.5. The bombarding electrons are now giving up some of their energy to accomplish this. Once excited, the mercury electrons drop back to their ground state by emitting a photon.

Increasing the accelerating potential further will enable the mercury electrons to reach other excited states. Each time a new quantum state is reached, the energy at the collecting plate drops again and additional photons are released. The photon energy is always just equal to the amount of energy the bombarding electrons lose between the gun and the collecting plate.

This sort of experiment has been done with many different kinds of atoms, and has been a useful way to measure the excited states.

EXAMPLE 16.4

In a Franck-Hertz type experiment using an unknown gas, the current cuts off first at 3.85 volts and then again at 5.20 volts. What is the longest wavelength of the spectrum produced by the gas?

SOLUTION

The longest wavelength is produced by the smallest energy jump of the three available, which are 3.85 eV, 5.20 eV, and the jump from the third to the second level, $5.20 - 3.85 = 1.35$ eV. From Equations 16.1 and 8.4, $f = E/h = c/\lambda$, so

$$\lambda = \frac{ch}{E} = \frac{(3.0 \times 10^8 \text{ m/s})(4.1 \times 10^{-15} \text{ eV} \cdot \text{s})}{1.35 \text{ eV}} = 9.1 \times 10^{-7} \text{ m}$$

which you will need an infrared detector to find.

CORE CONCEPT

Bombarding electrons can give up their energy to atoms only when they can provide enough energy to reach the quantum states of the atom.

TRY THIS
—5—

When electrons are fired into mercury vapor in a Franck-Hertz experiment, the energy at the collecting plate drops when the potential on the plate gets bigger than 4.86 volts. What are the energies of the photons emitted (consult Figure 16.5)?

Matter Waves

By the 1920s, so many strange ideas had emerged from the experiments in quantum theory that anything seemed possible. It was then that Louis de Broglie suggested that, if light could behave like particles, maybe particles could behave like waves.

Waves must have wavelength. In a stream of particles, as in a stream of photons, wavelength is inversely proportional to momentum. Equation 14.5 applies to particles as well as to waves. For particles, however, the speed is not c, and the momentum is mv. Therefore, when Equation 14.5 is applied to particles, the wavelength of a particle can be expressed as

$$\lambda = \frac{h}{mv} \qquad \textbf{(Equation 16.3)}$$

What distinguishes a wave from a stream of particles is the phenomenon of interference. If two streams of water from two hoses arrive simultaneously at the same spot, there will always be twice as much water as would come from one stream. However, two streams of photons, two light waves, can arrive at a spot and cancel each other. They will do so if they arrive out of phase. When a light beam passes through a pair of slits or a diffraction grating ("Measuring Wavelength" in Chapter 14), the light forms alternate light and dark bands. Where two parts of the original beam arrive out of phase, there is no light. Waves, but not particles, can cancel each other.

To find out whether a beam is a wave or a stream of particles, we must look for interference. For this purpose we need a diffraction grating. Equation 14.1b tells us what sort of diffraction angles to expect:

$$\sin \theta_n = \frac{n\lambda}{d}$$

To get any sizable diffraction angle, then, we need a grating whose spacing, d, is not much greater than the wavelength.

EXAMPLE 16.5

What is the wavelength of a stream of electrons accelerated in an electron gun through a potential difference of 100 volts?

SOLUTION

The kinetic energy of the electrons is equal to the electric energy they lose in the gun:

$$q\Delta V = \frac{mv^2}{2}$$

Multiplying both sides of the equation by $2m$ gives

$$2mq\,\Delta V = m^2 v^2$$

So the momentum of the electrons is

$$mv = \sqrt{2mq\Delta V} = \sqrt{2(9.1 \times 10^{-31}\ \text{kg})(1.6 \times 10^{-19}\ \text{C})(100\ \text{V})}$$

$$mv = 5.4 \times 10^{-24}\ \text{kg·m/s}$$

and the wavelength is

$$\lambda = \frac{h}{mv} = \frac{6.62 \times 10^{-34}\ \text{J·s}}{5.4 \times 10^{-24}\ \text{kg·m/s}} = 1.2 \times 10^{-10}\ \text{m}$$

The grating spacing required is the size of an atom. It would be quite hopeless to try to make such a grating. Fortunately, nature provides one, A crystal consists of atoms lined up in neat rows. These rows are extremely straight, and the surface of the crystal acts like a grating ruled at atomic dimensions. The experiment has been done; a beam of electrons reflected from the surface of a crystal does in fact form a series of alternate nodes and antinodes. This phenomenon has now provided a standard method of investigating the intimate structures of crystals.

The existence of matter waves suggests another interpretation of the atom of atomic hydrogen. Bohr got his spectacular result by assuming that the angular momenta of electrons are quantized:

$$mvr = \frac{nh}{2\pi}$$

This equation can be rearranged to give

$$n\frac{h}{mv} = 2\pi r$$

But h/mv is the wavelength of the electron matter wave, and $2\pi r$ is the circumference of the orbit, The equation is now telling us that the only orbits possible are those in which a whole number of wavelengths fit into the circumference. The electron matter waves form standing waves, ringing the nucleus. We are reminded of a guitar string, in which standing waves can form when a whole number of half waves fit into the string. The existence of matter waves produced the wave model of the atom, the third attempt to penetrate the mystery of the nature of matter.

If moving objects have wave properties, why do we never see interference under ordinary conditions? Consider what the wavelengths of ordinary objects are.

EXAMPLE 16.6   What is the wavelength of a 5-gram marble moving at 1 meter per second?

SOLUTION

$$\lambda = \frac{h}{mv} = \frac{6.62 \times 10^{-34}\ \text{J·s}}{(5 \times 10^{-3}\ \text{kg})(1\ \text{m/s})} = 1.3 \times 10^{-31}\ \text{m}$$

The wavelength of the marble is a billion billion times smaller than a proton. There is no hope of ever finding a grating that can diffract it. However, wave properties have been studied in ions and other kinds of atomic and subatomic particles, which have momenta small enough to reveal them.

CORE CONCEPT

Moving matter has wave properties; the wavelengths are inversely proportional to the momenta of the particles.

TRY THIS
—6—

What would the wavelength of the electrons in Example 16.5 be if the electrons were accelerated at 1 volt instead of 100 volts?

The Probability Wave

Can photons act like waves? An interesting experiment was done to investigate the relationship between wave and particle properties, The apparatus, illustrated in Figure 16.7, consists of a long, light-tight box with an extremely dim light at one end. At the other end is a photographic plate. The box is so long, and the light is so dim, that there is never more than one photon at a time in the box, going from the lamp to the plate.

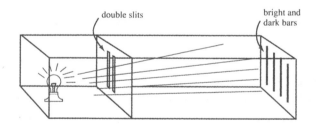

double slits
bright and
dark bars

FIGURE 16.7

Between the lamp and the plate is a pair of slits. A light wave passing through these slits forms a diffraction pattern. Constructive and destructive interferences alternate, producing a series of light and dark bars on the plate. The box was left in place for several months to get enough light on the plate to produce an image. When the plate was developed, the typical diffraction pattern was found.

It seems that one photon at a time came from the lamp, went through the slits, and landed on the plate. It always landed at a bright bar on the plate, never in a dark region. How did the photon "know" where it was supposed to go? It was the only photon in the box, so how could it behave in such a way as to form an interference pattern?

Furthermore, how does a single photon go through a double slit? If each photon goes through only one slit, why do the photons form the typical double-slit interference pattern?

Quantum effects such as this one call for a whole new way of thinking about the universe. Scientists are accustomed to thinking in terms of cause and effect. Newton taught us that, if you know the present conditions and the laws of nature, you can predict what will happen. Given the mass and present velocity of a tennis ball, you can tell exactly where it will go if you apply a known force to it. If you do the measurements, you will find that they obey your calculations. True, the measurements never match the calculations *exactly;* there is always some uncertainty because measurement is never perfect. However, by careful experimental design, it is always possible to reduce the uncertainty, to increase the accuracy of measurement with no ultimate limit.

It turns out that in the atomic domain this is not true. In dealing with tiny particles and minute amounts of energy, there is an ultimate limit to the accuracy of measurement that is available. To find out where an electron is, you will have to do something to it. Perhaps you will bombard it with a single photon. In the realm of the extremely small, the energy of a photon is enough to change things. The fact that you cannot find out where an electron is without disturbing it places an ultimate limit on the accuracy of your measurement of the electron. Uncertainty is built into the experiment and cannot be reduced below a certain level.

The limits on the accuracy of measurements are given by the *Heisenberg uncertainty principle,* which says that physical quantities come in pairs. There is always some uncertainty in measuring either member of the pair. If you measure both members of the pair, the product of the two uncertainties can never be less than the quantity $h/2\pi$. This is an extremely small limit, but in dealing with anything the size of an electron, it is significant.

Momentum and position make a Heisenberg pair. There is no theoretical limit to the accuracy of measurement of the position of an electron. But as the accuracy of position measurement increases, the accuracy of momentum measurement must decrease. The more you learn about one of the two quantities, the less you can find out about the other. If you know *exactly* what the momentum of an electron is, you can have no idea *where* it is.

Energy and time are another Heisenberg pair. An experiment designed to give exceptional accuracy in regard to the energy of a particle will sacrifice accuracy about the length of time it exists.

What happens when a photon goes through a double slit? Since there is no way of knowing which slit it went through, the distance between the slits is the uncertainty in its position. The uncertainty in its momentum is represented by its deviation from a straight-line path. It is this deviation that produces the diffraction pattern. The closer together the slits are, the less the uncertainty in position, and the more the uncertainty in momentum. This means that, when the slits are closer together, the diffraction pattern spreads out further. We have seen this before; Equation 14.4b tells us that the sines of the diffraction angles are inversely proportional to the slit spacing.

These rules tell us nothing about where any particular photon will go, It is possible, however, to calculate the *probability* that a photon will land in any given region. The probabilities are high that photons will land in the bright region, and very low for the dark ones. When many photons pass through the slits, most of them will land in regions of high probability and produce a bright bar. The relationship is something like what happens in throwing dice. You cannot predict how any given toss will turn out, but it is safe to say that the probability of coming up seven is just one out of six. The more tosses you make, the closer the actual ratio comes to one in six.

During the several months of the photon diffraction experiment, an enormous number of photons strike the plate. There are so many that, with the right mathematics, the distribution of the photons on the plate can be predicted with great accuracy. The equations that produce this prediction are the equations that describe waves. The location of the highest probabilities is found with the ordinary equations for the diffraction of light. In that sense, a light wave, or a matter wave, is a wave of probabilities. It describes the probable distribution of particles.

The laws of modern physics make no attempt to predict what will happen in any single event; they give only probabilities. If many identical processes take place, the laws can predict the overall effect with high precision. The predictions are never perfect; they tell us only probabilities. The old idea that the universe operates like a clockwork, with all future conditions determined by the present state, is gone forever.

The wave atom was a good model for its time. It described the electrons of the atom in terms of a matter wave. In the most modern atomic model, that wave is a probability wave. It gives the probabilities of the electrons being at various places. Electrons are not localized; the nucleus is surrounded by "electron clouds." Each cloud symbolizes the probabilities for the location of an electron.

CORE CONCEPT
Quantum theory predicts only probabilities of single events. Light and matter waves are waves of probability.

TRY THIS
—7—

Why is there no Heisenberg uncertainty in calculating the path of a bullet?

Experiment and Reality

Well, then, what *is* light, a wave or a stream of particles? Is it possible for it to be both at the same time?

No, it is not. Sometimes light is a wave, and sometimes a stream of particles. Which it is depends on the experiment in which it is being observed. It is the experimenter who decides.

The peculiarities of quantum theory were described with the coming of the *Copenhagen interpretation.* Applied to light, this interpretation says that light has either aspect, depending on the kind of experiment being done. In figuring out where the light is going to land in the diffraction-box experiment, light must be treated as a wave. No model describing it as a stream of photons will give the answer. On the other hand, if you want to investigate what the light is doing to the photographic emulsion, the light must be thought of as a stream of photons. Each photon raises the energy level of a silver atom, so that it can be developed into a black spot. No wave model can explain this.

Everyone is now tempted to ask, But what is light *really?* A scientist must beg off this question. Science does not offer answers to all questions. The scientist is limited to questions that can be answered by experiment or observation. There are many kinds of questions that cannot be answered by experimentation. Here are a few: Is honesty the best policy? Is the Sistine Chapel beautiful? Is a straight line the shortest distance between two points? Does God exist? These may or may not be meaningful questions, but in any case, science cannot answer them.

There is no experiment that can tell what light is like when it is not being investigated. In any experiment, it is either a wave or a particle—never both. The question posed above—What is light really, when it is not being observed?—has no meaning.

How about that other question—Which slit does a given photon go through? Or does it go through both? A photon in motion cannot be detected. Light is seen only when it strikes something. Then it is absorbed into matter and disappears. You might try covering one of the slits to see whether a photon goes through the other; if you do, the interference pattern disappears. Since there is no way to design an experiment to answer the question, the question is meaningless. Forget it. While traveling through the slits, light is a wave.

CORE CONCEPT

The question of whether a beam, whether of light or of electrons, is a wave or a stream of particles is meaningful only when it is detected, and the answer depends on the kind of experiment.

TRY THIS
—8—

Cite an experiment in which an electron is a wave and another in which it is a particle.

The New Hydrogen Atom

The Heisenberg uncertainty principle put an end to efforts to state precisely where an electron is located as it circles the nucleus. The discovery of matter waves necessarily led to a whole new model for the atom. A matter wave, remember, is a wave of probability. The new model of the atom makes no effort to locate the electron precisely. The position of the electron is represented as a kind of cloud of probability surrounding the nucleus. Mathematically, this cloud is called the *wave function* of the electron.

Still, the atom has a definite size. It can be calculated from the Bohr model, which is still useful in certain limited applications. The centripetal force on the electron is its electrical attraction to the nucleus, so (Equations 4.4 and 9.1)

$$\frac{mv^2}{r} = \frac{kq^2}{r^2}$$

Equation 16.2 tells us that the angular momentum of the electron is

$$mvr = \frac{nh}{2\pi}$$

Solving these two equations for v^2 gives

$$v^2 = \frac{kq^2}{mr} = \frac{n^2h^2}{4\pi^2m^2r^2}$$

And solving for r gives

$$r = \frac{n^2h^2}{4\pi^2mkq^2}$$

Putting in the known values of Planck's constant, the mass and charge of the electron, and the electric constant of free space:

$$r = (5 \times 10^{-11}\text{ m})n^2$$

The size of the hydrogen atom in its ground state is known from experimental data, and it agrees well with this value when $n = 1$.

The modern electron cloud model of the atom perceives the electron only in terms of its probable location in the atom. In the ground state, the probabilities, the wave function, form a kind of fuzzy hollow sphere, a spherical shell. This condition is the only possible configuration for the wave function when the electron is in the ground, or $n = 1$, state. The Bohr radius now becomes the most probable distance of the electron from the center.

EXAMPLE 16.7   What is the radius of a hydrogen atom in its fourth excited state?

SOLUTION

In the fourth excited state, $n = 5$ ($n = 1$ is the ground state). So

$$r = (5 \times 10^{-11} \text{ m})(5^2) = 1.3 \times 10^{-9} \text{ m}$$

CORE CONCEPT

The Bohr model of the hydrogen atom provides an explanation for the emission spectrum of hydrogen.

Other Elements

The Bohr model of the atom, useful as it was for atomic hydrogen, was completely useless as soon as a second electron came into the atom. Magnetic and electrical interactions between the electrons and the nucleus, and with each other, complicated the picture. Only the wave function model was able to produce anything meaningful.

Consider an atomic nucleus with two protons, giving it a charge of +2 elementary units. This is a helium nucleus. A neutral helium atom must have two electrons. The second electron can join the first in its $n \geq 1$ shell but only under a most important constraint. Its spin quantum number must be the opposite of that of the other electron. This is an example of the basic principle that accounts for the entire periodic table of the elements, as well as much of solid-state physics. Applied to an atom, the Pauli exclusion principle states that *no two electrons in an atom may have the same set of quantum numbers.*

In the modern model of the atom, the electrons form a series of concentric shells around the nucleus. The organization of these shells is determined by a set of four quantum numbers. Each electron has its own unique set of quantum numbers. These are the four numbers:

- the *principal quantum number*, n, is 1 for the first shell, 2 for the second shell, and so on.
- the *spin quantum number*, s, can be either $+\frac{1}{2}$ or $-\frac{1}{2}$. Thus, helium has two electrons in the $n = 1$ shell, and each must have a different s number.
- the *angular momentum quantum number*, l, which can have any integral value between 0 and $(n-1)$. An electron in the third shell ($n = 3$) for example, can have $l = 0, 1$ or 2.
- the *orbital magnetic quantum number*, m, can have any value between $+l$ and $-l$. When $l = 2$, m can be $+2, +1, 0, -1$ or -2.

Using this principle, there are only two kinds of atoms with only one shell; where $n = 1$, l and m are both zero, and there are two values of s. In the second shell, l can be 0 or 1; when l is 0, m is 0; when l is 1, there are 3 possible values of m; for each of these four, s has two values. Total: 8.

EXAMPLE 16.9

How many electrons are there in a complete fourth shell?

SOLUTION

When $n = 4$, l can be 0, 1, 2, or 3. When $l = 3$, there are seven values of m; with $l = 2$, five values for m; with $l = 1$, three values for m; with $l = 0$, one value for m. In each of these 16 values, there are two for s. Total: 32.

CORE CONCEPT

In an atom, the major electron shells are divided into subshells by the angular momentum quantum number and still further by the orbital and spin magnetic quantum numbers.

Making X-rays

The heart of the doctor's X-ray machine is an X-ray tube. This tube is an evacuated glass container in which electrons are released from a heated filament, accelerated through a potential difference of several thousand volts, and spewed onto a metal target. The electrons come abruptly to a stop, and their kinetic energy is turned into the energy of a stream of X-ray photons.

Each energetic electron may release one or more photons, but their combined energy must be equal to the kinetic energy lost by the bombarding electron. The photons will have a broad range of wavelengths. However, the spectrum of these X-ray photons has an absolute upper limit. The electron may put all its energy into a single photon. The most energetic released photon has all the kinetic energy that the electron lost.

EXAMPLE 16.10

What is the shortest wavelength in the X-ray spectrum that is produced when a beam of electrons that has been accelerated through 25 000 volts strikes a tungsten target?

SOLUTION

Tungsten is often used for various reasons, but the nature of the target does not matter. The most energy an emitted photon can have is 25 000 electron volts. Then, from Equation 14.2,

$$\lambda = \frac{hc}{E} = \frac{\left(4.13 \times 10^{-15} \text{ eV·s}\right)\left(3.0 \times 10^8 \text{ m/s}\right)}{25\,000 \text{ eV}} = 4.9 \times 10^{-11} \text{ m} = 0.049 \text{ nm}$$

The spectrum of emitted X-rays will contain several sharp peaks, photon wavelengths containing far more energy than others. The wavelengths of these *characteristic* X-rays are unique for each target; indeed, they have been used to identify the elements present in rocks on Mars. Characteristic X-rays result from the loss of energy when an electron in some inner shell drops down into a vacant space in a lower shell. This can only happen if an incoming, high-energy electron knocks an electron out of a lower shell, creating a place into which a higher-energy electron can drop.

To calculate the wavelength of the photon emitted in this circumstance, we have to know how much energy an electron has in any given shell. Equation 16.1 implies that in hydrogen, the energy of an electron in any shell is $(-13.6$ electron volts$)/n^2$. This equation must be modified if the nucleus has more than one positive charge; the energy then becomes $(-13.6$ electron volts$)(Z^2/n^2)$ where Z is the atomic number, the charge on the nucleus. This equation is very limited; it works as long as there is only one electron. It works, for example, for ionized helium, where $Z = 2$ and only one electron remains in the first shell.

For atoms with more electrons, the situation becomes more complicated. An electron in the second shell looks not only at a nucleus but also at 2 electrons in the first shell. These neutralize two units of the positive charge of the nucleus. The effective atomic number, for this purpose, is two less than the charge on the nucleus. Similarly, an electron in the third shell sees the 2 electrons in the first shell and 8 in the second. Therefore, the effective atomic number, for the purpose of calculating the energy of the third-shell electron, is 10 less than the charge on the nucleus. The energy of any electron is given by

$$E = \left(-13.6 \text{ eV}\right)\left(\frac{Z_{\text{eff}}^2}{n^2}\right) \qquad \textbf{(Equation 16.4)}$$

where Z_{eff} is the atomic number minus the number of electrons between the nucleus and the electron in question.

EXAMPLE 16.11 What is the wavelength of the characteristic X-ray produced when an electron in the third orbit of a tungsten atom ($Z = 74$) drops from the third shell to the first?

SOLUTION

To get the energy of an electron in the first shell, we must take into account the influence of the other electron. The effective nuclear charge for an electron in the first shell is thus $Z - 1 = 73$. The energy of an electron in the first shell is, from Equation 16.4,

$$E = (-13.6 \text{ eV})\left(\frac{73^2}{1^2}\right) = -72\ 500 \text{ eV}$$

An electron in the third shell can drop into the first shell only if one of the first shell electrons has been knocked out. An electron in the third shell looking toward the nucleus sees 8 electrons in the second shell and the one remaining in the first shell, so $Z_{\text{eff}} = 74 - 9 = 65$. Then

$$E = (-13.6 \text{ eV})\left(\frac{65^2}{3^2}\right) = -6\ 390 \text{ eV}$$

The difference between these energy levels is the amount that the energy of the electron dropped, which is 66 100 eV. Then, from Equation 14.2,

$$\lambda = \frac{hc}{E} = \frac{(4.14 \times 10^{-15} \text{ eV} \cdot \text{s})(3.0 \times 10^8 \text{ m/s})}{66\ 100 \text{ eV}} = 1.88 \times 10^{-11} \text{ m} = 0.0188 \text{ nm}$$

CORE CONCEPT

X-rays are produced by sudden deceleration of electrons; peaks in the spectrum come from change in the energy levels of inner electrons.

REVIEW EXERCISES FOR CHAPTER 16

Fill-In's. For each of the following, fill in the missing word or phrase.

1. A cathode ray is a stream of _____.

2. The nucleus of any atom contains some whole number of _____ charges.

3. The number of electrons surrounding the nucleus of an atom is equal to the _____ of the nucleus.

4. The _____ atom has an atomic number of 1.

5. In dropping from a high energy level to a lower one, the electron of a hydrogen atom emits a(n) _____.

6. When the principal quantum number of the electron energy of a hydrogen atom reaches infinity, the atom is _____.

7. Quantized energy levels of the hydrogen electron can be derived on the assumption that its _____ is quantized.

8. Angular momentum is always an integral multiple of _____.

9. In the wave model, electrons form _____ in their orbits.

10. All charged particles that go straight through crossed magnetic and electric fields have the same _____.

11. The kinetic energy of electrons emerging from an electron gun is equal to the charge on the electron times the _____ of the gun.

12. The existence of the atomic nucleus was established by bombarding gold foil with _____.

13. Classical physics could not account for the amount of energy produced by _____.

14. A line spectrum represents a set of _____, each with a definite energy.

15. To make a mercury vapor glow, colliding electrons must have enough energy to raise the mercury atoms to the first _____.

16. The force exerted by a light beam is determined by the _____ of photons.

17. As the wavelength of photons increases, their momentum _____.

18. The force exerted by a light beam on a silvered surface is _____ the force exerted on a black surface.

19. Diffraction effects cannot be observed with ordinary objects because their _____ is too large.

20. Unlike a wave, a stream of particles cannot exhibit the phenomenon of _____.

21. A(n) _____ provides a diffraction grating suitable for showing interference patterns of a beam of electrons.

22. The _____ model of light is necessary to account for the effect of light on a photographic plate.

23. Whether a beam of light is a wave or a stream of photons depends on the kind of _____ in which it exists.

24. There is a lower limit to _____ in measurement of momentum and position.

25. Quantum physics predicts only the _____ of the outcome of a single event.

26. The intellectual discipline known as _____ restricts itself to questions that can be answered by experiment or observation.

27. The wave function is a field of _____ of the position of an electron.

28. The radius of a hydrogen atom in its second excited state is _____ times as great as the radius in its first excited state.

Multiple Choice.

1. Under appropriate conditions, which of the following can be shown to act like waves?
   (1) photons only
   (2) photons and electrons only
   (3) electrons and protons only
   (4) photons and protons only
   (5) photons, electrons, and protons.

2. The criterion for demonstrating the existence of wave properties is
   (1) interference
   (2) velocity of light
   (3) absence of mass
   (4) photoelectric effect
   (5) refraction.

3. The wavelengths of electron beams vary according to
   (1) frequency and mass
   (2) mass and velocity
   (3) velocity only
   (4) rest mass only
   (5) mass and charge.

4. When alpha particles pass through a thin sheet of gold foil, the particles are deflected by
    (1) elastic collision with the nucleus
    (2) electrostatic repulsion from the nucleus
    (3) electrostatic attraction to the nucleus
    (4) inelastic collision with the nucleus
    (5) friction within the foil.

5. The planetary model of the atom cannot work because
    (1) Coulomb's law does not function at atomic distances
    (2) accelerated electrons must radiate energy
    (3) there can be no energy stored in a circular orbit
    (4) an orbiting electron has no momentum
    (5) gravity is too weak to keep the electrons in orbit.

6. In developing his theory of the atom, Bohr violated known physical law by assuming that
    (1) energy is not necessarily conserved
    (2) momentum is not necessarily conserved
    (3) angular momentum is not necessarily conserved
    (4) angular momentum is quantized
    (5) energy is quantized.

7. If an electron in a hydrogen atom drops from the fifth shell to the second shell, it emits a photon whose energy is
    (1) 0.54 electron volts   (4) 3.40 electron volts
    (2) 1.51 electron volts   (5) 4.08 electron volts.
    (3) 2.86 electron volts

8. A collision between an electron and a mercury atom will be elastic unless the energy of the electron is sufficient to
    (1) ionize the mercury atom   (4) penetrate the nucleus of the atom
    (2) produce a violet photon   (5) split the nucleus.
    (3) raise the mercury atom to
         the first excited state

9. If the radius of a helium atom in its ground state is r, its radius in the third excited state is
    (1) $3r$ (2) $9r$ (3) $16r$ (4) r^2 (5) r^3.

10. A beam of electrons can be deflected by
     (1) either an electric or a magnetic field   (4) electric fields only
     (2) crossed electric and magnetic fields only   (5) magnetic fields only.
     (3) parallel electric and magnetic fields only

11. The lines in the spectrum of a gas are measures of
     (1) the angular momentum of electrons
     (2) the atomic number of atoms
     (3) the energy of electrons in orbit
     (4) the energy change when an electron changes orbit
     (5) the strength of the magnetic fields of the electrons.

12. If the energy of the hydrogen electron is zero, this means that
     (1) the atom is ionized   (4) the electron cloud has collapsed
     (2) the electron is not in motion (5) the electron is in its ground state.
     (3) the electron charge has been
          neutralized

Problems.

(Hint: Many of these problems call for the use of certain constants, which are listed in Appendix 2.)

1. What is the velocity of a stream of charged particles if they pass straight through an electric field of 6 500 newtons per coulomb crossed with a magnetic field of 0.25 teslas?

2. What is the energy of a hydrogen electron in (a) its ground, or lowest energy, state; (b) its third quantum state; and (c) its ionized state?

3. A hydrogen electron drops from its third quantum state to the first (ground) state. Find (a) the energy of the photon emitted; and (b) its wavelength.

4. In a collision with another particle, a hydrogen atom is ionized, gaining 18 electron volts of energy in the process. What is the kinetic energy of the electron after it leaves the atom?

5. A beam of protons is accelerated through a potential difference of 50 volts. Find their (a) kinetic energy in joules; (b) velocity; (c) momentum; and (d) wavelength.

6. List the energies of all the photons that could be released from a mercury atom in dropping from level f to level b. (There are 10 of them.)

7. In a Franck-Hertz experiment, the electron-gun potential is 5.0 volts. What retarding potential on the collecting plate will cut off the current?

8. A beam of violet light of wavelength 460 nanometers strikes a black surface. If 7.0×10^{15} photons strike the surface every second, how much force does the beam exert?

9.   What potential difference would be needed to get an electron to a velocity so high that its mass doubles?

10. An ultraviolet photon of wavelength 55 nanometers enters a hydrogen atom and ionizes it. What is the kinetic energy of the emerging electron?

11. What is the wavelength of the light emitted when a hydrogen electron drops from the sixth to the third level?

12. A bottle full of atomic hydrogen atoms are excited to the fourth state. How many lines will appear in the spectrum? Which will have the shortest wavelength?

The Atomic Nucleus and Beyond

WHAT YOU WILL LEARN

The nature of the atomic nucleus and the subatomic particles; radioactivity and nuclear energy.

Within the Nucleus

Every nucleus possesses a positive charge that is an integral multiple of the elementary charge. This must be so, for the atom as a whole is electrically neutral, and the nuclear charge must exactly cancel out the negative charge on some whole number of electrons. This suggests that the nucleus contains some sort of particle possessing a single elementary unit of positive charge. This particle is called a *proton*.

The mass of the nucleus presents a problem. Chemists had long since discovered that the masses of most nuclei are very nearly whole-number multiples of the mass of the smallest atom, that of hydrogen. But this multiple is always larger than the atomic number. If the protons contain all the charge of the nucleus, there must be something else there to account for all the mass. A nucleus of gold, for example, has exactly 79 times the charge on a hydrogen nucleus (a proton). But its mass, on the average, is 195.4 times as much as the mass of a proton.

The problem was partially solved with the discovery of the *neutron*. This particle was found in one of the many alpha-particle bombardment experiments being done to investigate the atom. An alpha particle, on striking a beryllium nucleus, knocked out of it a neutral particle with a mass a little greater than that of a proton. Now it became clear that the extra mass of the nuclei was provided by these neutral particles. The neutron has no charge, but its mass turned out to be nearly the same as the mass of a proton.

There was still a problem. If the nucleus is made of some integral number of protons and neutrons, then the mass of each nucleus ought to be some whole-number multiple of the mass of one proton. Most nuclei came pretty close; but there were some disturbing deviations. How was it possible to account for the atomic mass of chlorine, for example, which is 35.5 times the mass of a proton?

The answer is that not all atoms of chlorine contain the same number of neutrons. In a mass spectrometer a beam of chlorine ions is divided into two parts with different masses. All have the same number of protons; that is what makes them chlorine atoms. But there are several *isotopes* of chlorine, containing different numbers of neutrons. The total number of particles in the nucleus, protons, and neutrons together, is called the *atomic mass number* of the particular atomic species.

The composition of a nucleus is thus represented by two numbers, written before the chemical symbol of the element as a subscript and superscript. Chlorine-35, for example, is symbolized in this way:

$$^{35}_{17}\text{Cl}$$

indicating that its atomic number (number of protons) is 17 and its atomic mass number (both kinds of particles) is 35. The other common isotope of chlorine is chlorine-37:

$$^{37}_{17}\text{Cl}$$

Still, there is a problem. The actual masses of the various isotopes are *still* not exact multiples of the mass of a proton, or of the separate masses of the protons and the neutrons in the nucleus! We measure the masses of nuclei in *daltons*; a dalton (Dl) is defined as $\frac{1}{12}$ of the mass of a nucleus of carbon-12. This, of course, is an extremely small unit of mass; it takes 6.02×10^{26} daltons to make up one kilogram. Isotope masses (except for carbon-12) are never exactly whole numbers of daltons. As the sample values of Appendix 4 show, they are always within a few hundredths of a dalton of the atomic mass number. The deviation from integral values lies at the heart of our knowledge of nuclear energy.

CORE CONCEPT
Nuclei are made of protons and neutrons; their masses are not quite whole numbers of daltons.

TRY THIS
—1—

The atomic number of iron is 26. State the number of protons and neutrons in a nucleus of iron-54.

Tracking the Particles

All of the heaviest metals—polonium, radium, uranium, thorium—seem to violate the law of energy conservation. They produce a steady stream of energetic particles, with no observable energy input. These elements are said to be *radioactive.*

This peculiar property was found when it was noted that an ore containing these metals fogged a photographic plate. A charged particle passing through a photographic emulsion gives up some of its energy in colliding with the atoms of the sensitive material. Its path is marked by a streak of ionized atoms, which are converted into visible black specks of metallic silver when the plate is developed. Using a stack of photographic plates, an investigator can trace the three-dimensional path of a charged particle. Since each speck of silver represents a definite amount of energy lost by the particle, it is possible to analyze the plates to study the energy of the particle.

In a *cloud chamber,* the entire track of a charged particle is visible as it is made. This device is a round box with a glass top, containing a little water. The air in the box is saturated with water vapor. A charged particle in the vapor leaves a trail of tiny water droplets, which are visible against a black background as a thin line of fog. If the particle is moving fast enough, it can be seen that the trail is made of separate droplets, each representing the ionization of a water molecule, with energy transferred from the moving charged particle.

Modern equipment produces charged particles of extremely high energy, and much more elaborate equipment is needed to trace their history. Most of them disintegrate into other particles within times as short as 10^{-20} seconds. One tool for studying the path and the interactions of these particles is the *bubble chamber.* Such a chamber may contain hundreds of gallons of liquid hydrogen at its boiling point of about 4 kelvins. A charged particle going through it leaves a trail of tiny hydrogen bubbles. A light flashes, and two cameras take pictures from different angles. The advantage of a bubble chamber is that its high density enables it to absorb the energy of the particle more effectively than water vapor; a high-energy particle could pass completely through a cloud chamber before anything interesting happened to it.

In a typical operation, particles of one kind or another enter the bubble chamber, which is in a strong magnetic field. Thousands of photographs are made of tracks of the particles. In the example in Figure 17.1, a charged particle is coming up from the bottom, leaving a very slightly curved track. The momentum of a particle in a magnetic field is equal to Bqr (see "Currents in Space" in Chapter 11). Since B (magnetic field) and q (charge) are known, the momentum can be found by measuring r, the radius of curvature of the path. With this magnetic field, the curvature to the right indicates that the charge is negative. The particle abruptly disintegrates into three others, one curving more sharply to the right and the others leaving no track at all. The other particles are gamma photons, which leave no track because they have no charge. There was a lead plate in the chamber; the gamma photons entered into it. Each photon decayed to form a positron and an electron. The tight spiral is just a stray electron.

FIGURE 17.1

Since each bubble represents a loss of energy, the energy of the particles can be determined from the spacing of the bubbles. With the momentum (mv) and the kinetic energy ($\frac{1}{2} mv^2$) both known, the mass and velocity of the particle can be found. The velocity and the length of the track, in turn, provide a measure of the time of survival of the particle.

CORE CONCEPT
The properties of charged particles can be found by studying the tracks they leave in various kinds of detection devices.

TRY THIS
—2—

Explain why the path of an electron in a bubble chamber is a spiral rather than a straight line or a circle.

Radioactive Elements

A piece of uranium or polonium emits three kinds of radiation. In a cloud chamber in a magnetic field, alpha rays curve gently in one direction; beta rays curve sharply in the other; gamma rays leave no track but can be detected by various kinds of devices; they pass straight through without deflection.

Alpha rays consist of massive particles, each composed of two protons and two neutrons. With atomic number 2, they are considered to be nuclei of helium, $^4_2$He.

Beta rays are electrons emitted with much higher energy than can be produced in any sort of chemical reaction.

Gamma rays are X-ray photons of extremely high frequency and energy. They give us an inkling of the difference between the energy of chemical reactions and of nuclear reactions. Electron transitions in the outer atom, as in ordinary chemical processes, produce visible photons with energies of a few electron volts; gamma photon energies are measured in hundreds of thousands of electron volts. Nuclear energies are related to the *strong nuclear force*. Electron transitions obey the rules of the much weaker electromagnetic force.

The strong nuclear force is what holds nuclei together. A group of protons, all with positive electric charge, would surely be expected to blow up unless some force stronger than the electric repulsion holds them together. The strong nuclear force is a strong force of attraction by which all protons and neutrons attract each other, regardless of electric charge. This force is effective only at extremely short ranges, such as the distance inside a nucleus. It is not strong enough to bind two protons to each other.

The smallest atomic nucleus is helium-3, $^3_2$He. It takes the additional strong force provided by the neutron to hold the two protons together. From $^4_2$He to $^{32}_{16}$S (sulfur), the numbers of protons and neutrons are equal. When the number of protons is more than 16, it takes more than 16 neutrons to make the nucleus stable. The largest stable nucleus is $^{207}_{82}$Pb (lead). More than 82 protons cannot be held together permanently by any number of neutrons. Every nucleus larger than this is *radioactive*. Sooner or later, it must emit bursts of energy and particles that reduce its size to a stable level.

One form of radioactivity is alpha emission, in which an alpha particle is emitted, for example:

$$^{238}_{92}U \rightarrow \,^{234}_{90}Th + \,^{4}_{2}He$$

A nucleus of uranium-238 has emitted an alpha particle (helium-4 nucleus), thus reducing its atomic number by 2 and its atomic mass number by 4 and becoming thorium-90.

EXAMPLE 17.1

Write the equation for the alpha decay of thorium-230.

SOLUTION

Consult Appendix 4; thorium-230 has an atomic number of 90, just like all the other forms of thorium. So

$$^{230}_{90}Th \rightarrow \,^{4}_{2}He + \,^{226}_{88}Ra$$

Element 88 is radium, according to the table.

Emitted alpha particles always have, in any particular reaction, certain definite energy values. Just like the definite energy values of the photons emitted in an electron transition, these values indicate that there are quantized energy levels within the nucleus. If the alpha particle energy is less than the maximum possible, the rest is released as a gamma photon.

Alpha emission decreases the charge on the nucleus. In other nuclei, stability is increased by *increasing* the nuclear charge. This is done by converting a neutron to a proton, with the negative charge being carried off in the form of an electron, a beta particle. A gamma photon may or may not come off also, but there is always another particle known as a neutrino, which carries off some of the excess energy. Thorium decay is an example:

$$^{234}_{90}Th \rightarrow \,^{234}_{91}Pa + \,^{0}_{-1}e + \,^{0}_{0}\nu$$

The neutrino, with no charge and negligible mass, appears in the equation as the Greek letter nu (ν). Note that, in beta decay, the atomic mass number does not change but the atomic number increases by 1. The neutrino, having no charge, makes no change in the atomic number. Whether it has rest mass is one of the still unsolved questions, but if it does, the rest mass is surely not much.

EXAMPLE 17.2

Write the equation for the beta decay of palladium-234.

SOLUTION

Palladium is element 91:

$$^{234}_{91}Pa \rightarrow \,^{234}_{92}U + \,^{0}_{-1}e + \,^{0}_{0}\nu$$

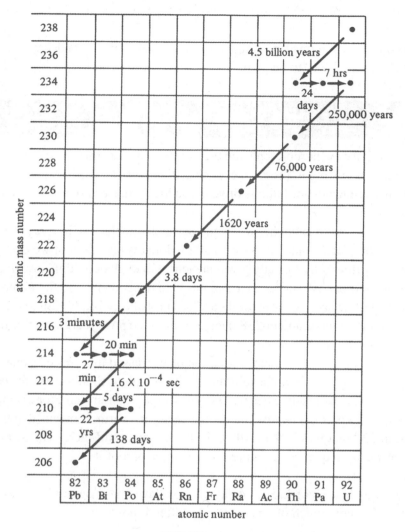

FIGURE 17.2

Palladium is still unstable and undergoes another beta decay. This is followed by a series of alpha and beta decays, until the nucleus finally becomes stable in the form of lead-206. The entire series is illustrated in Figure 17.2.

CORE CONCEPT

In natural radioactivity, nuclei become more stable by emitting alpha and beta particles.

TRY THIS
—3—

Write the equations for an alpha decay, followed by a beta decay, of polonium-214. (Atomic numbers are listed in Appendix 4.)

How Long Does It Take?

Each radioactivity event represents the transformation of a single nucleus. These nuclei have been resting in the earth ever since there was an earth. A nucleus, having been a uranium nucleus for billions of years, suddenly decides that now is the time to turn into thorium.

We know of no way to predict when any particular nucleus will decay. The timing seems to be completely random. As with all random events, the best we can do is to determine the *probability* that any given nucleus will explode within any given time period. Some radioactive nuclei are more unstable than others, and the probability that any nucleus will decay is greater for them. This probability is a billion to one *against* the decay of any given uranium-238 nucleus within the next year; for a thorium-234 nucleus, it is 40 000 to 1 *in favor.*

The stability of radioactive substances can be expressed in terms of a quantity called *half-life*. The half-life of a substance is the length of time it takes for half of any sample to undergo radioactive decay. The half-life of uranium-238, for example, is 4.5 billion years. Since this is approximately the age of the earth, it means that half of all the uranium-238 that was in the earth at the beginning is still there. The rest has gone through the decay sequence of Figure 17.2, or some part of it.

EXAMPLE 17.3 Thorium-234 has a half-life of 24 days. If you have a 1-kilogram block of thorium-234, how much of it is unchanged after 4 months?

SOLUTION

Four months is about 120 days, which is five half-lives of thorium. The mass of thorium then must drop to one-half five times, so it becomes

$$(1 \text{ kg})\left(\frac{1}{2}\right)^5 = 0.031 \text{ kg}$$

Half-lives vary enormously, as you can see by reference to Figure 17.2. The shortest in the sequence is that of polonium-214, which loses half of its nuclei every 160 microseconds. A lump of uranium contains all the members of this sequence, in various stages of decay, but the life of polonium-84 is so fleeting that there will be very little of it in the mix.

CORE CONCEPT

The half-life of a radioactive substance is the length of time it takes for half of its nuclei to disintegrate.

TRY THIS
—4—

If you start with 1 gram of polonium-214, how much will be left at the end of $\frac{1}{100}$ second?

The Energy of the Rays

What is the source of the energy of these emissions? The answer turns out to be intimately associated with a question we asked earlier: Why are all nuclear masses not equal to some integral multiple of the mass of a proton?

The answer lies in the famous Einstein mass–energy relationship:

$$E = mc^2$$

which tells us that mass and energy are simply two different ways of measuring the same thing.

This leads to a contradiction with the most basic rule of chemistry. Chemists tell us that 16 grams of oxygen and 2 grams of hydrogen form an explosive mixture. A spark ignites it, and the oxygen and hydrogen combine to form exactly 18 grams of water. Mass is conserved. But if mass is energy, we ought to expect that the mass of the water produced in the explosion will be *less* than 18 grams, since a lot of energy is given off in the explosion. In general, every molecule ought to have less mass than the separate atoms that make it up, since it always takes energy to separate the atoms from each other.

The reason that chemists have never noticed this missing mass is that it is far too small to measure. But with nuclei, the situation is different. The nuclear force is far stronger than the electromagnetic force that holds molecules together. Thus, the energy given up when protons and neutrons combine to form nuclei produces a *measurable* difference in mass. Every nucleus has a *mass deficit,* which is the difference between the mass of the nucleus and the masses of the separate particles of which it is composed. Consider, for example, iron-56, the most common isotope of iron:

mass of 26 protons

 26 × 1.007833   26.203658

mass of 30 neutrons

 30 × 1.008665   30.25995

total mass of separate particles   56.29653

mass of $^{56}_{26}$Fe   55.9349 daltons

mass deficit of $^{56}_{26}$Fe   0.3616 daltons

This mass deficit represents the energy that would be released if we could find a way to make 30 neutrons and 26 protons combine into a single nucleus. It is also the energy that would have to be added to separate the nucleus into its parts.

While energy and mass are completely interchangeable, we do not usually think of a dalton as a unit of energy. We can use Einstein's equation to express this energy in more familiar units. The result is

$$1 \text{ dalton (Dl)} = 931 \text{ million electron volts (MeV)}$$

EXAMPLE 17.4

How many million electron volts correspond to a mass of 1 dalton?

SOLUTION

From Equation 5.7,

$$E = mc^2 = (1 \text{ Dl})(2.998 \times 10^8 \text{ m/s})^2 \times \frac{\text{kg}}{6.025 \times 10^{26} \text{ Dl}}$$

$$E = (1.4918 \times 10^{-10} \text{ J}) \times \frac{\text{eV}}{1.602 \times 10^{-19} \text{ J}}$$

$$E = 9.31 \times 10^8 \text{ eV} = 931 \text{ MeV}$$

Thus the mass deficit of iron-56 (0.5143 daltons) corresponds to a nuclear binding energy of 479 million electron volts. Compare this with the chemical binding energy of the two atoms of hydrogen and one of oxygen that make up a water molecule. It is a mere 7 electron volts, corresponding to a few billionths of a dalton of mass.

EXAMPLE 17.5

Find (a) the mass deficit and (b) the binding energy of the nucleus of boron-11.

SOLUTION

 (a) From Appendix 4, the atomic number of boron is 5, so boron-11 contains five protons and six neutrons.

mass of five protons
  5 × 1.007833   5.039165 daltons
mass of six neutrons
  6 × 1.008665   6.051990
total separate masses   11.09116
mass of boron-11   11.00931
mass deficit   0.08185 daltons

(b) The binding energy is found by transforming mass units to energy units:

$$0.08185 \text{ D1}\left(\frac{931 \text{ MeV}}{\text{D1}}\right) = 76.2 \text{ MeV}$$

In any radioactive decay process, the mass deficit must increase; the total mass of the products must be less than the mass of the nucleus that exploded. The lost mass provides the energy of the emitted particles. Consider, for example, the alpha decay of the most stable isotope of radium, which has a half-life of 1 600 years:

$$^{226}_{88}\text{Ra} \rightarrow {}^{222}_{86}\text{Rn} + {}^{4}_{2}\text{He} \qquad (1\ 600 \text{ years})$$

It decays to form a nucleus of radon-222 and an alpha particle. Then here is the mass-energy budget:

mass of $^{226}_{88}\text{Ra}$ 226.0254 daltons
mass of $^{222}_{86}\text{Rn}$ 222.0175
mass of $^{4}_{2}\text{He}$ 4.0026
total mass of product nuclei   226.0201
additional mass deficit   0.0053 daltons

corresponding to a binding energy of 4.93 million electron volts.

Most of the alpha particles emitted have only 4.78 million electron volts of kinetic energy. The additional 0.15 million electron volts of energy is then given off as a gamma photon. It is an extremely high-energy photon, with a wavelength of only about 10^{-11} meters.

EXAMPLE 17.6

The alpha particle emitted in the decay of radium-226 sometimes has an energy of 4.340 million electron volts. What is the wavelength of the gamma photon that follows?

SOLUTION

The total increase in binding energy is still 4.93 MeV, so the energy of the photon must be $4.93 - 4.34 = 0.59$ MeV. The wavelength, from Equation 14.2, is

$$\lambda = \frac{hc}{E} = \frac{\left(4.14 \times 10^{-15} \text{ eV} \cdot \text{s}\right)\left(3.00 \times 10^{8} \text{ m/s}\right)}{0.59 \times 10^{6} \text{ eV}}$$

$$\lambda = 2.1 \times 10^{-12} \text{ m}$$

CORE CONCEPT

A nuclear reaction always results in increased mass deficit, with the lost mass appearing in the form of the energy of the emitted particles.

TRY THIS
—5—

What is the binding energy of an alpha particle? (All needed data appear in Appendix 4.)

Increasing the Energy

The energies of alpha particles emitted in radioactive decay run to about 5 million electron volts. If they are used to bombard gold foil, they cannot collide head-on with a nucleus because it takes more energy than this for the particle to overcome the electric repulsion. However, if these alpha particles are sent into a metal with a smaller nucleus, the nuclear charge will be smaller and the alpha particle might actually enter the nucleus. Usually, it forms an unstable combination that breaks up immediately. In fact, it was this kind of nuclear reaction that first revealed the existence of the neutron. Alpha particles were sent into a piece of beryllium, with this result:

$$^{9}_{4}\text{Be} + {}^{4}_{2}\text{He} \rightarrow ({}^{13}_{6}\text{C}) \rightarrow {}^{12}_{6}\text{C} + {}^{1}_{0}\text{n}$$

The carbon-13 formed was in a highly excited state and settled down by emitting a neutron (n).

To make alpha particles, or any other charged particles enter a large nucleus they must be given more energy. This is the function of a particle accelerator. Many types of accelerators have been designed, and the newest ones are the largest and most expensive pieces of research equipment ever produced.

A charged particle is accelerated by placing it in an electric field. Equation 9.3 tells us that, as a positive charge is allowed to fall in an electric field from high potential to low, it gains an amount of energy equal to $q\Delta V$. If the value of q is one elementary charge, the energy the particle gains, in electron volts, is equal to the difference of potential in volts.

The simplest particle accelerator is a Van de Graaff generator, such as that described in Chapter 9. A large Van de Graaff generator can produce potentials up to several million volts. To prevent sparking, they are made with smooth, round terminals and enclosed in a case filled with a special gas under high pressure. Protons have been accelerated up to 9 million electron volts with a device of this sort.

The key to producing particles with much higher energies than that is to accelerate them in steps rather than all at once. This can be done in a *linear accelerator,* which consists of a series of *drift tubes* in a linear array, as in Figure 17.3. There is no force on a particle while it is inside a tube. Acceleration occurs in the gaps between tubes and is produced by a difference of potential between the two tubes. Thus, if the potential of the second tube is 10 000 volts lower than that of the first, a positive particle will gain 10 000 electron volts of energy as it crosses the gap. The tubes are connected to carefully synchronized AC sources, so that the polarity switches while the particles are passing through the tubes. Then, when the particle reaches the next gap, it finds an electric field that will boost its energy by another 10 000 electron volts.

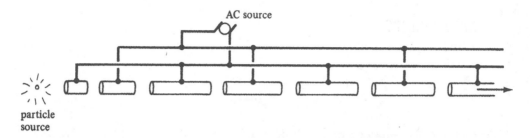

FIGURE 17.3

The largest linear accelerator is in a tunnel 2 miles long at Stanford University. Its 960 drift tubes accelerate electrons up 20 gigaelectron volts (1 GeV = 1 000 MeV). The lengths of the drift tubes must be carefully controlled; the first few must increase in length as the electrons speed up. Very soon, however, the electrons are going practically at the speed of light, and nothing can go faster. Therefore, all but the first few tubes are the same length. As the electrons gain kinetic energy, their speed cannot increase, but their mass does.

EXAMPLE 17.7

What is the mass of a proton that has been accelerated to a kinetic energy of 20 000 million electron volts?

SOLUTION

The kinetic energy has mass equal to

$$(20\ 000\ \text{MeV})\left(\frac{\text{D1}}{931\ \text{MeV}}\right) = 21\ \text{D1}$$

Added to the rest mass of 1 Dl, this gives a total of 22 Dl.

EXAMPLE 17.8

The mass of an electron at rest is 9.1×10^{-31} kilogram. By what factor does its mass increase when its energy is raised to 20 gigaelectron volts?

SOLUTION

From Equation 5.7,

$$m = \frac{E}{c^2} = \frac{\left(20 \times 10^9\ \text{eV}\right)\left(1.60 \times 10^{-19}\ \text{J/eV}\right)}{\left(3.00 \times 10^8\ \text{m/s}\right)^2}$$

$$m = 3.56 \times 10^{-26} \text{ kg}$$

To get the factor of the increase, divide this by the rest mass:

$$\frac{3.56 \times 10^{-26} \text{ kg}}{9.1 \times 10^{-31} \text{ kg}} = 39\ 000$$

CORE CONCEPT

High-energy accelerators use electric fields to boost the energy of charged particles, and magnetic fields to control their paths.

TRY THIS
—6—

What is the mass of a proton with 500 gigaelectron volts of kinetic energy? How does this compare with its rest mass?

Accelerating in Circles

The earliest device that boosted particle energy in steps was the *cyclotron,* invented in 1934. It works because of a simple mathematical relationship. The momentum of a particle moving in a magnetic field ("Measuring Mass" in Chapter 11) is

$$mv = Bqr$$

The velocity is the circumference of the circle divided by the period of revolution:

$$v = \frac{2\pi r}{T}$$

Substituting this value for v above and solving for T gives

$$T = \frac{2\pi m}{Bq}$$

The radius of the circle has dropped out of the equation. This means that all particles traveling in this circle will take the same length of time to make the trip. Those farther from the center will be going faster. The cyclotron was designed to take advantage of this fact.

A cyclotron (Figure 17.4) consists of two hollow d-shaped containers, called *dees*. The flat edges face each other across a gap. The whole circular apparatus is placed between the poles of an electromagnet. There is an alternating electric potential difference between the dees. (The circle with the little squiggle in it stands for an AC generator.) The period of the AC potential is the same as T in the equation above.

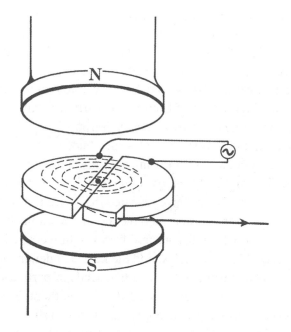

FIGURE 17.4

When a charged particle crosses the gap, it is accelerated by the potential difference between the dees. There is no electric field inside the dees; there never is inside a conductor. The magnetic field makes the particles go in a circular path. When they have gone halfway around, they reach the gap again, and by that time the polarity of the dees has reversed. The particles are accelerated again. They speed up and move farther from the center, but this does not change the time it takes them to pass through a dee. They go faster and faster, and farther from the center, until they shoot out an opening in one of the dees. In practice, a source of particles, such as a radioactive substance, is placed at the center of the dees and the particles spiral outward until they emerge.

In the first cyclotron, the dees were the size of a dinner plate, and the cyclotron produced protons with an energy of 80 kiloelectron volts—not much, even in those days. Within a few years, the concept was scaled up and protons and deuterons were accelerated to 20 million electron volts and more. There was, however, a severe limit on the cyclotron. The equation on which it is based includes the mass of the particles. As their energy increased, so did their mass. When their speed reached 10 percent of the speed of light, their mass was so great that they took too long to go around, and fell out of step with the AC on the dees.

The problem of the relativistic increase of mass was solved by a system that cycled the frequency of the applied AC. Protons were accelerated in bunches rather than in a continuous stream. As each bunch made its way farther from the center, the frequency of the AC dropped to allow for the longer transit time. A cyclotron using this principle had enormous dees, 4.6 meters across, and produced protons with over 200 million electron volts of energy.

Accelerators were pushed up into the gigaelectron volt (GeV) range by the use of a different principle. A *synchrotron,* in principle, is something like a linear accelerator bent into a circle. The drift tubes are surrounded by strong electromagnets, so that the particles go in a circular path between gaps, through the same tubes over and over again. The AC frequency drops, as the particles travel, to compensate for the mass increase. There is also the problem of keeping the particles centered in the tube as they go faster, instead of spiraling outward as in a cyclotron. This is done by increasing the strength of the magnetic fields as the particles speed up.

Synchrotrons have been the basic instrument in investigations of the most intimate structures of matter. The higher the energy of the bombarding particles, the more detail they can show. High-energy beams have revealed that there is structure within a proton. The energy of these beams has been converted into a whole assemblage of new kinds of particles, most of which disappeared from the universe within the first few seconds of its existence. These particles hold the key to understanding how the universe began. Theorists have predicted that certain kinds of particles can be found if only we can produce the particle energy needed to create them. Synchrotrons have been spectacularly successful in producing the predicted particles.

The search for ever higher levels of energy continues. The most ambitions project is the Large Hadron Collider, built into a 17-mile circular tunnel on the French-Swiss border. Protons will be accelerated by linear accelerators, and then injected into the tunnel, where they are accelerated by electric fields. They are kept in the circular path by magnetic fields, produced by powerful electromagnets that are kept at a temperature of few kelvins to make them superconductive. The particles will travel in opposite directions around the ring, and collide with energies of more than 10^{15} eV. At the present time the accelerator is operational, but some problems must be solved before the crucial collision experiments can be done. Then, it is expected that the collisions will produce many kinds of particles, hopefully including the hypothetical Higgs boson. This particle, if it exists, explains many of the remaining mysteries of the theory of fundamental particles.

CORE CONCEPT

Giant particle accelerators produce beams of particles with enormous energy by accelerating them between gaps as magnetic fields keep them in circular paths.

TRY THIS
—7—

A cyclotron that has been used to accelerate protons is now to be used with alpha particles. What changes will have to be made?

Transforming the Nucleus

Firing energetic bullets into nuclei can transform one element into another. In one common type of reaction, an alpha particle is absorbed by a nucleus, which forms a new and highly unstable nucleus. This immediately breaks down with emission of a proton:

$$_2^4\text{He} + _{13}^{27}\text{Al} \rightarrow (_{15}^{31}\text{P}) \rightarrow _{14}^{30}\text{Si} + _1^1\text{H}$$

The alpha bullet has transformed aluminum-27 into silicon-30.

In a similar reaction, the bombarded nucleus emits not a proton, but a neutron. We saw such a reaction earlier. For another example, see Example 17.9.

EXAMPLE 17.9 Fluorine-19 emits a neutron when bombarded by an alpha particle. Write the equation.

SOLUTION

$$_9^{19}\text{F} + _2^4\text{He} \rightarrow (_{11}^{23}\text{Na}) \rightarrow _{11}^{22}\text{Na} + _0^1\text{n}$$

The product is sodium-22.

Often the unstable nucleus produced by alpha bombardment does not break down immediately. It becomes an artificially produced radioactive substance with a half-life that might be many years. Carbon-14 is one such artificially radioactive nucleus. It can be produced by alpha bombardment of boron:

$$_2^4\text{He} + _5^{11}\text{B} \rightarrow (_7^{15}\text{N}) \rightarrow _6^{14}\text{C} + _1^1\text{H}$$

Carbon-14 is also produced naturally in the atmosphere, when a nitrogen nucleus absorbs a neutron:

$$_7^{14}\text{N} + _0^1\text{n} \rightarrow (_7^{15}\text{N}) \rightarrow _6^{14}\text{C} + _1^1\text{H}$$

The carbon-14 decays by beta emission with a half-life of 5 730 years:

$$_6^{14}\text{C} \rightarrow _7^{14}\text{N} + _{-1}^0\text{e} + _0^0\upsilon \quad (5\ 730\ \text{years})$$

Artificially induced radioactivity takes some forms that are quite different from the alpha and beta decay of naturally radioactive substances. In one form, a proton in the nucleus converts itself into a neutron by emitting a *positron*. This is a positive electron; it has all the usual properties of an electron, except that its charge is positive. An example is the decay of phosphorus-30, produced by alpha bombardment of aluminum. It decays by positron emission:

$$_{15}^{30}\text{P} \rightarrow _{14}^{30}\text{Si} + _{+1}^0\text{e} + _0^0\upsilon \quad (2.5\ \text{minutes})$$

EXAMPLE
17.10

When aluminum-27 is bombarded with an alpha particle, a neutron is released and a new kind of nucleus is formed. (a) Write the equation for this reaction. (b) Write the equation for the decay of the new nucleus by positron emission.

SOLUTION

(a)
$$^{27}_{13}\text{Al} + {}^4_2\text{He} \rightarrow ({}^{31}_{15}\text{P}) \rightarrow {}^1_0\text{n} + {}^{30}_{15}\text{P}$$

(b)
$$^{30}_{15}\text{P} \rightarrow {}^{30}_{14}\text{Si} + {}^0_{+1}\text{e} + {}^0_0\upsilon.$$

Nuclear transformation produces a whole string of new elements larger than uranium-238, which is the largest naturally occurring nucleus. In this reaction uranium-238 is bombarded with neutrons. The neutrons do not need high energy; since they have no charge, they are not repelled by the uranium nucleus and can simply fall inward. This is what happens:

$$^{238}_{92}\text{U} + {}^1_0\text{n} \rightarrow {}^{239}_{92}\text{U}$$

The new isotope of uranium undergoes beta decay:

$$^{239}_{92}\text{U} \rightarrow {}^{239}_{93}\text{Np} + {}^0_{-1}\text{e} + {}^0_0\upsilon \quad \text{(24 minutes)}$$

Neptunium (Np) is a new element. It also decays:

$$^{239}_{93}\text{Np} \rightarrow {}^{239}_{94}\text{Pu} + {}^0_{-1}\text{e} + {}^0_0\upsilon \quad \text{(2.4 days)}$$

The other new element is plutonium (Pu). It is much more stable than the others:

$$^{239}_{94}\text{Pu} \rightarrow {}^{235}_{92}\text{U} + {}^4_2\text{He} \quad \text{(24 360 years)}$$

As we shall see, plutonium is now available in large quantities. It is an extremely useful material—and the most dangerous ever to come out of a laboratory.

CORE CONCEPT

Nuclei can be transformed by bombarding them with particles.

TRY THIS
—8—

Complete the following nuclear equation:

$$^4_2\text{He} + {}^{39}_{19}\text{K} \rightarrow ({}^?_?\text{Sc}) \rightarrow {}^1_1\text{H} + {}^?_?\text{Ca}$$

Nuclei That Split

Certain nuclei, made unstable by absorption of a particle, will split nearly in half instead of just emitting a particle of some sort:

$$^{235}_{92}\text{U} + {}^1_0\text{n} \rightarrow ({}^{236}_{92}\text{U}) \rightarrow {}^{141}_{56}\text{Ba} + {}^{92}_{36}\text{Kr} + 3\,{}^1_0\text{n}$$

This is only one of a number of ways in which the uranium-235 nucleus can split when it is struck by a neutron. Fission produces a wide variety of fragments, always radioactive, medium-sized nuclei. Most very large nuclei can undergo fission this way, but many of them do not unless the incident neutron has a great deal of energy.

Medium-sized nuclei are more stable and more tightly bound than the real giants. This means that their mass deficits, for each particle, are much larger. That is why the fission of a big nucleus releases a lot of energy. If you calculate the additional binding energy produced in the fission of uranium-235 into barium-141 and krypton-92, you will find that it comes to 174 million electron volts, about 20 times as much as in an ordinary alpha decay.

EXAMPLE 17.11   How much additional binding energy is generated when a uranium-235 nucleus is struck by a neutron and splits to form barium-141, krypton-92, and three neutrons?

SOLUTION

The total mass going into the reaction—one uranium-235 plus one neutron—is

$$235.0439 + 1.008665 = 236.0526 \text{ Dl}$$

and coming out of the reaction:

$$140.9140 + 91.9261 + 3(1.008665) = 235.8661 \text{ Dl}$$

The difference is 0.1865 Dl. Multiplying by 931 MeV per Dl gives 174 MeV as the energy released.

It is another feature of the fission process shown above that provides the key to usable atomic energy. The process uses 1 neutron and produces 3. Each extra neutron can strike another nucleus and cause it to split, releasing still more neutrons. This *chain reaction* can build up rapidly, until enormous numbers of nuclei are splitting. Uncontrolled, this produces an atomic explosion, which releases thousands of times more energy than the largest TNT bomb ever made.

To sustain a chain reaction, more neutrons must be produced than consumed. Only three kinds of nuclei can be used: uranium-235, uranium-233, and plutonium-239. Of these, only uranium-235 is found naturally, and it is scarce. Natural uranium has 140 times as much uranium-238 as uranium-235. Uranium-238 cannot sustain a chain reaction because it undergoes fission only with extremely energetic neutrons, and when it splits it does not release more of them. To obtain nuclear fuel, it is necessary to separate uranium-235 from uranium-238. This cannot be done by chemical means, since both are chemically the same. The physical separation process, based on the slight difference in mass, is difficult and expensive.

The most significant bomb fuel today is plutonium-239, which is produced from the abundant uranium-238 by bombardment with slow neutrons. The equations for this reaction were given earlier.

To make a bomb, you have to have fairly pure fuel. Impurities soak up neutrons, so that they become unavailable for producing additional fissions. The fast neutrons produced in fission can knock as many as five additional neutrons out of a nucleus, but they must not be lost. When the mass of fuel is too small, many of the nuclei are near its surface and can escape. On the other hand, if the size of the fuel package is big enough, all it needs to set it off is one neutron. A few kilograms of uranium-235 or plutonium-239 will explode spontaneously.

The trick in making an atom bomb is to bring the fuel together and keep it in one place long enough for the reaction to take place—say, a few millionths of a second. In a bomb, the fuel is at the center, loosely packed. It is surrounded by TNT, and the whole thing is packaged within a strong, thick metal wall. The explosion of the TNT cannot breach the wall, so it actually implodes; it pushes toward the center, compressing the fuel into a critical mass and holding it there just long enough. Boom!

CORE CONCEPT

A nuclear chain reaction can occur when nuclear fission releases more neutrons than are consumed or lost.

TRY THIS
—9—

An atomic bomb contains 2.0 kilograms of fuel. What fraction of the fuel undergoes fission if the bomb produces energy equivalent to 25 kilotons of TNT? (A ton of TNT produces about 4×10^9 joules of energy.)

The Reactor

The fission of uranium-235 does not have to be used to make bombs. The reaction can be controlled, and when operated slowly can produce electricity. This is done in a *nuclear reactor.*

There are many kinds of reactors. Some are designed for research purposes, some for the production of artificial radioisotopes, some to drive ships, and others to operate electric power stations. A *breeder reactor* exposes uranium-238 to a flux of neutrons, converting it into fissionable plutonium.

Nuclear power plants in the United States are of the kind called boiling water reactors. The fuel is in the form of long rods containing uranium oxide pellets. A typical reactor has 560 fuel assemblies, each consisting of 64 rods several meters long. The rods are immersed in very pure water, and the whole assembly is housed inside a steel vessel about 10 meters tall. This is the reactor core. The uranium used is "enriched"; that is, the proportion of fissionable uranium-235 has been increased from its natural value of 0.7 percent to about 3 percent. This, of course, is far below the 90 percent purity needed to make it explode like a bomb.

With the fuel in long, thin rods, most of the neutrons produced in the fission of uranium-235 escape from the rod into the water. The water acts as a *moderator;* it absorbs most of the energy of the neutrons, slowing them down. This is important, since slow neutrons are captured for fission by uranium-235 and are not as likely to get lost inside other nuclei, such as uranium-238. The slow neutrons reach another rod, where they do their job properly. As fission products accumulate, they trap some neutrons, but the reaction continues as long as one neutron from each fission, on the average, manages to reach another fissionable nucleus.

The large amount of uranium-238 in the rods can catch fast neutrons before they leave the rod, converting their nuclei into uranium-239. Two beta decays turn the uranium-239 into fissionable plutonium-239—more nuclear fuel. Usually, the fission of this plutonium is a rather small part of the energy produced in the reactor. A breeder reactor is designed to produce plutonium, which can be extracted chemically from the spent fuel rods. It can be fed back into a reactor as fuel, or used to make bombs.

To control the rate of the reaction, a number of *control rods* project into the reactor core. These are made of a boron–carbon alloy. Their function is to absorb neutrons. As the accumulation of fission products slows the reaction down, the control rods are gradually withdrawn to keep the flow of neutrons from one fuel rod to another at the desired level. Inserting the control rods quickly into the reactor core will shut down the whole operation promptly.

The highly purified water that serves as a moderator has another function as well. It is the working fluid of the power generation system. The energy of the neutrons and other particles generated in the reaction heats up the water. It is kept under pressure, so that it boils at a temperature far above its usual value. The steam produced flows into the turbines that turn the electric generators.

Once a year, the reactor is shut down and about 30 percent of the fuel rods are replaced. The spent rods are highly radioactive. This makes them extremely dangerous to the public health, so they must be stored in isolation for thousands of years. Now, they are considered waste, but they are undoubtedly the plutonium mines of the future. Surely, many uses will be found for the other radioactive substances they also contain. Meanwhile, disposal of radioactive waste remains the thorniest problem for the nuclear energy industry.

CORE CONCEPT

In a reactor, fission of uranium-235 is accomplished by means of neutrons slowed down by passage through a moderator.

TRY THIS

—10—

Like uranium-238, thorium-232 is "fertile"; that is, it can be converted into a fissionable substance by bombardment with neutrons. Write the three nuclear reaction equations showing the effect of the thorium absorbing a neutron, followed by two beta decays to produce fissionable uranium-233.

Fusion

Energy is released by any nuclear reaction in which the binding energy, or mass deficit, of the products is more than that of the original nuclei. As we have seen, this situation exists whenever a very large nucleus is split into medium-sized ones. It also exists, however, when very small nuclei are combined to form medium-sized ones. This is called *nuclear fusion*. Medium-sized nuclei have the largest mass deficit per particle.

The fusion of small nuclei provides the energy radiated out by the sun and other stars. Many different processes go on; one of the most important, in the sun at least, is a three-step process that converts hydrogen to helium. It starts with two protons (nuclei of hydrogen-1) combining to form hydrogen-2, also known as deuterium, with emission of a positron:

$$_1^1\text{H} + {}_1^1\text{H} \rightarrow {}_1^2\text{H} + {}_{+1}^0\text{e} + {}_0^0\upsilon$$

Then deuterium reacts with another proton to form helium-3:

$$_1^2\text{H} + {}_1^1\text{H} \rightarrow {}_2^3\text{He} + {}_0^0\upsilon$$

Finally, two of these helium-3 nuclei combine to form helium-4 and a couple of protons:

$$_2^3\text{He} + {}_2^3\text{He} \rightarrow {}_2^4\text{He} + 2\,{}_1^1\text{H}$$

At each step, the total mass decreases and energy is released.

These reactions are called *thermonuclear* because they can occur only at extraordinarily high temperatures, about 40 million kelvins. The reason is that all the particles involved have positive charges, so they repel each other. Violent vibration at high temperatures is necessary to bring the nuclei close enough for the nuclear force to take over and draw the two nuclei together.

On earth, we have mastered nuclear physics well enough to produce our own thermonuclear reactions—in hydrogen bombs. One type of fuel used for this sort of bomb is the chemical lithium deuteride, a solid composed of atoms of lithium-6 and hydrogen-2 (deuterium) in equal numbers. The explosion of a hydrogen bomb is triggered by an ordinary atom bomb, in which uranium-235 undergoes fission. This reaction produces not only the necessary temperature for fusion, but also a tremendous flux of fast neutrons. These neutrons react with the lithium nuclei:

$$_3^6\text{Li} + {}_0^1\text{n} \rightarrow {}_2^4\text{He} + {}_1^3\text{H}$$

Now the nuclei of hydrogen-3 (tritium) combine in a thermonuclear reaction with the deuterium nuclei:

$$_1^3\text{H} + {}_1^2\text{H} \rightarrow {}_2^4\text{He} + {}_0^1\text{n}$$

This reaction releases an enormous amount of energy, but this bomb has still another trick. All fast neutrons strike the uranium-238 shell that holds the bomb together. The resulting fission of the uranium adds its energy to the mix. The overall output is enormous. The largest hydrogen bombs produce about 25 megatons of energy, a thousand times as much as that produced by the atom bomb that destroyed Hiroshima in World War II.

Efforts are being made to find ways to use nuclear fusion in a controlled way to produce electricity. If this can be done economically, it will put an end to our shortage of energy. The fuel will be deuterium, which is found as 1 part in 6 000 parts of all natural hydrogen and is thus available in endless supply. There is no fission involved, so very little radioactive waste will be produced. The process will be nonpolluting. Controlled fusion has already been achieved on a laboratory scale, but there are great obstacles in the way of making it commercially useful.

CORE CONCEPT
Fusion of small nuclei produces great amounts of nuclear energy.

TRY THIS
—11—

The hydrogen bomb thermonuclear fusion results in an overall effect, which is the conversion of lithium-6 and deuterium into two nuclei of helium-4. How much energy is released?

The Four Forces

Once upon a time, not so long ago, it was generally believed by scientists that the whole universe was made of only two kinds of fundamental particles: protons and electrons. It was thought that the nucleus contained both, and is positive because of an excess of protons. When the neutron was discovered, this theory fell apart. Then the positron came along. With the introduction of new detection devices and particle accelerators, well over a hundred particles were added to the list.

The discovery of the positron was a major breakthrough in the development of modern physics. The list of "fundamental" particles was expanded, but it now included a particle doomed to early annihilation. The positron is like an electron in every respect except that its charge is positive. Electrons are everywhere; when a positron meets one, both particles disappear in a burst of gamma photons.

The mutual annihilation of an electron and a positron is constrained by the conservation laws. Electric charge is conserved since the sum of the positive and the negative charge is zero. Mass-energy is conserved only if the energy of the photons adds up to the masses of the two particles. The necessity of momentum conservation calls for two photons to be made, moving in opposite directions so that their total momentum is zero.

> **EXAMPLE 17.12**
>
> What is the wavelength of the photons produced in the electron-positron annihilation?

SOLUTION

Each photon must have the same energy as is embodied in the mass of a single electron, which is 0.511 million electron volts. Then, from Equation 14.9,

$$\lambda = \frac{hc}{E} = \frac{(4.14 \times 10^{-15} \text{ eV} \cdot \text{s})(3.0 \times 10^8 \text{ m/s})}{0.511 \times 10^6 \text{ eV}} = 2.4 \times 10^{-12} \text{ m}$$

In the world of subnuclear particles, there are other conservation laws that do not appear in the physics of ordinary objects. Angular momentum is conserved in the ordinary world, but there is an additional constraint in the world of particles. Angular momentum can change only by the Planck quantum $h/2\pi$. An electron has $+\frac{1}{2}$ unit of angular momentum and a positron has $-\frac{1}{2}$ unit; when these two particles meet, the total angular momentum is 0. The resulting photons have no net spin; one has spin $+1$ and the other -1. Even a neutron has its antiparticle; these have no charge, but they can annihilate each other because they have opposite values of angular momentum.

The right gamma photon can disappear and reemerge as a positron-electron pair. This cannot happen in empty space. The photon has momentum, and the new pair, moving in opposite directions, has none. Pair production can take place in a bubble chamber, where much more massive atoms are available to absorb the momentum of the photon. The right photon projected into a bubble chamber suddenly appears as a pair of tight spiral trails, coiling in opposite directions in the magnetic field of the chamber. If the photon energy is more than the minimum needed to create the mass of the pair, the extra energy becomes the kinetic energy of the new particles.

Particles exert forces of several kinds on each other. A force, in this realm, is produced by an exchange particle, shuttling between the two particles that are attracting or repelling each other. Four kinds of force are recognized, each with its own characteristic exchange particle. The properties of each of these exchange particles, as of many other kinds of particles, have been predicted by theory. Here are the four forces:

Gravitation depends only on the masses of the objects; it acts at long range but is extremely weak. It is always too small to take into consideration in particle physics. Its exchange particle is the graviton, which has so little energy that it has never been detected. This particle responds only to the gravitational force.

Electromagnetism depends only on electric charge. Charged particles obey the well-known laws of electromagnetism. It is a long-range force, about 10^{36} times as strong as gravity. Its exchange particle is the virtual photon, which exists for such a short period of time that it can probably never be detected.

Strong nuclear force is the force that holds neutrons and protons together to form a nucleus. It is a hundred times as strong as the electromagnetic force, but it operates only at the tiny distance inside the nucleus. Protons and neutrons are composed of still smaller particles called *quarks*, and the nuclear force between quarks is mediated by exchange particles called *gluons*. Neither quarks nor gluons have ever been detected as separate particles, but are always part of other particles. Some of these are the kind called *pions*, or pi mesons, which act as the exchange particles that hold the nucleus together.

Weak nuclear force is another short-range force, only 10^{-13} times as strong as the strong nuclear force. It is responsible for much of the radioactive decay going on within nuclei. Its exchange particles, called W^+, W^-, and Z^0, are so massive that they could not be created until particle accelerators could produce protons with enormous energy. These exchange particles respond to the weak nuclear and electromagnetic forces but not to the strong nuclear force.

CORE CONCEPT

There are four fundamental forces recognized, and each is mediated by a particle that can disappear in interactions.

TRY THIS
—12—

What is the longest wavelength photon that can create an electron-positron pair?

The Particle Zoo

All the exchange particles have one important feature in common: their angular momentum is in whole numbers, not fractions. This means that their spin, under appropriate conditions, can become zero. Then, the particle ceases to exist, transferring its energy to some other particle. The more permanent particles, the fermions, have half-unit spins. Since spin can change only by whole numbers, they cannot disappear. Many fermions are unstable, disintegrating into several smaller particles. The spin of the decay products, can be $+\frac{1}{2}$ or $-\frac{1}{2}$ but never zero. A fermion cannot just disappear.

There are two categories of fermions:

Leptons generally have small mass and are affected by the weak interaction but not the strong nuclear force. When a neutron breaks down to form a proton and two leptons (an electron and a neutrino), the weak interaction is responsible. These are the leptons:

electron small mass, unit charge, a stable particle.

muon like an electron, with unit charge but 200 times as massive. It is formed in the breakdown of pions and other particles. It turns into an electron and neutrinos in a couple millionths of a second.

tauon a still heavier electron, over 3 000 times the electron mass and with a half-life less than a trillionth of a second.

antiparticles for the electron, muon, and tauon, differing only in having positive charge.

neutrinos and *antineutrinos* Each of the leptons is associated with a neutrino, making a total of 12 leptons. Neutrinos have no charge; whether they have any mass is questionable; they are stable. Originally postulated to make the angular momentum equation come out right, their existence has been substantiated experimentally.

Baryons are the more massive particles, and they alone (with the pions) respond to the strong nuclear force. In their interactions and breakdowns, the total number of baryons is conserved.

proton has about 1 800 times the mass of an electron, is positively charged, and is stable. It is the smallest baryon.

neutron is slightly heavier than the proton and has no charge. Outside the nucleus, it breaks down with a half-life of 15 minutes to form a proton, an electron, and an electron antineutrino. In the nucleus, the continual exchange of pions with protons and other neutrons keeps it stable.

hyperons are a variety of larger baryons, some more than twice the mass of a proton, that have been produced in the bubble chambers of particle accelerators. Typically, they last less than a billionth of a second and break down to form pions and other particles, finally becoming protons, electrons, photons, and neutrinos.

antiparticles Every baryon has its antiparticle.

CORE CONCEPT

The four physical forces each have their own exchange particle; the two classes of structural particles are the leptons and the baryons.

TRY THIS
—13—

Why is it impossible for a neutron to break down into a proton and an electron without also emitting some kind of lepton?

Inside the Hadrons

Hadrons are the particles that respond to the strong nuclear force. The smaller ones, up to about a thousand times the electron mass, are the three kinds of mesons; you have already met one, the pion. The larger hadrons are the baryons: proton, neutron, and a number of hyperons.

Many theories were advanced to explain why only certain kinds of hadrons exist. It was suggested that protons and neutrons are composed of still smaller particles. An experimental test was made by bombarding protons with high-energy electrons to observe the scattering pattern. Clear evidence showed that protons have some kind of internal structure.

A model has emerged in which all hadrons are made of still smaller particles called *quarks*. An exchange particle called a *gluon* binds quarks together to make mesons and baryons. The nature of the quark-binding force seems to be such that it becomes stronger as the quarks are separated, so that it will never be possible to separate the quarks and study them in isolation.

Every quark has an electric charge, but the charges are not integral multiples of the electron charge. The charges are either $\frac{1}{3}$ or $\frac{2}{3}$ of the fundamental charge, positive or negative. Other properties of quarks: their spin is always $+\frac{1}{2}$ unit; they have another conserved quantity called *baryon number* which has the value $+\frac{1}{3}$. There is also a full set of antiquarks, which have the opposite charge of the quarks and a baryon number of $-\frac{1}{3}$.

When quarks combine to form mesons and baryons, they must do so in such a way as to obey certain rules:
- the particle must have just 1 elemental charge or none;
- the spin of a baryon must be half-integral; a meson must have integral spin;
- a baryon must have baryon number 1 and a meson must have baryon number 0.

EXAMPLE 17.13   Using these rules, explain why it takes 3 quarks to make a baryon, and a quark and an antiquark to make a meson.

SOLUTION

Any baryon must have a baryon number of 1, which takes three quarks. For a meson, the zero baryon number must come from a quark and an antiquark.

The two quarks in ordinary matter are called, for no particular reason, *up* and *down*. The up (u) quark charge is $+\frac{2}{3}$ and the down (d) charge is $-\frac{1}{3}$. Each has an antiquark with opposite charge and a baryon number of $-\frac{1}{3}$; antiup ($\bar{u}$) has charge $-\frac{2}{3}$ and antidown ($\bar{d}$) has $+\frac{1}{3}$.

EXAMPLE 17.14

Show that the combination *uud* has all the properties of a proton.

SOLUTION

The charge is $2(+\frac{2}{3}) + (-\frac{1}{3}) = +1$; the baryon number is $3(+\frac{1}{3}) = 1$; the spin is $3(\frac{1}{2})$ which is half-integral. If one of the quarks is in the opposite orientation, its spin would be $-\frac{1}{2}$, making the total spin $+\frac{1}{2}$.

EXAMPLE 17.15

What kind of particle has the quark formula $u\bar{d}$?

SOLUTION

The charge is $(+\frac{2}{3}) + (+\frac{1}{3}) = +1$; the baryon number is $(+\frac{1}{3}) + (-\frac{1}{3}) = 0$; the spin is $2(\frac{1}{2}) = 1$, a whole number, or it could be 0. A pion!

So this is it. Our world consists of the first family of particles: two leptons (electron and its neutrino), two quarks (up and down), and their antiparticles, bound together by gluons, photons, gravitons, and W-Z bosons. This model works well, and will stand until something better comes along.

There is another world, a world of particles created by protons with speeds approaching the speed of light, crashing into targets and spewing particles created by their energy into bubble chambers. They decay into first-family particles in times that make "instantaneous" seem forever. The fundamental units of this second family are many times more massive than those of the first family, but they have corresponding natures: two leptons (muon and its neutrino) and two quarks, called strange and charmed. In forming their combinations, they obey the same rules as those of the first family and an additional one as well. A property called strangeness is added to those that must be conserved, except in weak interactions. Typical members of this family are the K$^+$ meson with formula $u\bar{s}$ and the Σ^0 baryon, *uds*.

There is yet a third family with its own two leptons (tauon and its neutrino) and two quarks (top and bottom), and an additional conservation law of its own. Only a few particles of this family have been found. The energies of these particles are so great that only the most powerful particle accelerators can push protons to the energy needed to create them. The top quark has 25 times the mass of a proton, the size of a sodium nucleus, and was not found until the energy of accelerated protons reached an enormous level in 1995. Theory has been highly successful in predicting the existence of these quarks, and it tells us that there is no fourth family.

The mysteries keep coming. Astronomers have found that the observable matter in the universe—stars, planets, galaxies—does not produce enough gravitation to account for the way these objects move. They have concluded that most of the universe is made of some kind of "dark matter." No one knows what this stuff might be. It might be made of a whole now class of particles, entirely different from those that make up the visible world. Stay tuned.

CORE CONCEPT

Hadrons are formed by combinations of quarks, each with fractional electric charge, spin, and baryon number.

The search for ever higher levels of energy continues. The most ambitious project is the Large Hadron Collider, built into a 17-mile circular tunnel on the French-Swiss border, financed by a consortium of 19 countries. Heavy ion particles are accelerated by linear accelerators, and then injected into the tunnel, where they are accelerated by electric fields. They are kept in the circular path by magnetic fields, produced by powerful electromagnets kept at a few kelvins to make them superconductive. The particles travel in opposite directions around the ring, and collide with energies of more than 10^{15} eV. When it goes operational in 2008, it is hoped that the collisions will produce new particles that will solve some of the most perplexing mysteries in the world of particle physics.

TRY THIS
—14—

What is the quark formula for a neutron?

REVIEW EXERCISES FOR CHAPTER 17

Fill-In's. For each of the following, fill in the missing word or phrase.

1. The atomic nucleus was discovered by bombarding gold foil with _____.

2. The nucleus designated as $^{95}_{40}$Zr contains _____ neutrons.

3. The radius of curvature of the track of a particle in a bubble chamber is measured in order to determine the _____ of the particle.

4. Gamma photons leave no track in a bubble chamber because they have no_____.

5. The operation of a cyclotron depends on the fact that all the circulating particles have the same _____.

6. The relativistic increase in the _____ of the particles imposes a limit on the energy obtained in a cyclotron.

7. The collision of electrons with _____ produces the greatest energy yet obtained in particle accelerators.

8. To obtain nuclear fuel, it is necessary to separate the natural isotopes of _____.

9. A chain reaction can be produced by the splitting of a nucleus that releases _____.

10. The positively charged particle in the nucleus is called a(n) _____.

11. The number of protons plus the number of _____ equals the atomic mass number of the nucleus.

12. _____ of the same element differ only in the number of neutrons in their nuclei.

13. Nuclear masses are measured in _____.

14. Charged particles can be tracked through detection devices because they lose energy in _____ the material through which they pass.

15. The tracks of charged particles are curved as a result of the influence of a(n) _____.

16. Charged particles passing through liquid hydrogen at its boiling point leave a trail of _____.

17. The photons emitted by radioactive materials are called _____ rays.

18. Beta rays are _____.

19. Alpha particles are nuclei of _____.

20. The process of _____ decay increases the atomic number by 1.

21. In alpha decay, the atomic mass number decreases by _____.

22. If a gram of a radioactive substance is reduced to $\frac{1}{2}$ gram in 6 hours, then the _____ of that substance is 6 hours.

23. The mass of a nucleus is always _____ the mass of the separate particles of which it is composed.

24. Mass deficit is _____ energy.

25. In any spontaneous nuclear process, mass deficit always _____ .

26. Charged particles are increased in energy by the action of a(n) _____ field.

27. In a linear accelerator, energy is added to particles as they pass through the gap between the _____.

28. In a synchrotron, a(n) _____ field bends the path of the particles.

29. A positron is a positive _____.

30. Small radioactive nuclei can be made artificially by bombardment with _____.

31. _____ with very low energy can enter nuclei.

32. Plutonium is produced by reacting uranium with a(n) _____.

33. A chain reaction can be sustained only by a fuel that produces _____ when it undergoes fission.

34. The preferred fuel for atom bombs is _____.

35. In a nuclear reactor, it is the function of the _____ to slow down the neutrons.

36. In a reactor core, neutrons are absorbed by the _____.

37. Temperatures in the millions of kelvins are needed to produce nuclear _____.

38. The largest mass deficit per particle is found in _____ nuclei.

39. The boson of the electromagnetic force is the _____.

40. A proton is made of three _____.

41. The force that binds nucleons together in the nucleus is supplied by an exchange of _____.

42. The _____ is the charged lepton in the first family of fundamental particles.

Multiple Choice.

1. One reason atomic masses are not integral multiples of the mass of a proton is that nuclei of an element differ in numbers of
   (1) quarks   (2) protons   (3) neutrons   (4) pions   (5) electrons.

2. A particle that will not leave a curved track in a bubble chamber is the
   (1) neutron   (2) proton   (3) positron   (4) electron   (5) alpha particle.

3. Emission of an electron from a nucleus results in
   (1) an increase in atomic number
   (2) an increase in atomic mass number
   (3) a decrease in atomic number
   (4) a decrease in atomic mass number
   (5) no change in either atomic mass number or atomic number.

4. When $^{234}_{90}$Th emits an alpha particle, it becomes
   (1) $^{230}_{88}$Th   (2) $^{230}_{89}$Ac   (3) $^{230}_{89}$Th   (4) $^{230}_{88}$Ra   (5) $^{235}_{90}$Pa.

5. The mass deficit of a nucleus is an indication of its
   (1) atomic mass number   (4) total energy
   (2) total nuclear mass   (5) radioactive decay mode.
   (3) binding energy

6. At higher energies, the proton beam of a cyclotron goes out of synchronization because
   (1) the magnetic field is not regular
   (2) the electric field breaks down
   (3) the protons gain mass
   (4) the protons lose some of their charge
   (5) the protons lose some of their mass.

7. A nuclear chain reaction can take place if the splitting of a nucleus
   (1) releases energy   (4) raises the pressure
   (2) produces neutrons   (5) raises the temperature.
   (3) increases binding energy

8. A nuclear reactor is said to be a breeder if it
   (1) produces plutonium
   (2) produces large amounts of energy
   (3) generates large amounts of radioactive waste
   (4) is in danger of exploding
   (5) releases neutrons.

9. If a neutron and an antineutron combine, the result is
   (1) a helium nucleus   (4) two photons
   (2) a proton and an electron   (5) two protons.
   (3) a photon

10. The exchange particle of the strong nuclear force is the
    (1) photon   (2) neutron   (3) pion   (4) W boson   (5) nucleon.

11. Particles with unit spin have the property that
    (1) they react only with the strong force
    (2) they have no charge
    (3) they have zero mass
    (4) they cannot react with protons
    (5) they can be absorbed and disappear.

Problems.

1. An atomic species is represented as $^{190}_{77}\text{Ir}$. State the number of (a) electrons in the rings; (b) neutrons in the nucleus; and (c) protons in the nucleus.

2. (a) What is the momentum of a proton in a magnetic field of 2.5 teslas if it travels in a circle of radius 7.5 centimeters? (b) How fast is it traveling?

3. A gamma photon strikes a lead plate, and its energy is converted into an electron and a positron. (a) How much energy, in joules, must the photon have? (b) What is the largest wavelength of a gamma photon that can do this?

4. What is the frequency of an AC that will accelerate protons in a cyclotron if the magnetic field is 1.8 teslas?

5. What combination of quarks is present in (a) an antiproton; (b) a negative pion?

6. How much energy is released when two nuclei of helium-3 combine to form a nucleus of helium-4 and two protons?

7. Determine the mass deficit of phosphorus-30 and the corresponding binding energy.

8. (a) Write the equation for the alpha decay of polonium-218; and (b) calculate the combined energy of the alpha particle and the gamma photon released.

9. How long would it take for a sample of polonium-218 to be reduced to $\frac{1}{16}$ of its original mass?

10. (a) Write the equation for the beta decay of lead-210; and (b) calculate the combined energy of the electron and the neutrino produced.

11. What is the kinetic energy of an alpha particle that has been accelerated in an electric field by a potential difference of 650 000 volts?

12. Through what potential difference would a proton have to be accelerated to double its mass?

13. An alpha particle strikes the nucleus of boron-10 and combines with it. The compound nucleus stabilizes by emitting a neutron. Write the equation.

14. Write the equation for the decay of sodium-22 by emission of a positron.

15. Plutonium-239 receives a slow neutron and splits, releasing 3 neutrons and a nucleus of xenon-142. Determine the other fission product and write the complete equation for the reaction.

16. Determine the energy of the gamma photon released in the following fusion reaction, which takes place in the sun:

$$\mathrm{^{1}_{1}H + {}^{12}_{6}C \rightarrow {}^{13}_{7}N + gamma}$$

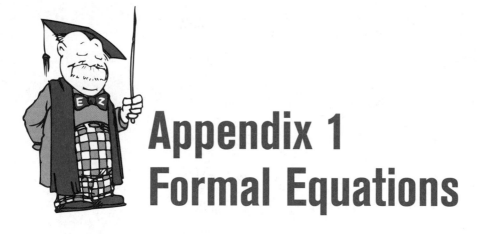

Appendix 1
Formal Equations

Numbers refer to the section of the text in which the equation is introduced.

2.1.	Definition of average speed	$v_{av} = \dfrac{\Delta s}{\Delta t}$
2.2.	Component of a vector	$P_x = \mathbf{P} \cos \theta_x$
2.3.	Average acceleration	$\mathbf{a}_{av} = \dfrac{\Delta \mathbf{v}}{\Delta t}$
2.4.	Uniformly accelerated motion	(a) $v_f = v_i + at$
		(b) $s = v_i t + \dfrac{1}{2} a t^2$
		(c) $v_f^2 = v_i^2 + 2as$
2.5.	Motion in a circular path	(a) $v = 2\pi r / T$
		(b) $a_c = v^2/r$
2.6.	Galilean velocity transformation	$v' = v + u$
3.1.	Weight and mass	$\mathbf{w} = m\mathbf{g}$
3.2.	Sliding friction	$\mathbf{F}_{fr} = \mu \mathbf{F}_{nor}$
3.3.	Definition of torque	$\tau = rF \sin \theta$
4.1.	Impulse	(a) $\mathbf{P} = \mathbf{F}\Delta t$
		(b) $\Delta \mathbf{v} \propto \mathbf{F}\Delta t$
4.2.	Newton's law of inertia	$\mathbf{F} = m\mathbf{a}$
4.3.	Impulse and momentum	$\mathbf{F}\Delta t = \Delta(m\mathbf{v})$
4.4.	Centripetal force	$F_c = \dfrac{mv^2}{r}$
4.5.	Gravitational field and altitude	$\dfrac{g}{g_0} = \left(\dfrac{r_0}{r}\right)^2$
4.6.	Universal gravitation	$F_{grav} = \dfrac{Gm_1 m_2}{r^2}$
5.1.	Definition of work	$W = \mathbf{F}\Delta \mathbf{s} \cos \theta$

5.2.	Gravitational potential energy	$E_{\text{grav}} = mgh$

5.3.   Kinetic energy $E_{\text{kin}} = \frac{1}{2}mv^2$

5.4.   Gravitational energy in space $E_{\text{grav}} = -\dfrac{Gm_1 m_2}{r}$

5.5.   Energy of a spring $E_{\text{spr}} = \frac{1}{2}k\Delta l^2$

5.6.   Definition of power $P = \dfrac{\Delta E}{\Delta t}$

5.7.   Mass-energy $E = mc^2$

6.1.   Definition of density $D = \dfrac{m}{V}$

6.2   Definition of pressure $p = \dfrac{F}{A}$

6.3.   Pressure in a liquid $p = hDg$
6.4.   Buoyancy $B = VDg$
7.1.   Temperature conversion $T_{\text{K}} = T_{\text{C}} + 273.15 \text{ K}$
7.2.   Specific heat $\Delta H = cm\Delta T$
7.3.   Method of mixtures $(mc\Delta T)_{\text{water}} = (mc\Delta T)_{\text{other}}$
7.4.   Latent heat $\Delta H = mL$
7.5.   Ideal gas law $pV \propto T_{\text{K}}$
7.6.   Boltzman's constant $pV = NkT_{\text{K}}$

7.7.   Molecular temperature $\overline{E} = \frac{3}{2}kT$

7.8.   Ideal heat engines $\varepsilon = 1 - \dfrac{T_{\text{o}}}{T_{\text{i}}}$

8.1.   Period of a pendulum $T = 2\pi\sqrt{\dfrac{l}{g}}$

8.2.   Simple harmonic motion $T = 2\pi\sqrt{\dfrac{x}{a}}$

8.3.   Definition of frequency $f = 1/T$
8.4.   Velocity of a wave $v = f\lambda$
8.5.   Speed of sound in air $v_{\text{air}} = 331 \text{ m/s} + (0.6 \text{ m/s})T_{\text{C}}$

8.6. Loudness of sound

(a) $\beta = 10 \log\left(\dfrac{I}{I_0}\right)$

(b) $I = \dfrac{P}{4\pi r^2}$

9.1. Coulombs's law

$F_{el} = \dfrac{kq_1q_2}{r^2}$

9.2. Definition of electric field

$\mathscr{E} = \dfrac{F_{el}}{q}$

9.3. Definition of electric potential difference

$\Delta V = \dfrac{\Delta E_{el}}{q}$

9.4. Electric field between parallel plates

$\mathscr{E} = \dfrac{\Delta V}{\Delta s}$

9.5. Potential at a point near a charge

$V = k\dfrac{q}{r}$

10.1. Definition of electric current

$I = \dfrac{\Delta q}{\Delta t}$

10.2. Definition of resistance

$R = \dfrac{\Delta V}{I}$

10.3. Resistance of a wire

$R = \dfrac{\rho l}{A}$

10.4. Electric power consumed

(a) $P = I\,\Delta V$
(b) $P = I^2 R$

10.5. Battery with internal resistance

(a) $I = \dfrac{emf}{R_{internal} + R_{external}}$

(b) $\Delta V_{terminal} = emf - IR_{internal}$

10.6. Series circuit

(a) $\Delta V_{total} = \Delta V_1 + \Delta V_2 + \Delta V_2 + \ldots$

(b) $R_s = R_1 + R_2 + R_3 + \ldots$

10.7. Parallel circuit

$\dfrac{1}{R_p} = \dfrac{1}{R_1} + \dfrac{1}{R_2} + \dfrac{1}{R_3} + \ldots$

11.1. Magnetic force on a current

$\mathbf{F}_{mag} = IlB\sin\theta$

11.2. Torque on a loop in a magnetic field

$\tau = NBIA\sin\theta$

11.3. Magnetic field around a current

$B = 2k'\dfrac{I}{r}$

11.4. Magnetic force between currents

$\dfrac{F_{mag}}{l} = \dfrac{2k'I_1I_2}{r}$

11.5. Field in a solenoid $\qquad$ $B = 4\pi k'I\dfrac{n}{l}$

11.6. Magnetic force on a charged particle $\qquad$ $F_{mag} = Bqv$

11.7. Induced emf $\qquad$ $emf = Blv$

11.8. Magnetic flux in a loop $\qquad$ $\phi = BA$

12.1. Alternating current $\qquad$ (a) $P_{av} = I_{eff}^2 R$

(b) $\bar{I} = \dfrac{I_{max}}{\sqrt{2}}$

(c) $\Delta\bar{V} = \dfrac{\Delta V_{max}}{\sqrt{2}}$

12.2. Transformer rules $\qquad$ (a) $\dfrac{\Delta\bar{V}_s}{\Delta\bar{V}_P} = \dfrac{n_s}{n_P}$

(b) $\dfrac{\bar{I}_P}{\bar{I}_s} = \dfrac{n_s}{n_P}$

13.1. Definition of capacitance $\qquad$ $C = \dfrac{q}{\Delta V}$

13.2. Definition of capacitive reactance $\qquad$ $X_C = \dfrac{\Delta\bar{V}}{\bar{I}} = \dfrac{1}{2\pi fC}$

13.3. Self-inductance $\qquad$ $\Delta V_L = -L\dfrac{\Delta I}{\Delta t}$

13.4. Inductive reactance $\qquad$ $X_L = 2\pi fL$

13.5. Impedance $\qquad$ $Z = \sqrt{R^2 + X^2}$

13.6. Series R-L-C circuit $\qquad$ $Z = \sqrt{R^2 + (X_L - X_C)^2}$

13.7. Resonant frequency $\qquad$ $f = \dfrac{1}{2\pi\sqrt{LC}}$

14.1. Diffraction grating interference pattern $\qquad$ (a) $s_2 - s_1 = n\lambda$

(b) $\sin\theta_n = \dfrac{n\lambda}{d}$

(c) $\lambda = d\dfrac{x}{L}$

14.2. Quantum of energy $\qquad$ $\Delta E = hf = hc/\lambda$

14.3. Energy of a photon $\qquad$ $E_{photon} = hf$

14.4. Photoelectric effect $\qquad$ $q\Delta V = hf - W$

14.5. Momentum of a photon $\qquad$ $p = h/\lambda$

15.1. Illumination $\qquad I = \dfrac{F}{4\pi r^2}$

15.2. Refraction $\qquad \dfrac{\sin\theta_A}{\sin\theta_B} = \dfrac{n_B}{n_A}$

15.3. Image formation $\qquad$ (a) $\dfrac{1}{f} = \dfrac{1}{D_o} + \dfrac{1}{D_i}$

$\qquad$ (b) $\dfrac{D_o}{D_i} = \dfrac{S_o}{S_i}$

15.4. Magnifying glass $\qquad M_{max} = 1 + \dfrac{25\text{ cm}}{f}$

16.1. Photon energy of hydrogen spectrum $\qquad E = hf = 13.6\text{ eV}\left(\dfrac{1}{n_1^2} - \dfrac{1}{n_2^2}\right)$

16.2. Angular momentum of hydrogen electron $\qquad mvr = \dfrac{nh}{2\pi}$

16.3. Wavelength of a particle $\qquad \lambda = \dfrac{h}{mv}$

16.4. Energy of an electron in an atom $\qquad E = (-13.6\text{ eV})\left(\dfrac{Z_{eff}^2}{n^2}\right)$

17.1. Rate of radioactive decay $\qquad$ (a) $\dfrac{\Delta N}{\Delta t} = \gamma N$

$\qquad$ (b) $T\dfrac{1}{2} = \dfrac{0.693}{\gamma}$

Appendix 2
Physical Constants

Acceleration due to gravity on earth (average): $g_o = 9.81$ m/s^2

Atomic mass unit: 6.023×10^{26} daltons $= 1$ kg

Boltzmann gas constant: $k = 1.380 \times 10^{-23}$ J/mlc·K

Electric constant of free space: $k = 8.987 \times 10^9$ N·m^2/C^2

Elementary unit of charge: $e = 1.602 \times 10^{-19}$ C

Magnetic constant of free space: $k' = 10^{-7}$ N/A^2 exactly

Mass of the earth $= 5.976 \times 10^{24}$ kg

Mass–energy conversion: 1 dalton $= 931$ MeV

Planck's quantum constant: $h = 6.624 \times 10^{-34}$ J·s
$\qquad\qquad\qquad\qquad\quad = 4.135 \times 10^{-15}$ eV·s

Radius of the earth (average) $= 6\ 370$ km

Rest mass of electron $= 9.110 \times 10^{-31}$ kg
$\qquad\qquad\qquad\quad = 5.486 \times 10^{-4}$ Dl
$\qquad\qquad\qquad\quad = 0.511$ MeV

Rest mass of proton $= 1.673 \times 10^{-27}$ kg

Speed of electromagnetic wave in vacuum: c $= 2.998 \times 10^8$ m/s

Universal gravitation constant: $G = 6.672 \times 10^{-11}$ N·m^2/kg^2

Appendix 3
Values of the
Trigonometric Functions

Angle	Sin	Cos	Tan
1°	.0175	.9998	.0175
2°	.0349	.9994	.0349
3°	.0523	.9986	.0524
4°	.0698	.9976	.0699
5°	.0872	.9962	.0875
6°	.1045	.9945	.1051
7°	.1219	.9925	.1228
8°	.1392	.9903	.1405
9°	.1564	.9877	.1584
10°	.1736	.9848	.1763
11°	.1908	.9816	.1944
12°	.2079	.9781	.2126
13°	.2250	.9744	.2309
14°	.2419	.9703	.2493
15°	.2588	.9659	.2679
16°	.2756	.9613	.2867
17°	.2924	.9563	.3057
18°	.3090	.9511	.3249
19°	.3256	.9455	.3443
20°	.3420	.9397	.3640
21°	.3584	.9336	.3839
22°	.3746	.9272	.4040
23°	.3907	.9205	.4245
24°	.4067	.9135	.4452
25°	.4226	.9063	.4663
26°	.4384	.8988	.4877
27°	.4540	.8910	.5095
28°	.4695	.8829	.5317
29°	.4848	.8746	.5543
30°	.5000	.8660	.5774
31°	.5150	.8572	.6009
32°	.5299	.8480	.6249

Angle	Sin	Cos	Tan
33°	.5446	.8387	.6494
34°	.5592	.8290	.6745
35°	.5736	.8192	.7002
36°	.5878	.8090	.7265
37°	.6018	.7986	.7536
38°	.6157	.7880	.7813
39°	.6293	.7771	.8098
40°	.6428	.7660	.8391
41°	.6561	.7547	.8693
42°	.6691	.7431	.9004
43°	.6820	.7314	.9325
44°	.6947	.7193	.9657
45°	.7071	.7071	1.0000
46°	.7193	.6947	1.0355
47°	.7314	.6820	1.0724
48°	.7431	.6691	1.1106
49°	.7547	.6561	1.1504
50°	.7660	.6428	1.1918
51°	.7771	.6293	1.2349
52°	.7880	.6157	1.2799
53°	.7986	.6018	1.3270
54°	.8090	.5878	1.3764
55°	.8192	.5736	1.4281
56°	.8290	.5592	1.4826
57°	.8387	.5446	1.5399
58°	.8480	.5299	1.6003
59°	.8572	.5150	1.6643
60°	.8660	.5000	1.7321
61°	.8746	.4848	1.8040
62°	.8829	.4695	1.8807
63°	.8910	.4540	1.9626
64°	.8988	.4384	2.0503
65°	.9063	.4226	2.1445
66°	.9135	.4067	2.2460
67°	.9205	.3907	2.3559
68°	.9272	.3746	2.4751
69°	.9336	.3584	2.6051
70°	.9397	.3420	2.7475
71°	.9455	.3256	2.9042
72°	.9511	.3090	3.0777
73°	.9563	.2924	3.2709
74°	.9613	.2756	3.4874
75°	.9659	.2588	3.7321
76°	.9703	.2419	4.0108
77°	.9744	.2250	4.3315
78°	.9781	.2079	4.7046
79°	.9816	.1908	5.1446
80°	.9848	.1736	5.6713
81°	.9877	.1564	6.3138
82°	.9903	.1392	7.1154
83°	.9925	.1219	8.1443
84°	.9945	.1045	9.5144

Angle	Sin	Cos	Tan
85°	.9962	.0872	11.4301
86°	.9976	.0698	14.3007
87°	.9986	.0523	19.0811
88°	.9994	.0349	28.6363
89°	.9998	.0175	57.2900
90°	1.0000	.0000	

Appendix 4
Properties of
Selected Nuclides

Atomic number	Name	Symbol	Atomic mass number	Atomic mass (DI)	Half-life
0	Electron	$_{-1}e$	0	5.49×10^{-4}	Stable
0	Neutron	n	1	1.008665	11 minutes
1	Hydrogen	H	1	1.007825	Stable
	(deuterium)		2	2.0140	Stable
	(tritium)		3	3.01605	12 years
2	Helium	He	3	3.01603	Stable
			4	4.00260	Stable
3	Lithium	Li	6	6.01512	Stable
			7	7.01600	Stable
4	Beryllium	Be	9	9.01218	Stable
5	Boron	B	10	10.0129	Stable
			11	11.00931	Stable
6	Carbon	C	12	12 exactly	Stable
			14	14.0032	5730 years
7	Nitrogen	N	13	13.0057	10 minutes
			14	14.00307	Stable
9	Fluorine	F	19	18.99840	Stable
10	Neon	Ne	22	21.99138	Stable
11	Sodium	Na	22	21.9944	2.6 years
13	Aluminum	Al	27	26.98153	Stable
14	Silicon	Si	30	29.97376	Stable
15	Phosphorus	P	30	29.9783	2.5 minutes
17	Chlorine	Cl	35	34.96885	Stable
			37	36.9659	Stable
19	Potassium	K	39	38.96371	Stable
20	Calcium	Ca	42	41.95863	Stable
26	Iron	Fe	54	53.9396	Stable
			56	55.9349	Stable
36	Krypton	Kr	92	91.9261	3.0 seconds
40	Zirconium	Zr	95	91.224	65 days

Atomic number	Name	Symbol	Atomic mass number	Atomic mass (DI)	Half-life
54	Xenon	Xe	142	131.293	1.5 seconds
56	Barium	Ba	141	140.9140	18 minutes
82	Lead	Pb	206	205.9745	Stable
			210	209.9848	21 years
			214	213.9998	27 minutes
83	Bismuth	Bi	210	209.9840	5 days
84	Polonium	Po	214	213.9952	1.6×10^{-4} second
			218	218.0089	3 minutes
86	Radon	Rn	222	222.0175	38 seconds
88	Radium	Ra	226	226.0254	1,600 years
90	Thorium	Th	230	230.0331	80,000 years
			234	234.0436	24 days
91	Palladium	Pa	234	234.0433	6.7 hours
92	Uranium	U	233	233.0395	160,000 years
			235	235.0439	710 million years
			238	238.0508	4.5 billion years
			239	239.05433	24 minutes
93	Neptunium	Np	239	239.05295	2.4 days
94	Plutonium	Pu	239	239.0522	24,000 years

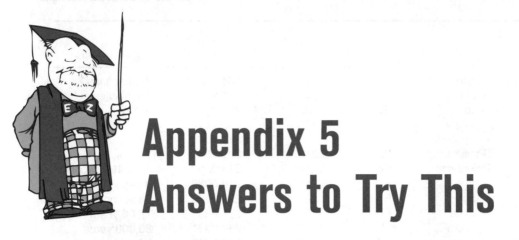

Appendix 5
Answers to Try This

Chapter 1

1. For international trade and science, it is important to have universally accepted standards.
2. 2/55 is about 4 percent.
3. $87.0 + 9.136 = 96.14$, carrying one nonsignificant digit. Then $96.14 \div 22 = 4.4$, rounding off to only two digits.
4. 1. 1.4×10^7.
   2. 2.34×10^{-1}.
   3. 4.19×10^3; if the final zero is significant, the answer is 4.190×10^3.
   4. 3.98×10^{-7}.
5. 1. 16×10^6, which is 1.6×10^7.
   2. 34×10^3, or 3.4×10^4.
   3. 2.0×10^{-4}.
   4. The exponent of 10 must be made the same:

$$3.75 \times 10^4 + 0.191 \times 10^4 = 3.94 \times 10^4.$$

   5. $0.395 \times 10^5 - 5.17 \times 10^5 = -4.78 \times 10^5$.
6. 1. 75 oz.
   2. 31 in.
7. 1. yes
   2. yes
   3. no
   4. no
   5. no
8. 1. 4.2 lb/in^2.
   2. 320 ft-lb.
   3. $(100 \text{ cm})^3 = 10^6 \text{ cm}^3$.
9. Kilograms per second (kg/s).

Chapter 2

1. 3 min 30 s is 210 s:

$$v_{av} = \frac{\Delta s}{\Delta t} = \frac{1\ 500\ m}{210\ s} = 7.1\ m/s$$

2. Since east and south are at right angles to each other, their sum is the hypotenuse of a right triangle. Using the theorem of Pythagoras, we have $3.5^2 + 5.8^2 = s^2$, so $s = 6.8$ m. The direction will be south of east at an angle whose tangent is 5.8/3.5, or 59°.

3. To compensate for that easterly wind, he will have to head the plane in a direction east of north. His air velocity and the eastward wind velocity produce a sum directly north. He will achieve this if the angle has a sine equal to 110/620, so his heading is north 10° east.

4. a. Since the angle is 22° with the horizontal, the horizontal component is (40 mi/hr)(cos 22°) = 37 mi/hr.

   b. The other rectangular component is (40 mi/hr)(sin 22°) = 15 mi/hr.

5. Acceleration is change of speed per unit time; the change of velocity is $(2\ 840 - 2\ 800) = 40$ m/s. Then

$$a = \frac{\Delta v}{\Delta t} = \frac{40\ m/s}{25\ s} = 1.6\ m/s^2$$

6. You are given time, distance, and initial speed, so Equation 2.4b will do the job. Substituting in the known values,

$$360\ m = (22\ m/s)(8.3\ s) + \frac{1}{2}a(8.3\ s)^2$$

   from which $a = 5.2\ m/s^2$.

7. a. Starting from rest with uniform acceleration:
   $s = 1/2\ at^2 = 1/2(9.8\ m/s^2)(2.3\ s)^2 = 26$ m.

   b. $v = at = (9.8\ m/s^2)(2.3\ s)^2 = 23$ m/s.

8. From Equation 2.4a, taking the upward direction as positive,

$$-15\ m/s = (22\ m/s) + (-9.8\ m/s^2)t$$

   from which $t = 3.8$ s.

9. Find out how long it took for the trip using the vertical motion and Equation 2.4b:

$$t = \sqrt{\frac{2s}{g}} = \sqrt{\frac{2(32\ m)}{(9.8\ m/s^2)}} = 2.56\ s$$

   Then find the horizontal speed from Equation 2.1:

$$v = s\ /\ t = \frac{135\ m}{2.56\ s} = 53\ m/s$$

10. a. He travels the circumference of the circle in 30 s, so

$$v = \frac{2\pi r}{T} = \frac{2\pi(150 \text{ m})}{30 \text{ s}} = 31 \text{ m/s}$$

At the southernmost point, he is going west.

b. $a = \frac{v^2}{r} = \frac{(31.4 \text{ m/s})^2}{150 \text{ m}} = 6.6 \text{ m/s}^2$

The direction of the acceleration is toward the center of the circle, northward.

11. a. In the elevator's frame of reference, the watch starts at zero velocity with an acceleration of 9.8 m/s².

b. In the building's frame of reference, the initial velocity is 3.0 m/s, and the acceleration is still 9.8 m/s².

Chapter 3

1. Two forces pushing in opposite directions cancel each other, so the direction matters.

2. $w = mg$, so

$$m = \frac{w}{g} = \frac{340 \text{ N}}{9.8 \text{ m/s}^2} = 35 \text{ kg}$$

3. Since the surface is horizontal, the normal force is the weight of the chair, $mg = (65 \text{ kg})(9.8 \text{ m/s}^2) = 637 \text{ N}$. Then

$$\mu = \frac{F_{\text{fr}}}{F_{\text{nor}}} = \frac{220 \text{ N}}{637 \text{ N}} = 0.35$$

4. The weight of the object is

$$(6 \text{ kg})(9.8 \text{ m/s}^2) = 59 \text{ N}$$

Since this is less than the buoyant force, the object will float.

5. a. 6   b. 6   c. 1   d. 5   e. 2   f. 8

6. The force of your tires pulling the earth along with the car.

7. a. The friction is the force that maintains horizontal equilibrium, so it must equal the scale reading, 4 N.

b. The elastic recoil of the tabletop maintains vertical equilibrium, so it must be equal to the weight of the brick; $w = mg = (1.6 \text{ kg})(9.8 \text{ m/s}^2) = 16 \text{ N}$.

8. a. Since the 30° is the angle with the horizontal, the horizontal component is $(160 \text{ N})(\cos 30°) = 139 \text{ N}$.

b. The vertical component is $(160 \text{ N})(\sin 30°) = 80 \text{ N}$.

9. First, find the horizontal and vertical components of the tension in the string. Since the 20° angle is with the vertical, $T_{vert} = (16 \text{ N})(\cos 20°) = 15.0 \text{ N}$. The horizontal component, acting toward the left, is $(16 \text{ N})(\sin 20°) = 5.5 \text{ N}$.

   Now set the upward forces equal to the downward forces: $B = 15.0 \text{ N} + 25 \text{ N} = 40 \text{ N}$. And the forces toward the right equal those toward the left: $W = 5.5 \text{ N}$.

10. 1. a. The angle θ in Figure 3.13 is equal to the slope of the surface, which is 20°. Therefore, the component of weight parallel to the plane is $(120 \text{ kg})(9.8 \text{ m/s}^2)(\sin 20°) = 400 \text{ N}$, and this is just balanced by the tension in the rope.

      b. The component of weight perpendicular to the plane is $(120 \text{ kg})(9.8 \text{ m/s}^2)(\cos 20°) = 1\,100 \text{ N}$.

    2. The weight of the crate is $(60 \text{ kg})(9.8 \text{ N/kg}) = 588 \text{ N}$. The component of weight parallel to the ramp is $(588 \text{ N})(\sin 36°) = 346 \text{ N}$. The rope supplies 240 N, so friction must provide the remaining 106 N to keep the crate in equilibrium. The component perpendicular to the ramp is $(588 \text{ N})(\cos 36°) = 476 \text{ N}$. Therefore

$$\mu = \frac{F_f}{F_N} = \frac{106 \text{ N}}{476 \text{ N}} = 0.22$$

11. Clockwise and counterclockwise torques around any selected pivot must be equal. All distances must be measured from that pivot. In this case, the rock is the most convenient pivot. The woman, her weight pushing (say) clockwise, is 1.0 m from the rock; then the scale must be pushing counterclockwise at a distance of 3.5 m from the rock. Therefore, $w(1.0 \text{ m}) = (160 \text{ N})(3.5 \text{ m})$, and $w = 560 \text{ N}$.

Chapter 4

1. A force of 520 N for 10 s provides an impulse of 5 200 N·s; 130 N for 4 s gives 520 N·s, only $\frac{1}{10}$ as much. Since the first blast slowed the craft down by 50 m/s, the second slows it by an additional 5 m/s ($\frac{1}{10}$ as much) to bring it to 295 m/s.

2. $F = ma$, so

$$m = \frac{F}{a} = \frac{130 \text{ N}}{3.0 \text{ m/s}^2} = 43 \text{ kg}$$

3. The net force is 8 500 N – 6 200 N = 2 300 N. Then

$$a = \frac{F}{m} = \frac{2\ 300\ \text{N}}{1\ 200\ \text{kg}} = 1.9\ \text{m/s}^2$$

4. a. Momentum $= mv = (0.30\ \text{kg})(35\ \text{m/s}) = 10.5\ \text{kg·m/s}$.
   b. Impulse is the change in momentum; since the ball started at rest, its change in momentum $= 10.5\ \text{kg·m/s}$, and the impulse is 10.5 N·s.
   c. Impulse $= F\,\Delta t$;

$$F = \frac{10.5\ \text{N} \cdot \text{s}}{0.5\ \text{s}} = 21\ \text{N}$$

5. Total momentum before the collision must equal total momentum after. Taking east as the positive direction:

$$(2\ 000\ \text{kg})(22\ \text{m/s}) + (1\ 200\ \text{kg})(-30\ \text{m/s}) = (2\ 000\ \text{kg} + 1\ 200\ \text{kg})v$$

from which $v = 2.5$ m/s.

6. $F = ma = \dfrac{mv^2}{r} = \dfrac{(55\ \text{kg})(6.0\ \text{m/s})^2}{30\ \text{m}} = 66\ \text{N}$

7. Its speed must be just enough to make its centripetal acceleration equal to the acceleration due to gravity. Then, from Equation 2.5b,

$$v = \sqrt{ar} = \sqrt{(1.67\ \text{m/s}^2)(3.5 \times 10^6\ \text{m})} = 2\ 400\ \text{m/s}$$

8. $F_{\text{grav}} = \dfrac{Gm_1m_2}{r^2} = \dfrac{(6.67 \times 10^{-11}\ \text{N·m}^2/\text{kg}^2)(5\ 700\ \text{kg})(14\ 000\ \text{kg})}{(4.0 \times 10^5\ \text{m})^2}$

$$= 3.3 \times 10^{14}\ \text{N}$$

9. As long as the spacecraft is in free fall, whether or not it is near a planet, everything in it falls in the same way and everything is weightless. If you see something fall within the spaceship, that object is not falling the same way the spaceship is. Therefore, the ship is not in free fall. Its engines are on.

Chapter 5

1. 1. 0
   2. 1030 J
   3. 0
2. Since it takes work to move the charges apart, their potential energy increases as they separate and decreases as they approach each other.

3. Since $W = mgh$,

$$h = \frac{W}{mg} = \frac{12\ 000\ \text{J}}{(280\ \text{kg})(9.8\ \text{m/s}^2)} = 4.4\ \text{m}$$

4. $E_{\text{kin}} = \frac{1}{2}mv^2 = \frac{1}{2}(1\ 200\ \text{kg})(22\ \text{m/s})^2 = 2.9 \times 10^5\ \text{J}$.

5. a. $E_{\text{kin}} = \frac{1}{2}mv^2 = \frac{1}{2}(30\ \text{kg})(6.0\ \text{m/s})^2 = 540\ \text{J}$.

   b. All this kinetic energy turns to gravitational energy ($= mgh$), so $540\ \text{J} =$ $(30\ \text{kg})(9.8\ \text{m/s}^2)h$, and $h = 1.8\ \text{m}$.

6. Its energy drops from 0 to $-Gm_1m_2/r$, so the loss is

$$\frac{\left(6.67 \times 10^{-11}\ \text{N}\cdot\text{m}^2/\text{kg}^2\right)\left(6.0 \times 10^{24}\ \text{kg}\right)(60\ \text{kg})}{6.4 \times 10^6\ \text{m}} = 3.8 \times 10^9\ \text{J}$$

7. The escape velocity equation still holds:

$$v_{\text{escape}} = \sqrt{\frac{2Gm_2}{r}} = \sqrt{\frac{2\left(6.67 \times 10^{-11}\ \text{N}\cdot\text{m}^2/\text{kg}^2\right)\left(6.0 \times 10^{24}\ \text{kg}\right)}{2\left(6.4 \times 10^6\ \text{m}\right)}} = 7\ 900\ \text{m/s}$$

8. The increase in internal energy is equal to the kinetic energy lost:

$$\Delta H = \frac{mv_2^2}{2} - \frac{mv_1^2}{2} = \frac{2\ 200\ \text{kg}}{2}\left[(26\ \text{m/s})^2 - (11\ \text{m/s})^2\right] = 6.1 \times 10^5\ \text{J}$$

9. The gravitational energy lost by the rock is equal to the deformation energy gained by the spring:

$$mgh = \frac{1}{2}k\Delta l^2$$

Since $h = \Delta l$, the equation becomes

$$\Delta l = \frac{2\ mg}{k} = \frac{2(2.0\ \text{kg})(9.8\ \text{m/s}^2)}{150\ \text{N/m}} = 0.26\ \text{m}$$

10. The work that must be done is

$$(40\ \text{kg})(9.8\ \text{N/kg})(25\ \text{m}) = 9\ 800\ \text{J}$$

From Equation 5.6,

$$p = \frac{\Delta E}{\Delta t} = \frac{9\ 800}{30s} = 330\ \text{W}$$

11. Its kinetic energy is $\frac{1}{2}mv^2 = \frac{1}{2}(1\ 400\ \text{kg})(30\ \text{m/s})^2 = 6.3 \times 10^5\ \text{J}$.

This corresponds to an increase in mass:

$$\frac{E}{c^2} = \frac{6.3 \times 10^5\ \text{J}}{(3.0 \times 10^8\ \text{m/s})^2} = 7 \times 10^{-12}\ \text{kg}$$

12. The engine continually turns chemical energy into internal energy, and may produce other kinds as well. As the car speeds up, it produces kinetic energy. In climbing the hill, it produces gravitational energy. When the car goes downhill, some of its gravitational energy may become kinetic. When it comes to rest, the brakes and tires convert any remaining kinetic energy into internal energy.

Chapter 6

1. a. Solid, liquid, gas.
   b. Liquid, solid.
   c. Solid, liquid, gas.
   d. Solid.
   e. Solid, liquid.
   f. Solid.
   g. Solid, liquid, gas.
   h. Solid, liquid, gas.
   i. Solid.

2. $D = m/V$, so

$$V = \frac{m}{D} = \frac{650\ \text{g}}{0.62\ \text{g/cm}^3} = 1\ 050\ \text{cm}^3$$

3. $p = \frac{F}{A} = \frac{50\ \text{N}}{3(0.2\ \text{cm}^2)} = 83\ \text{N/cm}^2$

4. $p = F/A$, so

$$F = pA = 1.01 \times 10^5\ \frac{\text{N}}{\text{m}^2} \times 2\ \text{m}^2 = 2 \times 10^5\ \text{N}$$

5. $p = hDg = (6\ \text{m})(0.9 \times 10^3\ \text{kg/m}^3)(9.8\ \text{m/s}^2) = 5.3 \times 10^4\ \text{N/m}^2$. This is the excess over the atmospheric pressure.

6. The mass of fluid displaced is the difference between the scale reading in air and in fluid, which is 10 g. The volume of fluid displaced is 8 cm³. Therefore

$$D = \frac{m}{V} = \frac{10\ \text{g}}{8\ \text{cm}^3} = 1.3\ \text{g/cm}^3$$

7. The mass of the ice is equal to the mass of the displaced water. The volume of ice submerged is $(2 \text{ cm})(2 \text{ cm})(1.7 \text{ cm}) = 6.8 \text{ cm}^3$, so the mass of displaced water is 6.8 g—and this is also the mass of the ice cube. The volume of the ice cube is 8 cm³. Therefore

$$D = \frac{m}{V} = \frac{6.8 \text{ g}}{8 \text{ cm}^3} = 0.85 \text{ g/cm}^3$$

8. Try putting soap in the water.
9. a. The parachute presents a broad surface to the air, and causes the air to flow around it turbulently, thus increasing the drag.
   b. A fish is streamlined so that the flow of water over its body is laminar.
10. In a swollen part of an artery, blood flow slows down, so the pressure increases. The artery could burst.
11. The extra drag slows it down, and the extra lift allows it to remain airborne at a slower speed.

Chapter 7

1. When the level of mercury in the thermometer stops changing, the thermometer is at the same temperature as your mouth.
2. It is silvered to prevent transfer of heat in or out by radiation. It is evacuated so that no convective transfer can take place.
3. The amount of column length above the zero mark is proportional to the Celsius temperature, so

$$\frac{11.5 \text{ cm}}{15.0 \text{ cm}} = \frac{T}{100°C}; \quad T = 76.7°C$$

4. When the pressure drops to half, the Kelvin temperature must also. The original temperature is $273 \text{ K} + 40°C = 313 \text{ K}$, so the new temperature is 157 K, or $-116°C$.
5. $\Delta H = cm \, \Delta T = (0.90 \text{ J/g·K})(250 \text{ g})(65 \text{ K}) = 15\,000 \text{ J}$.
6. The heat lost by the pot must equal the heat gained by the water. Both the water and the pot must finish at 75°C:

$$(cm \, \Delta T)_{pot} = (cm \, \Delta T)_{water}$$
$$(0.48 \text{ J/g·K})(350 \text{ g})(120 \text{ K}) = (4.19 \text{ J/g·K})(m)(55K)$$

so

$$m_{water} = 87 \text{ g}$$

7. The heat lost by the ball is equal to the heat that melts the ice. Since the cake of ice is very large, the temperature of the ball will continue to drop until it reaches 0°C:

$$(cm\ \Delta T)_{iron} = (mL)_{ice}$$
$$(0.48\ \text{J/g·K})(600\ \text{g})(240\ \text{K}) = (m)(335\ \text{J/g})$$

so

$$m_{ice} = 210\ \text{g}$$

8. The ratio pV/T does not change, since no gas has been added or removed. The initial temperature is $20°C + 273\ \text{K} = 293\ \text{K}$, and the final temperature is $100°C + 273\ \text{K} = 373\ \text{K}$. Therefore

$$\frac{(1\ \text{atm})(300\ \text{cm3})}{293\ \text{K}} = \frac{p(100\ \text{cm3})}{373\ \text{K}}; \quad p = 3.8\ \text{atm}$$

This is the absolute pressure. The gauge reads 2.8 atm.

9. $pV = nkT$, so $p = nkT/V$. To find the answer in SI units, all quantities in the equation must be in Sl units. The temperature is $T_k = 30°C + 273\ \text{K} = 303\ \text{K}$. Since there are 10^3 liters in a cubic meter, the volume is $5.0 \times 10^{-4}\ \text{m}^3$. Then

$$p = \frac{(5.0 \times 10^{22}\ \text{mcl})(1.38 \times 10^{-23}\ \text{J/mcl} \cdot \text{K})(303\ \text{K})}{5.0 \times 10^{-4}\ \text{m}^3} = 4.2 \times 10^5\ \text{Pa}$$

10.

$$E_{kin} = \frac{3}{2}kT; \quad T = \frac{2}{3}\frac{E}{k} = \frac{2}{3}\left(\frac{3.5 \times 10^{-21}\ \text{J}}{1.38 \times 10^{-23}\ \text{J/mcl} \cdot \text{K}}\right)$$

169 K, or −104°C

11. From Equation 7.8,

$$0.25 = 1 - \frac{293\ \text{K}}{T_i}$$

So $T_1 = 390\ \text{K}$, or 118°C.

Chapter 8

1. A seconds pendulum makes a tick and a tock on each full cycle, so its period is 2 s. The length must be, from Equation 8.1,

$$l = \frac{T^2 g}{4\pi^2} = \frac{(2\ \text{s})^2 (9.8\ \text{m/s}^2)}{4\pi^2} = 0.99\ \text{m}$$

2. When the object is being held back, it is in equilibrium. Therefore, the force holding it back is equal to the force that will accelerate it when it is released. Its acceleration, from Equation 8.2, is

$$a = \frac{4\pi^2 x}{T^2} = \frac{4\pi^2 (0.05 \text{ m})}{(1.5 \text{ s})^2} = 0.88 \text{ m/s}^2$$

Then, from Equation 4.2,

$$m = \frac{F}{a} = \frac{40 \text{ N}}{0.88 \text{ m/s}^2} = 45 \text{ kg}$$

3. If the frequency is 7.5 Hz, the period is $1/f = 0.133$ s. The phase difference is therefore $(0.05/0.133)360° = 135°$.

4. Two points 90° out of phase are a quarterwavelength apart so the wavelength is 100 cm. Then $v = f\lambda = (6.0 \text{ Hz})(1.00 \text{ m}) = 6.0 \text{ m/s}$.

5. $v = 331 \text{ m/s} + (0.6)24 = 345 \text{ m/s}$. Then

$$\lambda = \frac{v}{f} = \frac{345 \text{ m/s}}{320 \text{ Hz}} = 1.08 \text{ m}$$

6. The second overtone has three times the frequency of the fundamental, or $3 \times 264 \text{ Hz} = 792 \text{ Hz}$. The velocity is $331 + (0.6)22 = 344$ m/s. Then

$$\lambda = \frac{v}{f} = \frac{344 \text{ m/s}}{792 \text{ Hz}} = 0.43 \text{ m}$$

7. You will not hear the high frequencies as well, since they do not bend around the edges of the doorway.

8. Wavelength changes only if the source is moving.

9. If they arrive at the listener in phase, the path difference must be one wavelength, so the second speaker is 14.5 m from the listener.

10. The wavelength is

$$\frac{v}{f} = \frac{15 \text{ m/s}}{5.0 \text{ Hz}} = 3.0 \text{ m}$$

There must be a node at the attached end.
   a. The first node is a halfwavelength from the end, or 1.5 m.
   b. The first antinode is a quarterwavelength from the end, or 0.75 m.

11. At the fundamental, the wavelength is twice the length of the string:

$$f = \frac{v}{\lambda} = \frac{280 \text{ m/s}}{0.60 \text{ m}} = 467 \text{ Hz}$$

The first overtone is twice this, or 933 Hz, and the second overtone is three times the fundamental, or 1 400 Hz.

12. The velocity is $331 + (0.6)22 = 344$ m/s. The wavelength of the fundamental is twice the length of the pipe, so

$$f = \frac{v}{\lambda} = \frac{344 \text{ m/s}}{0.44 \text{ m}} = 782 \text{ Hz}$$

The first overtone is twice this, or 1 565 Hz, and the second overtone is three times the fundamental, or 2 347 Hz.

13. The intensity of the sound at that distance, from Equation 8.6b, is

$$I = \frac{P}{4\pi r^2} = \frac{3 \times 10^{-5} \text{ W}}{4\pi (20 \text{ m})^2} = 6.0 \times 10^{-9} \text{ W/m}^2$$

Then, from Equation 8.6a,

$$\beta = 10 \log \left(\frac{I}{I_0}\right) = 10 \log \left(\frac{6.0 \times 10^{-9} \text{ W/m}^2}{1 \times 10^{-12} \text{ W/m}^2}\right) = 38 \text{ dB}$$

14. From a node at the handle to an antinode at the tip is a quarter wavelength, so the whole wavelength is 72 cm. Overtones are at $\frac{3}{4}\lambda$ and $\frac{5}{4}\lambda$. For the fundamental,

$$f = \frac{v}{\lambda} = \frac{420 \text{ m/s}}{0.72 \text{ m}} = 583 \text{ Hz}$$

and the overtones are $3 \times 583 = 1\,750$ Hz and $5 \times 583 = 2\,917$ Hz.

Chapter 9

1. It must be the same as the charge on the rubber, since it repels an object charged from the rubber rod.
2. Steel is a conductor, and the charges will repel each other and spread throughout the rod.
3. All the charge is outside, on the surface.
4. The charges will be equal. Use Coulomb's law: $F = kq_1q_2/r^2$. With $q_1 = q_2 = q$, this becomes

$$q = \sqrt{\frac{Fr^2}{k}} = \sqrt{\frac{(6.0 \times 10^{-5} \text{ N})(0.15 \text{ m})^2}{9.0 \times 10^9 \text{ N} \cdot \text{m}^2/\text{C}^2}} = 1.2 \times 10^{-8} \text{ C}$$

5. With three times the charge, the force will be three times as great, or 6×10^{-5} N. A negative charge is pushed the opposite way, to the east.

6. On a sphere, the curvature is the same everywhere, so the charges are uniformly spaced. The field lines are radial because they must be perpendicular to the surface. They point inward because they terminate on negative charges.

7. The field may induce a separation of the charges, but there will be just as much positive charge on one end as negative on the other. Thus, in a uniform field, the two forces on the object will be equal in magnitude and opposite in direction. Unless they are in line, there will be a torque.

8. Only motion in the direction of the field counts. So the work done in moving the charge is $F \, \Delta s = (6.5 \times 10^{-5} \, \text{N})(0.05 \, \text{m}) = 3.3 \times 10^{-6} \, \text{J}$.

9. From Equation 9.3,

$$q = \frac{E}{\Delta V} = \frac{850 \, \text{J}}{12 \, \text{V}} = 71 \, \text{C}$$

10. The number of electrons will be the total charge divided by the charge on each electron:

$$\frac{8.0 \times 10^{-19} \, \text{C}}{1.60 \times 10^{-19} \, \text{C}} = 5$$

11. First, find the total charge, from Equation 9.5:

$$q = \frac{Vr}{k} = \frac{(2\,500 \, \text{V})(0.15 \, \text{m})}{\left(9.0 \times 10^9 \, \text{N} \cdot \text{m}^2/\text{C}^2\right)} = 42 \, \text{nC}$$

Then divide the total charge by the surface area of the sphere:

$$\text{charge density} = \frac{42 \, \text{nC}}{4\pi(0.15 \, \text{m})^2} = 150 \, \text{nC/m}^2$$

Chapter 10

1. Each cell provides 1.5 V, so the number of cells is 22.5 V/1.5 V = 15.
2. From Equation 10.1, $\Delta q = I \, \Delta t = (50 \, \text{A})(300 \, \text{S}) = 15\,000 \, \text{C}$.
3. From Equation 10.2, $\Delta V = IR = (0.15 \, \text{A})(20 \, \Omega) = 3.0 \, \text{V}$. The current is therefore flowing to a point 3.0 V lower than 45 V, or 42 V.
4. The resistance of the wire is 30 V/16 A = 1.88 Ω. Then

$$\rho = \frac{RA}{l} = \frac{(1.88\Omega)(0.025 \, \text{cm}^2)}{1\,000 \, \text{cm}} = 4.7 \times 10^{-5} \, \Omega \cdot \text{cm}$$

5. From Equation 10.4a,

$$I = \frac{P}{\Delta V} = \frac{600 \, \text{W}}{120 \, \text{V}} = 5.0 \, \text{A}$$

6. It could save so much electricity that the need for imported oil would be greatly reduced.

7. The total resistance is $1.0 \, \Omega + 0.20 \, \Omega = 1.2 \, \Omega$, so the current is, from Equation 10.2,

$$I = \frac{\Delta V}{R} = \frac{3.0 \text{ V}}{1.2 \, \Omega} = 2.5 \text{ A}$$

8. From Equation 10.2,

$$R = \frac{\Delta V}{I} = \frac{4.5 \text{ V}}{0.12 \text{ A}} = 38 \, \Omega$$

9. I_1 and I_2 (the currents in R_1 and R_2) must both be 12 A, since there is no branching between them and A_2. Of the 20 A flowing out of the battery, 12 A go into R_1, so the other 8 A must be I_4. Of the 12 A in R_2, 9 A go into R_5 and R_6, so the other 3 A must be in R_3. I_4 ($= 8$ A) and R_3 ($=3$ A) both flow into R_7, so $I_7 = 11$ A.

10. The potential differences through any path out of the positive end of the battery and back into the negative end must total 50 V. Thus, $\Delta V_6 + 20 \text{ V} = 50 \text{ V}$ and $\Delta V_6 = 30 \text{ V}$. Also, $20 \text{ V} + \Delta V_3 + 4 \text{ V} + 18 \text{ V} = 50 \text{ V}$, so $\Delta V_3 = 8 \text{ V}$. And $\Delta V_7 + 5 \text{ V} + 4 \text{ V} + 18 \text{ V} = 50 \text{ V}$, so $\Delta V_7 = 23 \text{ V}$.

11. a. The total resistance is $20 \, \Omega + 40 \, \Omega + 60 \, \Omega = 120 \, \Omega$. Then

$$I = \frac{\Delta V}{R} = \frac{24 \text{ V}}{120} = 0.20 \text{ A}$$

   b. $\Delta V = IR = (0.20 \text{ A})(20 \, \Omega) = 4 \text{ V}$.
   c. $P = I^2 R = (0.20 \text{ A})^2 (60 \, \Omega) = 2.4 \text{ W}$.

12. a. $\dfrac{1}{R} = \dfrac{1}{20 \, \Omega} + \dfrac{1}{40 \, \Omega} + \dfrac{1}{60 \, \Omega} = \dfrac{6 + 3 + 2}{120 \, \Omega}$

   so $R = 120 \, \Omega / 11 = 11 \, \Omega$.

   b. $I = \dfrac{\Delta V}{R} = \dfrac{24 \text{ V}}{11 \, \Omega} = 2.2 \text{ A}$

   c. The current in the 40-Ω resistance is

$$\frac{\Delta V}{R} = \frac{24 \text{ V}}{40 \, \Omega} = 0.60 \text{ A}$$

   Then $P = I \, \Delta V = (0.60 \text{ A})(24 \text{ V}) = 14 \text{ W}$.

13.   First, find the equivalent resistance of each of the parallel combinations:

$$\frac{1}{R} = \frac{1}{40\ \Omega} + \frac{1}{60\ \Omega} = \frac{3+2}{120\ \Omega};\quad R = 24\ \Omega$$

$$\frac{1}{R} = \frac{1}{10\ \Omega} + \frac{1}{20\ \Omega} = \frac{2+1}{20\ \Omega};\quad R = 6.7\ \Omega$$

Since both parallel combinations are in series with the fifth resistor, their resistances can just be added: $24\ \Omega + 12\ \Omega + 6.7\ \Omega = 43\ \Omega$.

Chapter 11

1.   Since it attracts an N-pole, it must be an S-pole.
2.   Since the current is perpendicular to the field, the force is $IlB =$ $(10\ \text{A})(0.15\ \text{m})(2.0 \times 10^{-3}\ \text{T}) = 3 \times 10^{-3}\ \text{N}$. Based on the right-hand rule, its direction is east.
3. $\tau = NBIA \sin \theta = (1)(0.030\ \text{T})(5\ \text{A})(0.02\ \text{m}^2) = 0.003\ \text{N·m}$
     Based on the righthand rule, it is directed northward.
4.   Use a finer wire; add additional layers of wire; increase the current. All three methods call for a larger potential difference. In the first two cases, this is needed in order to keep the current constant in spite of the increased resistance produced by the longer wire.
5.   From Equation 11.5,

$$I = \frac{B}{4\pi k'(n/l)}$$

If the diameter of the wire is 0.2 mm, there are 5 wires per millimeter, or 5 000 wires per meter, in each layer, for a total value $n/l = 20\ 000/\text{m}$. Then

$$I = \frac{1.5 \times 10^{-3}\ \text{T}}{4\pi (10^{-7}\ \text{N/A}^2)(20\ 000/\text{m})} = 0.060\ \text{A}$$

6.   See Figure A5.1.

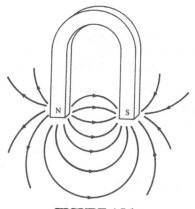

FIGURE A5.1

7. The dip of 76° means that the field vector points 76° from the horizontal. So, from Equation 2.2, $B_{hor} = (5.3 \times 10^{-5}\ \text{T})(\cos 76°) = 1.3 \times 10^{-5}\ \text{T}$.

8. From Equation 11.6,

$$B = \frac{F}{qv} = \frac{2.0 \times 10^{-12}\ \text{N}}{(1.60 \times 10^{-19}\ \text{C})(2.5 \times 10^8\ \text{m/s})} = 0.050\ \text{Ts}$$

9. a. The kinetic energy it gains equals the potential energy it lost; from Equation 9.3, $\Delta E = q\ \Delta V = (1.60 \times 10^{-19}\ \text{C})(75\ \text{V}) = 1.20 \times 10^{-17}\ \text{J}$.

   b. $mv = Bqr = (3.7 \times 10^{-3}\ \text{T})(1.60 \times 10^{-19}\ \text{C}) \times (0.30\ \text{m}) = 1.78 \times 10^{-22}\ \text{kg·m/s}$.

   c. $E_{kin} = \frac{1}{2}mv^2 = \frac{1}{2}(mv)^2/m$; so

   $$m = \frac{(mv)^2}{2E_{kin}} = \frac{(1.78 \times 10^{-22}\ \text{kg · m/s})^2}{2(1.20 \times 10^{-17}\ \text{J})} = 1.3 \times 10^{-27}\ \text{kg}$$

10. $\Delta V = Blv = (50 \times 10^{-3}\ \text{T})(0.15\ \text{m})(2.5\ \text{m/s}) = 0.0188\ \text{V}$. The current is

$$\frac{\Delta V}{R} = \frac{0.0188\ \text{V}}{10\Omega} = 1.9 \times 10^{-3}\ \text{A}$$

The right-hand rule shows that the force on positive charge in the rod is downward, and this is the way the current travels.

11. The maximum flux passing through the loop is $\phi = BA = (50 \times 10^{-3}\ \text{T})(150 \times 10^{-4}\ \text{m}^2) = 7.5 \times 10^{-4}\ \text{Wb}$. Since this changes to 0 in 0.05 s, the average induced emf is

$$\frac{7.5 \times 10^{-4}\ \text{Wb}}{0.05\ \text{s}} = 0.015\ \text{V}$$

for each turn, for a total of 0.75 V.

12. Either connect it to the other terminals of the battery or rewind it in the opposite direction.

Chapter 12

1. You would need a device that produces a current proportional to velocity. There is one in the speedometer of your car.

2. If it operates at 78% efficiency, 22% of the power input is converted to heat:

$$(0.22)(2.0\ \text{A})(240\ \text{V}) = 106\ \text{J/s}$$

3. As the loop rotates, the direction in which the magnetic flux passes through it changes every half-cycle.

4. a. Maximum current is $I_{eff}\sqrt{2} = (3.0 \text{ A})(\sqrt{2}) = 4.2 \text{ A}$
    b. $P = I_{eff}^{2}R = (3.0 \text{ A})^2(20 \text{ } \Omega) = 180 \text{ W}.$

5. a. $\dfrac{\Delta V_P}{\Delta V_s} = \dfrac{n_P}{n_s} = \dfrac{120 \text{ V}}{15 \text{ V}} = 8$

    b. $I = P/\Delta V$. In the primary, $I = 3 \text{ W}/120 \text{ V} = 25 \text{ mA}$. In the secondary, $I = 3 \text{ W}/15 \text{ V} = 0.20 \text{ A}.$

6. The current entering the school is

    $$\frac{P}{\Delta V} = \frac{140 \times 10^3 \text{ W}}{120 \text{ V}} = 1.17 \times 10^3 \text{ A}$$

    This must pass through lines with a total resistance of 0.004 Ω, so the power lost in the lines is $I^2R = (1.17 \times 10^3 \text{ A})^2(0.004 \text{ } \Omega) = 5.5 \text{ kW}.$

7. #8.

8. The grounding wire is to protect against current accidentally delivered to the case. A plastic case is an insulator and cannot carry current.

9. There are many—20 or more, probably.

Chapter 13

1. This is straightforward Ohm's Law:

    $$\Delta V = IR = (2.5 \times 10^{-3} \text{ A})(16 \text{ } 000 \text{ } \Omega) = 40 \text{ V}$$

2. The rate at which the resistor dissipates heat is

    $$P = I^2R = (20 \times 10^{-3} \text{ A})^2(2 \text{ } 000 \text{ } \Omega) = 0.8 \text{ W}$$

    A 1-watt resistor will do. In a resistor, the frequency does not matter.

3. From Equation 13.1,

    $$q = C\Delta V = (30 \times 10^{-12} \text{ F})(80 \text{ V}) = 2.4 \text{ nC}$$

4. The reactance of a capacitor is inversely proportional to the frequency. At four times the frequency, the reactance is one-fourth as great, and the current is four times as great, 160 mA.

5. At the instant the switch is closed, the entire 24 volts is applied to the inductance. At that instant, the current is increasing at a rate of 20 milliamperes per microsecond, or 20 000 amperes per second. Equation 13.3 gives an estimate of the inductance that results in this situation:

    $$L = \frac{\Delta V}{\Delta I / \Delta t} = \frac{24 \text{ V}}{20 \times 10^3 \text{ A/s}} = 1.2 \text{ mH}$$

    Of course, this rate of current increase does not continue.

6. We need the current; to find it, we need the reactance:

$$X_L = 2\pi f L = 2\pi(600 \text{ Hz})(3.0 \text{ H}) = 11\ 300\ \Omega$$

This is essentially the impedance, since the 120 ohms of resistance will have negligible effect. Then the current is

$$I = \frac{\Delta V}{Z} = \frac{40 \text{ V}}{11\ 300\ \Omega} = 3.54 \text{ mA}$$

Then the power loss is

$$P = I^2 R = (3.54 \times 10^{-3} \text{ A})^2(120\ \Omega) = 1.5 \times 10^{-3}\text{W}$$

7. From Equation 13.7,

$$C = \frac{1}{4\pi^2 f^2 L} = \frac{1}{4\pi^2 \left(1.56 \times 10^6 \text{ Hz}\right)^2 \left(5.0 \times 10^{-3} \text{ H}\right)} = 2.1 \text{ pF}$$

8. The speed of the AC wave is essentially the speed of light; from Equation 8.4,

$$f = \frac{v}{\lambda} = \frac{3.0 \times 10^8 \text{ m/s}}{0.10 \text{ m}} = 3.0 \times 10^9 \text{ Hz}$$

9. With a very high potential, electrons can be driven across the base into the n-type semiconductor, against the opposing field set up by the excess electrons.

10. Both batteries would have to be reversed.

Chapter 14

1.
$$f = \frac{c}{\lambda} = \frac{3.00 \times 10^8 \text{ m/s}}{5.5 \times 10^{-7} \text{ m}} = 5.4 \times 10^{14} \text{ Hz}$$

2. The wavelength is

$$\lambda = \frac{c}{f} = \frac{3.00 \times 10^8 \text{ m/s}}{4\ 500 \times 10^6 \text{ Hz}} = 6.67 \times 10^{-2} \text{ m}$$

The antenna should be a halfwavelength long, or 3.3 cm.

3. Longer waves diffract around the object; since the object will cast no shadow, it cannot be seen.

4.
$$\lambda = d\frac{x}{L} = \frac{\text{mm}}{640}\left(\frac{14.3 \text{ cm}}{38 \text{ cm}}\right) = 5.9 \times 10^{-4} \text{ mm, or 590 nm}$$

5.
$$f = \frac{c}{\lambda} = \frac{3.00 \times 10^8 \text{ m/s}}{6.9 \times 10^{-7} \text{ m}} = 4.3 \times 10^{14} \text{ Hz}$$

6. Observe it through a polarizing filter. Rotate the filter until the light is blocked out; its axis is then perpendicular to the direction of polarization of the light.

7. From Figure 14.15, the magenta filter seems to cut out everything from about 470 nm to 600 nm.

8. The color will be missing if the film is a halfwavelength thick. Then the missing wavelength is 480 nm. In air, this wave would have a wavelength of

$$\lambda = (480 \text{ mm}) \frac{3.0 \times 10^8 \text{ m/s}}{2.2 \times 10^8 \text{ m/s}} = 654 \text{ nm}$$

which is in the middle of the red.

9.
$$E = hf = \frac{hc}{\lambda} = \frac{\left(4.14 \times 10^{-15} \text{ eV·s}\right)\left(3.00 \times 10^8 \text{ m/s}\right)}{240 \times 10^{-9} \text{ m}} = 5.2 \text{ eV}$$

10. From Equation 14.4,
$$W = hf - q\,\Delta V = (4.14 \times 10^{-15} \text{ eV·s})(3.51 \times 10^{15} \text{ Hz}) - 9.2 \text{ eV} = 5.3 \text{ eV}$$

11. $E = hc/\lambda$, so
$$\lambda = \frac{hc}{E} = \frac{\left(4.14 \times 10^{-15} \text{ eV·s}\right)\left(3.00 \times 10^8 \text{ m/s}\right)}{22\,000 \text{ eV}} = 5.6 \times 10^{-11} \text{ m}$$

12. From Equations 14.2 and 8.4, the energy of a single photon is
$$E = hf = \frac{hc}{\lambda} = \frac{\left(6.6 \times 10^{-34} \text{ J·s}\right)\left(3.0 \times 10^8 \text{ m/s}\right)}{250 \times 10^{-9} \text{ m}} = 7.9 \times 10^{-19} \text{ J}$$

To deliver a microjoule, the number needed is
$$\frac{10^{-6} \text{ J}}{7.9 \times 10^{-19} \text{ J}} = 1.3 \times 10^{12}$$

Chapter 15

1. The rays from a bulb are arrayed radially outward; those from a laser are parallel.

2. The visible power produced by a 40-W bulb operating at 16% efficiency is (0.16)(40 W) = 6.4 W. This produces (6.4 W)(500 lum/W) = 3 200 lum. The illumination at 3.5 m is

$$\frac{F}{4\pi r^2} = \frac{3\,200 \text{ lum}}{4\pi(3.5 \text{ m})^2} = 21 \text{ lux}$$

3. The ray coming from your feet to the mirror and to your eye strikes the mirror at half the distance from your eye to the floor, and you do not use any of the mirror below that point. The smallest mirror is thus half your height.

4. The size of the image is always the same as the size of the object.

5. $\dfrac{\sin \theta_A}{\sin \theta_B} = \dfrac{n_B}{n_A};$ $\dfrac{\sin 55°}{\sin r} = \dfrac{2.42}{1.33};$ $r = 27°$

6. $\dfrac{\sin \theta_A}{\sin \theta_B} = \dfrac{n_B}{n_A};$ $\dfrac{\sin i_c}{\sin 90°} = \dfrac{1}{2.42};$ $i_c = 24°$

   It sparkles because the light cannot escape unless it is coming to the surface at a small angle of incidence.

7. For red:

$$\frac{\sin 65.0°}{\sin r} = \frac{1.61}{1}; \quad r = 34.3°$$

   For violet:

$$\frac{\sin 65.0°}{\sin r} = \frac{1.66}{1}; \quad r = 33.1°$$

8. In a +20 cm lens, rays entering parallel to the principal axis converge at 20 cm from the lens. In a –30 cm lens, rays entering parallel to the principal axis diverge as though coming from a point 30 cm behind the lens.

9. See Figure A5.2.

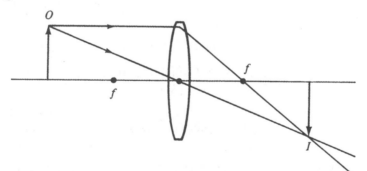

FIGURE A5.2

10. See Figure A5.3.

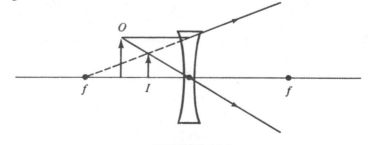

FIGURE A5.3

11.   a.   First find the object distance:

$$\frac{D_o}{D_i} = \frac{S_o}{S_i} \qquad \frac{D_o}{3.0 \text{ m}} = \frac{3.5 \text{ cm}}{50 \text{ cm}}; \qquad D_o = 0.21 \text{ m}$$

b.   Now use this value to find the focal length:

$$\frac{1}{f} = \frac{1}{D_o} + \frac{1}{D_i} = \frac{1}{0.21 \text{ m}} + \frac{1}{3.0 \text{ m}}$$

$$\frac{1}{f} = \frac{(3.0 \text{ m}) + (0.21 \text{ m})}{(0.21 \text{ m})(3.0 \text{ m})}$$

$$f = 0.20 \text{ m}$$

12.   This doubling of the diameter of the lens opening increases the area of the opening by a factor of 4, thus letting in four times as much light.

13.   The image distance is negative, so

$$\frac{1}{f} = \frac{1}{3.0 \text{ m}} + \frac{1}{-0.45 \text{ m}}$$

$$\frac{1}{f} = \frac{(3.0 \text{ m})(-0.45 \text{ m})}{-0.45 \text{ m} + 3.0 \text{ m}} = -0.52 \text{ m}$$

The radius of curvature is twice this, or 1.04 m.

Chapter 16

1.   This would increase the velocity of the electrons. The beam would curve because the magnetic force increases, but the electric force does not.

2.   This is 50 elementary charges, or $50(1.60 \times 10^{-19} \text{ C}) = 8.0 \times 10^{-18} \text{ C}$.

3.   At the highest frequency of the visible spectrum, $n_1 = 2$ and n_2 is infinity. Therefore, $E = 13.6 \text{ eV}/2^2 = 3.4 \text{ eV}$. Then

$$f = \frac{E}{h} = \frac{3.4 \text{ eV}}{4.15 \times 10^{-15} \text{ eV} \cdot \text{s}} = 8.1 \times 10^{14} \text{ Hz}$$

4.   Subtract the energy at f from the energy at c:

$$-2.68 \text{ eV} - (-5.52 \text{ eV}) = 2.84 \text{ eV}$$

$$\lambda = \frac{hc}{E} = \frac{(4.15 \times 10^{-15} \text{ eV} \cdot \text{s})(3.0 \times 10^8 \text{ m/s})}{2.84 \text{ eV}} = 4.4 \times 10^{-7} \text{ m}$$

5. The energy loss is 10.38 eV–5.52 eV = 4.86 eV. The electron was at the c level. It could drop to ground state, or to the b level and then to the ground state, so there are three possible energies: 4.86 eV, 0.22 eV, and 4.64 eV.

6. The velocity is proportional to the square root of the accelerating potential difference, so the momentum would be one-tenth as great. Wavelength is inversely proportional to momentum, so it would be 10 times as much.

7. The wavelength is extremely small—smaller than a proton.

8. An electron beam is a wave in the experiment in which diffraction effects are seen when the beam is reflected from a crystal. It is a particle when its mass and charge are being measured, as in Thomson's experiment or Millikan's oil drop.

Chapter 17

1. 26 protons and 28 neutrons.

2. It bends into a circular path because of the centripetal force exerted by the magnetic field. It spirals inward because it is constantly giving up energy in collisions with hydrogen molecules.

3.
$$^{214}_{84}\text{Po} \rightarrow {}^{4}_{2}\text{He} + {}^{210}_{82}\text{Pb}$$

$$^{210}_{82}\text{Pb} \rightarrow {}^{210}_{83}\text{Bi} + {}^{0}_{-1}\text{e} + {}^{0}_{0}$$

4. From Appendix 4 the half-life of polonium 214 is 1.6×10^{-4} s. The number of half-lives in 0.01 s is thus $0.01/(1.6 \times 10^{-4})$ = about 6. Since the amount of polonium drops to half six times, the amount remaining is $(1 \text{ g}) \left(\dfrac{1}{2}\right)^6 =$ 0.016 g.

5.
Mass of protons:	$2 \times 1.007825 = 2.01565$
Mass of neutrons:	$2 \times 1.008665 = \underline{2.01733}$
Mass of separate particles	4.03298
Mass of the helium nucleus	4.00260 Dl
Mass deficit of helium-4	0.03038 Dl

The binding energy is found by multiplying the mass deficit by 931 MeV/Dl, which gives 28 MeV.

6. $\left(500 \times 10^9 \text{ eV}\right)\left(\dfrac{\text{D1}}{931 \times 10^6 \text{ eV}}\right) = 540 \text{ D1}$

compared with about 1 Dl when it is at rest.

7. Alpha particles have 4 times the mass and twice the charge of protons, so the ratio m/q will double. To maintain synchronism, either the period or the magnetic field (or at any rate their product) will have to double.

8.

$$_2^4He + _{19}^{39}K \rightarrow (_{21}^{43}Sc) \rightarrow _1^1H + _{20}^{42}Ca$$

9. The energy released is $(25 \times 10^3 \text{ tons})(4 \times 10^9 \text{ J/ton}) = 1.0 \times 10^{14} \text{ J}$. The corresponding mass is

$$\frac{E}{c^2} = \frac{1.0 \times 10^{14} \text{ J}}{\left(3.00 \times 10^8 \text{ m/s}\right)^2} = 1.1 \times 10^{-3} \text{ kg}$$

10.

$$_{90}^{232}Th + _0^1n \rightarrow _{90}^{233}Th$$

$$_{90}^{233}Th \rightarrow _{91}^{233}Pa + _{-1}^0e + _0^0v$$

$$_{91}^{233}Pa \rightarrow _{92}^{233}U + _{-1}^0e + _0^0v$$

11. Mass of lithium-6   6.01512 Dl
     Mass of deuterium   2.0140
     Original total mass   8.02912
     Mass of two helium nuclei: $2 \times 4.00260 = $ 8.00520
     Additional mass deficit   0.02392 Dl

     Energy released per fusion reaction is $(0.02392 \text{ Dl})(931 \text{ MeV/Dl}) = 22 \text{ MeV}$.

12. The photon must have enough energy to create 2 electrons, one negative and the other positive. The energy needed is $2 \times 0.511 \text{ MeV} = 1.02 \text{ MeV}$. The necessary wavelength is

$$\lambda = \frac{hc}{E} = \frac{\left(4.14 \times 10^{-15} \text{ eV} \cdot \text{s}\right)\left(3.0 \times 10^8 \text{ m/s}\right)}{1.02 \times 10^6 \text{ eV}} = 1.22 \times 10^{-12} \text{ m}$$

14. Neutrons, protons, and electrons each have spin $+\frac{1}{2}$. To balance the spin equation, something with a spin $-\frac{1}{2}$ must come out of it. That's an antineutrino.

15. A neutron is a baryon and needs baryon number 1, which can only be supplied by 3 quarks. An up quark with charge $+\frac{2}{3}$ and two down quarks, each with $-\frac{1}{3}$ will supply the necessary zero charge, so the formula is *udd*. If one of the quarks is reversed, the spin is $+\frac{1}{2}$ as required.

Appendix 6
Answers to Chapter Review Exercises

Chapter 1

FILL IN'S

1. measurement
2. kilograms
3. percent
4. accuracy
5. beginning
6. decimal point
7. addition; subtraction
8. 10
9. scientific notation
10. add
11. dimensionality
12. multiplied; divided
13. liters or cubic meters
14. multiplication; division
15. second

MULTIPLE CHOICE

1. (5)
2. (5)
3. (4)
4. (5)
5. (1)
6. (3)
7. (3)
8. (3)
9. (3)
10. (1)
11. (5)
12. (3)
13. (4)
14. (1)

PROBLEMS

1. $(0.2/35) \times 100\% = 0.6\%$.
2. $0.03 \times 655 \text{ kg} = 20 \text{ kg}$.
3. 3; the first zero is not significant, but the zero after a decimal point is.
4. 5 100; adding $231 + 45.7 + 6.195 = 282.8$; the last significant digit is in the units place. Since 0.056 has only two digits, the answer must be rounded off to two digits.
5. 4; since there are no digits beyond the units place in 122, there can be none in the answer.
6. 5.5×10^{-3}; two significant digits,
7. 1.92×10^{-3}.

8. 5.58×10^4; before adding, convert 8.02×10^3 to 0.802×10^4.
9. 75 m^3.
10. 0.275 kg/s.
11. $3.1 \times 10^2 \text{ m/s}^2$.
12. $1.8 \times 10^{-14} \text{ m}^3/\text{kg}$.

Chapter 2

FILL IN'S

1. average speed
2. meter per second
3. scalar
4. vector
5. 4; 8
6. cosine
7. acceleration
8. meter per second squared
9. square
10. acceleration
11. south
12. west
13. vertical
14. air resistance, friction
15. centripetal
16. earth
17. 2π
18. the radius of the wheel
19. the speed of light
20. shorter

MULTIPLE CHOICE

1. (2)   3. (3)   5. (5)   7. (1)   9. (4)
2. (2)   4. (3)   6. (2)   8. (2)   10. (1)

PROBLEMS

1. Since $v_{av} = \Delta s/\Delta t$, $\Delta s = v_{av}\,\Delta t = (4.5 \text{ m/s})(600 \text{ s}) = 2\,700 \text{ m}$.
2. The time in SI units:

$$(1 \text{ hr})\left(\frac{60 \text{ min}}{\text{hr}}\right)\left(\frac{60 \text{ s}}{\text{min}}\right) = 3\,600 \text{ s}$$

Then

$$v_{av} = \frac{4.6 \times 10^4 \text{ m}}{3\,600 \text{ s}} = 12.8 \text{ m/s}$$

3. For two velocities at right angles to each other, the vector sum is the hypotenuse of a triangle. Therefore $v = \sqrt{3.6^2 + 12.0^2} = 12.5 \text{ m/s}$. The direction will be south of east at an angle whose tangent is 3.6/12.0, or 17°.

4. The plane must go at an angle to the north such that the northward component of its velocity is 55 mi/hr, just enough to compensate for the wind. That angle will be the angle whose sine is 55/230, which is 13.8°.

5. Your displacement has two components: 4.5 mi east and $(6.2 - 2.0) = 4.2$ mi south. The magnitude of this displacement is the vector sum of these components, $\sqrt{4.2^2 + 4.5^2} = 6.2$ mi. The direction is south of east at arctangent $4.2/4.5 = 43°$.

6. This direction is 60° north of west, so the westward component is $(45 \text{ mi})(\cos 60°) = 23$ mi.

7. The 35 mi is the northward component of his displacement; $35 \text{ mi} = \Delta s \cos 55°$, so his displacement is $(35 \text{ mi})/\cos 55° = 61$ mi. Then

$$v_{av} = \frac{\Delta s}{\Delta t} = \frac{61 \text{ mi}}{6.0 \text{ hr}} = 10.1 \text{ mi/hr}$$

8. $a = \Delta v/\Delta t$. The increase in speed is $(30 \text{ m/s } 12 \text{ m/s}) = 18$ m/s. Then

$$a = \frac{18 \text{ m/s}}{15 \text{ s}} = 1.2 \text{ m/s}^2$$

9. $a = \Delta v/\Delta t$, so

$$\Delta t = \frac{\Delta v}{a} = \frac{22 \text{ m/s} - 15 \text{ m/s}}{3.2 \text{ m/s}^2} = 2.2 \text{ s}$$

10. When starting from rest and accelerating uniformly, the average velocity is half the final velocity. Since $v_{av} = \Delta s/\Delta t$, $s = v_{av} \Delta t = (11 \text{ m/s})(15 \text{ s}) = 165$ m.

11. When starting from rest and accelerating uniformly, the standard formulas apply, so $v = \sqrt{2as}$. Then

$$a = \frac{v^2}{2s} = \frac{(18 \text{ m/s})^2}{2 \times 240 \text{ m}} = 0.68 \text{ m/s}^2$$

12. In free fall, the acceleration is 9.8 m/s²; there is uniform acceleration starting from rest, so $s = \frac{1}{2}at^2$ and

$$t = \sqrt{\frac{2s}{a}} = \sqrt{\frac{2 \times 24 \text{ m}}{9.8 \text{ m/s}^2}} = 2.2 \text{ s}$$

13. It comes to rest uniformly, so $v = \sqrt{2as}$ and

$$s = \frac{v^2}{2a} = \frac{(15 \text{ m/s})^2}{2 \times 9.8 \text{ m/s}^2} = 11.5 \text{ m}$$

14. There is uniform acceleration from rest, so

$$v = \sqrt{2as} = \sqrt{2(9.8 \text{ m/s}^2)(34 \text{ m})} = 26 \text{ m/s}$$

15. $v = \sqrt{2as}$, so

$$s = \frac{v^2}{2a} = \frac{(15 \text{ m})^2}{2 \times 9.8 \text{ m/s}^2} = 11.5 \text{ m}$$

16. $a = \Delta v/\Delta t$, so $\Delta v = a\,\Delta t = (1.5 \text{ m/s}^2)(2.0 \text{ s}) = 3.0 \text{ m/s}$. Upward acceleration for a descending elevator means it is slowing down, so its final speed is $4.4 - 3.0 = 1.4 \text{ m/s}$.

17. a. $v = \dfrac{2\pi r}{T} = \dfrac{2\pi(2.0 \text{ m})}{4.5 \text{ s}} = 2.8 \text{ m/s}.$

   b. $a_c = \dfrac{v^2}{r} = \dfrac{(2.79 \text{ m/s})^2}{2.0 \text{ m}} = 3.9 \text{ m/s}^2.$

18. $a_c = v^2/r$, so $v = \sqrt{ar} = \sqrt{(3.0 \text{ m/s}^2)(6.5 \text{ m})} = 4.4 \text{ m/s}.$

19. $a_c = v^2/r$, so

$$r = \frac{v^2}{a} = \frac{(20 \text{ m/s})^2}{3.5 \text{ m/s}^2} = 114 \text{ m}$$

20. $a_c = v^2/r$, so $v = \sqrt{ar} = \sqrt{(75 \text{ m/s}^2)(0.30 \text{ m})} = 4.74 \text{ m/s}.$ If a point on the basket makes a complete revolution in time T, its speed is

$$v = 2\pi r/T$$

So $T = 2\pi r/v = 2\pi(0.30 \text{ m})/(4.74 \text{ m/s}) = 0.398 \text{ s}$ and the number of complete revolutions in 60 s is 60 s/(0.398 s per revolution) = 151 rpm.

21. a. Vertical motion is free fall starting from rest; $s = \dfrac{1}{2}at^2$, so

$$t = \sqrt{\frac{2s}{a}} = \sqrt{\frac{2 \times 22 \text{ m}}{9.8 \text{ m/s}^2}} = 2.12 \text{ s}$$

   b. Horizontal motion is uniform speed, so $v = \Delta s/\Delta t$ and $\Delta s = v\,\Delta t = (18 \text{ m/s})(2.12 \text{ s}) = 38 \text{ m}.$

22. In a frame of reference attached to the first train, the velocity of the second train is $22 + 14 = 36 \text{ m/s}$. Then the vector sum of the velocity of the train and that of the airplane is

$$v = \sqrt{(36 \text{ m/s})^2 + (8 \text{ m/s})^2} = 37 \text{ m/s}$$

23. The thief runs a distance s in time t at 6.0 m/s; the policeman runs $(s + 15 \text{ m})$ in the same time, going 8.5 m/s. For both, time is distance over speed, so

$$t = \frac{s}{6.0 \text{ m/s}} = \frac{(s + 15 \text{ m})}{8.5 \text{ m/s}}$$

from which $s = 36 \text{ m}$. The policeman runs 51 m.

24. At 36 m/s, the car does the first 5 km in

$$t = \frac{s}{v} = \frac{5 \times 10^3 \text{ m}}{36 \text{ m/s}} = 139 \text{ s}$$

It must complete the last 5 000 m in 101 s by going 49 m/s.

25. Having lost 10 percent of its speed, the bullet enters the sandbag at $0.9 \times 340 = 306$ m/s. Coming from that speed to rest, its average speed must have been half of 306 m/s, or 153 m/s. Then

$$t = \frac{s}{v} = \frac{0.12 \text{ m}}{153 \text{ m/s}} = 8 \times 10^{-4} \text{ s}$$

26. Both vehicles travel the same distance. For the car, going at constant speed, the distance is vt; for the motorcycle, it is $\frac{1}{2}at^2$. Then

$$s = (24 \text{ m/s})t = \frac{1}{2}(3.2 \text{ m/s}^2)t^2$$

from which $t = 15$ s.

27. Once your foot hits the brake, your car has to get from 26 m/s to zero; from Equation 2.7c the distance it travels is

$$s = \frac{v^2}{2a} = \frac{(26 \text{ m/s})^2}{2(2.5 \text{ m/s}^2)} = 135 \text{ m}$$

You have 15 m of space in which to get your foot on the brake; the time it takes your car to go that distance is

$$t = \frac{s}{v} = \frac{15 \text{ m}}{26 \text{ m/s}} = 0.6 \text{ s}$$

which is just about enough if you are sober.

28. The time it takes the ball to get to the plate, at constant horizontal speed, is $(18.3 \text{ m})/(42.1 \text{ m/s}) = 0.435$ s. During that time, gravity makes it drop a distance of

$$s = \frac{1}{2}at^2 = \frac{1}{2}(9.8 \text{ m/s}^2)(0.435 \text{ s})^2 = 0.93 \text{ m}$$

29. The vertical component of the rock's velocity when it hits the ocean is

$$v = \sqrt{2as} = \sqrt{2(9.8 \text{ m/s}^2)(45 \text{ m})} = 29.7 \text{ m/s}$$

The horizontal component is still 18 m/s, so the angle the trajectory makes with the water is arctan $(29.7/18) = 59°$.

30. The point makes a complete revolution in one day, which is (24 hrs)(3 600 s/hr) $= 8.6 \times 10^4$ s. Then the speed of the point is $2\pi r/T =$ $2\pi(6.4 \times 10^6$ m$)/(8.6 \times 10^4$ s$) = 468$ m/s. Its centripetal acceleration is

$$a_c = \frac{v^2}{r} = \frac{(468 \text{ m/s})^2}{6.4 \times 10^6 \text{ m}} = 0.034 \text{ m/s}^2$$

31. To find its speed when its centripetal acceleration is 1.2 meters per second squared, Equation 2.5b converts to

$$v = \sqrt{ar} = \sqrt{1.2 \text{ m/s}^2 \times \frac{1200 \text{ m}}{2\pi}} = 15.2 \text{ m/s}$$

To find how long it takes for the car to get up to this speed, use Equation 2.3; $v_i = 0$, so

$$t = \frac{v}{a} = \frac{15.1 \text{ m/s}}{1.2 \text{ m/s}^2} = 13 \text{ s}$$

Chapter 3

FILL IN'S

1. velocity
2. opposite
3. equilibrium
4. newton
5. pound
6. viscous drag
7. weight
8. mass
9. weight
10. newton
11. elastic recoil
12. tension
13. buoyancy
14. 30
15. elastic recoil
16. 25 pounds
17. cosine
18. 200 newtons

MULTIPLE CHOICE

1. (5)
2. (3)
3. (2)
4. (1)
5. (4)
6. (3)
7. (5)
8. (4)
9. (3)
10. (2)

PROBLEMS

1. $w = mg$, so

$$m = \frac{w}{g} = \frac{75 \text{ N}}{9.8 \text{ m/s}^2} = 7.7 \text{ kg}$$

2. $w = mg = (95 \text{ kg})(1.67 \text{ m/s}^2) = 159 \text{ N}$.

3. Downward forces (weight and viscous drag) must equal upward forces (buoyancy and tension). Weight is $mg = (35 \text{ kg})(9.8 \text{ m/s}^2) = 343 \text{ N}$. Therefore $343 \text{ N} + 25 \text{ N} = 50 \text{ N} + T$, and $T = 320 \text{ N}$.

4. Force forward (engine thrust) = force backward (friction + viscous drag). Therefore, $31\,000 \text{ N} = 22\,000 \text{ N} + D$, and $D = 9\,000 \text{ N}$.

5. The horizontal component of the tension in the rope is $T \cos 25°$. Forward force = retarding force, so $85 \text{ N} = T \cos 25°$, and $T = 94 \text{ N}$.

6. a. Since the whole system is symmetrical, each half of the rope must be supporting half the weight, or $\frac{1}{2}(32 \text{ kg})(9.8 \text{ m/s}^2) = 157 \text{ N}$.

   b. The rope makes an angle of 50° with the vertical, so the vertical component of the tension in the rope is $T \cos 50°$ (or $T \sin 40°$). The vertical component of one side of the rope supports half the weight of the knapsack, so $157 \text{ N} = T \sin 40°$ and $T = 240 \text{ N}$.

7. The vertical component of the tension is $T \cos 25°$, and each rope supports half the weight. Therefore $T \cos 25° = \frac{1}{2}(35 \text{ kg})(9.8 \text{ m/s}^2)$, and $T = 190 \text{ N}$.

8. The angle the rope makes with the vertical is 40°, so the vertical component of the tension is $T \cos 40°$. Since this is the only upward force, it must be equal to the weight; $w = (350 \text{ N})(\cos 40°) = 270 \text{ N}$.

9. The angle the weight vector makes with the perpendicular to the surface is 20°, so the force the books exert against the plank is $mg \cos 20° = (20 \text{ kg})(9.8 \text{ m/s}^2)(\cos 20°) = 180 \text{ N}$.

10. The angle the weight vector makes with the surface is $90° - 35°$, so the component of weight acting parallel to the slope is $w \sin 35°$. This is the force needed to push the (frictionless) wagon up the slope, so $40 \text{ lb} = w \sin 35°$, and $w = 70 \text{ lb}$.

11. a. The component of weight parallel to the ramp is $mg \sin \theta = (7.0 \text{ kg})(9.8 \text{ m/s}^2)(\sin 15°) = 18 \text{ N}$.

    b. The component of weight normal to the surface is $mg \cos \theta = (7.0 \text{ kg})(9.8 \text{ m/s}^2)(\cos 15°) = 66 \text{ N}$.

12. The force needed is enough to balance the friction, and the normal force is the weight of the crate:

$$F_{\text{fr}} = \mu m g = (0.18)(62 \text{ kg})(9.8 \text{ m/s}^2) = 110 \text{ N}$$

13. George's and Janet's forces add up to $\sqrt{(85^2 + 62^2)} = 105$ N. Henry must pull this hard in the opposite direction.

14. Presumably, the floor is horizontal, so the weight of the table is a normal force. Then

$$\mu = F_f/FN = 170 \text{ N}/(45 \text{ kg} \times 9.8 \text{ N/kg}) = 0.39$$

15. a. The nail supports the whole weight of the picture,

$$(18 \text{ kg})(9.8 \text{ N/kg}) = 176 \text{ N}$$

   b. Since the situation is symmetrical, half the weight of the picture (88 N) is supported by each half of the wire; this is the vertical component of the tension in the wire. The angle the wire makes with the vertical is arcsin $(22.5/31) = 46.5°$. Therefore $(88 \text{ N})/T = \cos 46.5°$, and $T = 128$ N.
   Incidentally, it is not a good idea to suspend a picture, particularly a heavy one, from a nail. Get a picture hook.

16. Equation 3.1 tells us that the gravitational field, equal to the acceleration due to gravity, is $w/m = (240 \text{ N})/(65 \text{ kg}) = 3.7 \text{ m/s}^2$. Then, from Equation 2.4c,

$$v = \sqrt{2as} = \sqrt{2(3.7 \text{ m/s}^2)(8.5 \text{ m})} = 7.9 \text{ m/s}$$

17. a. $\tau = Fr = (250 \text{ N})(1.5 \text{ m})$, measuring r from the pivot. Then $Fr = 375$ N·m.
   b. The torque applied by the rock must be the same, so

$$F = \frac{\tau}{r} = \frac{375 \text{ N·m}}{0.5 \text{ m}} = 750 \text{ N}$$

18. When the left end is taken as the pivot, the torque applied by the weight of the painter is $Fr = (600 \text{ N})(1.0 \text{ m}) = 600$ N·m, clockwise. The other rope exerts a counterclockwise torque of $T_R(2.5 \text{ m})$. Since there are no other torques, these must be equal, so 600 N·m $= T_R(2.5 \text{ m})$, and $T_R = 240$ N. The upward forces must equal the weight of the painter, so $T_R + T_L = 600$ N, and $T_L = 360$ N.

19. The clockwise torque exerted by the rope must equal the counterclockwise torque at the handle, so $(1\,200 \text{ N})(1.5 \text{ cm}) = F(45 \text{ cm})$, and $F = 40$ N.

20. The component of weight acting down the slope is $(450 \text{ N})(\sin 25°) = 190$ N, and this must be equal to the sum of the uphill components. Two forces pull uphill, friction and the tension in the rope. Therefore, 190 N $= 75 \text{ N} + T$, and $T = 115$ N.

21. Clockwise torque exerted by the girl around the rock must be equal to the counterclockwise torque exerted by the rope. Thus

$$(55 \text{ kg}) \times (9.8 \text{ m/s}^2)r = (350 \text{ N})(4.0 \text{ m}), \text{ so } r = 2.6 \text{ m}.$$

22. a. Force (a) is the component of weight parallel to the plane:

$$F_{par} = mg \sin \theta = (45 \text{ kg})(9.8 \text{ m/s}^2)(\sin 38°) = 272 \text{ N}$$

  b. Force (b) is the component of weight perpendicular to the plane:

$$F_{per} = mg \cos \theta = (45 \text{ kg})(9.8 \text{ m/s}^2)(\cos 38°) = 348 \text{ N}$$

  c. $F_{fr} = \mu F_{nor} = 0.26 \times 348 \text{ N} = 90 \text{ N}.$

  d. The friction must be added to the parallel component of weight:
     272 N + 90 N = 360 N (rounded off).

23. Since segment A is horizontal, it cannot be supporting any part of the weight. The vertical component of F_B is the full weight of the goodies, (28 kg)(9.8 N/kg) = 274 N. This tension is directed along the rope, at an angle of arcsin (3.0 m/11.5 m) = 15.1° with the horizontal. Then the tension in segment B is (274 N)/(sin 15.1°) = 1 050 N. Since the system is in equilibriuim horizontally, the tension in segment A must be equal to the horizontal component of the tension in segment B, which is (1 050 N)(cos 15.1°) = 1 020 N.

24. The line makes an angle of arcsin (5.6 m/17.0 m) = 19.2°. The net downward force on the instrument is 85 N – 22 N = 63 N. Viscous drag is the only horizontal force, so (63 N)/T = tan 19.2°, and T = 180 N.

Chapter 4

FILL IN'S

1. time
2. impulse
3. mass
4. newton
5. net unbalanced
6. velocity
7. impulse
8. momentum
9. perpendicular
10. centripetal
11. gravity
12. masses
13. distance between them
14. mass, radius
15. free fall
16. velocity
17. altitude
18. gravity
19. angular momentum
20. 9
21. northwest

MULTIPLE CHOICE

1. (2)   4. (2)   6. (3)   8. (2)   10. (5)
2. (5)   5. (1)   7. (1)   9. (1)   11. (2)
3. (5)

PROBLEMS

1. Impulse $= F\,\Delta t$,

$$\Delta t = \frac{10^6\ \text{N}\cdot\text{s}}{45\ 000\ \text{N}} = 22\ \text{s}$$

2. The same impulse is needed in both cases, so $F_1\,\Delta t_1 = F_2\,\Delta t_2$ and

$$F_2 = F_1\left(\frac{\Delta t_1}{\Delta t_2}\right) = (3\ 500\ \text{N})\left(\frac{5.0\ \text{s}}{2.0\ \text{s}}\right) = 8\ 750\ \text{N}$$

3. Acceleration is

$$\frac{\Delta v}{\Delta t} = \frac{6.5\ \text{m/s} - 4.0\ \text{m/s}}{3.0\ \text{s}} = 0.83\ \text{m/s}^2$$

   Then $F = ma = (60\ \text{kg})(0.83\ \text{m/s}^2) = 50\ \text{N}$.

4. a. From Equation 2.4c

$$a = \frac{v^2}{2s} = \frac{(25\ \text{m/s})^2}{2(0.40\ \text{m})} = 780\ \text{m/s}^2$$

   b. $F = ma = (1\ 400\ \text{kg})(780\ \text{m/s}^2) = 1.1 \times 10^6\ \text{N}$.

5. $a = \dfrac{v}{t} = \dfrac{30\ \text{m/s}}{8.0\ \text{s}} = 3.75\ \text{m/s}^2$

$$F = ma = (2\ 800\ \text{kg})(3.75\ \text{m/s}^2) = 10\ 500\ \text{N}$$

6. The downward force on it is its weight $= mg = (680\ \text{kg})(1.67\ \text{m/s}^2) = 1\ 136\ \text{N}$.
   If F is the thrust of the engines, the net upward force will be $F - 1\ 136\ \text{N} = ma$;
   then $F = (680\ \text{kg})(2.0\ \text{m/s}^2) + 1\ 136\ \text{N} = 2\ 500\ \text{N}$.
7. The net force is $220\ \text{N} - F$ (the friction), so
   $220\ \text{N} - F = (75\ \text{kg})(2.0\ \text{m/s}^2)$; $F = 70\ \text{N}$.
8. The total momentum starts at zero and does not change, so
   $(35\ \text{kg})v + (6\ \text{kg})(3.5\ \text{m/s}) = 0$, and $v = -0.6\ \text{m/s}$.
9. The total momentum is zero and does not change:

$$(7.5\ \text{kg})v + (0.0080\ \text{kg})(640\ \text{m/s}) = 0; v = 0.68\ \text{m/s}.$$

10. a. $\Delta(mv) = (45 \text{ kg})(85 \text{ m/s}) = 3\ 825 \text{ kg·m/s}$.

    b. $\Delta(mv)$ for the rocket is $-3\ 825$ kg·m/s; since it starts at rest, this is also its final mv. Its final mass is 705 kg, so

    $$v = \frac{3\ 825 \text{ kg·m/s}}{705 \text{ kg}} = 5.4 \text{ m/s}$$

11. a. From rest in the engine, the gas increases its momentum to $mv =$

    $$(60 \text{ kg})(95 \text{ m/s}) = 5\ 700 \text{ kg·m/s}.$$

    b. Impulse is change of momentum, both for the gas and for the ship. Therefore $F\ \Delta t = 5\ 700 \text{ N·s} = F(20 \text{ s})$, and $F = 290$ N.

12. The total momentum before equals the total momentum after. Since they are going in opposite directions, one of them has negative momentum. Then $(62 \text{ kg})(3.5 \text{ m/s}) - (53 \text{ kg})(5.0 \text{ m/s}) = (53 \text{ kg} + 62 \text{ kg})v$ and $v = 0.42$ m/s.

13. $F = ma = mv^2/r$, so

    $$r = \frac{mv^2}{F} = \frac{(240 \text{ kg})(35 \text{ m/s})^2}{6\ 200 \text{ N}} = 47 \text{ m}$$

14. Function is the centripetal force, so

    $$F = \frac{mv^2}{r} = \frac{(45 \text{ kg})(4.5 \text{ m/s})^2}{2.5 \text{ m}} = 365 \text{ N}$$

15. The acceleration is due to gravity alone, so $g \propto 1/r^2$ and gr^2 is a constant. We know g and r at the surface of the earth. For the meteor r is its altitude plus the radius of the earth: $55\ 000 \text{ km} + 6\ 400 \text{ km} = 61\ 400 \text{ km}$. Then $g_1 r_1^{\,2} = g_2 r_2^{\,2}$, or

    $$g_2 = g_1 \left(\frac{r_1}{r_2}\right)^2 = (9.8 \text{ m/s}^2)\left(\frac{6\ 400 \text{ km}}{61\ 400 \text{ km}}\right)^2 = 0.11 \text{ m/s}^2$$

16. First find the acceleration due to gravity, as in Problem 15:

    $$g = (9.8 \text{ m/s}^2)\left(\frac{6\ 400 \text{ km}}{6\ 400 \text{ km} + 4\ 200 \text{ km}}\right)^2 = 3.57 \text{ m/s}^2$$

    Then $w = mg = (3\ 500 \text{ kg})(3.57 \text{ m/s}^2) = 12\ 500$ N.

17.

$$F = \frac{Gm_1 m_2}{r^2} = \frac{\left(6.67 \times 10^{-11} \text{ N·m}^2/kg^2\right)\left(4.5 \times 10^5 \text{ kg}\right)\left(7 \times 10^5 \text{ kg}\right)}{(22\ 000 \text{ m})^2} = 4.3 \times 10^{-8} \text{ N}.$$

18. The desired increase in speed, Δv, is 4.0 m/s; the net force producing that increase is $510 \text{ N} - 240 \text{ N} = 270 \text{ N}$. Therefore, from Equation 4.3,

$$\Delta t = \frac{m\Delta v}{F} = \frac{(64 \text{ kg})(4.0 \text{ m/s})}{270 \text{ N}} = 0.9 \text{ s}$$

19. At the earth's surface, 6 400 km from its center, a kilogram weighs 9.8 N. Then, from Equation 4.6,

$$m_1 = \frac{F_{\text{grav}}r^2}{Gm_2} = \frac{(9.8 \text{ N})(6.4 \times 10^6 \text{ m})^2}{(6.67 \times 10^{-11} \text{ N} \cdot \text{m}^2 / \text{kg}^2)(1 \text{ kg})} = 6.0 \times 10^{24} \text{ kg}$$

20. The period of revolution of the satellite must be exactly one day, or 86 400 s. From Equations 2.5, the centripetal acceleration of the satellite must be $4\pi^2 r/T^2$. From Equation 4.5, the gravitational field must be $g = g_0(r_0/r)^2$. In free fall, $a = g$, so

$$\frac{4\pi^2 r}{T^2} = g_0\left(\frac{r_0}{r}\right)^2 \quad \text{so} \quad r^3 = \frac{g_0 r_0{}^2 T^2}{4\pi^2}$$

$$r = \sqrt[3]{\frac{(9.8 \text{ m/s}^2)(6.4 \times 10^6 \text{ m})^2 (86\ 400 \text{ s})^2}{4\pi^2}} = 4.16 \times 10^7 \text{ m}$$

To get the altitude, subtract the radius of the earth. The satellite must be at an altitude of 35 000 km.

21. The craft is going in a circle of radius $6\ 500 + 2\ 100 = 8\ 600$ km. Its centripetal acceleration is thus

$$a_c = \frac{v^2}{r} = \frac{(1\ 200 \text{ m/s})^2}{8.6 \times 10^6 \text{ m}} = 0.167 \text{ m/s}^2$$

This is the value of g at altitude. At the surface, it will be

$$g_0 = g\left(\frac{r}{r_0}\right)^2 = (0.167 \text{ m/s}^2)\left(\frac{8\ 600 \text{ km}}{2\ 100 \text{ km}}\right)^2 = 2.80 \text{ m/s}^2$$

Then the weight of the pack at the surface is

$$w = mg = (60 \text{ kg})(2.80 \text{ N/kg}) = 170 \text{ N}$$

Chapter 5

FILL IN'S

1. work
2. increases
3. kinetic
4. 4
5. scalar
6. gravitational potential; kinetic
7. total
8. altitude
9. internal
10. energy
11. chemical
12. mass
13. the speed of light
14. internal
15. zero

MULTIPLE CHOICE

1. (2) 3. (3) 5. (2) 7. (2) 9. (1)
2. (1) 4. (5) 6. (3) 8. (1) 10. (5)

PROBLEMS

1. $\Delta E = W = F\,\Delta s = (150\ \text{N})(22\ \text{m}) = 3\,300\ \text{J}.$

2. a. The only form of energy added is kinetic, so

$$\Delta E_{\text{kin}} = W = F\,\Delta s = (75\ \text{N})(15\ \text{m}) = 1\,130\ \text{J}.$$

   b. $E_{\text{kin}} = \dfrac{1}{2}mv^2 \qquad v = \sqrt{\dfrac{2E_{\text{kin}}}{m}} = \sqrt{\dfrac{2 \times 1\,130\ \text{J}}{380\ \text{kg}}} = 2.4\ \text{m/s}.$

3. All the kinetic energy it loses does work, so

$$W = \frac{1}{2}mv^2 = \frac{1}{2}(20\ \text{kg}) \times (10.0\ \text{m/s})^2 = 1\,000\ \text{J}.$$

4. The gravitational potential energy it loses turns into kinetic energy, so

$$(E_{\text{grav}})_{\text{top}} = (E_{\text{kin}})_{\text{bottom}}: \qquad mgh_{\text{top}} = \frac{1}{2}mv^2{}_{\text{bottom}}, \quad \text{and}$$

$$v = \sqrt{2gh} = \sqrt{2(9.8\ \text{m/s}^2)(0.15\ \text{m})} = 1.7\ \text{m/s}$$

5. a. $W = F\,\Delta s = (6.5\ \text{N})(25\ \text{m}) = 163\ \text{J}.$
   b. 163 J.
   c. $E_{\text{grav}} = mgh$; $163\ \text{J} = (0.30\ \text{kg})(9.8\ \text{m/s}^2)h$, so $h = 55\ \text{m}$. This is the additional height gained after the rocket burns out, so it comes to rest at 80 m.

6. a. $mgh_1 = mgh_2 + E_{int}$; $E_{int} = mg(h_1 - h_2) =$

    $$(0.16 \text{ kg})(9.8 \text{ m/s}^2)(5.0 \text{ m} - 4.2 \text{ m}) = 1.3 \text{ J}$$

    b. As it leaves the ground, its energy is all kinetic and equal to mgh_2. Therefore

    $\frac{1}{2}mv^2 = mgh_2$, and $v = \sqrt{2gh_2} = \sqrt{2(9.8 \text{ m/s}^2)(4.2 \text{ m})} = 9.1 \text{ m/s}$.

7. a. $W = F \Delta s = (250 \text{ N})(12 \text{ m}) = 3\,000 \text{ J}$.
    b. $E_{grav} = mgh = (30 \text{ kg})(9.8 \text{ m/s}^2)(6.0 \text{ m}) = 1\,760 \text{ J}$.
    c. Internal energy is the energy produced by friction, so it is 10% of
        3 000 J, or 300 J.
    d. The work produced three kinds of energy: $W = E_{int} + E_{kin} + E_{grav}$;
        3 000 J = 300 J + E_{kin} + 1 760 J, so E_{kin} = 940 J.

    e. $E_{kin} = \frac{1}{2}mv^2$, so 940 J = $\frac{1}{2}(30 \text{ kg})v^2$; $v = 7.9 \text{ m/s}$.

8. Internal energy created = loss of kinetic energy = $(\frac{1}{2}mv^2)_{initial} - (\frac{1}{2}mv^2)_{final} =$

    $\frac{1}{2}(1\,200 \text{ kg})(22 \text{ m/s})^2 - \frac{1}{2}(1\,200 \text{ kg})(15 \text{ m/s})^2 = 1.55 \times 10^5 \text{ J}$.

9. $E_{grav} = mgh = (1\,800 \text{ kg})(9.8 \text{ m/s}^2)(15 \text{ m}) = 2.6 \times 10^5 \text{ J}$.

10. $m = \dfrac{E_{grav}}{gh} = \dfrac{20\,000 \text{ J}}{(9.8 \text{ m/s}^2)(12 \text{ m})} = 170 \text{ kg}$.

11. a. $(mv)_{before} = (mv)_{after}$; $(1.5 \text{ kg})(4.0 \text{ m/s}) = (4.0 \text{ kg})v$, so v, after the impact, is
        1.5 m/s.
    b. The internal energy created is the difference between the kinetic energy of
        the rock before the impact and the kinetic energy of the combination after:

    $\frac{1}{2}(1.5 \text{ kg})(4.0 \text{ m/s})^2 - \frac{1}{2}(4.0 \text{ kg})(1.5 \text{ m/s})^2 = E_{int}$, therefore $E_{int} = 7.5 \text{ J}$.

12. $E = W = F \Delta s = (60 \text{ N})(0.35 \text{ m}) = 21 \text{ J}$.
13. $P = E/\Delta t$, so

    $$E = P\Delta t = (10^9 \text{ W})(365 \text{ days})\left(\frac{24 \text{ hours}}{\text{day}}\right)\left(\frac{3\,600 \text{ s}}{\text{hour}}\right) = 3.2 \times 10^{16} \text{ J}$$

    $E = mc^2$, so

    $$m = \frac{E}{c^2} = \frac{3.2 \times 10^{16} \text{ J}}{(3.00 \times 10^8 \text{ m/s})^2} = 0.35 \text{ kg}$$

14. $E = mc^2 = (10^{-7} \text{ kg})(3.0 \times 10^8 \text{ m/s})^2 = 9 \times 10^9 \text{ J}$.

15.   The work that has to be done is

$$(500 \text{ kg})(9.8 \text{ N/kg})(75 \text{ m}) = 368\,000 \text{ J}$$

From Equation 5.6

$$\Delta t = \frac{W}{P} = \frac{380 \text{ kJ}}{20 \text{ kW}} = 19 \text{ s}$$

16.   a. $E = \dfrac{Gm_1m_2}{r} = \dfrac{(6.67 \times 10^{-11} \text{ N·m}^2/\text{kg}^2)(6.0 \times 10^{24} \text{ kg})(250 \text{ kg})}{6.4 \times 10^6 \text{ m}}$

$$E = 1.56 \times 10^{10} \text{ J}$$

b. $W = F\,\Delta s$; 1.56×10^{10} J $= (640\,000$ N$)(\Delta s)$, so the height $\Delta s = 24\,000$ m.

17.   The gravitational energy it loses is

$$\frac{Gm_1m_2}{r} = \frac{(6.67 \times 10^{-11} \text{ N·m}^2/\text{kg}^2)(7.48 \times 10^{22} \text{ kg})(60 \text{ kg})}{1.73 \times 10^6 \text{ m}} = 1.73 \times 10^8 \text{ J}$$

18.

$$v_{\text{escape}} = \sqrt{\frac{2Gm_2}{r}} = \sqrt{\frac{2(6.67 \times 10^{-11} \text{ N·m}^2/\text{kg}^2)(7.48 \times 10^{22} \text{ kg})}{1.73 \times 10^6 \text{ m}}} = 2\,400 \text{ m/s}$$

19. $E_{\text{spr}} = \dfrac{1}{2} k\Delta l^2 = \dfrac{1}{2}(75 \text{ N/m})(0.165 \text{ m})^2 = 1.02$ J.

20.   The energy stored in the spring is

$$E_{\text{spr}} = \frac{1}{2} k\Delta l^2 = \frac{1}{2}(28 \text{ N/m})(0.65 \text{ m})^2 = 0.059 \text{ J}$$

The dart gets half of this as kinetic energy ($\frac{1}{2}mv^2$) so then

$$v = \sqrt{\frac{2E}{m}} = \sqrt{\frac{2(0.0295 \text{ J})}{0.012 \text{ kg}}} = 2.2 \text{ m/s}$$

Chapter 6

FILL IN'S

1. gas
2. shape
3. density
4. density
5. area
6. smaller
7. density
8. 760 mm
9. 3 atm
10. 2 200 mb
11. buoyancy
12. weight
13. weight
14. density
15. surface tension
16. viscosity
17. velocity
18. weight
19. laminar
20. increases
21. decreases
22. faster
23. speed
24. angle of attack
25. remains constant
26. decreases
27. increases

MULTIPLE CHOICE

1. (4)	3. (2)	5. (5)	7. (3)	9. (2)
2. (1)	4. (3)	6. (5)	8. (3)	10. (4)

PROBLEMS

1. The mass of the liquid is 265 g − 120 g = 145 g.

$$D = \frac{m}{V} = \frac{145 \text{ g}}{100 \text{ cm}^3} = 1.45 \text{ g/cm}^3$$

2. $D = m/V$, so $m = DV = (0.83 \text{ g/cm}^3)(1\,000 \text{ cm}^3) = 830 \text{ g}$.

3. Its density must be 1.20 g/cm^3, and

$$V = \frac{m}{D} = \frac{500 \text{ g}}{1.20 \text{ g/cm}^3} = 417 \text{ cm}^3$$

4. $$p = \frac{F}{A} = \frac{2\,500 \text{ lb}}{4 \times 4 \text{ in}^2} = 156 \text{ lb/in.}^2$$

5. $$F = pA = (8\tfrac{1}{2} \text{ in})(11 \text{ in}) \left(14.7 \frac{\text{lb}}{\text{in}^2}\right)$$

$$F = 1\,370 \text{ lb}$$

6. $$\left(62 \text{ lb/in.}^2\right)\left(\frac{1 \text{ atm}}{14.7 \text{ lb/in}^2}\right) = 4.2 \text{ atm.}$$

7. $p = hDg$. To put everything in SI units, we have to convert:

$$D = \left(1\frac{\text{g}}{\text{cm}^3}\right)\left(\frac{10^2 \text{ cm}}{\text{m}}\right)^3\left(\frac{\text{kg}}{10^3 \text{ g}}\right) = 10^3 \text{ kg/m}^3$$

So $p = (65 \text{ m})(10^3 \text{ kg/m}^3)(9.8 \text{ m/s}^2) = 6.37 \times 10^5 \text{ N/m}^2$. Adding the pressure at the surface, $1.0 \times 10^5 \text{ N/m}^2$, gives an answer of $7.4 \times 10^5 \text{ N/m}^2$, or 7.4 atm.

8. $P = hDg = (110 \text{ m})(10^3 \text{ kg/m}^3)(9.8 \text{ m/s}^2) = 1.08 \times 10^6 \text{ Pa}$. It is not necessary to add the atmospheric pressure, since the atmosphere is also pressing on the other side of the dam.

9. $P = hDg = (7.5 \text{ m})(880 \text{ kg/m}^3)(9.8 \text{ m/s}^2) = 6.5 \times 10^4 \text{ N/m}^2 = 0.65 \text{ atm}$. This is the gauge pressure.

10. It displaces 650 cm³ of water, and its buoyancy is the weight of this water. Then $w = mg$ and $m = DV$, so

$$w = DVg = \left(1\frac{\text{g}}{\text{cm}^3}\right)(650 \text{ cm}^3)(9.8 \text{ m/s}^2)\left(\frac{\text{kg}}{10^3 \text{ g}}\right) = 6.4 \text{ N}$$

11. Buoyancy is $(350 \text{ N} - 245 \text{ N}) = 105 \text{ N}$. The volume of the anchor is the volume of water that weighs that much; $D = m/V$ and $w = mg$, so

$$V = \frac{w}{Dg} = \frac{105 \text{ N}}{(10^3 \text{ kg/m}^3)(9.8 \text{ m/s}^2)} = 1.1 \times 10^{-2} \text{ m}^3$$

12. The buoyancy on the rock is the weight of $(45 \text{ g} - 32 \text{ g}) = 13 \text{ g}$ of water, so its volume is 13 cm³. Then

$$D = \frac{m}{V} = \frac{45 \text{ g}}{13 \text{ cm}^3} = 3.5 \text{ g/cm}^3$$

13. a. The mass of the 18 cm³ of water is 18 g; since the block displaces its own weight, its mass is also 18 g.

   b. $D = \dfrac{m}{V} = \dfrac{18 \text{ g}}{(3.0 \text{ cm})^3} = 0.67 \text{ g/cm}^3$

14. The plank floats with 5/6.5 of its volume under water, so the water it displaces must have just 5/6.5 as much volume as the plank. Therefore, the density of the plank must be 5/6.5 that of water, or 0.77 g/cm³.

15. Its density is 86% that of water, or 0.86 g/cm³.

Chapter 7

FILL IN'S

1. temperature
2. heat
3. thermal equilibrium
4. a vacuum
5. metals
6. warmer
7. expands
8. ice
9. heat
10. specific heat
11. water
12. phase
13. 2 260
14. pressure
15. 273 K
16. kelvin
17. 1 atm
18. pressure; volume
19. low P; high T
20. size
21. temperature
22. work
23. efficiency

MULTIPLE CHOICE

1. (1)	3. (5)	5. (2)	7. (4)	9. (2)
2. (3)	4. (2)	6. (5)	8. (1)	10. (3)

PROBLEMS

1. $\dfrac{T}{100°C} = \dfrac{12.2 \text{ cm} - 3.5 \text{ cm}}{19.0 \text{ cm} - 3.5 \text{ cm}}$ $T = 56°C$.

2. The heat generated must be equal to the kinetic energy lost:

$$\frac{1}{2}mv^2 = cm\Delta T$$

In SI, the specific heat of lead is 130 J/kg·K, so

$$\Delta T = \frac{v^2}{2c} = \frac{(460 \text{ m/s})^2}{2(130 \text{ J/kg} \cdot \text{K})} = 810 \text{ K}$$

3. Lost potential energy becomes heat:

$$mgh = cm\ \Delta T$$

In SI, the specific heat of water is 4 185 J/kg·K, so

$$\Delta T = \frac{gh}{c} = \frac{(9.8 \text{ m/s}^2)(110 \text{ m})}{(4\ 185 \text{ J/kg} \cdot \text{K})} = 0.26 \text{ K}$$

4. $\Delta H = cm\Delta T = (0.92 \text{ J/g·K})(580 \text{ g})(160 \text{ K}) = 85\ 000 \text{ J}$.

5. Heat lost equals heat gained, so $(cm\Delta T)_{metal} = (cm\Delta T)_{water}$

$$c_{metal} = \frac{(cm\Delta T)_{water}}{(m\Delta T)_{metal}} = \frac{(4.19 \text{ J/g·K})(300 \text{ g})(23 \text{ K})}{(760 \text{ g})(67 \text{ K})} = 0.57 \text{ J/g·K}$$

6. $(cm\Delta T)_{iron} = (cm\Delta T)_{water}$, so

$$m_{iron} = \frac{(cm\Delta T)_{water}}{(c\Delta T)_{iron}} = \frac{(4.19 \text{ J/g·K})(220 \text{ g})(30 \text{ K})}{(0.48 \text{ J/g·K})(160 \text{ K})} = 360 \text{ g}$$

7. First melt the ice:

$$\Delta H = mL_{fusion} = (200 \text{ g})(335 \text{ J/g}) = 67\ 000 \text{ J}$$

Then heat the water:

$$\Delta H = cm\Delta T = (4.19 \text{ J/g·K})(200 \text{ g})(100 \text{ K}) = 83\ 800 \text{ J}$$

And then boil the water:

$$\Delta H = mL_{vaporization} = (200 \text{ g})(2\ 260 \text{ J/g}) = 452\ 000 \text{ J}$$

Total heat adds up to 603 000 J.

8. The final temperature is 0°C, and $(mc\Delta T)_{copper} = mL_{ice}$

$$\Delta T_{copper} = \frac{mL_{ice}}{(mc)_{copper}} = \frac{(75 \text{ g})(335 \text{ J/g})}{(350 \text{ g})(0.38 \text{ J/g·K})} = 190 \text{ K}$$

So the initial temperature of the copper was 190°C.

9. Let T_f = the final temperature

$$(cm\Delta T)_{water} = (cm\Delta T)_{copper}$$

Since the water warms up, $\Delta T_{water} = T_f - 15°C$. The copper cools down, so $\Delta T_{copper} = 145°C - T_f$. Then

$$(4.19 \text{ J/g·K})(300 \text{ g})(T_f - 15°C) = (0.38 \text{ J/g·K})(580 \text{ g})(145°C - T_f)$$

Since (J/g·K)(g) appears in every term, it can be eliminated. Clearing parentheses gives

$$1257T_f - 18\ 855°C = 31\ 958°C - 220T_f$$
$$T_f = 34°C$$

10. a. $p \propto T$, so p/T is constant. For ice water, $T = 0°C + 273 \text{ K} = 273 \text{ K}$. Then

$$\frac{p_1}{T_1} = \frac{p_2}{T_2} \qquad \frac{95\text{kPa}}{273 \text{ K}} = \frac{62\text{kPa}}{T_2} \qquad T_2 = 178\text{K}$$

b. $T_e = T_k - 273 \text{ K}$; $178 \text{ K} - 273 \text{ K} = -95°C$.

11. For ice water, $T = 273$ K; for boiling water, $T = 373$ K.

$$\frac{p_1}{T_1} = \frac{p_2}{T_2} \qquad \frac{1\ 400\ \text{mb}}{273\ \text{K}} = \frac{p_2}{373\ \text{K}} \qquad p_2 = 1\ 910\ \text{mb}$$

12. $V \propto 1/p$, so pV is constant. Then $p_1V_1 = p_2V_2$: (1 atm)(2.5 L) $= p_2$(1.0 L); and $p_2 = 2.5$ atm. This is the absolute pressure; the gauge pressure is 1.5 atm.

13. The initial temperature is 295 K. $pV \propto T$, so

$$\frac{p_1V_1}{T_1} = \frac{p_2V_2}{T_2} \qquad \frac{(1\ \text{atm})(600\ \text{cm}^3)}{295\ \text{K}} = \frac{(40\ \text{atm})(50\ \text{cm}^3)}{T_2}$$

and $T_2 = 980$ K.

14. p_1 (absolute) $= 3$ atm; $T_1 = 295$ K;

$$\frac{p_1V_1}{T_1} = \frac{p_2V_2}{T_2} \qquad \frac{(3\ \text{atm})(1\ \text{L})}{295\ \text{K}} = \frac{(6\ \text{atm})(2\ \text{L})}{T_2}$$

and $T_2 = 1\ 180$ K.

15. $pV = nkT$, so $n = PV/kT$; 1 liter $= 10^{-3}$ m^3 and 20°C $= 293$ K.

$$n = \frac{(0.3 \times 10^5\ \text{N/m}^2)(10^{-3}\ \text{m}^3)}{(1.39 \times 10^{-23}\ \text{J/mcl·K})(293\ \text{K})} = 7.4 \times 10^{21}\ \text{molecules}$$

16. At constant temperature, PV/n is constant;

$$\frac{p_1V_1}{n_1} = \frac{p_2V_2}{n_2} \qquad \frac{(1\ \text{atm})(500\ \text{cm}^3)}{n} = \frac{p_2(200\ \text{cm}^3)}{2n}$$

and $p_2 = 5$ atm.

17. $E_{\text{kin}} = \frac{3}{2}kT = \frac{3}{2}\left(1.38 \times 10^{-23}\ \frac{\text{J}}{\text{mcl·K}}\right)(373\ \text{K}) = 7.7 \times 10^{-21}$ J

18. Since both have the same temperature, both have the same kinetic energy, and $m_1v_1^2 = m_2v_2^2$. Then

$$\frac{v_2}{v_1} = \sqrt{\frac{m_1}{m_2}} = \sqrt{\frac{32}{2}} = \frac{4}{1}$$

19. First, find the amount of heat put in.

$$0.25 = 1 - \frac{20\ 000\ \text{J}}{H_i}$$

so $H_i = 27\ 000$ J. The difference between this and the heat out is the work done, 7 000 J.

20. It must be taking in an amount of heat equal to (540 J)/0.30 $= 1\ 800$ J.

21. $$\frac{H_o}{H_i} = \frac{T_o}{T_i}$$

Therefore, using kelvin temperatures,

$$\frac{H_o}{2\ 400\ \text{J}} = \frac{293\ \text{K}}{513\ \text{K}}$$

and $H_o = 1\ 370$ J. Subtracting this from the heat absorbed gives the work done, 1 030 J.

22.   From Equation 7.8, using kelvin temperatures,

$$T_0 = T_i\,(1 - \varepsilon)$$
$$= 1\ 023\ \text{K}\,(1 - .7)$$

so the output temperature must be 307 K, or 34°C.

Chapter 8

FILL IN'S

1. period	17. wavelength
2. frequency	18. point source
3. frequency	19. frequency
4. 180	20. source
5. amplitude	21. out of phase
6. damping	22. difference
7. wavelength	23. node
8. velocity	24. halfwavelength
9. parallel	25. node
10. velocity	26. one half
11. rarefactions	27. node
12. temperature	28. natural or resonant frequencies
13. frequency	29. acceleration
14. 110	30. maximum
15. tone quality	31. 20
16. overtones	32. 30

MULTIPLE CHOICE

1. (1)	3. (2)	5. (5)	7. (5)	9. (5)
2. (4)	4. (3)	6. (2)	8. (3)	10. (3)

PROBLEMS

1. a. $v = f\lambda$, so

$$f = \frac{v}{\lambda} = \frac{12 \text{ m/s}}{1.2 \text{ m}} = 10 \text{ Hz}$$

   b. $T = \frac{1}{f} = \frac{1}{10 \text{ s}^{-1}} = 0.10 \text{ s}$

2. $12/30 = \phi/360°$; $\phi = 144°$.

3. From a compression to the nearest rarefaction is 1/2 wavelength, so $\lambda = 0.70$ m. Then $v = f\lambda = (4.0 \text{ Hz})(0.70 \text{ m}) = 2.8$ m/s.

4. $v = 331 \text{ m/s} + (0.6 \text{ m/s})(-6°\text{C}) = 327$ m/s.

5. $v = 331 \text{ m/s} + (0.6)(26°\text{C}) = 347$ m/s. Then

$$\lambda = \frac{v}{f} = \frac{347 \text{ m/s}}{640 \text{ Hz}} = 0.54 \text{ m}$$

6. Divide by 2 twice to get 110 Hz.

7. $v = 331 \text{ m/s} + (0.6 \text{ m/s·°C})(10°\text{C}) = 337$ m/s. Then

$$\lambda = \frac{v}{f} = \frac{337 \text{ m/s}}{680 \text{ Hz}} = 0.50 \text{ m}$$

   The other speaker is one-half wavelength farther than the near one, at 16.3 m.

8. Tightening the matching string raises its frequency, so it must have been lower than 220 Hz to start. With 4 beats per second, it was 4 Hz lower, at 216 Hz.

9. $\lambda = \frac{v}{f} = \frac{8.0 \text{ m/s}}{6.0 \text{ Hz}} = 1.33$ m

   The first two antinodes are $\frac{1}{4}$ and $\frac{3}{4}$ wavelength from the end, at 0.33 m and 1.00 m.

10. At the fundamental, it is $\frac{1}{2}$ wavelength long, so

$$\lambda = 0.70 \text{ m}; f = \frac{v}{\lambda} = \frac{180 \text{ m/s}}{0.70 \text{ m}} = 257 \text{ Hz}$$

   At the first overtone, the string is one wavelength long;

$$f = \frac{180 \text{ m/s}}{0.35 \text{ m}} = 514 \text{ Hz}$$

At the second overtone, $0.35\ \text{m} = \dfrac{3}{2}\ \lambda$, so

$$\lambda = 0.233\ \text{m}; f = \frac{180\ \text{m/s}}{0.233\ \text{m}} = 771\ \text{Hz}$$

11. $v = 331\ \text{m/s} + (0.6\ \text{m/s})(20°\text{C}) = 343\ \text{m/s}$. At the fundamental, the pipe is

$\dfrac{1}{4}$ wavelength, so $\lambda = 1.40\ \text{m}$; then $f = v/\lambda = 245\ \text{Hz}$. At the first overtone,

$0.35\ \text{m} = \dfrac{3}{4}\ \lambda$, so $\lambda = 0.47\ \text{m}$ and $f = 730\ \text{Hz}$. At the second overtone, $0.35\ \text{m}$

$= \dfrac{5}{4}\lambda$ and $f = 1\ 225\ \text{Hz}$.

12. At the fundamental, the fork is $\dfrac{1}{2}$ wavelength long, so

$$f = \frac{60\ \text{m/s}}{0.36\ \text{m}} = 167\ \text{Hz}$$

At the first overtone, $0.18\ \text{m} = \dfrac{3}{2}\lambda$ and $f = 500\ \text{Hz}$.

13. The frequency of the second overtone is $3 \times 440\ \text{Hz} = 1\ 320\ \text{Hz}$, so the beat frequency is $1318.5 - 1320 = 1.5\ \text{Hz}$.

14. From Equation 8.6a, the intensity of the sound you hear is

$$15\ \text{db} = 10\ \log\left(\frac{I}{1 \times 10^{-12}\ \text{W/m}^2}\right)$$

$$\frac{I}{1 \times 10^{-12}\ \text{W/m}^2} = 10^{1.5} = 32$$

$$I = 32(1 \times 10^{-12}\ \text{W/m}^2) = 3.2 \times 10^{-11}\ \text{W/m}^2$$

From Equation 8.6b, the power at the source must be

$$P = 4\pi r^2 I = 4\pi(30\ \text{m})^2(3.2 \times 10^{-11}\ \text{W/m}^2) = 3.6 \times 10^{-7}\ \text{W}$$

15. Equation 8.6 says that the intensity of a sound is inversely proportional to the square of the distance from the source. The threshold of pain is 10^{12} times as great as the threshold of hearing. Therefore

$$\frac{I_1}{I_2} = 10^{12} = \left(\frac{2\ \text{m}}{r_1}\right)^2$$

$$r_1 = (2\ \text{m})\sqrt{10^{12}} = 2 \times 10^6\ \text{m}$$

16. The intensity required would be the threshold of hearing, which is 1×10^{-12} W/m². Then the power needed is

$$p = 4\pi r^2 I = 4\pi (10^3 \text{ m})^2 (1 \times 10^{-12} \text{ W/m}^2) = 1.3 \times 10^{-5} \text{ W}$$

17. The period of the pendulum on the other planet is

$$T = \frac{1}{f} = \frac{1}{2.4 \text{ Hz}} = 0.42 \text{ s}$$

Equation 8.1 tells us that, for a given length of pendulum, $T \propto 1/g$, so

$$g_2 = g_1 \left(\frac{T_1}{T_2}\right)^2 = (9.8 \text{ m/s}^2)\left(\frac{2.0 \text{ s}}{0.42 \text{ s}}\right)^2 = 220 \text{ m/s}^2$$

Chapter 9

FILL IN'S

1. repel
2. electric
3. negative
4. electrons
5. conductors
6. surface
7. insulator
8. curvature
9. coulomb
10. charges
11. one-fourth
12. closest together
13. negative
14. charge
15. positive
16. perpendicular
17. uniform
18. electric field
19. equipotential
20. field lines
21. volts
22. coulomb
23. electron
24. electrons
25. spherical
26. 4 000

MULTIPLE CHOICE

1. (3)
2. (5)
3. (1)
4. (2)
5. (2)
6. (4)
7. (1)
8. (4)
9. (1)
10. (3)

PROBLEMS

1. The total charge is +20 nC; each has +10 nC.

2. a.

$$F = \frac{kq_1q_2}{r^2} = \frac{(9.0 \times 10^9 \text{ N·m}^2 / \text{C}^2)(25 \times 10^{-9} \text{ C})(15 \times 10^{-9} \text{ C})}{(0.15 \text{ m})^2} = 1.5 \times 10^{-4} \text{ N}$$

   b. Same.

3. The ratio of force to charge is the field, so

$$\frac{3.0 \times 10^{-3} \text{ N}}{12 \text{ nC}} = \frac{2.5 \times 10^{-5} \text{ N}}{q}$$

   and $q = 0.10$ nC.

4. Charge is conserved, so the other has +12 nC.

5. $\Delta E = q \, \Delta V = (220 \times 10^{-9} \text{ C})(30 \text{ V} - 5 \text{ V}) = 5.5 \times 10^{-6}$ J.

6. $\Delta E = q \, \Delta V = (6.0 \text{ C})(24 \text{ V}) = 144$ J.

7. $\Delta E = q \, \Delta V = (240 \text{ nC})(110 \ 000 \text{ V}) = 0.026$ J.

8. a. $\Delta E = q \, \Delta V = (2.0 \times 10^{-9} \text{ C})(240 \text{ V}) = 4.8 \times 10^{-7}$ J.
   b. $\Delta E = F \Delta s$; 4.8×10^{-7} J $= F(0.12$ m$)$, so $F = 4.0 \times 10^{-6}$ N.

9. $\dfrac{8.0 \times 10^{-9} \text{ C}}{1.60 \times 10^{-19} \text{ C}} = 5 \times 10^{10}$ excess electrons.

10. $\Delta E = q \, \Delta V = (1.60 \times 10^{-19} \text{ C})(120 \ 000 \text{ V}) = 1.9 \times 10^{-14}$ J.

11. $\mathscr{E} = \dfrac{\Delta V}{\Delta s} = \dfrac{12 \text{ V}}{2.5 \text{ m}} = 4.8$ V/m, or 4.8 N/C.

12. The upward force produced by the electric field must be equal to the weight of the ball:

$$mg = \mathscr{E}q$$

$$\mathscr{E} = \frac{mg}{q} = \frac{(0.0035 \text{ kg})(9.8 \text{ m/s}^2)}{8.0 \times 10^{-9} \text{ C}} = 4.3 \times 10^6 \text{ N/C}$$

13. The electric energy lost by the ball must be equal to the kinetic energy it gained:

$$q\Delta V = \frac{1}{2}mv^2$$

$$v = \sqrt{\frac{2q\Delta V}{m}} = \sqrt{\frac{2(22 \times 10^{-9} \text{ C})(12 \text{ V})}{0.25 \times 10^{-3} \text{ kg}}} = 4.6 \times 10^{-2} \text{ m/s}$$

14. Between parallel plates, the field is uniform and equal to

$$\mathscr{E} = \frac{\Delta V}{\Delta s} = \frac{120 \text{ V}}{0.04 \text{ m}} = 3 \ 000 \text{ V/m, or } 3 \ 000 \text{ N/C}$$

So the force on the charge is

$$F = \mathscr{E}q = (3\,000 \text{ N/C})(15 \times 10^{-9} \text{ C}) = 4.5 \times 10^{-5} \text{ N}$$

15. Divide the total charge by the charge on a single electron:

$$\frac{0.5 \times 10^{-9} \text{ C}}{1.60 \times 10^{-19} \text{ C/electron}} = 3 \times 10^9 \text{ electrons}$$

16. The strength of the field must be

$$\mathscr{E} = \frac{4 \times 10^{-17} \text{ N}}{1.60 \times 10^{-19} \text{ C}} = 250 \text{ N/C}$$

Then the required potential difference is

$$\Delta V = \mathscr{E}\,\Delta s = (250 \text{ V/m})(0.10 \text{ m}) = 25 \text{ V}$$

17. From Equation 9.3,

$$\Delta E_{el} = q\Delta V = (4.5 \times 10^5 \text{ C})(12 \text{ V}) = 5.4 \times 10^6 \text{ J}$$

To get the number of electrons, divide the total charge by the charge on an electron:

$$\frac{4.5 \times 10^5 \text{ C}}{1.60 \times 10^{-19} \text{ C}} = 2.8 \times 10^{24}$$

18. The electric energy lost must equal the kinetic energy gained:

$$q\Delta V = \frac{1}{2}mv^2$$

$$\Delta V = \frac{mv^2}{2q} = \frac{(9.1 \times 10^{-31} \text{ kg})(1.5 \times 10^8 \text{ m/s})^2}{2(1.6 \times 10^{-19} \text{ C})} = 64\,000 \text{ V}$$

19. The charge, from Equation 9.5, is

$$q = \frac{Vr}{k} = \frac{(1\,200 \text{ V})(0.20 \text{ m})}{(9.0 \times 10^9 \text{ N·m}^2 / \text{C}^2)} = 27 \text{ nC}$$

To get the surface charge density, divide the charge by the area of the terminal:

$$\text{charge density} = \frac{27 \text{ nC}}{4\pi(0.20 \text{ m})^2} = 53 \text{ nC/m}^2$$

20. The pith ball comes from a position of zero potential up to the potential of the terminal, so

$$\Delta E_{el} = q\Delta V = (6.0 \times 10^{-9} \text{ C})(1\,200 \text{ V}) = 7.2 \times 10^{-6} \text{ J}$$

21. The total charge of 27 nC is divided in proportion to the radius of the two spheres. Then the pith ball gets 1/21 of the total and the terminal gets 20/21. Therefore the charge on the pith ball is (27 nC)/21 = 1.3 nC.

22. The charge on a proton is the same as on the electron, only positive; from Equation 9.5,

$$V = k\frac{q}{r} = \left(9.0 \times 10^9 \text{ N·m}^2/\text{C}^2\right)\frac{\left(+1.60 \times 10^{-19} \text{ C}\right)}{(0.02 \text{ m})} = +7.2 \times 10^{-8} \text{ V}$$

23. Just calculate the two potentials separately and add them.

$$V_1 + V_2 = \left(9.0 \times 10^9 \text{ N·m}^2/\text{C}^2\right)\left(\frac{+6.0 \times 10^{-9} \text{ C}}{0.06 \text{ m}} + \frac{-3.5 \times 10^{-9} \text{ C}}{0.06 \text{ m}}\right) = 380 \text{ V}$$

Chapter 10

FILL IN'S

1. separation
2. emf
3. volt
4. current
5. ampere
6. ammeter
7. voltmeter
8. watts
9. power
10. resistance
11. ohm
12. superconductive
13. current
14. zero
15. current
16. potential differences
17. series
18. largest
19. potential difference
20. current; power
21. smallest
22. resistivity
23. heat
24. 160
25. decreases
26. internal resistance
27. 2

MULTIPLE CHOICE

1. (2)
2. (4)
3. (4)
4. (1)
5. (3)
6. (5)
7. (3)
8. (3)
9. (1)
10. (1)

PROBLEMS

1. $12 \times 2.2 \text{ V} = 26 \text{ V}$.

2. a. 26 J, by definition of potential difference.
   b. $I = q/\Delta t$, so $q = I\,\Delta t = (4.0 \text{ A})(300 \text{ s}) = 1\,200 \text{ C}$.
   c. $\Delta E = q\,\Delta V = (1\,200 \text{ C})(26 \text{ V}) = 31\,000 \text{ J}$.

   d. $P = \dfrac{\Delta E}{\Delta t} = \dfrac{31\,200 \text{ J}}{300 \text{ s}} = 104 \text{ W}$.

3. a. P
   b. V
   c. T
   d. Q

4. Electric energy delivered $= I\,\Delta V\,\Delta t$, which must equal the heat energy generated. Then $I\,\Delta V\,\Delta t = cm\,\Delta T$. In SI units, $c = 4\,190 \text{ J/kg·K}$. Then

$$\Delta t = \frac{cm\Delta T}{I\Delta V} = \frac{\left(4\,190 \text{ J/kg·K}\right)(0.25 \text{ kg})(80 \text{ K})}{(11 \text{ A})(115 \text{ V})} = 66 \text{ s}$$

5. $P = I\,\Delta V$, so

$$I = \frac{P}{\Delta V} = \frac{60 \text{ W}}{115 \text{ V}} = 0.52 \text{ A}$$

   Then

$$R = \frac{\Delta V}{I} = \frac{115 \text{ V}}{0.52 \text{ A}} = 220\ \Omega$$

6. $P = I\,\Delta V = (4.0 \text{ A})(120 \text{ V}) = 480 \text{ W}$. In running for 8 hours, the energy it uses is $(480 \text{ W})(8 \text{ hours}) = 3\,840$ watt-hours, or 3.84 kWh. Multiply by 8¢/kWh to get 31¢.

7. $I = \dfrac{\Delta V}{R} = \dfrac{(180 \text{ V} - 30 \text{ V})}{1\,200\ \Omega} = 0.13 \text{ A}$.

8. $\Delta V = IR = (90 \times 10^{-3} \text{ A})(40\ \Omega) = 3.6 \text{ V}$.

9. $P = I\,\Delta V$; $900 \text{ W} = I(115 \text{ V})$, so $I = 7.8\text{A}$.

10. A_6 reads the same as A_2, 3 A. Also, $A_1 = A_5 + A_4 = 10$ A. And $A_1 = A_2 + A_3 + A_4$, so $A_3 = 5\text{A}$.

11. $V_4 + V_5 = V_3$, so $V_4 = 5$ V. Also, $V_4 + V_5 + V_6 = V_7$, so $V_6 = 12$ V. And $V_1 = V_2 + V_7 = 26$ V.

12. a. $12\ \Omega + 18\ \Omega = 30\ \Omega$.

b. $I = \dfrac{\Delta V}{R} = \dfrac{24\ \text{V}}{30\ \Omega} = 0.80\ \text{A}$.

c. $P = I^2R = (0.80\ \text{A})^2(12\ \Omega) = 7.7\ \text{W}$.
    $P = (0.80\ \text{A})^2(18\ \Omega) = 11.5\ \text{W}$.

13. a. $P = I^2R$, so

$$I = \sqrt{\frac{P}{R}} = \sqrt{\frac{850\ \text{W}}{20\ \Omega}} = 6.5\ \text{A}$$

b. The total resistance needed is

$$R = \frac{\Delta V}{I} = \frac{160\ \text{V}}{6.5\ \text{A}} = 25\ \Omega$$

so the variable resistor must supply $5\ \Omega$.

14. a. $I = P/\Delta V =$ (toaster) $600\ \text{W}/120\ \text{V} = 5.0\ \text{A}$; (lamp) $150\ \text{W}/120\ \text{V} = 1.3\ \text{A}$; (radio) $40\ \text{W}/120\ \text{V} = 0.33\ \text{A}$.

b. Add them up to get $6.6\ \text{A}$.

15. $\dfrac{1}{R_P} = \dfrac{1}{R_1} + \dfrac{1}{R_2} + \dfrac{1}{R_3} = \dfrac{1}{100\ \Omega} + \dfrac{1}{250\ \Omega} + \dfrac{1}{400\ \Omega}$

$$\frac{1}{R_P} = \frac{60 + 24 + 15}{6\ 000\ \Omega}$$

so

$$R_P = \frac{6\ 000\ \Omega}{99} = 61\ \Omega$$

16. First determine the combined resistance of the parallel combination:

$$\frac{1}{R_P} = \frac{1}{20\ \Omega} + \frac{1}{60\ \Omega} + \frac{1}{80\ \Omega} = \frac{12 + 4 + 3}{240\ \Omega};\quad R_P = 13\ \Omega$$

Then add this to the others in series to get $50\ \Omega$.

17. The current in each bulb is, from Equation 10.4a,

$$I = \frac{P}{\Delta V} = \frac{50\ \text{W}}{120\ \text{V}} = 0.42\ \text{A}$$

so the maximum number of bulbs is

$$\frac{15\ \text{A}}{0.42\ \text{A}} = 35$$

This is one case in which you round off 35.7 to 35, not 36.

18. The saving is (45 W)(8 hrs)(30 days) = 10 800 watt-hours or 10.8 kWh. You saved 97¢.

19. Apply Equation 7.4; it takes 2 200 J to vaporize a gram (= 1 cm^3) of water. Then the rate at which electric energy is supplied is

$$\left(2\ 200\frac{J}{g}\right)\left(\frac{500\ g}{1\ hour}\right)\left(\frac{1\ hour}{3\ 600\ s}\right) = 306\ W$$

Then by Equation 10.4a, the current is

$$I = \frac{P}{\Delta V} = \frac{306\ W}{120\ V} = 2.6\ A$$

And from Equation 10.3,

$$R = \frac{\Delta V}{I} = \frac{120\ V}{2.6\ A} = 46\ \Omega$$

20. The resistance of the wire, by Equation 10.4, is

$$R = \frac{\rho l}{A} = \frac{(1.72 \times 10^{-8}\ \Omega\cdot m\ )(150\ m)}{5 \times 10^{-6}\ m^2} = 0.52\ \Omega$$

The current in the wire is

$$I = \frac{\Delta V}{R} = \frac{12\ V}{0.52\ \Omega} = 23\ A$$

Then the power consumed, which is equal to the rate of heat production, is

$$P = I^2R = (23\ A)^2(0.52\ \Omega) = 275\ W$$

21. The current is 0.060 coulombs per second; divide by the charge on an electron:

$$\frac{0.060\ C/s}{1.60 \times 10^{-19}\ C/electron} = 3.8 \times 10^{17}\ electrons/s$$

22. The potential drop is uniform throughout the length of the wire; the voltmeter is marking off 60/100 of the length of the wire. The drop over that distance is thus (0.6)(12 V) = 7.2 V.

23. The electric power input is

$$P = I\,\Delta V = (6.5\ A)(120\ V) = 780\ W$$

If only 500 W turns up as useful output, the fraction of power wasted is 280 W/780 W = 35%.

24. The emf of the 3-cell battery is 4.5 volts; its terminal difference is the potential difference across the bulb, which is 6.0 W/1.5 A = 4.0 V. Then the potential drop due to the internal resistance of the battery is 0.5 V, so its internal resistance is 0.5 V/1.5 A = 0.33 Ω, or 0.11 Ω for each cell.

25. From its rated values, the motor ordinarily draws 60 W/24 V = 2.5 A, so its resistance is 24 V/2.5 A = 9.6 Ω. With this battery, the total resistance of the circuit is 11.6 Ω, so the current will be 24 V/11.6 Ω = 2.1 A.

Chapter 11

FILL IN'S

1. attract
2. magnetic
3. north
4. S
5. field; current
6. downward
7. electric
8. circle
9. mass
10. currents
11. 2
12. clockwise
13. increase
14. ferromagnetic
15. electrons
16. alloys or oxides
17. north magnetic
18. moving
19. length and velocity
20. magnetic flux
21. work done
22. radius
23. circles
24. eastward
25. magnetic
26. iron
27. S
28. parallel
29. four times
30. south

MULTIPLE CHOICE

1. (3)
2. (5)
3. (2)
4. (2)
5. (2)
6. (3)
7. (1)
8. (4)
9. (1)
10. (5)

PROBLEMS

1. $B = \dfrac{F}{Il} = \dfrac{2.5 \times 10^{-4} \text{ N}}{(5.0 \text{ A})(1 \text{ m})} = 5 \times 10^{-5}$ T

2. The force per unit length exerted by the field is the weight per unit length of the wire. Thus, $F/l = mg/l$. Since $B = F/Il$,

$$B = \frac{mg}{Il} = \left(1.50 \times 10^{-3} \frac{\text{kg}}{\text{m}}\right)\left(\frac{9.8 \text{ m/s}^2}{2.0 \text{ A}}\right) = 7.4 \text{ mT}$$

3. a. East.
   b. $F = Bqv = (1.5 \text{ T})(1.60 \times 10^{-19} \text{ C})(1.8 \times 10^7 \text{ m/s}) = 4.3 \times 10^{-12} \text{ N}$.

4. $Bqv = mv^2/r$, so

$$m = \frac{Bqr}{v} = \frac{(0.75 \text{ T})(1.60 \times 10^{-19} \text{ C})(2.4 \text{ m})}{2.6 \times 10^7 \text{ m/s}} = 1.10 \times 10^{-26} \text{ kg}$$

5. $mv = Bqr$, so

$$v = \frac{Bqr}{m} = \frac{(0.25 \text{ T})(1.60 \times 10^{-19} \text{ C})(0.40 \text{ m})}{1.67 \times 10^{-27} \text{ kg}} = 9.6 \times 10^6 \text{ m/s}$$

6. $F = BIl = (0.055 \times 10^{-3} \text{ T})(1.0 \text{ m})(8.0 \text{ A}) = 4.4 \times 10^{-4} \text{ N}$. The direction is northward, up at an angle of 18° from the horizontal.

7. $E = Blv = (1.2 \text{ T})(0.20 \text{ m})(3.5 \text{ m/s}) = 0.84 \text{ V}$.

8. $\phi = BA = (350 \times 10^{-3} \text{ T})\pi(0.10 \text{ m})^2 = 1.1 \times 10^{-2} \text{ Wb}$.

9. $\Delta V = \dfrac{\Delta \phi}{\Delta t} = \dfrac{(150)(1.1 \times 10^{-2} \text{ Wb})}{0.10 \text{ s}} = 17 \text{ V}$.

10. a. $B = \dfrac{2k'I}{r} = \dfrac{2(10^{-7} \text{ N/A}^2)(20 \text{ A})}{0.15 \text{ m}} = 2.7 \times 10^{-5} \text{ T}$.

    b. According to the right-hand rule, east.

11. $I = \dfrac{Br}{2k'} = \dfrac{(0.050 \times 10^{-3} \text{ T})(0.02 \text{ m})}{2 \times 10^{-7} \text{ N/A}^2} = 5.0 \text{ A}$.

12. $B = 4\pi k'I(n/l)$. There are 10 turns per millimeter in two layers, so $n/l = 20\,000$/m. Then, $B = 4\pi(10^{-7} \text{ N/A}^2)(0.50 \text{ A})(20\,000\text{/m}) = 12.6 \text{ mT}$.

13. a. $I = \dfrac{\Delta V}{R} = \dfrac{12 \text{ V}}{30 \text{ } \Omega} = 0.40 \text{ A}$

    Then

$$B = 4\pi k'I\frac{n}{l} = \frac{4\pi(10^{-7} \text{ N/A}^2)(0.40 \text{ A})(200)}{0.15 \text{ m}} = 6.7 \times 10^{-4} \text{ T}$$

    b. Nothing changes except that the number of turns per unit length doubles, so the field also doubles.

14. It will still be 15 mT. Doubling the amount of wire doubles its resistance, so the current will drop to half, just balancing the effect of the increased number of windings.

15. First determine how much current is needed. The number of turns of wire per meter is 4(1 000 mm/0.71 per mm) = 5 600 turns per meter. Then, from Equation 11.5,

$$I = \frac{B}{(4\pi k'n/l)} = \frac{8.5 \times 10^{-3} \text{ Ts}}{4\pi(10^{-7} \text{ N/A}^2)(5 \text{ 600/m})} = 1.21 \text{ A}$$

The number of turns is (5 600 turns/m)(0.15 m) = 840 turns, and the total length of the wire is π(0.30 m)(840) = 790 m. The resistance of the wires is thus (790 m)(0.053 Ω/m) = 42 Ω. Then from Equation 10.2,

$$\Delta V = IR = (1.21 \text{ A})(42 \text{ Ω}) = 51 \text{ V}$$

16. Any time you know both the kinetic energy and the momentum of something, you can find its velocity. In this problem, you can eliminate the velocity from the calculation. The kinetic energy of the proton is provided by the accelerating electric field, so it is

$$\frac{1}{2}mv^2 = q\Delta V$$

As it travels in a circle in the magnetic field, its momentum is

$$mv = Bqr$$

Solve the two equations simultaneously, eliminating v and solving for r. Then

$$r = \sqrt{\frac{2m\Delta V}{B^2 q}} = \sqrt{\frac{2(1.67 \times 10^{-27} \text{ kg})(1 \text{ 500 V})}{(0.80 \text{ T})^2 (1.60 \times 10^{-19} \text{ C})}} = 7.0 \times 10^{-3} \text{ m}$$

17. In order for any particle to go straight through without deflection, the force exerted upward (say) by the electric field must be just balanced by the force of the magnetic field pushing it downward. Then $\mathscr{E}q = Bqv$. Note that the charge does not matter. Then

$$v = \frac{\mathscr{E}}{B} = \frac{22 \text{ 000 NC}}{35 \times 10^{-3} \text{ T}} = 6.3 \times 10^5 \text{ m/s}$$

18. From Equation 11.2

$$I = \frac{\tau}{vBA \cos \theta} = \frac{3.2 \times 10^{-3} \text{ N·m}}{(60)(0.35 \text{ T})(\pi)(0.025 \text{ m})^2 (\sin 20°)} = 0.23 \text{ A}$$

Chapter 12

FILL IN'S

1. currents
2. voltmeter
3. torque
4. shunts
5. commutator
6. magnetic
7. commutator
8. AC
9. zero
10. effective

11. smaller
12. magnetic field
13. potentials
14. transformer
15. 230 000
16. fuse; circuit breaker
17. increases
18. short circuits
19. green
20. synchronous

MULTIPLE CHOICE

1. (1)
2. (3)

3. (2)
4. (5)

5. (3)
6. (5)

7. (4)
8. (2)

9. (2)
10. (3)

PROBLEMS

1. a. $P_{max} = I_{max}{}^2 R = (0.50 \text{ A})^2 (60 \text{ }\Omega) = 15 \text{ W}.$

    b. $\Delta V_{eff} = \dfrac{\Delta V_{max}}{\sqrt{2}} = \dfrac{I_{max}R}{\sqrt{2}} = \dfrac{(0.50 \text{ A})(60 \text{ }\Omega)}{\sqrt{2}} = 21 \text{ V}.$

2. a. $I_{eff} = \dfrac{P_{av}}{\Delta V_{eff}} = \dfrac{300 \text{ W}}{120 \text{ V}} = 2.5 \text{ A}.$

    b. Twice $P_{av} = 600 \text{ W}.$

3. $\dfrac{\Delta V_P}{\Delta V_s} = \dfrac{n_P}{n_S} \qquad \dfrac{120 \text{ V}}{6 \text{ V}} = \dfrac{2\,400}{N_S} \qquad n_S = 120.$

4. a. $\bar{I} = \dfrac{P}{\Delta \bar{V}} = \dfrac{60 \text{ W}}{12 \text{ V}} = 5 \text{ A}.$

    b. $\dfrac{\Delta \bar{V}_P}{\Delta \bar{V}_S} = \dfrac{\bar{I}_S}{\bar{I}_P}; \qquad \dfrac{120 \text{ V}}{12 \text{ V}} = \dfrac{5 \text{ A}}{I_P}; \qquad I_P = 0.5 \text{ A}.$

    c. $\Delta V_{max} = \Delta V_{eff}\sqrt{2} = (120 \text{ V})\sqrt{2} = 170 \text{ V}.$

5. At 120 V:

$$\bar{I} = \frac{6.5 \times 10^3 \text{ W}}{120 \text{ V}} = 54 \text{ A}$$

Power loss in the line $= \bar{I}^2 R = (54 \text{ A})^2 (0.20 \text{ }\Omega) = 580 \text{ W}.$
At 1 500 V:

$$\bar{I} = \frac{6.5 \times 10^3 \text{ W}}{1 \text{ 500 V}} = 4.3 \text{ A}$$

Power loss in the line $= \bar{I}^2 R = (4.3 \text{ A})^2 (0.20 \text{ }\Omega) = 3.7 \text{ W}.$ The saving is 576 W.

6. $\bar{I} = \dfrac{P}{\Delta V} = \dfrac{45 \times 10^3 \text{ W}}{560 \text{ V}} = 80 \text{ A}.$

The power loss in the line $= \bar{I}^2 R = (80 \text{ A})^2 (1.5 \text{ }\Omega) = 9 \text{ 600 W}.$

7. #14 wire is limited to 15 A, so the most power that can be delivered is $(15 \text{ A})(120 \text{ V}) = 1 \text{ 800 W}.$

8. a. In the primary:

$$\bar{I} = \frac{P}{\Delta \bar{V}} = \frac{2 \text{ 500 000 W}}{325 \text{ 000 V}} = 7.7 \text{ A}$$

b. In the secondary:

$$\frac{2 \text{ 500 000 W}}{7 \text{ 500 V}} = 330 \text{ A}$$

9. $\Delta V_{max} = \Delta V_{eff} \sqrt{2} = \left(30 \text{ 000 V}\right)\sqrt{2} = 43 \text{ 000 V}.$

10. Since the galvanometer is in parallel with the shunt, the potential difference across the shunt and the galvanometer must be the same. Therefore

$$(IR)_{galv} = (IR)_{shunt}$$

$$R_{shunt} = \frac{(IR)_{galv}}{I_{shunt}} = \frac{\left(0.60 \times 10^{-3} \text{ A}\right)\left(70 \text{ }\Omega\right)}{\left(10 \text{ A}\right)} = 4.2 \times 10^{-3} \text{ }\Omega$$

11. Since the resistance is in series with the galvanometer, both must carry the same current, 0.60 milliamperes. The potential difference across the galvanometer is $(0.60 \times 10^{-3} \text{ A})(70 \text{ }\Omega) = 0.042 \text{ V};$ 99.96 volts are across the resistor. Therefore

$$R_{res} = \frac{\Delta V}{I} = \frac{100 \text{ V}}{0.60 \times 10^{-3} \text{ A}} = 1.67 \times 10^5 \text{ }\Omega$$

Chapter 13

FILL IN'S

1. information
2. filament
3. wattage
4. resistor
5. capacitor
6. capacitance
7. electric field
8. breakdown voltage
9. inductance
10. changing
11. resistance
12. inductance
13. capacitance
14. resistance
15. four
16. drift velocity
17. longitudinal
18. diode
19. base
20. p-type
21. higher
22. zero
23. remains the same
24. increases
25. impedance
26. decreases

MULTIPLE CHOICE

1. (2)
3. (2)
5. (3)
7. (3)
9. (1)
2. (5)
4. (5)
6. (4)
8. (1)
10. (5)

PROBLEMS

1. To dissipate 8 times as much heat, the larger resistor needs 8 times the surface area. It is twice as long, so it needs only 4 times the circumference. Cirumference is proportional to diameter, so it must be 16 millimeters wide.

2. The largest charge the capacitor can tolerate without breaking down is, from Equation 13.1

$$q = C\,\Delta V = (40 \times 10^{-6}\,\text{F})(12\,\text{V}) = 4.8 \times 10^{-4}\,\text{C}$$

The charge is flowing in at 2.0×10^{-3} coulomb per second, so the breakdown voltage will be reached in

$$\frac{4.0 \times 10^{-4}\,\text{C}}{2.0 \times 10^{-3}\,\text{C/s}} = 0.2\,\text{s}$$

3. The amount of charge transferred is

$$C = I\Delta t = (200 \times 10^3 \, \text{A})(20 \times 10^{-6} \, \text{s}) = 4 \, \text{C}$$

The potential difference at breakdown is

$$(10^6 \, \text{V/m})(5 \times 10^3 \, \text{m}) = 5 \times 10^9 \, \text{V}$$

And the capacitance is

$$C = \frac{q}{\Delta V} = \frac{4 \, \text{C}}{5 \times 10^9 \, \text{V}} = 800 \, \text{pF}$$

4. The greatest current this resistor can tolerate, from Equation 10.4b, is

$$I = \sqrt{\frac{P}{R}} = \sqrt{\frac{0.25 \, \text{W}}{200 \, \Omega}} = 35 \, \text{mA}$$

The potential difference producing this current is

$$\Delta V = IR = (35 \times 10^{-3} \, \text{A})(200 \, \Omega) = 7 \, \text{V}$$

5. The charge being added to the capacitor is 3.6 coulombs every second. Therefore, from Equation 13.1, the capacitor potential difference increases in every second by

$$\Delta V = \frac{q}{C} = \frac{3.6 \, \text{C}}{2 \times 10^{-6} \, \text{F}} = 1.8 \times 10^6 \, \text{V}$$

The rate of increase is 1.8 million volts per second. It surely will not keep rising at that rate.

6. The resistance of the coil is

$$R = \frac{\Delta V}{I} = \frac{12 \, \text{V}}{0.50 \, \text{A}} = 24 \, \Omega$$

When the current is 0.20 ampere, the potential difference across the resistance is

$$\Delta V = IR = (0.20 \, \text{A})(24 \, \Omega) = 4.8 \, \text{V}$$

which leaves 7.2 volts for the inductance. Then, from Equation 13.3,

$$\frac{\Delta I}{\Delta t} = \frac{V_\text{L}}{L} = \frac{7.2 \, \text{V}}{40 \times 10^{-3} \, \text{H}} = 180 \, \text{A/s}$$

7. From Equation 13.3, when the current is increasing at 400 amperes per second, the potential difference across the inductance is

$$\Delta V = L\frac{\Delta I}{\Delta t} = \left(25 \times 10^{-3} \text{ H}\right)\left(400 \text{ A/s}\right) = 10 \text{ V}$$

The other 14 volts are across the resistance, which must be

$$R = \frac{\Delta V}{I} = \frac{14 \text{ V}}{0.25 \text{ A}} = 56 \text{ }\Omega$$

At steady current, the current is

$$I = \frac{\Delta V}{R} = \frac{24 \text{ V}}{56 \text{ }\Omega} = 0.43 \text{ A}$$

So the rate of power loss is

$$P = I^2R = (0.43 \text{ A})^2(56 \text{ }\Omega) = 10 \text{ W}$$

8. Since the current is proportional to the frequency, the current at 1 000 hertz is 20 times as great as at 50 hertz.

9. From Equation 13.2,

$$C = \frac{\overline{I}}{2\pi f \Delta \overline{V}} = \frac{2.5 \times 10^{-3} \text{ A}}{2\pi(1 \text{ 000 Hz})(120 \text{ V})} = 3.3 \times 10^{-9} \text{ F}$$

10. The combined impedance of the capacitor and resistor must be

$$Z = \frac{\Delta \overline{V}}{\overline{I}} = \frac{60 \text{ V}}{30 \times 10^{-3} \text{ A}} = 2 \text{ 000 }\Omega$$

From Equation 13.2, the capacitive reactance is

$$X = \frac{1}{2\pi f C} = \frac{1}{2\pi\left(25 \times 10^6 \text{ Hz}\right)\left(4.0 \times 10^{-12} \text{ F}\right)} = 1 \text{ 590 }\Omega$$

Then, from Equation 13.5,

$$R = \sqrt{Z^2 - X^2} = \sqrt{(2 \text{ 000 }\Omega)^2 - (1 \text{ 590 }\Omega)^2} = 1 \text{ 200 }\Omega$$

11. The reactance of the coil is

$$X_L = 2\pi f L = 2\pi(1.2 \times 10^6 \text{ Hz})(20 \times 10^{-6} \text{ H}) = 151 \text{ }\Omega$$

The impedance of the coil is

$$Z = \sqrt{R^2 + X^2} = \sqrt{(75 \text{ }\Omega)^2 + (151 \text{ }\Omega)^2} = 169 \text{ }\Omega$$

Then the required potential difference is

$$\Delta \bar{V} = \bar{I}Z = (0.25 \text{ A})(169 \ \Omega) = 42 \text{ V}$$

12. From Equation 13.2, the capacitive reactance is

$$X_C = \frac{1}{2\pi f C} = \frac{1}{2\pi(1.8 \times 10^6 \text{ Hz})(550 \times 10^{-12} \text{ F})} = 161 \ \Omega$$

From Equation 13.4, the inductive reactance is

$$X_L = 2\pi f L = 2\pi(1.8 \times 10^6 \text{ Hz})(5.0 \times 10^{-6} \text{ H}) = 56 \ \Omega$$

From Equation 13.6, the impedance is

$$Z = \sqrt{R^2 + (X_L - X_C)^2} = \sqrt{(40 \ \Omega)^2 + (56 \ \Omega - 161 \ \Omega)^2} = 112 \ \Omega$$

Then the current is

$$\bar{I} = \frac{\Delta \bar{V}}{Z} = \frac{12 \text{ V}}{112 \ \Omega} = 0.11 \text{ A}$$

13. From Equation 13.7,

$$C = \frac{1}{4\pi^2 f^2 L} = \frac{1}{4\pi^2 (94 \times 10^6 \text{ Hz})^2 (1.2 \times 10^{-6} \text{ H})} = 2.3 \text{ pF}$$

14. The required impedance is

$$Z = \frac{\Delta \bar{V}}{\bar{I}} = \frac{100 \text{ V}}{0.22 \text{ A}} = 454 \ \Omega$$

The inductive reactance is

$$X_L = 2\pi f L = 2\pi(400 \text{ Hz})(0.25 \text{ H}) = 628 \ \Omega$$

The capacitive reactance is

$$X_C = \frac{1}{2\pi f C} = \frac{1}{2\pi(400 \text{ Hz})(1.0 \times 10^{-6} \text{ F})} = 398 \ \Omega$$

The net reactance is

$$X = X_L - X_C = 628 \ \Omega - 398 \ \Omega = 230 \ \Omega$$

From Equation 13.6

$$R = \sqrt{Z^2 - X^2} = \sqrt{(454 \ \Omega)^2 - (230 \ \Omega)^2} = 390 \ \Omega$$

Chapter 14

FILL IN'S

1. accelerated
2. speed
3. half-wavelength
4. perpendicular
5. X-rays or gamma rays
6. shorter
7. shorter
8. diffraction
9. wavelengths
10. antinode
11. wavelength
12. red
13. laser
14. polarized
15. perpendicular
16. three
17. yellow
18. short
19. infrared
20. frequency; power
21. quantized
22. frequency
23. light
24. frequency
25. work function
26. energy
27. short
28. horizontal

MULTIPLE CHOICE

1. (5)
2. (3)
3. (3)
4. (1)
5. (4)
6. (2)
7. (2)
8. (1)
9. (2)
10. (3)
11. (1)
12. (4)

PROBLEMS

1. a. Wave
    b. Electric field
    c. Nothing
    d. Wave
    e. Electric and magnetic fields.

2. a. $\lambda = \dfrac{c}{f} = \dfrac{3.00 \times 10^8 \text{ m/s}}{106 \times 10^6 \text{ Hz}} = 2.8 \text{ m.}$

    b. Half a wavelength, 1.4 m.

3. $\lambda = \dfrac{c}{f} = \dfrac{3.00 \times 10^8 \text{ m/s}}{7.1 \times 10^{14} \text{ Hz}} = 4.2 \times 10^{-7} \text{ m, or 420 nm.}$

    Figure 14.15 shows this to be blue.

4. a. 10λ, or 5 300 nm.
    b. $\sin \theta = n\lambda/d$, so

$$d = \frac{n\lambda}{\sin \theta} = \frac{5\ 300 \times 10^{-9}\ \text{m}}{\sin 12°} = 2.5 \times 10^{-5}\ \text{m}$$

5. $\sin \theta = \dfrac{\lambda}{d} \qquad d = \dfrac{1\ \text{m}}{580 \times 10^3}$

For the shortest wavelength,

$$\sin \theta = \left(\frac{580 \times 10^3}{\text{m}}\right)(380 \times 10^{-9}) \qquad \theta = 12.7°$$

For the longest wavelength,

$$\sin \theta = \left(\frac{580 \times 10^3}{\text{m}}\right)(760 \times 10^{-9}\ \text{m}) \qquad \theta = 26.2°$$

6. The shortest wavelength for red light is about 700 nm;

$$\sin \theta = \left(\frac{530 \times 10^3}{\text{m}}\right)(3)(700 \times 10^{-9}\ \text{m}) = 1.1$$

No angles have sines greater than 1.

7. Those two frequencies act on the red and green pigments, with very little in the blue. It appears yellow.

8. Everything below about 650 nm.

9. About 600 nm.

10. $E = hf = (6.62 \times 10^{-34}\ \text{J·s})(10^{14}\ \text{Hz}) = 6.6 \times 10^{-20}\ \text{J}$, or

$$(4.14 \times 10^{-15}\ \text{eV·s})(10^{14}\ \text{Hz}) = 0.41\ \text{eV}.$$

11. $E = hf = \dfrac{hc}{\lambda} = \dfrac{(4.14 \times 10^{-15}\ \text{eV·s})(3.00 \times 10^8\ \text{m/s})}{140 \times 10^{-9}\ \text{m}} = 8.9\ \text{eV}.$

12. This is the photon that has just that energy. $E = hc/\lambda$, so

$$\lambda = \frac{hc}{E} = \frac{(6.62 \times 10^{-34}\ \text{J·s})(3.00 \times 10^8\ \text{m/s})}{3.1 \times 10^{-19}\ \text{J}} = 6.4 \times 10^{-7}\ \text{m, or 640 nm}$$

13. The photon energy is $E = hf = (4.14 \times 10^{-15}\ \text{eV·s})(6.8 \times 10^{15}\ \text{Hz}) = 28.2\ \text{eV}$. Since the electrons leave the surface with 22.0 eV of energy, the work function is $28.2 - 22.0 = 6.2\ \text{eV}$.

14. The photon energy is

$$\frac{hc}{\lambda} = \frac{\left(4.14 \times 10^{-15} \text{ eV·s}\right)\left(3.00 \times 10^8 \text{ m/s}\right)}{580 \times 10^{-9} \text{ m}} = 2.1 \text{ eV}$$

The electron energy is 2.1 eV − 1.6 eV = 0.5 eV.

15. The photon energy is

$$\frac{hc}{\lambda} = \frac{\left(4.14 \times 10^{-15} \text{ eV·s}\right)\left(3.00 \times 10^8 \text{ m/s}\right)}{470 \times 10^{-9} \text{ m}} = 2.6 \text{ eV}$$

The maximum electron energy is 2.6 − 2.2 eV = 0.4 eV, and the current will be cut off by 0.4 V.

16. From Figure 14.15, the purest green light has a wavelength of about 530 nanometers. Then each photon has a momentum of

$$p = \frac{h}{\lambda} = \frac{6.6 \times 10^{-34} \text{ J·s}}{530 \times 10^{-9} \text{ m}} = 1.25 \times 10^{-27} \text{ kg·m/s}$$

Since the photons are reflected from the mirror, their change of momentum is twice this, 2.50×10^{-27} kg · m/s. This is the impulse each photon applies to the mirror. In each second, the total impulse is 3.8×10^{-9} N·s. Therefore, the number of photons striking the mirror in each second is

$$\frac{3.8 \times 10^{-9} \text{ N·s}}{2.50 \times 10^{-27} \text{ N·s}} = 1.5 \times 10^{18}$$

17. The energy of each photon is

$$E = \frac{hc}{\lambda} = \frac{(6.6 \times 10^{-34} \text{ J·s})(3 \times 10^8 \text{ m/s})}{530 \times 10^{-9} \text{ m}} = 3.7 \times 19^{-19} \text{ J}$$

Multiply by the number that strike the black surface in 10 minutes:

$$(3.7 \times 10^{-19} \text{ J})(1.5 \times 10^{18})(600) = 330 \text{ J}$$

18. The frequency of the light cannot change as it enters the glass. Since $v = f\lambda$, $v \propto \lambda$, the wavelength in the glass is (0.65)(530 nm) = 345 nm. Therefore, the number of wavelengths in the glass is

$$\frac{2.0 \times 10^{-3} \text{ m}}{345 \times 10^{-9} \text{ m}} = 5\,800$$

Chapter 15

FILL IN'S

1. diffraction
2. photon
3. lumens
4. green
5. lux
6. normal
7. angle of incidence
8. long
9. specular
10. virtual
11. the same
12. toward
13. index of refraction
14. speed
15. sines
16. critical angle of incidence
17. lower
18. frequency or wavelength
19. violet
20. converging
21. principal focus
22. principal focus
23. real
24. the focal length
25. two
26. diverging or concave
27. focal length
28. size; size
29. virtual
30. infinity
31. dispersion
32. focal ratio or f-number
33. long
34. circle of confusion
35. concave
36. 20
37. eye
38. near point

MULTIPLE CHOICE

1. (4)	3. (5)	5. (2)	7. (3)	9. (3)
2. (5)	4. (2)	6. (1)	8. (1)	10. (5)

PROBLEMS

1. $I = F/4\pi r^2$, so $F = 4\pi r^2 I = 4\pi(1.5\text{ m})^2(20\text{ lux}) = 570$ lum.
2. The visible power output is 4% of 200 W = 8 W. Since a visible watt corresponds to roughly 400 lum, the luminous flux is (400 lum/W) × (8W) = 3 200 lum.
3. The size does not change, but the distance drops to 6 m.

4. $n = \dfrac{c}{v} = \dfrac{3.00 \times 10^8 \text{ m/s}}{2.00 \times 10^8 \text{ m/s}} = 1.50$.

5. $n = \dfrac{\sin i}{\sin r} = \dfrac{\sin 45°}{\sin 32°} = 1.33$.

6. The first refraction occurs when the beam enters the benzene (B) from the air (A):

$$\frac{\sin A}{\sin B} = \frac{n_B}{n_A}; \qquad \frac{\sin 55°}{\sin B} = \frac{1.50}{1}$$

so $B = 33°$. This is the angle at which it enters the water, so

$$\frac{\sin 33°}{\sin W} = \frac{1.33}{1.50}; \qquad W = 38°$$

7. We need a material whose critical angle is 45°, so

$$\frac{\sin 45°}{\sin 90°} = \frac{1}{n}; \qquad n = 1.41$$

8. $n = \dfrac{\sin 62°}{\sin 47°} = \dfrac{\sin 90°}{\sin i_C}; \qquad i_C = 56°$.

9. Both make the same angle (A) in air, but different angles B_v and B_r in the glass. Then, since $n = \sin A/\sin B$, $\sin A = n \sin B$ is the same for both. Therefore $1.590 \sin 65.0° = 1.566 \sin B_r$, and $B_r = 67.0°$.

10. $\dfrac{1}{f} = \dfrac{1}{D_o} + \dfrac{1}{D_i}$, so

$$\frac{1}{D_i} = \frac{1}{f} - \frac{1}{D_o} = \frac{1}{5.0 \text{ cm}} - \frac{1}{2.0 \text{ cm}} = \frac{2 - 5}{10.0 \text{ cm}}$$

and $D_i = -3.33$ cm, a virtual image. To find the size:

$$\frac{D_o}{D_i} = \frac{S_o}{S_i} \qquad \frac{2.0 \text{ cm}}{-3.33 \text{ cm}} = \frac{1.2 \text{ cm}}{S_i}$$

and $S_i = -2.0$ cm.

11. $\dfrac{1}{f} = \dfrac{1}{D_o} + \dfrac{1}{D_i} = \dfrac{1}{9.0 \text{ cm}} + \dfrac{1}{6.5 \text{ cm}} = \dfrac{6.5 + 9.0}{(9.0)(6.5) \text{ cm}}$

and $f = 3.8$ cm.

12. $$\frac{1}{D_i} = \frac{1}{f} - \frac{1}{D_o} = \frac{1}{20 \text{ cm}} - \frac{1}{50 \text{ cm}} = \frac{5-2}{100 \text{ cm}}$$

and $D_i = 33.3$ cm. To find its size:

$$\frac{D_o}{D_i} = \frac{S_o}{S_i} \qquad \frac{50 \text{ cm}}{33.3 \text{ cm}} = \frac{40 \text{ cm}}{S_i}$$

which gives $S_i = 27$ cm.

13. In this setup, $D_i = (3.0 \text{ m} - D_o)$, and we are trying to find values for D_o. Then

$$\frac{1}{f} = \frac{1}{D_o} + \frac{1}{D_i}$$

becomes

$$\frac{1}{0.50 \text{ m}} = \frac{1}{D_o} + \frac{1}{3.0 \text{ m} - D_o}$$

Multiplying through by $(0.50 \text{ m})(D_o)(3.0 \text{ m} - D_o)$ gives $D_o(3.0 - D_o) = 0.50(3.0 - D_o) + 0.50 D_o$, which simplifies to $D_o^2 - 3.0 D_o + 1.5 = 0$. Now we apply the quadratic equation formula

$$x = \frac{-b \pm \sqrt{b^2 - 4ac}}{2a}$$

to get

$$D_o = \frac{3.0 \pm \sqrt{3.0^2 - 4(1.5)}}{2}$$

so $D_o = 2.36$ m or 0.63 m.

14. $$\frac{D_o}{D_i} = \frac{S_o}{S_i} = \frac{1.00 \text{ m}}{0.065 \text{ m}} \qquad D_i = 0.065 \, D_o$$

Therefore, the $1/f$ equation becomes

$$\frac{1}{80 \text{ mm}} = \frac{1}{D_o} + \frac{1}{0.065 \, D_o}$$

Then

$$\frac{1}{80 \text{ mm}} = \frac{0.065 + 1}{0.065 \, D_o}$$

and $D_o = 1300$ mm, or 1.3 m.

15. $3.5 = \dfrac{\text{focal length}}{\text{diameter}} = \dfrac{80 \text{ mm}}{\text{diameter}}$.

So the diameter $= 23$ mm.

16. If the radius of curvature of a convex mirror is 40 cm, its focal length is -20 cm. First find the image distance:

$$\frac{1}{D_i} = \frac{1}{f} - \frac{1}{D_o}$$

$$D_i = \frac{f D_o}{D_o - f} = \frac{(-20 \text{ cm})(150 \text{ cm})}{150 \text{ cm} - (-20 \text{ cm})} = -17.6 \text{ cm}$$

Then use the distance ratio to find the image size:

$$S_i = S_o \left(\frac{D_i}{D_o} \right) = 12 \text{ cm} \left(\frac{-17.6 \text{ cm}}{35 \text{ cm}} \right) = -3.0 \text{ cm}$$

The negative sign shows that the image is virtual.

17. Since the image is real and one-half the size of the object, $S_i = \dfrac{1}{2} S_o$, so $D_i = \dfrac{1}{2} D_o$. The focal length is half the radius of curvature, $+39$ cm. It is positive for a concave mirror. Then

$$\frac{1}{D_o} + \frac{1}{D_i} = \frac{1}{f}$$

$$\frac{1}{D_o} + \frac{1}{\frac{1}{2} D_o} = \frac{1}{39 \text{ cm}}$$

$$\frac{3}{D_o} = \frac{1}{39 \text{ cm}}$$

and $D_o = 117$ cm.

18. From Equation 15.3b, the image distance is

$$D_i = D_o \left(\frac{S_i}{S_o} \right) = (30 \text{ cm}) \left(\frac{6.5 \text{ cm}}{10 \text{ cm}} \right) = 19.5 \text{ cm}$$

Then, from Equation 15.3a, the focal length is

$$f = \frac{D_o D_i}{D_o + D_i} = \frac{(30 \text{ cm})(19.5 \text{ cm})}{30 \text{ cm} + 19.5 \text{ cm}} = 11.8 \text{ cm}$$

At the new position,

$$D_i = \frac{fD_o}{D_o - f} = \frac{(11.8 \text{ cm})(5.0 \text{ cm})}{5.0 \text{ cm} - 11.8 \text{ cm}} = -8.7 \text{ cm}$$

So the image is virtual. Its size is

$$S_i = S_o\left(\frac{D_i}{D_o}\right) = \frac{(10.0 \text{ cm})(8.7 \text{ cm})}{5.0 \text{ cm}} = 17 \text{ cm}$$

19. The oncoming traffic can be considered essentially at infinity, so the image is at the principal focus. The focal length must be half the radius of curvature, which is therefore 30 cm.

20. Since maximum magnification is needed, the image will be placed at the near point of the eye. Therefore, from Equation 15.4,

$$f = \frac{25 \text{ cm}}{M - 1} = \frac{25 \text{ cm}}{5 - 1} = 6.3 \text{ cm}$$

Chapter 16

FILL IN'S

1. electrons
2. positive elementary
3. atomic number
4. hydrogen
5. photon
6. ionized
7. angular momentum
8. $h/2\pi$
9. standing waves
10. velocity
11. potential difference or voltage
12. alpha particles
13. the sun or radioactive rocks
14. photons
15. excited state
16. momentum
17. decreases
18. twice
19. wavelength
20. interference
21. crystal
22. photon or particle
23. experiment
24. uncertainty
25. probability
26. science
27. probabilities
28. nine

MULTIPLE CHOICE

1. (5)	4. (2)	7. (3)	9. (3)	11. (4)
2. (1)	5. (2)	8. (3)	10. (1)	12. (1)
3. (3)	6. (4)			

PROBLEMS

1. If they pass straight through, the electric and magnetic forces are equal:
$Bqv = \mathscr{E}q$, and

$$v = \frac{\mathscr{E}}{B} = \frac{6\ 500\ \text{N/C}}{0.25\ \text{T}} = 2.6 \times 10^4\ \text{m/s}$$

2. a. $E = -13.6\ \text{eV}/n^2$. Since $n = 1, E = -13.6\ \text{eV}$.
   b. With $n = 3, E = -1.51\ \text{eV}$.
   c. n is indefinitely large and $E = 0$.

3. a. Using the answers to Problem 2, the energy drops from $-1.51\ \text{eV}$ to -13.6 eV, so $\Delta E = 12.1\ \text{eV}$, and that is the energy of the photon.
   b. $E = hc/\lambda$, so

$$\lambda = \frac{hc}{E} = \frac{\left(4.14 \times 10^{-15}\ \text{eV} \cdot \text{s}\right)\left(3.00 \times 10^8\ \text{m/s}\right)}{12.1\ \text{eV}} = 1.03 \times 10^{-7}\ \text{m}$$

4. Its total energy is then $18\ \text{eV} - 13.6\ \text{eV} = 4.4\ \text{eV}$.

5. a. $E_{\text{kin}} = \Delta E_{\text{elec}} = q\ \Delta V$

$$= (1.60 \times 10^{-19}\ \text{C})(50\ \text{V}) = 8.0 \times 10^{-18}\ \text{J}.$$

   b. $v = \sqrt{\dfrac{2\ E_{\text{kin}}}{m}} = \sqrt{\dfrac{2\left(8.0 \times 10^{-18}\right)\text{J}}{1.67 \times 10^{-27}\ \text{kg}}} = 9.8 \times 10^4\ \text{m/s}$

   c. $p = mv = (1.67 \times 10^{-27}\ \text{kg})(9.8 \times 10^4\ \text{m/s}) = 1.63 \times 10^{-22}\ \text{kg·m/s}$.

   d. $\lambda = \dfrac{h}{p} = \dfrac{6.62 \times 10^{-34}\ \text{J} \cdot \text{s}}{1.63 \times 10^{-22}\ \text{kg} \cdot \text{m/s}} = 4.0 \times 10^{-12}\ \text{m}$.

6. 3.06 eV, 2.84 eV, 2.27 eV, 2.03 eV, 1.81 eV, 1.24 eV, 1.03 eV, 0.79 eV, 0.57 eV, 0.22 eV.

7. About 0.14 V, since electrons will be raised to the c level by 4.86 eV.

8. The momentum of one photon is

$$p = \frac{h}{\lambda} = \frac{6.24 \times 10^{-34} \text{ J} \cdot \text{s}}{460 \times 10^{-9} \text{ m}} = 1.36 \times 10^{-27} \text{ kg} \cdot \text{m/s}$$

The photons are absorbed, so they transfer their entire momentum to the surface. The force is the change of momentum per second:

$$F = \frac{\Delta p}{\Delta t} = \left(1.36 \times 10^{-27} \text{ kg} \cdot \text{m/s}\right)\left(7.0 \times 10^{15} \text{ m/s}\right) = 9.5 \times 10^{-12} \text{ N}$$

9. The amount of mass increase is 9.1×10^{-31} J. Then the additional energy is

$$E = mc^2 = (9.1 \times 10^{-31} \text{ J})(3.0 \times 10^8 \text{ m/s})^2 = 8.2 \times 10^{-14} \text{ J}$$

The required potential difference is

$$\Delta V = \frac{E_{elec}}{q} = \frac{8.2 \times 10^{-14} \text{ J}}{1.60 \times 10^{-19} \text{ C}} = 5.1 \times 10^5 \text{ V}$$

10. From Equation 14.2, the energy of this photon is

$$E = \frac{hc}{\lambda} = \frac{\left(4.14 \times 10^{-15} \text{ eV} \cdot \text{s}\right)\left(3.0 \times 10^8 \text{ m/s}\right)}{55 \times 10^{-9} \text{ m}} = 22.6 \text{ eV}$$

Subtracting the 13.6 eV needed to ionize the hydrogen atom, it leaves 9.0 eV for the electron.

11. Equation 16.1 gives the change in the energy of the electron:

$$\Delta E = -(13.6 \text{ ev})\left(\frac{1}{n_1{}^2} - \frac{1}{n_2{}^2}\right) = -(13.6 \text{ ev})\left(\frac{1}{6^2} - \frac{1}{3^2}\right) = -1.13 \text{ eV}$$

The energy lost by the electron becomes the energy of the photon. From Equation 14.2, the wavelength of this photon is

$$\lambda = \frac{hc}{E} = \frac{\left(4.14 \times 10^{-15} \text{ eV} \cdot \text{s}\right)\left(3.0 \times 10^8 \text{ m/s}\right)}{1.13 \text{ eV}} = 1.10 \text{ } \mu\text{m}$$

which is invisible, in the infrared.

12.   Six. The drop from level 2 to level 1 is by far the largest energy drop of all those possible, so this line will have the shortest wavelength.

Chapter 17

FILL IN'S

1. alpha particles
2. 55
3. momentum or mass
4. electric charge
5. period of revolution
6. mass
7. positrons or antielectrons
8. uranium
9. neutrons
10. proton
11. neutrons
12. isotopes
13. daltons
14. ionizing
15. magnetic field
16. bubbles
17. gamma
18. electrons
19. helium
20. beta
21. 4
22. half-life
23. smaller than
24. binding
25. increases
26. electric
27. drift tubes
28. magnetic
29. electron
30. neutrons or protons
31. neutrons
32. neutrons
33. neutrons
34. plutonium-239
35. moderator
36. control rods
37. fusion
38. medium-sized
39. photon
40. quarks
41. pions
42. electron

MULTIPLE CHOICE

1. (3)
2. (1)
3. (1)
4. (4)
5. (3)
6. (3)
7. (2)
8. (1)
9. (4)
10. (3)
11. (5)

PROBLEMS

1.   a. 77.
     b. 113 (atomic mass number minus atomic number).
     c. 77 (atomic number).

2. a. $mv = Bqr = (2.5 \text{ Ts})(1.60 \times 10^{-19} \text{ C})(0.075 \text{ m})$

$$mv = 3.0 \times 10^{-20} \text{ kg·m/s}$$

b. $v = \dfrac{p}{m} = \dfrac{3.0 \times 10^{-20} \text{ kg} \cdot \text{m/s}}{1.67 \times 10^{-27} \text{ kg}} = 1.8 \times 10^7 \text{ m/s.}$

3. a. It needs the energy of two electrons:

$$E = mc^2 = 2(9.1 \times 10^{-31} \text{ kg})(3.0 \times 10^8 \text{ m/s})^2 = 1.64 \times 10^{-13} \text{ J}$$

b. $\lambda = \dfrac{hc}{E} = \dfrac{(6.24 \times 10^{-34} \text{ J} \cdot \text{s})(3.0 \times 10^8 \text{ m/s})}{1.64 \times 10^{-13} \text{ J}} = 1.1 \times 10^{-12} \text{ m.}$

4. $f = \dfrac{1}{T} = \dfrac{Bq}{2\pi m} = \dfrac{(1.8 \text{ T})(1.60 \times 10^{-19} \text{ C})}{2\pi (1.67 \times 10^{-27} \text{ kg})} = 2.7 \times 10^7 \text{ Hz}$

5. a. 2 *antiup* + 1 *antidown*.
    b. 1 *antiup* + 1 *down*.

6. mass of two helium-3 nuclei: 2×3.01603 Dl =   6.03206 Dl
    mass of helium-4:   4.00260 D1
    two protons: 2×1.007825 =   2.015650
    Subtract the product nuclei: 6.01825 D1
    Mass deficit: 0.01381 D1
    Multiply by 931 MeV/Dl to get 12.9 MeV.

7. From Appendix 4:
    protons: $15 \times 1.007825 = 15.11738$
    neutrons: $15 \times 1.008665 = 15.12998$
    mass of separate particles   30.24735
    mass of phosphorus-30   29.9783 D1
    mass deficit   0.2691 D1
    binding energy: (0.2691 Dl) (931 MeV/Dl) = 250 MeV

8. a. $^{218}_{84}\text{Po} \rightarrow \, ^{214}_{82}\text{Pb} + \, ^{4}_{2}\text{He}$.

    b. The mass of polonium-218 is 218.0089. Subtract mass of helium-4 (4.00260) and lead-214 (213.9998) to find the additional mass deficit of 0.0065 Dl; multiply by 931 MeV/Dl to get 6.1 MeV.

9. It must drop to half four times; with a half-life of 3 min, this would take 12 min.

10. a. $^{210}_{82}\text{Pb} \rightarrow \, ^{210}_{83}\text{Bi} + \, ^{0}_{-1}\text{e} + \, ^{0}_{0}\text{v}$.

    b. The mass of lead-210 is 209.9848. Subtract the mass of bismuth-210 (209.9840) and an electron (0.00055) to get 0.0002 Dl; multiplying by 931 gives 0.2 MeV.

11. $E = q \, \Delta V = \text{(two elementary charges)}(6.5 \times 10^5 \text{ V}) = 1.3 \text{ MeV}$.

12. It must acquire an additional mass equal to its own mass, 1.67×10^{-27} kg. Then $E = mc^2 = (1.67 \times 10^{-27} \text{ kg}) (3.00 \times 10^8 \text{ m/s})^2 = 1.51 \times 10^{-10}$ J. To find the potential difference, $\Delta E = q \, \Delta V$ and

$$\Delta V = \frac{\Delta E}{q} = \frac{1.51 \times 10^{-10} \text{ J}}{1.60 \times 10^{-19} \text{ C}} = 9.4 \times 10^8 \text{ V}$$

which is not practical.

13. $^{4}_{2}\text{He} + \, ^{10}_{5}\text{B} \rightarrow (^{14}_{7}\text{N}) \rightarrow \, ^{13}_{7}\text{N} + \, ^{1}_{0}\text{n}$.

14. $^{22}_{11}\text{Na} \rightarrow \, ^{22}_{10}\text{Ne} + \, ^{0}_{+1}\text{e} + \, ^{0}_{0}\text{v}$.

15. $^{239}_{94}\text{Pu} + \, ^{1}_{0}\text{n} \rightarrow \, ^{142}_{54}\text{Xe} + \, ^{94}_{40}\text{Zr} + 3 \, ^{1}_{0}\text{n}$.

16.

mass of hydrogen-1 (proton)	1.007825 Dl
mass of carbon-12	<u>12.00000</u>
	13.007825
mass of nitrogen-13	<u>13.0057</u>
mass deficit of reaction	0.0021 Dl
$\times$ 931 MeV/DI = 1.96 MeV	

Index

Success on Advanced Placement Tests Starts with Help from Barron's

Each May, thousands of college-bound students take one or more Advanced Placement Exams to earn college credits—and for many years, they've looked to Barron's, the leader in Advanced Placement test preparation. You can get Barron's user-friendly manuals for 22 different AP subjects, most with optional CD-ROMs. Every Barron's AP manual gives you—

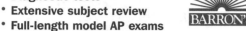

* **Diagnostic tests**
* **Extensive subject review**
* **Full-length model AP exams**
* **Study help and test-taking advice**

Model exams are designed to reflect the actual AP exams in question types, subject matter, length, and degree of difficulty. All questions come with answers and explanations. All books are paperback.

AP: Art History, w/optional CD-ROM
Book: ISBN 978-0-7641-3737-2, 448 pp.
Book w/CD-ROM: ISBN 978-0-7641-9463-4

AP: Biology, 2nd Ed., w/optional CD-ROM
Book: ISBN 978-0-7641-3677-1, 496 pp.
Book w/CD-ROM: ISBN 978-0-7641-9327-9

AP: Biology Flash Cards
ISBN 978-0-7641-7868-9, 504 cards

AP: Calculus, 9th Ed., w/optional CD-ROM
Book: ISBN 978-0-7641-3679-5, 672 pp.
Book w/CD-ROM: ISBN 978-0-7641-9328-6

AP: Calculus Flash Cards
ISBN 978-0-7641-9421-4, 300 cards

AP: Chemistry, 4th Ed., w/optional CD-ROM
Book: ISBN 978-0-7641-3685-6, 688 pp.
Book w/CD-ROM: ISBN 978-0-7641-9329-3

AP: Chemistry Flash Cards
ISBN 978-0-7641-6116-2, 500 cards

AP: Chinese Language and Culture
ISBN 978-0-7641-9400-9, 592 pp.

AP: Computer Science Levels A and AB, 4th Ed., w/optional CD-ROM
Book: ISBN 978-0-7641-3709-9, 736 pp.
Book w/CD-ROM: ISBN 978-0-7641-9350-7

AP: English Language and Composition, 2nd Ed., w/optional CD-ROM
Book: ISBN 978-0-7641-3690-0, 224 pp.
Book w/CD-ROM: ISBN 978-0-7641-9330-9

AP: English Literature and Composition, 2nd Ed., w/optional CD-ROM
Book: ISBN 978-0-7641-3682-5, 416 pp.
Book w/CD-ROM: ISBN 978-0-7641-9331-6

AP: Environmental Science, 3rd Ed.
ISBN 978-0-7641-4052-5, 504 pp.
Book w/CD-ROM: ISBN 978-0-7641-9565-5

AP: Environmental Science Flash Cards
ISBN 978-0-7641-9599-0, 500 cards

AP: European History, 4th Ed., w/optional CD-ROM
Book: ISBN 978-0-7641-3680-1, 336 pp.
Book w/CD-ROM: ISBN 978-0-7641-9332-3

AP: French w/Audio CDs, 3rd Ed., w/optional CD-ROM
Book w/3 audio CDs: ISBN 978-0-7641-9337-8, 464 pp.
Book w/3 audio CDs and CD-ROM:
ISBN 978-0-7641-9336-1

AP: Human Geography, 2nd Ed.
ISBN 978-0-7641-3817-1, 336 pp.

AP: Human Geography Flash Cards
ISBN 978-0-7641-9598-3, 500 cards

AP: Italian Language and Culture, w/Audio CDs, 2nd Ed.
ISBN 978-0-7641-9368-2, 448 pp.

AP: Microeconomics/Macroeconomics, 3rd Ed.
ISBN 978-0-7641-3930-7, 360 pp.

AP: Physics B, 4th Ed., w/optional CD-ROM
Book: ISBN 978-0-7641-3706-8, 476 pp.
Book w/CD-ROM: ISBN 978-0-7641-9351-4

AP: Physics C, 2nd Ed.
ISBN 978-0-7641-3710-5, 768 pp.

AP: Psychology, 3rd Ed., w/optional CD-ROM
Book: ISBN 978-0-7641-3665-8, 320 pp.
Book w/CD-ROM: ISBN 978-0-7641-9324-8, 320 pp.

AP: Psychology Flash Cards
ISBN 978-0-7641-9613-3, 400 cards

AP: Spanish w/Audio CDs, 6th Ed., and w/optional CD-ROM
Book w/3 audio CDs: ISBN 978-0-7641-9406-1, 448 pp.
Book w/3 audio CDs and CD-ROM:
ISBN 978-0-7641-9405-4

AP: Statistics, 4th Ed., w/optional CD-ROM
Book: ISBN 978-0-7641-3683-2, 560 pp.
Book w/CD-ROM: ISBN 978-0-7641-9333-0

AP: Statistics Flash Cards
ISBN 978-0-7641-9410-8, 500 cards

AP: U.S. Government and Politics, 8th Ed., w/optional CD-ROM
Book: ISBN 978-0-7641-3820-1, 510 pp.
Book w/CD-ROM: ISBN 978-0-7641-9404-7

AP: U.S. Government and Politics Flash Cards
ISBN 978-0-7641-6121-6, 400 cards

AP: U.S. History Flash Cards
ISBN 978-0-7641-7837-5, 500 cards

AP: U.S. History, 8th Ed., w/optional CD-ROM
Book: ISBN 978-0-7641-3684-9, 496 pp.
Book w/CD-ROM: ISBN 978-0-7641-9334-7

AP: World History, 8th Ed., w/optional CD-ROM
Book: ISBN 978-0-7641-3822-5, 512 pp.
Book w/CD-ROM: ISBN 978-0-7641-9403-0

AP: World History Flash Cards
ISBN 978-0-7641-7906-8, 400 cards

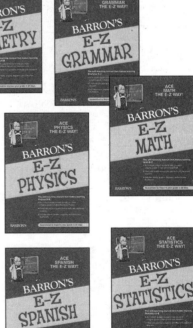